NCERT EXEMPLAR

Problems - Solutions

Mathematics 12th

BEST Solution
Guaranteed •••

NCERT EXEMPLAR

Problems – Solutions

Mathematics 12th

Author
Dr Sunil Kumar Goel

ARIHANT PRAKASHAN (Series), MEERUT

卐 **Administrative & Production Offices**

Regd. Office

'Ramchhaya' 4577/15, Agarwal Road, Darya Ganj, New Delhi -110002
Tele: 011- 47630600, 43518550

卐 **Head Office**

Kalindi, TP Nagar, Meerut (UP) - 250002
Tel: 0121-7156203, 7156204

卐 **Sales & Support Offices**

Agra, Ahmedabad, Bengaluru, Bareilly, Chennai, Delhi, Guwahati, Hyderabad, Jaipur, Jhansi, Kolkata, Lucknow, Nagpur & Pune.

卐 **PRICE** ₹385.00

卐 **PO No. :** TXT-59-T071864-7-26

Published by Arihant Publications (India) Ltd.

For further information about the books published by Arihant, log on to www.arihantbooks.com or e-mail at info@arihantbooks.com

Follow us on   

PREFACE

NCERT Exemplar Problems, published by the NCERT, are designed to provide students with a diverse range of quality questions in formats such as Short Answers Type Questions, Long Answer Type Questions and Objective Type Questions catering to varying levels of difficulty.

These problems are essential for success in school and board exams, as well as competitive exams like engineering and medical entrances. Regular practice with these problems prepares students to tackle challenging questions effectively.

This book, **NCERT Exemplar Problems-Solutions Mathematics 12th**, offers precise and detailed solutions to all the questions in the NCERT Exemplar Mathematics textbook. Each solution is complemented by explanatory notes and authentic explanations, enabling students to deepen their conceptual understanding and enhance their analytical skills.

Key features

- **Thinking Process** Provides simple steps to approach and solve the question, clearing up any confusion.
- **Concept Insight** Explains the main ideas and concepts needed to answer the questions correctly.
- **Mistake Alert!** Points out common mistakes students might make and helps them avoid these errors.
- **Notes** Highlights important points and useful tips to make learning easier.

I hope this book proves to be a valuable resource, aiding students in their academic journey and leading them to great success.

Author

CONTENTS

Chapter

01

Relations and Functions

Short Answer Type Questions

Q.1 Let $A = \{a, b, c\}$ and the relation R be defined on A as follows

$$R = \{(a, a), (b, c), (a, b)\}$$

Then, write minimum number of ordered pairs to be added in R to make R reflexive and transitive.

Ans. Given relation, $R = \{(a, a), (b, c), (a, b)\}$.

To make R is reflexive we must add (b, b) and (c, c) to R. Also, to make R is transitive we must add (a, c) to R.

So, minimum number of ordered pair is to be added are $(b, b), (c, c), (a, c)$.

Q.2 Let D be the domain of the real valued function f defined by $f(x) = \sqrt{25 - x^2}$. Then, write D.

Ans. Given function, $f(x) = \sqrt{25 - x^2}$

For real valued of $f(x)$, $25 - x^2 \geq 0$

$x^2 \leq 25$ [on multiplying in inequality by $-ve$ sign, then sign of inequality will change]

$$-5 \leq x \leq +5$$

$$\therefore \quad D = [-5, 5]$$

Q.3 If $f, g: R \to R$ be defined by $f(x) = 2x + 1$ and $g(x) = x^2 - 2$, $\forall\, x \in R$, respectively. Then, find gof.

Thinking Process If $f, g: R \to R$ be two functions, then $gof(x) = g\{f(x)\}$, $\forall x \in R$.

Ans. Given that, $f(x) = 2x + 1$ and $g(x) = x^2 - 2, \forall x \in R$

$$\therefore \quad gof = g\{f(x)\}$$

$$= g(2x + 1) = (2x + 1)^2 - 2$$

$$= 4x^2 + 4x + 1 - 2$$

$$= 4x^2 + 4x - 1$$

Q.4 Let $f: R \rightarrow R$ be the function defined by $f(x) = 2x - 3, \forall\, x \in R$. Write f^{-1}.

Ans. Given that, $f(x) = 2x - 3, \forall\, x \in R$

Now, let

$$y = 2x - 3$$

$$2x = y + 3$$

$$x = \frac{y+3}{2}$$

$$\therefore \quad f^{-1}(x) = \frac{x+3}{2}$$

Q.5 If $A = \{a, b, c, d\}$ and the function $f = \{(a, b), (b, d), (c, a), (d, c)\}$, write f^{-1}.

Ans. Given that, $A = \{a, b, c, d\}$

and $f = \{(a, b), (b, d), (c, a), (d, c)\}$

$$f^{-1} = \{(b, a), (d, b), (a, c), (c, d)\}$$

Q.6 If $f: R \rightarrow R$ is defined by $f(x) = x^2 - 3x + 2$, write $f\{f(x)\}$.

Thinking Process To solve this problem use the formula *i.e.*,
$(a + b + c)^2 = (a^2 + b^2 + c^2 + 2ab + 2bc + 2ca)$

Ans. Given that, $f(x) = x^2 - 3x + 2$

$$\therefore \quad f\{f(x)\} = f(x^2 - 3x + 2)$$

$$= (x^2 - 3x + 2)^2 - 3(x^2 - 3x + 2) + 2$$

$$= x^4 + 9x^2 + 4 - 6x^3 - 12x + 4x^2 - 3x^2 + 9x - 6 + 2$$

$$= x^4 + 10x^2 - 6x^3 - 3x$$

$$f\{f(x)\} = x^4 - 6x^3 + 10x^2 - 3x$$

Q.7 Is $g = \{(1, 1), (2, 3), (3, 5), (4, 7)\}$ a function? If g is described by $g(x) = \alpha x + \beta$, then what value should be assigned to α and β?

Ans. Given that, $g = \{(1, 1), (2, 3), (3, 5), (4, 7)\}$.

Here, each element of domain has unique image. So, g is a function.

Now given that, $g(x) = \alpha x + \beta$

$$g(1) = \alpha + \beta$$

$$\alpha + \beta = 1 \quad \ldots\text{(i)}$$

$$g(2) = 2\alpha + \beta$$

$$2\alpha + \beta = 3 \quad \ldots\text{(ii)}$$

From Eqs. (i) and (ii), we get

$$2(1 - \beta) + \beta = 3$$

$$\Rightarrow \quad 2 - 2\beta + \beta = 3$$

$$\Rightarrow \quad 2 - \beta = 3$$

$$\beta = -1$$

If $\beta = -1$, then $\alpha = 2$

Hence, $\alpha = 2, \beta = -1$

Q.8 Are the following set of ordered pairs functions? If so examine whether the mapping is injective or surjective.

(i) $\{(x, y) : x$ is a person, y is the mother of $x\}$.

(ii) $\{(a, b) : a$ is a person, b is an ancestor of $a\}$.

Ans. (i) Given set of ordered pair is $\{(x, y) : x$ is a person, y is the mother of $x\}$.

It represents a function. Here, the image of distinct elements of x under f are not distinct, so it is not a injective but it is a surjective.

(ii) Set of ordered pairs $= \{(a, b) : a$ is a person, b is an ancestor of $a\}$

Here, each element of domain does not have a unique image. So, it does not represent function.

Q.9 If the mappings f and g are given by $f = \{(1, 2), (3, 5), (4, 1)\}$ and $g = \{(2, 3), (5, 1), (1, 3)\}$, write fog.

Ans. Given that, $f = \{(1, 2), (3, 5), (4, 1)\}$

and $g = \{(2, 3), (5, 1), (1, 3)\}$

Now,

$$fog(2) = f\{g(2)\} = f(3) = 5$$

$$fog(5) = f\{g(5)\} = f(1) = 2$$

$$fog(1) = f\{g(1)\} = f(3) = 5$$

$$fog = \{(2, 5), (5, 2), (1, 5)\}$$

Q.10 Let C be the set of complex numbers. Prove that the mapping $f : C \to R$ given by $f(z) = |z|, \forall z \in C$, is neither one-one nor onto.

Ans. The mapping $f : C \to R$

Given, $f(z) = |z|, \forall z \in C$

$$f(1) = |1| = 1 \Rightarrow f(-1) = |-1| = 1 \Rightarrow f(1) = f(-1)$$

But $1 \neq -1$

So, $f(z)$ is not one-one. Also, $f(z)$ is not onto as there is no pre-image for any negative element of R under the mapping $f(z)$.

Q.11 Let the function $f : R \to R$ be defined by $f(x) = \cos x, \forall x \in R$. Show that f is neither one-one nor onto.

Ans. Given function, $f(x) = \cos x, \forall x \in R$

Now, $f\left(\frac{\pi}{2}\right) = \cos\frac{\pi}{2} = 0$

$\Rightarrow f\left(\frac{-\pi}{2}\right) = \cos\frac{\pi}{2} = 0$ $[\cos(-\theta) = \cos\theta]$

$\Rightarrow f\left(\frac{\pi}{2}\right) = f\left(\frac{-\pi}{2}\right)$

But $\frac{\pi}{2} \neq \frac{-\pi}{2}$

So, $f(x)$ is not one-one.

Now, $f(x) = \cos x, \forall x \in R$ is not onto as there is no pre-image for any real number, which does not belongs to the intervals $[-1, 1]$, the range of $\cos x$.

$\therefore$ f is not onto.

Q.12 Let $X = \{1, 2, 3\}$ and $Y = \{4, 5\}$. Find whether the following subsets of $X \times Y$ are functions from X to Y or not.

(i) $f = \{(1, 4), (1, 5), (2, 4), (3, 5)\}$ (ii) $g = \{(1, 4), (2, 4), (3, 4)\}$

(iii) $h = \{(1, 4), (2, 5), (3, 5)\}$ (iv) $k = \{(1, 4), (2, 5)\}$

Ans. Given that, $X = \{1, 2, 3\}$ and $Y = \{4, 5\}$

$$X \times Y = \{(1, 4), (1, 5), (2, 4), (2, 5), (3, 4), (3, 5)\}$$

(i) $f = \{(1, 4), (1, 5), (2, 4), (3, 5)\}$, f is not a function because f has not unique image.

(ii) $g = \{(1, 4), (2, 4), (3, 4)\}$, g is a function as each element of the domain has unique image.

(iii) $h = \{(1, 4), (2, 5), (3, 5)\}$. It is clear that h is a function.

(iv) $k = \{(1, 4), (2, 5)\}$, k is not a function as 3 does not have any image under the mapping.

Q.13 If functions $f : A \to B$ and $g : B \to A$ satisfy $gof = I_A$, then show that f is one-one and g is onto.

Ans. Given that, $f : A \to B$ and $g : B \to A$ satisfy $gof = I_A$

Let $f(x_1) = f(x_2)$ for any $x_1, x_2 \in A$

Then, $g(f(x_1)) = g(f(x_2))$ [by applying g on both sides]

$\Rightarrow$ $gof(x_1) = gof(x_2)$

$\Rightarrow$ $I_A(x_1) = I_A(x_2)$ [$\because gof = I_A$]

$\therefore$ $x_1 = x_2$

Hence, f is a one-one function.

Now, let $x \in A$ (codomain of g).

Then, $f(x) \in B$

On applying g to $f(x)$, we get

$$g(f(x)) = gof(x) = I_A(x) = x$$

$\therefore$ For all $x \in A$, there exists an element $f(x) \in B$ such that $g(f(x)) = x$.

Hence, g is onto.

Q.14 Let $f : R \to R$ be the function defined by $f(x) = \dfrac{1}{2 - \cos x}$, $\forall\, x \in R$. Then, find the range of f.

Thinking Process Range of $f = \{y \in Y : y = f(x) :$ for all values of in $x\}$ and use range of $\cos x$ is $[-1, 1]$.

Ans. Given function, $f(x) = \dfrac{1}{2 - \cos x}$, $\forall\, x \in R$

Let $y = \dfrac{1}{2 - \cos x}$

As the value of $\cos x$ lies in between -1 and 1.

Therefore, $2y - y\cos x = 1 \Rightarrow y\cos x = 2y - 1$

$\Rightarrow$ $\cos x = \dfrac{2y - 1}{y} = 2 - \dfrac{1}{y} \Rightarrow \cos x = 2 - \dfrac{1}{y}$

$\Rightarrow$ $-1 \le \cos x \le 1 \Rightarrow -1 \le 2 - \dfrac{1}{y} \le 1$

$$\Rightarrow \quad -3 \le -\frac{1}{y} \le -1 \quad \Rightarrow 1 \le \frac{1}{y} \le 3$$

$$\Rightarrow \quad \frac{1}{3} \le y \le 1$$

So, range of f is $\left[\frac{1}{3}, 1\right]$.

Q.15 Let n be a fixed positive integer. Define a relation R in Z as follows $\forall\, a, b \in Z$, aRb if and only if $a - b$ is divisible by n. Show that R is an equivalence relation.

Ans. Given that, $\forall\, a, b \in Z$, aRb if and only if $a - b$ is divisible by n.

Now,

I. **Reflexive** $aRa \Rightarrow (a - a)$ is divisible by n, which is true for any integer a as '0' is divisible by n.

Hence, R is reflexive.

II. **Symmetric**

$$aRb$$

$\Rightarrow a - b$ is divisible by $n \Rightarrow -(b - a)$ is divisible by n.

$\Rightarrow (b - a)$ is divisible by n.

$\Rightarrow bRa$

Hence, R is symmetric.

III. **Transitive** Let aRb and bRc

$\Rightarrow (a - b)$ is divisible by n and $(b - c)$ is divisible by n

$\Rightarrow (a - b) + (b - c)$ is divisibly by $n \Rightarrow (a - c)$ is divisible by n

$\Rightarrow aRc$

Hence, R is transitive.

So, R is an equivalence relation.

Long Answer Type Questions

Q.16 If $A = \{1, 2, 3, 4\}$, define relations on A which have properties of being

(i) reflexive, transitive but not symmetric.

(ii) symmetric but neither reflexive nor transitive.

(iii) reflexive, symmetric and transitive.

Ans. Given that, $A = \{1, 2, 3, 4\}$

(i) Let $R_1 = \{(1, 1), (1, 2), (2, 3), (2, 2), (1, 3), (3, 3), (4, 4)\}$

R_1 is reflexive, since, (1, 1) (2, 2) (3, 3) lie in R_1.

Now, $(1, 2) \in R_1, (2, 3) \in R_1 \Rightarrow (1, 3) \in R_1$

Hence, R_1 is also transitive but $(1, 2) \in R_1 \Rightarrow (2, 1) \notin R_1$.

So, it is not symmetric.

(ii) Let $R_2 = \{(1, 2), (2, 1)\}$

Now, $(1, 2) \in R_2, (2, 1) \in R_2$

So, it is symmetric but neither reflexive nor transitive.

(iii) Let $R_3 = \{(1, 2), (2, 1), (1, 1), (2, 2), (3, 3), (1, 3), (3, 1), (2, 3), (3, 2), (4, 4)\}$

R_3 is reflexive as (1, 1), (2, 2) and (3, 3), $(4, 4) \in R_3$

R_3 is symmetric as (1, 2), (1, 3), $(2, 3) \in R_3 \Rightarrow (2, 1), (3, 1), (3, 2) \in R_3$

R_3 is transitive as $(1, 2) \in R_3$, $(2, 3) \in R_3 \Rightarrow (1, 3) \in R_3$

Hence, R_3 is reflexive, symmetric and transitive.

Q.17 Let R be relation defined on the set of natural number N as follows, $R = \{(x, y) : x \in N, y \in N, 2x + y = 41\}$. Find the domain and range of the relation R. Also verify whether R is reflexive, symmetric and transitive.

Ans. Given that, $R = \{(x, y) : x \in N, y \in N, 2x + y = 41\}$.

$$y = 41 - 2x$$

As x is natural number

$$x = 1, 2, 3.......20$$
$$y = 39, 37, 35...1$$
$$R = \{(1, 39), (2, 37), (3, 35), \ldots, (19, 3), (20, 1)\}$$
$$\text{Domain} = \{1, 2, 3, \ldots, 20\}$$
$$\text{Range} = \{1, 3, 5, 7, \ldots, 39\}$$

R is not reflexive as $(2, 2) \notin R$

$$2 \times 2 + 2 \neq 41$$

As $(1, 39) \in R$ but $(39, 1) \notin R$

So, R is not symmetric.

As $(11, 19) \in R, (19, 3) \in R$

But $(11, 3) \notin R$

So, R is not transitive.

Hence, R is neither reflexive, nor symmetric and nor transitive.

Q.18 Given, $A = \{2, 3, 4\}$, $B = \{2, 5, 6, 7\}$. Construct an example of each of the following

(i) an injective mapping from A to B.

(ii) a mapping from A to B which is not injective.

(iii) a mapping from B to A.

Ans. Given that, $A = \{2, 3, 4\}$, $B = \{2, 5, 6, 7\}$

(i) Let $f : A \to B$ denote a mapping

$$f = \{(x, y) : y = x + 3\}$$

i.e., $f = \{(2, 5), (3, 6), (4, 7)\}$, which is an injective mapping.

(ii) Let $g : A \to B$ denote a mapping such that $g = \{(2, 2), (3, 5), (4, 5)\}$, which is not an injective mapping, becaue 5 have two pre-image.

(iii) Let $h : B \to A$ denote a mapping such that $h = \{(2, 2), (5, 3), (6, 4), (7, 4)\}$, which is a mapping from B to A.

Q.19 Give an example of a map

(i) which is one-one but not onto.

(ii) which is not one-one but onto.

(iii) which is neither one-one nor onto.

Ans. (i) Let $f : N \to N$, be a mapping defined by $f(x) = 2x$

which is one-one.

For $\quad f(x_1) = f(x_2) \Rightarrow 2x_1 = 2x_2 \Rightarrow x_1 = x_2$

Further f is not onto, as for $1 \in N$, there does not exist any x in $N(A)$.

(ii) Let $f : N \to N$, given by $f(1) = f(2) = 1$ and $f(x) = x - 1$ for every $x > 2$ is onto but not one-one. f is not one-one as $f(1) = f(2) = 1$. But f is onto.

(iii) The mapping $f : R \to R$ defined as $f(x) = x^2$, is neither one-one nor onto.

Q.20 Let $A = R - \{3\}$, $B = R - \{1\}$. If $f : A \to B$ be defined by $f(x) = \dfrac{x-2}{x-3}$, $\forall x \in A$. Then, show that f is bijective.

Thinking Process A function $f : X \to Y$ is said to be bijective, if f is both one-one (*i.e.*, injective) and onto (*i.e.*, surjective).

Ans. Given that, $\quad A = R - \{3\}, \; B = R - \{1\}$.

$f : A \to B$ is defined by $\quad f(x) = \dfrac{x-2}{x-3}, \forall x \in A$

For injectivity,

Let $\quad f(x_1) = f(x_2) \Rightarrow \dfrac{x_1 - 2}{x_1 - 3} = \dfrac{x_2 - 2}{x_2 - 3}$

$\Rightarrow \quad (x_1 - 2)(x_2 - 3) = (x_2 - 2)(x_1 - 3)$

$\Rightarrow \quad x_1x_2 - 3x_1 - 2x_2 + 6 = x_1x_2 - 3x_2 - 2x_1 + 6$

$\Rightarrow \quad -3x_1 - 2x_2 = -3x_2 - 2x_1$

$\Rightarrow \quad -x_1 = -x_2 \Rightarrow x_1 = x_2$

So, $f(x)$ is an injective function or one-one.

For surjectivity,

Let $\quad y = \dfrac{x-2}{x-3} \Rightarrow x - 2 = xy - 3y$

$\Rightarrow \quad x(1 - y) = 2 - 3y \Rightarrow x = \dfrac{2 - 3y}{1 - y}$

$\Rightarrow \quad x = \dfrac{3y - 2}{y - 1} \in A, \forall y \in B \quad$ [codomain]

So, $f(x)$ is surjective function.

Hence, $f(x)$ is a bijective function.

Q.21 Let $A = [-1, 1]$, then discuss whether the following functions defined on A are one-one onto or bijective.

(i) $f(x) = \dfrac{x}{2}$ (ii) $g(x) = |x|$ (iii) $h(x) = x|x|$ (iv) $k(x) = x^2$

Ans. Given that, $\quad A = [-1, 1]$

(i) $f(x) = \dfrac{x}{2}$

Let $\quad f(x_1) = f(x_2)$

$\Rightarrow \quad \dfrac{x_1}{2} = \dfrac{x_2}{2} \Rightarrow x_1 = x_2$

So, $f(x)$ is one-one.

Now, let $y = \frac{x}{2}$

$\Rightarrow$ $x = 2y \notin A, \forall\, y \in A$

As for $y = 1 \in A, x = 2 \notin A$

So, $f(x)$ is not onto.

Also, $f(x)$ is not bijective as it is not onto.

(ii) $g(x) = |x|$

Let $g(x_1) = g(x_2)$

$\Rightarrow$ $|x_1| = |x_2| \Rightarrow x_1 = \pm x_2$

So, $g(x)$ is not one-one.

Now, $y = |x| \Rightarrow x = \pm y \notin A, \forall\, y \in A$

So, $g(x)$ is not onto, also, $g(x)$ is not bijective.

(iii) $h(x) = x|x|$

Let $h(x_1) = h(x_2)$

$\Rightarrow$ $x_1|x_1| = x_2|x_2| \Rightarrow x_1 = x_2$

So, $h(x)$ is one-one.

Now, let $y = x|x|$

$\Rightarrow$ $y = x^2$, if $0 \le x < 1$

$= -x^2$, if $-1 \le x < 0$

So, $h(x)$ is onto also, $h(x)$ is a bijective.

(iv) $k(x) = x^2$

Let $k(x_1) = k(x_2)$

$\Rightarrow$ $x_1^2 = x_2^2 \Rightarrow x_1 = \pm x_2$

Thus, $k(x)$ is not one-one.

Now, let $y = x^2$

$\Rightarrow$ $x = \sqrt{y} \notin A, \forall\, y \in A$

As for $y = -1, x = \sqrt{-1} \notin A$

Hence, $k(x)$ is neither one-one nor onto.

Q.22 Each of the following defines a relation of N

(i) x is greater than y, $x, y \in N$. (ii) $x + y = 10$, $x, y \in N$.

(iii) xy is square of an integer $x, y \in N$.

(iv) $x + 4y = 10$, $x, y \in N$

Determine which of the above relations are reflexive, symmetric and transitive.

Ans. (i) x is greater than y, $x, y \in N$

$(x, x) \in R$

For xRx $x > x$ is not true for any $x \in N$.

Therefore, R is not reflexive.

Let $(x, y) \in R \Rightarrow xRy$

$x > y$

but $y > x$ is not true for any $x, y \in N$

Thus, R is not symmetric.

Let xRy and yRz

$$x > y \text{ and } y > z \Rightarrow x > z \Rightarrow xRz$$

So, R is transitive.

(ii) $x + y = 10, x, y \in N$

$$R = \{(x, y); x + y = 10, x, y \in N\}$$

$$R = \{(1, 9), (2, 8), (3, 7), (4, 6), (5, 5), (6, 4), (7, 3), (8, 2), (9, 1)\}$$

$$(1, 1), (2, 2) \notin R$$

So, R is not reflexive.

$$(x, y) \in R \Rightarrow (y, x) \in R$$

Therefore, R is symmetric.

Now, $(1, 9) \in R, (9, 1) \in R$ but $(1, 1) \notin R$

Hence, R is not transitive.

(iii) Given xy, is square of an integer $x, y \in N$.

$$\Rightarrow \quad R = \{(x, y) : xy \text{ is a square of an integer } x, y \in N\}$$

$$(x, x) \in R, \forall x \in N$$

As x^2 is square of an integer for any $x \in N$.

Hence, R is reflexive.

If $(x, y) \in R \Rightarrow (y, x) \in R$

Therefore, R is symmetric.

If $(x, y) \in R, (y, z) \in R$

So, xy is square of an integer and yz is square of an integer.

Let $xy = m^2$ and $yz = n^2$ for some $m, n \in Z$

$$x = \frac{m^2}{y} \text{ and } z = \frac{n^2}{y}$$

$$xz = \frac{m^2n^2}{y^2}, \text{ which is square of an integer.}$$

So, R is transitive.

(iv) $x + 4y = 10, \; x, y \in N$

$$R = \{(x, y) : x + 4y = 10, x, y \in N\}$$

$$R = \{(2, 2), (6, 1)\}$$

$$(1, 1), (3, 3), \ldots, \notin R$$

Thus, R is not reflexive.

Now, $(6, 1) \in R$ but $(1, 6) \notin R$

Hence, R is not symmetric.

$$(x, y) \in R \Rightarrow x + 4y = 10 \text{ and } (y, z) \in R$$

$$y + 4z = 10 \Rightarrow (x, z) \in R$$

So, R is transitive.

Q.23 Let $A = \{1, 2, 3, \ldots, 9\}$ and R be the relation in $A \times A$ defined by $(a, b) R (c, d)$ if $a + d = b + c$ for $(a, b), (c, d)$ in $A \times A$. Prove that R is an equivalence relation and also obtain the equivalent class $[(2, 5)]$.

Ans. Given that, $A = \{1, 2, 3, \ldots, 9\}$ and $(a, b) R(c, d)$ if $a + d = b + c$ for $(a, b) \in A \times A$ and $(c, d) \in A \times A$.

Let $(a, b)\ R\ (a, b)$

$\Rightarrow \quad a + b = b + a, \forall\ a, b \in A$

which is true for any $a, b \in A$.

Hence, R is reflexive.

Let $(a, b)\ R\ (c, d) \quad a + d = b + c$

$c + b = d + a \Rightarrow (c, d)\ R\ (a, b)$

So, R is symmetric.

Let $(a, b)\ R\ (c, d)$ and $(c, d)\ R\ (e, f)$

$$a + d = b + c \text{ and } c + f = d + e$$
$$a + d = b + c \text{ and } d + e = c + f$$
$$(a + d) - (d + e) = (b + c) - (c + f)$$
$$(a - e) = b - f$$
$$a + f = b + e$$
$$(a, b)\ R\ (e, f)$$

So, R is transitive.

Hence, R is an equivalence relation.

Now, equivalence class containing $[(2, 5)]$ is $\{(1, 4), (2, 5), (3, 6), (4, 7), (5, 8), (6, 9)\}$.

Q.24 Using the definition, prove that the function $f : A \to B$ is invertible if and only if f is both one-one and onto.

Ans. A function $f : X \to Y$ is defined to be invertible, if there exist a function $g = Y \to X$ such that $gof = I_X$ and $fog = I_Y$. The function is called the inverse of f and is denoted by f^{-1}.

A function $f = X \to Y$ is invertible iff f is a bijective function.

Q.25 Functions $f, g : R \to R$ are defined, respectively, by $f(x) = x^2 + 3x + 1$, $g(x) = 2x - 3$, find

(i) fog (ii) gof (iii) fof (iv) gog

Ans. Given that, $f(x) = x^2 + 3x + 1$, $g(x) = 2x - 3$

(i) $fog(x) = f\{g(x)\} = f(2x - 3)$
$$= (2x - 3)^2 + 3(2x - 3) + 1$$
$$= 4x^2 + 9 - 12x + 6x - 9 + 1 = 4x^2 - 6x + 1$$

(ii) $gof(x) = g\{f(x)\} = g(x^2 + 3x + 1)$
$$= 2(x^2 + 3x + 1) - 3$$
$$= 2x^2 + 6x + 2 - 3 = 2x^2 + 6x - 1$$

(iii) $fof(x) = f\{f(x)\} = f(x^2 + 3x + 1)$
$$= (x^2 + 3x + 1)^2 + 3(x^2 + 3x + 1) + 1$$
$$= x^4 + 9x^2 + 1 + 6x^3 + 6x + 2x^2 + 3x^2 + 9x + 3 + 1$$
$$= x^4 + 6x^3 + 14x^2 + 15x + 5$$

(iv) $gog(x) = g\{g(x)\} = g(2x - 3)$
$$= 2(2x - 3) - 3$$
$$= 4x - 6 - 3 = 4x - 9$$

***Q.26** Let $*$ be the binary operation defined on Q. Find which of the following binary operations are commutative

(i) $a * b = a - b, \forall a, b \in Q$ (ii) $a * b = a^2 + b^2, \forall a, b \in Q$

(iii) $a * b = a + ab, \forall a, b \in Q$ (iv) $a * b = (a - b)^2, \forall a, b \in Q$

Ans. Given that $*$ be the binary operation defined on Q.

(i) $a * b = a - b, \forall a, b \in Q$ and $b * a = b - a$

So, $a * b \neq b * a$ $[\because b - a \neq a - b]$

Hence, $*$ is not commutative.

The commutative is the binary operation in which operand is changed then result will not change.

(ii) $a * b = a^2 + b^2$

$b * a = b^2 + a^2$

$\because$ $a^2 + b^2 = b^2 + a^2$

So, $*$ is commutative. $[\because a * b = b * a]$

(iii) $a * b = a + ab$

$b * a = b + ab$

Clearly, $a + ab \neq b + ab$

So, $*$ is not commutative.

(iv) $a * b = (a - b)^2, \forall a, b \in Q$

$b * a = (b - a)^2$

$\because$ $(a - b)^2 = (b - a)^2$

Hence, $*$ is commutative. $[\because a * b = b * a]$

***Q.27** If $*$ be binary operation defined on R by $a * b = 1 + ab, \forall a, b \in R$. Then, the operation $*$ is

(i) commutative but not associative.
(ii) associative but not commutative.
(iii) neither commutative nor associative.
(iv) both commutative and associative.

Ans. (i) Given that, $a * b = 1 + ab, \forall a, b \in R$

$a * b = ab + 1 = b * a$

So, $*$ is a commutative binary operation.

Also, $a * (b * c) = a * (1 + bc) = 1 + a(1 + bc)$

$a * (b * c) = 1 + a + abc$ ...(i)

$(a * b) * c = (1 + ab) * c = 1 + (1 + ab)c = 1 + c + abc$ ...(ii)

From Eqs. (i) and (ii), we get

$a * (b * c) \neq (a * b) * c$

So, $*$ is not associative

Hence, $*$ is commutative but not associative.

Note (✱) Topics related to these question have been removed from latest NCERT.

Objective Type Questions

Choose the correct answer out of the given four options in each of the exercises from 28 to 47.

Q.28 Let T be the set of all triangles in the Euclidean plane and let a relation R on T be defined as aRb, if a is congruent to b, $\forall\ a, b \in T$. Then, R is

(a) reflexive but not transitive
(b) transitive but not symmetric
(c) equivalence
(d) None of these

Ans. ***(c)*** Consider that aRb, if a is congruent to b, $\forall\ a, b \in T$.

Then, $aRa \Rightarrow a \cong a$,

which is true for all $a \in T$

So, R is reflexive, ...(i)

Let $aRb \Rightarrow a \cong b$

$\Rightarrow b \cong a \Rightarrow b \cong a \Rightarrow bRa$

So, R is symmetric. ...(ii)

Let aRb and bRc

$\Rightarrow a \cong b$ and $b \cong c$

$\Rightarrow a \cong c \Rightarrow aRc$

So, R is transitive. ...(iii)

Hence, R is equivalence relation.

Q.29 Consider the non-empty set consisting of children in a family and a relation R defined as aRb, if a is brother of b. Then, R is

(a) symmetric but not transitive
(b) transitive but not symmetric
(c) neither symmetric nor transitive
(d) both symmetric and transitive

Ans. ***(b)*** Given, $aRb \Rightarrow a$ is brother of b

$\therefore$ $aRa \Rightarrow a$ is brother of a, which is not true.

So, R is not reflexive.

$aRb \Rightarrow a$ is brother of b.

This does not mean b is also a brother of a and b can be a sister of a.

Hence, R is not symmetric.

$aRb \Rightarrow a$ is brother of b

and $bRc \Rightarrow b$ is a brother of c.

So, a is brother of c.

Hence, R is transitive.

Q.30 The maximum number of equivalence relations on the set $A = \{1, 2, 3\}$ are

(a) 1
(b) 2
(c) 3
(d) 5

Ans. ***(d)*** Given that, $A = \{1, 2, 3\}$

Now, number of equivalence relations as follows.

$$R_1 = \{(1, 1), (2, 2), (3, 3)\}$$
$$R_2 = \{(1, 1), (2, 2), (3, 3), (1, 2), (2, 1)\}$$
$$R_3 = \{(1, 1), (2, 2), (3, 3), (1, 3), (3, 1)\}$$
$$R_4 = \{(1, 1), (2, 2), (3, 3), (2, 3), (3, 2)\}$$
$$R_5 = \{(1, 2, 3) \Leftrightarrow A \times A = A^2\}$$

$\therefore$ Maximum number of equivalence relation on the set $A = \{1, 2, 3\} = 5$

Q.31 If a relation R on the set $\{1, 2, 3\}$ be defined by $R = \{(1, 2)\}$, then R is

(a) reflexive (b) transitive (c) symmetric (d) None of these

Ans. ***(b)*** R is not reflexive.

Because $(1, 1), (2, 2), (3, 3) \notin R$

R is not symmetric.

Because $(1, 2) \in R$ but $(2, 1) \notin R$

R is transitive.

Because single element relation is always transitive.

Q.32 Let us define a relation R in R as aRb if $a \geq b$. Then, R is

(a) an equivalence relation
(b) reflexive, transitive but not symmetric
(c) symmetric, transitive but not reflexive
(d) neither transitive nor reflexive but symmetric

Ans. ***(b)*** Given that, aRb if $a \geq b$

$\Rightarrow$ $aRa \Rightarrow a \geq a$ which is true.

Let aRb, $a \geq b$, then $b \geq a$ which is not true R is not symmetric.

But aRb and bRc

$\Rightarrow$ $a \geq b$ and $b \geq c$

$\Rightarrow$ $a \geq c$

Hence, R is transitive.

Q.33 If $A = \{1, 2, 3\}$ and consider the relation

$R = \{(1, 1), (2, 2), (3, 3), (1, 2), (2, 3), (1, 3)\}$

Then, R is

(a) reflexive but not symmetric (b) reflexive but not transitive
(c) symmetric and transitive (d) neither symmetric nor transitive

Ans. ***(a)*** Given that, $A = \{1, 2, 3\}$

and $R = \{(1, 1), (2, 2), (3, 3), (1, 2), (2, 3), (1, 3)\}$

$\because$ $(1, 1), (2, 2), (3, 3) \in R$

Hence, R is reflexive.

$$(1, 2) \in R \text{ but } (2, 1) \notin R$$

Hence, R is not symmetric.

$$(1, 2) \in R \text{ and } (2, 3) \in R$$

$$\Rightarrow \quad (1, 3) \in R$$

Hence, R is transitive.

***Q.34** The identity element for the binary operation $*$ defined on $Q - \{0\}$ as $a * b = \dfrac{ab}{2}, \forall\, a, b \in Q - \{0\}$ is

(a) 1 (b) 0 (c) 2 (d) None of these

Thinking Process Q stands for rational number. For given binary operation $* : A \times A \to A$, an element $e \in A$, if it exists, is called identity for the operation $*$, if $a * e = a = e * a, \forall\, a \in A$.

Ans. *(c)* Given that, $a * b = \dfrac{ab}{2}, \forall\, a, b \in Q - \{0\}$.

Let e be the identity element for $*$.

$$\therefore \quad a * e = \frac{ae}{2} \qquad [\because a * e = e * a = a]$$

$$\Rightarrow \quad a = \frac{ae}{2} \Rightarrow e = 2$$

Q.35 If the set A contains 5 elements and the set B contains 6 elements, then the number of one-one and onto mappings from A to B is

(a) 720 (b) 120 (c) 0 (d) None of these

Ans. *(c)* We know that, if A and B are two non-empty finite set containing m and n elements respectively, then the number of one-one and onto mapping from A to B is

$$n!, \text{ if } m = n$$
$$0, \text{ if } m \neq n$$

Given that, $m = 5$ and $n = 6$

$\therefore$ $m \neq n$

Number of mapping = 0

Q.36 If $A = \{1, 2, 3, \ldots, n\}$ and $B = \{a, b\}$. Then, the number of surjections from A into B is

(a) ${}^nP_2$ (b) $2^n - 2$

(c) $2^n - 1$ (d) None of these

Ans. *(b)* Given that, $A = \{1, 2, 3, \ldots, n\}$ and $B = \{a, b\}$.

Total number of functions from A to $B = 2^n$

and number of elements in set $B = 2$

Then, total number of onto (surjections) functions from A into $B = 2^n - 2$

Note (*) Topics related to these question have been removed from latest NCERT.

Q.37 If $f: R \rightarrow R$ be defined by $f(x) = \frac{1}{x}, \forall x \in R$. Then, f is

(a) one-one (b) onto (c) bijective (d) f is not defined

Thinking Process In the given function at $x = 0$, $f(x) = \infty$. So, the function is not define.

Ans. **(d)** Given that, $f(x) = \frac{1}{x}, \forall x \in R$

For $x = 0, \quad f(x) = \infty$

$\therefore f(x)$ is not defined.

Hence, $f(x)$ is a not defined function.

Q.38 If $f: R \rightarrow R$ be defined by $f(x) = 3x^2 - 5$ and $g: R \rightarrow R$ by $g(x) = \frac{x}{x^2+1}$.

Then, gof is

(a) $\frac{3x^2 - 5}{9x^4 - 30x^2 + 26}$ (b) $\frac{3x^2 - 5}{9x^4 - 6x^2 + 26}$

(c) $\frac{3x^2}{x^4 + 2x^2 - 4}$ (d) $\frac{3x^2}{9x^4 + 30x^2 - 2}$

Ans. **(a)** Given that, $f(x) = 3x^2 - 5$ and $g(x) = \frac{x}{x^2 + 1}$

$$gof = g\{f(x)\} = g(3x^2 - 5)$$

$$= \frac{3x^2 - 5}{(3x^2 - 5)^2 + 1} = \frac{3x^2 - 5}{9x^4 - 30x^2 + 25 + 1}$$

$$= \frac{3x^2 - 5}{9x^4 - 30x^2 + 26}$$

Q.39 Which of the following functions from Z into Z are bijections?

(a) $f(x) = x^3$ (b) $f(x) = x + 2$ (c) $f(x) = 2x + 1$ (d) $f(x) = x^2 + 1$

Ans. **(b)** Here, $f(x) = x + 2 \quad \Rightarrow \quad f(x_1) = f(x_2)$

$x_1 + 2 = x_2 + 2 \quad \Rightarrow \quad x_1 = x_2$

Let $y = x + 2$

$x = y - 2 \in Z, \forall y \in x$

Hence, $f(x)$ is one-one and onto.

Q.40 If $f: R \rightarrow R$ be the functions defined by $f(x) = x^3 + 5$, then $f^{-1}(x)$ is

(a) $(x + 5)^{\frac{1}{3}}$ (b) $(x - 5)^{\frac{1}{3}}$

(c) $(5 - x)^{\frac{1}{3}}$ (d) $5 - x$

Ans. **(b)** Given that, $f(x) = x^3 + 5$

Let $y = x^3 + 5 \quad \Rightarrow \quad x^3 = y - 5$

$x = (y - 5)^{\frac{1}{3}} \quad \Rightarrow \quad f(x)^{-1} = (x - 5)^{\frac{1}{3}}$

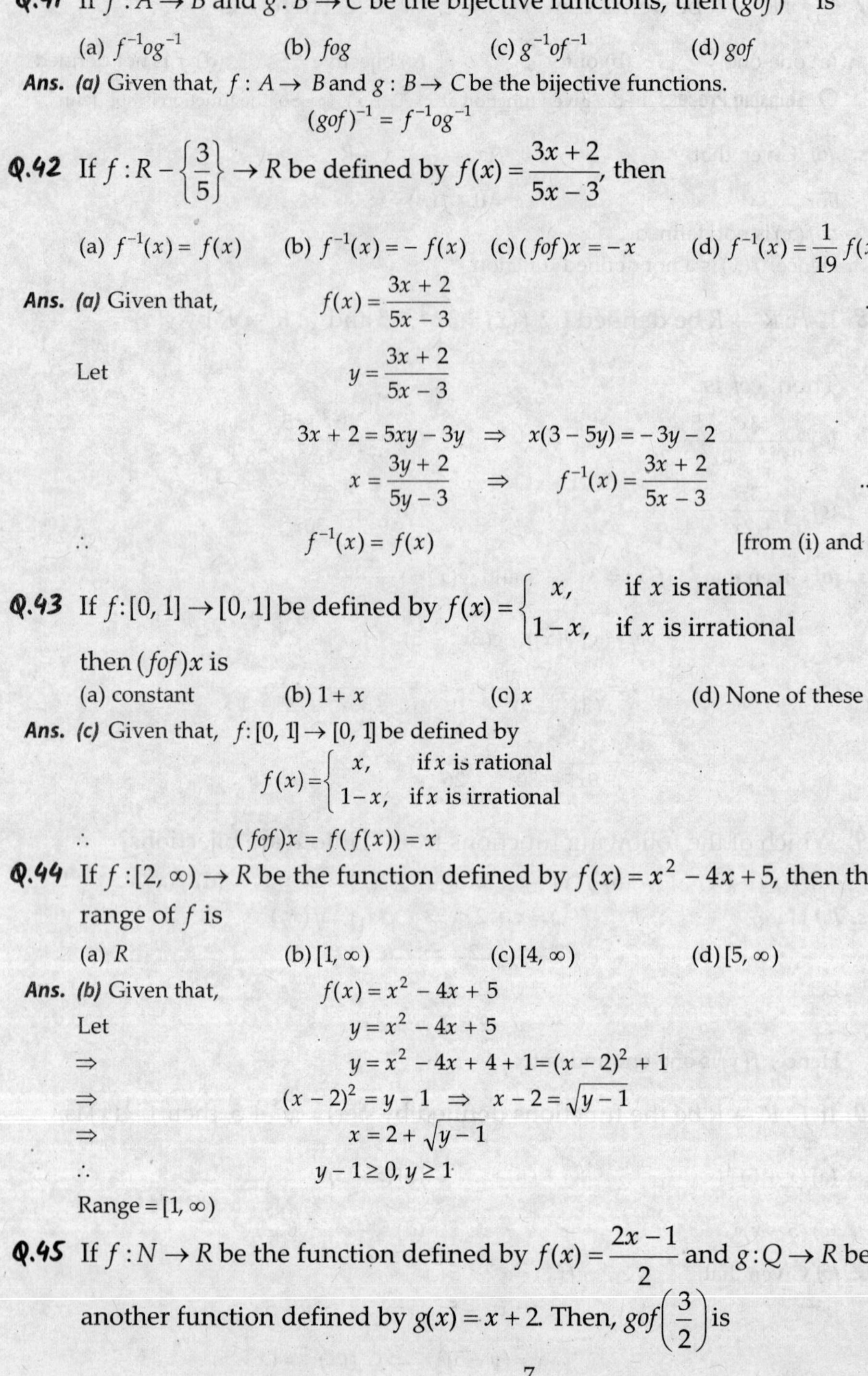

Q.41 If $f: A \to B$ and $g: B \to C$ be the bijective functions, then $(gof)^{-1}$ is

(a) $f^{-1}og^{-1}$ (b) fog (c) $g^{-1}of^{-1}$ (d) gof

Ans. **(a)** Given that, $f: A \to B$ and $g: B \to C$ be the bijective functions.

$$(gof)^{-1} = f^{-1}og^{-1}$$

Q.42 If $f: R - \left\{\frac{3}{5}\right\} \to R$ be defined by $f(x) = \frac{3x+2}{5x-3}$, then

(a) $f^{-1}(x) = f(x)$ (b) $f^{-1}(x) = -f(x)$ (c) $(fof)x = -x$ (d) $f^{-1}(x) = \frac{1}{19}f(x)$

Ans. **(a)** Given that, $f(x) = \dfrac{3x+2}{5x-3}$ ...(i)

Let $y = \dfrac{3x+2}{5x-3}$

$$3x + 2 = 5xy - 3y \Rightarrow x(3 - 5y) = -3y - 2$$

$$x = \frac{3y+2}{5y-3} \Rightarrow f^{-1}(x) = \frac{3x+2}{5x-3} \quad \text{...(ii)}$$

$\therefore$ $f^{-1}(x) = f(x)$ [from (i) and (ii)]

Q.43 If $f:[0, 1] \to [0, 1]$ be defined by $f(x) = \begin{cases} x, & \text{if } x \text{ is rational} \\ 1-x, & \text{if } x \text{ is irrational} \end{cases}$

then $(fof)x$ is

(a) constant (b) $1 + x$ (c) x (d) None of these

Ans. **(c)** Given that, $f:[0, 1] \to [0, 1]$ be defined by

$$f(x) = \begin{cases} x, & \text{if } x \text{ is rational} \\ 1-x, & \text{if } x \text{ is irrational} \end{cases}$$

$\therefore$ $(fof)x = f(f(x)) = x$

Q.44 If $f:[2, \infty) \to R$ be the function defined by $f(x) = x^2 - 4x + 5$, then the range of f is

(a) R (b) $[1, \infty)$ (c) $[4, \infty)$ (d) $[5, \infty)$

Ans. **(b)** Given that, $f(x) = x^2 - 4x + 5$

Let $y = x^2 - 4x + 5$

$\Rightarrow$ $y = x^2 - 4x + 4 + 1 = (x-2)^2 + 1$

$\Rightarrow$ $(x-2)^2 = y - 1 \Rightarrow x - 2 = \sqrt{y-1}$

$\Rightarrow$ $x = 2 + \sqrt{y-1}$

$\therefore$ $y - 1 \geq 0, y \geq 1$

Range $= [1, \infty)$

Q.45 If $f: N \to R$ be the function defined by $f(x) = \dfrac{2x-1}{2}$ and $g: Q \to R$ be another function defined by $g(x) = x + 2$. Then, $gof\left(\frac{3}{2}\right)$ is

(a) 1 (b) 1 (c) $\frac{7}{2}$ (d) None of these

Ans. ***(d)*** Given that, $f(x) = \dfrac{2x-1}{2}$ and $g(x) = x + 2$

$$gof\left(\frac{3}{2}\right) = g\left[f\left(\frac{3}{2}\right)\right] = g\left(\frac{2 \times \frac{3}{2} - 1}{2}\right) = g(1) = 1 + 2 = 3$$

Q.46 If $f : R \to R$ be defined by $f(x) = \begin{cases} 2x : x > 3 \\ x^2 : 1 < x \le 3 \\ 3x : x \le 1 \end{cases}$

Then, $f(-1) + f(2) + f(4)$ is

(a) 9 (b) 14 (c) 5 (d) None of these

Ans. ***(a)*** Given that, $f(x) = \begin{cases} 2x : x > 3 \\ x^2 : 1 < x \le 3 \\ 3x : x \le 1 \end{cases}$

$$f(-1) + f(2) + f(4) = 3(-1) + (2)^2 + 2 \times 4$$
$$= -3 + 4 + 8 = 9$$

Q.47 If $f : R \to R$ be given by $f(x) = \tan x$, then $f^{-1}(1)$ is

(a) $\dfrac{\pi}{4}$ (b) $\left\{n\pi + \dfrac{\pi}{4} : n \in Z\right\}$

(c) Does not exist (d) None of these

Ans. ***(a)*** Given that, $f(x) = \tan x$

Let $y = \tan x$

$\Rightarrow$ $x = \tan^{-1} y$

$\Rightarrow$ $f^{-1}(x) = \tan^{-1} x$

$\Rightarrow$ $f^{-1}(1) = \tan^{-1} 1$

$\Rightarrow$ $= \tan^{-1}\left(\tan\dfrac{\pi}{4}\right)$ $\left[\because \tan\dfrac{\pi}{4} = 1\right]$

$= \dfrac{\pi}{4}$ $[\because \tan^{-1}(\tan\theta) = \theta]$

Fill in the blanks in each of the exercises 48 to 52.

Q.48 Let the relation R be defined in N by aRb, if $2a + 3b = 30$. Then, $R = \ldots\ldots$.

Ans. Given that, $2a + 3b = 30$

$$3b = 30 - 2a$$
$$b = \frac{30 - 2a}{3}$$

For $a = 3, b = 8$

$a = 6, b = 6$

$a = 9, b = 4$

$a = 12, b = 2$

$R = \{(3, 8), (6, 6), (9, 4), (12, 2)\}$

Q.49 If the relation R be defined on the set $A = \{1, 2, 3, 4, 5\}$ by $R = \{(a, b) : |a^2 - b^2| < 8\}$. Then, R is given by ....

Ans. Given,
$$A = \{1, 2, 3, 4, 5\},$$
$$R = \{(a, b) : |a^2 - b^2| < 8\}$$
$$R = \{(1, 1), (1, 2), (2, 1), (2, 2), (2, 3), (3, 2), (3, 3), (4, 3), (3, 4), (4, 4), (5, 5)\}$$

Q.50 If $f = \{(1, 2), (3, 5), (4, 1)\}$ and $g = \{(2, 3), (5, 1), (1, 3)\}$, then $gof = ...$ and $fog = ...$.

Ans. Given that, $f = \{(1, 2), (3, 5), (4, 1)\}$ and $g = \{(2, 3), (5, 1), (1, 3)\}$
$$gof(1) = g\{f(1)\} = g(2) = 3$$
$$gof(3) = g\{f(3)\} = g(5) = 1$$
$$gof(4) = g\{f(4)\} = g(1) = 3$$
$$gof = \{(1, 3), (3, 1), (4, 3)\}$$
Now,
$$fog(2) = f\{g(2)\} = f(3) = 5$$
$$fog(5) = f\{g(5)\} = f(1) = 2$$
$$fog(1) = f\{g(1)\} = f(3) = 5$$
$$fog = \{(2, 5), (5, 2), (1, 5)\}$$

Q.51 If $f : R \to R$ be defined by $f(x) = \dfrac{x}{\sqrt{1 + x^2}}$, then $(fofof)(x) = ...$.

Ans. Given that,
$$f(x) = \frac{x}{\sqrt{1 + x^2}}$$
$$(fofof)(x) = f[f\{f(x)\}]$$
$$= f\left[f\left(\frac{x}{\sqrt{1 + x^2}}\right)\right] = f\left(\frac{\frac{x}{\sqrt{1 + x^2}}}{\sqrt{1 + \frac{x^2}{1 + x^2}}}\right)$$
$$= f\left(\frac{x}{\sqrt{1 + 2x^2}}\right)$$
$$= \frac{\frac{x}{\sqrt{1 + 2x^2}}}{\sqrt{1 + \frac{x^2}{1 + 2x^2}}} = \frac{x\sqrt{1 + 2x^2}}{\sqrt{1 + 2x^2}\sqrt{1 + 3x^2}} = \frac{x}{\sqrt{1 + 3x^2}}$$

Q.52 If $f(x) = [4 - (x - 7)^3]$, then $f^{-1}(x) = ...$.

Ans. Given that,
$$f(x) = \{4 - (x - 7)^3\}$$
Let
$$y = [4 - (x - 7)^3]$$
$$(x - 7)^3 = 4 - y$$
$$(x - 7) = (4 - y)^{1/3}$$
$$\Rightarrow \quad x = 7 + (4 - y)^{1/3}$$
$$f^{-1}(x) = 7 + (4 - x)^{1/3}$$

State True or False for the statements in each of the exercises 53 to 63.

Q.53 Let $R = \{(3, 1), (1, 3), (3, 3)\}$ be a relation defined on the set $A = \{1, 2, 3\}$. Then, R is symmetric, transitive but not reflexive.

Ans. *False*

Given that, $R = \{(3, 1), (1, 3), (3, 3)\}$ be defined on the set $A = \{1, 2, 3\}$

$(1, 1) \notin R$

So, R is not reflexive. $(3, 1) \in R, (1, 3) \in R$

Hence, R is symmetric.

Since, $(1, 3) \in R, (3, 1) \in R$

But $(1, 1) \notin R$

Hence, R is not transitive.

Q.54 If $f : R \to R$ be the function defined by $f(x) = \sin(3x + 2)\ \forall\ x \in R$. Then, f is invertible.

Ans. *False,*

Given that, $f(x) = \sin(3x + 2), \forall\ x \in R$ is not one-one function for all $x \in R$.

So, f is not invertible.

Q.55 Every relation which is symmetric and transitive is also reflexive.

Ans. *False*

For example Let R be a relation defined by

$R = \{(1, 2), (2, 1), (1, 1), (2, 2)\}$ on the set $A = \{1, 2, 3\}$

It is clear that $(3, 3) \notin R$. So, it is not reflexive.

and R is symmetric.

Because $(1, 2) \in R \Rightarrow (2, 1) \in R$

and R is transitive. $(1, 2) \in R$ and $(2, 1) \in R \Rightarrow (1, 1) \in R$

Q.56 An integer m is said to be related to another integer n, if m is a integral multiple of n. This relation in Z is reflexive, symmetric and transitive.

Ans. *False*

The given relation is reflexive and transitive but not symmetric.

Q.57 If $A = \{0, 1\}$ and N be the set of natural numbers. Then, the mapping $f : N \to A$ defined by $f(2n - 1) = 0, f(2n) = 1, \forall\ n \in N$, is onto.

Ans. *True*

Given, $A = \{0, 1\} \Rightarrow f(2n - 1) = 0, f(2n) = 1, \forall\ n \in N$

So, $A =$ Range

So, the mapping $f : N \to A$ is onto.

Q.58 The relation R on the set $A = \{1, 2, 3\}$ defined as $R = \{(1, 1), (1, 2), (2, 1), (3, 3)\}$ is reflexive, symmetric and transitive.

Ans. *False*

Given that, $R = \{(1, 1), (1, 2), (2, 1), (3, 3)\}$

$(2, 2) \notin R$

So, R is not reflexive.

Q.59 The composition of function is commutative.

Ans. *False*

Let $f(x) = x^2$

and $g(x) = x + 1$

$$fog(x) = f\{g(x)\} = f(x + 1)$$
$$= (x + 1)^2 = x^2 + 2x + 1$$
$$gof(x) = g\{f(x)\} = g(x^2) = x^2 + 1$$

$\therefore$ $fog(x) \neq gof(x)$

Therefore, the composition of function is not commutative.

Q.60 The composition of function is associative.

Ans. *True*

Let $f(x) = x,\ g(x) = x + 1$

and $h(x) = 2x - 1$

Then, $fo\{goh(x)\} = f[g\{h(x)\}]$

$$= f\{g(2x - 1)\}$$
$$= f(2x - 1 + 1)$$
$$= f(2x) = 2x$$

$\therefore$ $(fog)\,oh(x) = (fog)\{h(x)\}$

$$= (fog)(2x - 1)$$
$$= f\{g(2x - 1)\}$$
$$= f(2x - 1 + 1)$$
$$= f(2x) = 2x$$

Q.61 Every function is invertible.

Ans. *False*

Only bijective functions are invertible.

***Q.62** A binary operation on a set has always the identity element.

Ans. *False*

'+' is a binary operation on the set N but it has no identity element.

Note (✱) Topics related to these question have been removed from latest NCERT.

Chapter 02

Inverse Trigonometric Functions

Short Answer Type Questions

Q.1 Find the value of $\tan^{-1}\left(\tan\frac{5\pi}{6}\right)+\cos^{-1}\left(\cos\frac{13\pi}{6}\right)$.

Thinking Process Use the property, $\tan^{-1}(\tan x)=x, x\in\left(-\frac{\pi}{2},\frac{\pi}{2}\right)$ and $\cos^{-1}(\cos x)=x$, $x\in[0,\pi]$ to get the answer.

Ans. We know that, $\tan^{-1}(\tan x)=x;\ x\in\left(-\frac{\pi}{2},\frac{\pi}{2}\right)$ and $\cos^{-1}\cos x=x;\ x\in[0,\pi]$

We have, $\tan^{-1}\left(\tan\frac{5\pi}{6}\right)+\cos^{-1}\left(\cos\frac{13\pi}{6}\right)$

$$=\tan^{-1}\left[\tan\left(\pi-\frac{\pi}{6}\right)\right]+\cos^{-1}\left[\cos\left(\pi+\frac{7\pi}{6}\right)\right]$$

$$=\tan^{-1}\left(-\tan\frac{\pi}{6}\right)+\cos^{-1}\left(-\cos\frac{7\pi}{6}\right)$$

[$\because$ $\tan(\pi-\theta)=-\tan\theta$ and $\cos(\pi+\theta)=-\cos\theta$]

Let $\tan^{-1}\left(-\tan\frac{\pi}{6}\right)=x$ and $\cos^{-1}\left(-\cos\frac{7\pi}{6}\right)=y$

$\therefore\quad \tan^{-1}\left(\tan\frac{5\pi}{6}\right)+\cos^{-1}\left(\cos\frac{13\pi}{6}\right)=x+y$ ...(i)

Now, $\tan^{-1}\left(-\tan\frac{\pi}{6}\right)=x$

$\Rightarrow\quad -\tan\frac{\pi}{6}=\tan x$

$\Rightarrow\quad \tan\left(-\frac{\pi}{6}\right)=\tan x$ [$\because$ $\tan(-\theta)=-\tan\theta$]

$\Rightarrow\quad x=\frac{-\pi}{6}\in\left(-\frac{\pi}{2},\frac{\pi}{2}\right)$ [$\because$ principal value of $\tan^{-1}$ is $\left(-\frac{\pi}{2},\frac{\pi}{2}\right)$]

Also, $\cos^{-1}\left(-\cos\frac{7\pi}{6}\right)=y$

$\Rightarrow\quad -\cos\frac{7\pi}{6}=\cos y$

$\Rightarrow\quad -\cos\left(\pi+\frac{\pi}{6}\right)=\cos y$

$\Rightarrow\quad \cos\frac{\pi}{6}=\cos y$ [$\because$ $\cos(\pi+\theta)=-\cos\theta$]

$$\Rightarrow \quad y = \frac{\pi}{6} \in [0, \pi] \qquad [\because \text{principal value of } \cos^{-1} \text{ is } [0, \pi]]$$

$$\therefore \tan^{-1}\left(\tan\frac{5\pi}{6}\right) + \cos^{-1}\left(\cos\frac{13\pi}{6}\right) = \frac{-\pi}{6} + \frac{\pi}{6} = 0 \qquad \text{[from Eq. (i)]}$$

Mistake Alert! *Students may mistakenly take* $\tan^{-1}\left(\tan\frac{5\pi}{6}\right) = \frac{5\pi}{6}$ *and* $\cos^{-1}\left(\cos\frac{13\pi}{6}\right) = \frac{13\pi}{6}$ *without checking the principal value of* $\tan^{-1}$ *and* $\cos^{-1}$.

Q.2 Evaluate $\cos\left[\cos^{-1}\left(\frac{-\sqrt{3}}{2}\right) + \frac{\pi}{6}\right]$.

Ans. We have, $\cos\left[\cos^{-1}\left(\frac{-\sqrt{3}}{2}\right) + \frac{\pi}{6}\right] = \cos\left[\cos^{-1}\left(\cos\frac{5\pi}{6}\right) + \frac{\pi}{6}\right]$

$$\left[\because \cos\frac{5\pi}{6} = \cos\left(\pi - \frac{\pi}{6}\right) = \frac{-\sqrt{3}}{2}\right]$$

$$= \cos\left(\frac{5\pi}{6} + \frac{\pi}{6}\right) \qquad [\because \cos^{-1}(\cos x) = x;\ x \in [0, \pi]]$$

$$= \cos\left(\frac{6\pi}{6}\right)$$

$$= \cos(\pi) = -1$$

Q.3 Prove that $\cot\left(\frac{\pi}{4} - 2\cot^{-1} 3\right) = 7$.

Ans. We have to prove that, $\cot\left(\frac{\pi}{4} - 2\cot^{-1} 3\right) = 7$

$$\text{LHS} = \cot\left(\frac{\pi}{4} - 2\cot^{-1} 3\right)$$

Let $\quad 2\cot^{-1} 3 = x \Rightarrow \cot^{-1} 3 = \frac{x}{2} \Rightarrow \cot\frac{x}{2} = 3$

$\therefore \quad \tan\frac{x}{2} = \frac{1}{3}$

Now, $\quad \text{LHS} = \cot\left(\frac{\pi}{4} - x\right)$

$$= \frac{\cot\frac{\pi}{4}\cot x + 1}{\cot x - \cot\frac{\pi}{4}} = \frac{\cot x + 1}{\cot x - 1}$$

$$= \frac{\frac{1}{\tan x} + 1}{\frac{1}{\tan x} - 1} = \frac{1 + \tan x}{1 - \tan x}$$

$$= \frac{\frac{1+2\tan\frac{x}{2}}{1-\tan^2\frac{x}{2}}}{1-\frac{2\tan\frac{x}{2}}{1-\tan^2\frac{x}{2}}} \qquad \left[\because \tan 2\theta = \frac{2\tan\theta}{1-\tan^2\theta}\right]$$

$$= \frac{1+\frac{2\times\frac{1}{3}}{1-\left(\frac{1}{3}\right)^2}}{1-\frac{2\times\frac{1}{3}}{1-\left(\frac{1}{3}\right)^2}} \qquad \left[\because \tan\frac{x}{2} = \frac{1}{3}\right]$$

$$= \frac{1+\frac{\frac{2}{3}}{\frac{8}{9}}}{1-\frac{\frac{2}{3}}{\frac{8}{9}}} = \frac{1+\frac{3}{4}}{1-\frac{3}{4}} = 7 = \text{RHS}$$

$$\therefore \qquad \cot\left(\frac{\pi}{4} - 2\cot^{-1}3\right) = 7$$

Hence proved.

Q.4 Find the value of $\tan^{-1}\left(-\frac{1}{\sqrt{3}}\right) + \cot^{-1}\left(\frac{1}{\sqrt{3}}\right) + \tan^{-1}\left[\sin\left(\frac{-\pi}{2}\right)\right]$.

Ans. We have, $\tan^{-1}\left(-\frac{1}{\sqrt{3}}\right) + \cot^{-1}\left(\frac{1}{\sqrt{3}}\right) + \tan^{-1}\left[\sin\left(\frac{-\pi}{2}\right)\right]$

$$= \tan^{-1}\left(-\tan\frac{\pi}{6}\right) + \cot^{-1}\left(\cot\frac{\pi}{3}\right) + \tan^{-1}\left[-\sin\frac{\pi}{2}\right]$$

$$= \tan^{-1}\left[\tan\left(\frac{-\pi}{6}\right)\right] + \frac{\pi}{3} + \tan^{-1}(-1)$$

$$= \frac{-\pi}{6} + \frac{\pi}{3} + \tan^{-1}\left(-\tan\frac{\pi}{4}\right) = \frac{-\pi}{6} + \frac{\pi}{3} + \tan^{-1}\left[\tan\left(\frac{-\pi}{4}\right)\right]$$

$$= -\frac{\pi}{6} + \frac{\pi}{3} - \frac{\pi}{4} = \frac{-2\pi + 4\pi - 3\pi}{12} = \frac{-5\pi + 4\pi}{12} = -\frac{\pi}{12}$$

Q.5 Find the value of $\tan^{-1}\left(\tan\frac{2\pi}{3}\right)$.

Ans. We have, $\tan^{-1}\left(\tan\frac{2\pi}{3}\right) = \tan^{-1}\tan\left(\pi - \frac{\pi}{3}\right)$

$$= \tan^{-1}\left(-\tan\frac{\pi}{3}\right) \qquad [\because \tan(\pi - x) = -\tan x]$$

$$= \tan^{-1}\left[\tan\left(-\frac{\pi}{3}\right)\right] = -\frac{\pi}{3} \in \left(-\frac{\pi}{2}, \frac{\pi}{2}\right)$$

Note Remember that, $\tan^{-1}\left(\tan\frac{2\pi}{3}\right) \neq \frac{2\pi}{3}$

Since, $\tan^{-1}(\tan x) = x$, if $x \in \left(-\frac{\pi}{2}, \frac{\pi}{2}\right)$ and $\frac{2\pi}{3} \notin \left(\frac{-\pi}{2}, \frac{\pi}{2}\right)$

***Q.6** Show that $2\tan^{-1}(-3) = \frac{-\pi}{2} + \tan^{-1}\left(\frac{-4}{3}\right)$.

Ans. LHS $= 2\tan^{-1}(-3) = -2\tan^{-1}3$ $\quad [\because \tan^{-1}(-x) = -\tan^{-1}x, x \in R]$

$$= -\left[\cos^{-1}\frac{1-3^2}{1+3^2}\right] \quad \left[\because 2\tan^{-1}x = \cos^{-1}\frac{1-x^2}{1+x^2}, x \geq 0\right]$$

$$= -\left[\cos^{-1}\left(\frac{-8}{10}\right)\right] = -\left[\cos^{-1}\left(\frac{-4}{5}\right)\right]$$

$$= -\left[\pi - \cos^{-1}\left(\frac{4}{5}\right)\right] \quad \{\because \cos^{-1}(-x) = \pi - \cos^{-1}x, x \in [-1, 1]\}$$

$$= -\pi + \cos^{-1}\left(\frac{4}{5}\right)$$

Now let $\cos^{-1}\left(\frac{4}{5}\right) = \theta \Rightarrow \cos\theta = \frac{4}{5}$

$\Rightarrow \quad \tan\theta = \frac{3}{4} \Rightarrow \theta = \tan^{-1}\frac{3}{4}$

5 3 θ 4

$$= -\pi + \tan^{-1}\left(\frac{3}{4}\right) = -\pi + \left[\frac{\pi}{2} - \cot^{-1}\left(\frac{3}{4}\right)\right] \quad \left[\because \tan^{-1}x + \cot^{-1}x = \frac{\pi}{2}\right]$$

$$= -\frac{\pi}{2} - \cot^{-1}\frac{3}{4} = -\frac{\pi}{2} - \tan^{-1}\frac{4}{3} \quad \left[\because \cot^{-1}x = \tan^{-1}\frac{1}{x}\right]$$

$$= -\frac{\pi}{2} + \tan^{-1}\left(\frac{-4}{3}\right) \quad [\because \tan^{-1}(-x) = -\tan^{-1}x]$$

$= \text{RHS}$ **Hence proved.**

Q.7 Find the real solution of

$$\tan^{-1}\sqrt{x(x+1)} + \sin^{-1}\sqrt{x^2+x+1} = \frac{\pi}{2}.$$

Ans. We have, $\tan^{-1}\sqrt{x(x+1)} + \sin^{-1}\sqrt{x^2+x+1} = \frac{\pi}{2}$ ...(i)

Let $\tan^{-1}\sqrt{x(x+1)} = \alpha \Rightarrow \sqrt{x^2+x} = \tan\alpha$

$\Rightarrow \quad \tan^2\alpha = x^2 + x$ ...(ii)

Again let $\sin^{-1}\sqrt{x^2+x+1} = \beta \Rightarrow \sqrt{x^2+x+1} = \sin\beta$

$\Rightarrow \quad \sin^2\beta = x^2 + x + 1$ ...(iii)

From Eq. (i), we get

$$\alpha + \beta = \frac{\pi}{2} \Rightarrow \beta = \frac{\pi}{2} - \alpha \quad \text{...(iv)}$$

From Eqs.(ii) and (iii), we get

$$\sin^2\beta = \tan^2\alpha + 1 \Rightarrow \sin^2\beta = \sec^2\alpha \quad [\because 1 + \tan^2\theta = \sec^2\theta]$$

Note (✱) Topics related to these question have been removed from latest NCERT.

$\Rightarrow \quad \sin^2\beta\cos^2\alpha = 1 \Rightarrow \sin^2\left(\frac{\pi}{2}-\alpha\right)\cos^2\alpha = 1$ [from Eq.(iv)]

$\Rightarrow \quad \cos^2\alpha\cos^2\alpha = 1 \Rightarrow \cos^4\alpha = \cos 0 \Rightarrow \alpha = 0$

On putting value of α in Eq. (ii), we get

$$\tan^2(0) = x^2 + x \Rightarrow x(x+1) = 0 \Rightarrow x = 0, -1$$

Q.8 Find the value of $\sin\left(2\tan^{-1}\frac{1}{3}\right) + \cos(\tan^{-1}2\sqrt{2})$.

Ans. We have, $\sin\left(2\tan^{-1}\frac{1}{3}\right) + \cos(\tan^{-1}2\sqrt{2})$

Let $\tan^{-1}\frac{1}{3} = x \Rightarrow \tan x = \frac{1}{3}$ ...(i)

Again, let $\tan^{-1}2\sqrt{2} = y \Rightarrow \tan y = 2\sqrt{2}$ ...(ii)

$\therefore \sin\left(2\tan^{-1}\frac{1}{3}\right) + \cos(\tan^{-1}2\sqrt{2})$

$$= \sin(2x) + \cos(y) = \frac{2\tan x}{1+\tan^2 x} + \frac{1}{\sec y} \quad \left[\because \sin 2\theta = \frac{2\tan\theta}{1+\tan^2\theta}\right]$$

$$= \frac{2\times\frac{1}{3}}{1+\left(\frac{1}{3}\right)^2} + \frac{1}{\sqrt{1+\tan^2 y}} \quad [\text{from Eq.(i) and } 1+\tan^2\theta = \sec^2\theta]$$

$$= \frac{\frac{2}{3}}{\frac{10}{9}} + \frac{1}{\sqrt{1+(2\sqrt{2})^2}} \quad [\text{from Eq.(ii)}]$$

$$= \frac{3}{5} + \frac{1}{\sqrt{9}} = \frac{3}{5} + \frac{1}{3} = \frac{9+5}{15} = \frac{14}{15}$$

Q.9 If $2\tan^{-1}(\cos\theta) = \tan^{-1}(2\operatorname{cosec}\theta)$, then show that $\theta = \frac{\pi}{4}$, where n is any integer.

- **Concept Insight** In this question, we will use the identities, $\tan 2\theta = \frac{2\tan\theta}{1-\tan^2\theta}$ and $\sin^2\theta + \cos^2\theta = 1$.

Ans. We have, $2\tan^{-1}(\cos\theta) = \tan^{-1}(2\operatorname{cosec}\theta)$ ...(i)

Let $\tan^{-1}(\cos\theta) = \alpha$

$\Rightarrow \quad \tan\alpha = \cos\theta$ ...(ii)

$\therefore \quad 2\alpha = \tan^{-1}(2\operatorname{cosec}\theta)$ [from Eq.(i)]

$\Rightarrow \quad \tan(2\alpha) = 2\operatorname{cosec}\theta$

$\Rightarrow \quad \frac{2\tan\alpha}{1-\tan^2\alpha} = \frac{2}{\sin\theta}$

$\Rightarrow \quad \frac{\cos\theta}{1-\cos^2\theta} = \frac{1}{\sin\theta}$ [from Eq.(ii)]

$\Rightarrow \quad \frac{\cos\theta}{\sin^2\theta} = \frac{1}{\sin\theta} \Rightarrow \frac{\cos\theta}{\sin\theta} = 1$

$\Rightarrow \quad \cot\theta = 1$

$\Rightarrow \quad \cot\theta = \cot\frac{\pi}{4}$

$\therefore \quad \theta = \frac{\pi}{4}$

Q.10 Show that $\cos\left(2\tan^{-1}\frac{1}{7}\right)=\sin\left(4\tan^{-1}\frac{1}{3}\right)$.

Ans. We have to show that, $\cos\left(2\tan^{-1}\frac{1}{7}\right)=\sin\left(4\tan^{-1}\frac{1}{3}\right)$

$$\text{LHS} = \cos\left(2\tan^{-1}\frac{1}{7}\right)$$

Let $\tan^{-1}\frac{1}{7}=\alpha \Rightarrow \tan\alpha=\frac{1}{7}$

$\therefore$
$$\text{LHS}=\cos(2\alpha)=\frac{1-\tan^2\alpha}{1+\tan^2\alpha} \qquad \left[\because \cos 2\theta=\frac{1-\tan^2\theta}{1+\tan^2\theta}\right]$$

$$=\frac{1-\left(\frac{1}{7}\right)^2}{1+\left(\frac{1}{7}\right)^2}=\frac{\frac{49-1}{49}}{\frac{49+1}{49}}=\frac{48}{50}=\frac{24}{25}$$

Now, $\text{RHS}=\sin\left(4\tan^{-1}\frac{1}{3}\right)$

Let $\tan^{-1}\frac{1}{3}=\beta\Rightarrow\tan\beta=\frac{1}{3}$

$\therefore$
$$\text{RHS}=\sin(4\beta)=\sin(2.2\beta)$$
$$=2\sin 2\beta\cos 2\beta \qquad [\because \sin 2\theta=2\sin\theta\cos\theta]$$
$$=2\cdot\frac{2\tan\beta}{1+\tan^2\beta}\cdot\frac{1-\tan^2\beta}{1+\tan^2\beta}$$
$$=2\cdot\left(\frac{2\times\frac{1}{3}}{1+\left(\frac{1}{3}\right)^2}\right)\cdot\left(\frac{1-\left(\frac{1}{3}\right)^2}{1+\left(\frac{1}{3}\right)^2}\right)$$
$$=2\cdot\left(\frac{\frac{2}{3}}{\frac{10}{9}}\right)\left(\frac{\frac{8}{9}}{\frac{10}{9}}\right)$$
$$=2\times\frac{2}{3}\times\frac{9}{10}\times\frac{8}{9}\times\frac{9}{10}=\frac{24}{25}$$

$\therefore$ LHS = RHS **Hence proved.**

Mistake Alert! *Students may incorrectly apply the identity* $\cos 2\theta=\frac{1+\tan^2\theta}{1-\tan^2\theta}$ *instead of* $\cos 2\theta=\frac{1-\tan^2\theta}{1+\tan^2\theta}$.

Q.11 Solve the following equation $\cos(\tan^{-1}x)=\sin\left(\cot^{-1}\frac{3}{4}\right)$.

Ans. We have, $\cos(\tan^{-1}x)=\sin\left(\cot^{-1}\frac{3}{4}\right)$ ...(i)

Let $\tan^{-1}x=\theta_1 \Rightarrow \tan\theta_1=\frac{x}{1}$

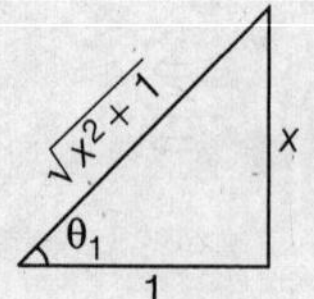

$\Rightarrow \quad \cos\theta_1 = \frac{1}{\sqrt{x^2+1}} \Rightarrow \theta_1 = \cos^{-1}\frac{1}{\sqrt{x^2+1}}$

Again, let $\quad \cot^{-1}\frac{3}{4} = \theta_2 \Rightarrow \cot\theta_2 = \frac{3}{4}$

$\Rightarrow \quad \sin\theta_2 = \frac{4}{5}$

$\Rightarrow \quad \theta_2 = \sin^{-1}\frac{4}{5}$

$\cos(\theta_1) = \sin(\theta_2)$ ...(ii)

On substituting the values of θ_1 and θ_2 in Eq. (i), we get [from Eq. (i)]

$$\cos\left(\cos^{-1}\frac{1}{\sqrt{x^2+1}}\right) = \sin\left(\sin^{-1}\left(\frac{4}{5}\right)\right)$$

$$\Rightarrow \quad \frac{1}{\sqrt{x^2+1}} = \frac{4}{5}$$

$[\because \cos(\cos^{-1}x) = x, x \in [-1, 1] \text{ and } \sin(\sin^{-1}x) = x, x \in [-1, 1]]$

On squaring both sides, we get

$$16(x^2+1) = 25$$

$$\Rightarrow \quad 16x^2 = 9$$

$$\Rightarrow \quad x^2 = \left(\frac{3}{4}\right)^2$$

$$\therefore \quad x = \pm\frac{3}{4}$$

Long Answer Type Questions

Q.12 Prove that $\tan^{-1}\left(\frac{\sqrt{1+x^2}+\sqrt{1-x^2}}{\sqrt{1+x^2}-\sqrt{1-x^2}}\right) = \frac{\pi}{4}+\frac{1}{2}\cos^{-1}x^2$.

Ans. We have to prove that,

$$\tan^{-1}\left(\frac{\sqrt{1+x^2}+\sqrt{1-x^2}}{\sqrt{1+x^2}-\sqrt{1-x^2}}\right) = \frac{\pi}{4}+\frac{1}{2}\cos^{-1}x^2$$

Now, $\quad \text{LHS} = \tan^{-1}\left(\frac{\sqrt{1+x^2}+\sqrt{1-x^2}}{\sqrt{1+x^2}-\sqrt{1-x^2}}\right)$ ...(i)

Let $x^2 = \cos 2\theta$

$\Rightarrow \quad \cos^{-1}x^2 = 2\theta \Rightarrow \theta = \frac{1}{2}\cos^{-1}x^2$ ...(ii)

$\therefore \quad \sqrt{1+x^2} = \sqrt{1+\cos 2\theta}$

$= \sqrt{1+2\cos^2\theta - 1} = \sqrt{2}\cos\theta$ $[\because \cos 2\theta = 2\cos^2\theta -`1]$

and $\quad \sqrt{1-x^2} = \sqrt{1-\cos 2\theta}$

$= \sqrt{1-1+2\sin^2\theta} = \sqrt{2}\sin\theta$

$[\because \cos 2\theta = 1 - 2\sin\sin^2\theta]$

On substituting the values of $\sqrt{1+x^2}$ and $\sqrt{1-x^2}$ in Eq. (i), we get

$$\text{LHS} = \tan^{-1}\left(\frac{\sqrt{2}\cos\theta + \sqrt{2}\sin\theta}{\sqrt{2}\cos\theta - \sqrt{2}\sin\theta}\right)$$

$$= \tan^{-1}\left(\frac{\cos\theta + \sin\theta}{\cos\theta - \sin\theta}\right)$$

On dividing both numerator and denominator by $\cos\theta$, we get

$$= \tan^{-1}\left(\frac{1+\tan\theta}{1-\tan\theta}\right) = \tan^{-1}\left(\frac{\tan\frac{\pi}{4} + \tan\theta}{1-\tan\frac{\pi}{4}\cdot\tan\theta}\right)$$

$$= \tan^{-1}\left[\tan\left(\frac{\pi}{4}+\theta\right)\right] \qquad \left[\because \tan(x+y) = \frac{\tan x + \tan y}{1-\tan x\cdot\tan y}\right]$$

$$= \frac{\pi}{4} + \theta = \frac{\pi}{4} + \frac{1}{2}\cos^{-1}x^2 \qquad \text{[from Eq. (ii)]}$$

$= \text{RHS}$ **Hence proved.**

Mistake Alert! *Students may incorrectly apply the identity* $\cos 2\theta = 1 - 2\cos^2\theta$ *instead of* $\cos 2\theta = 2\cos^2\theta - 1$ *and* $\cos 2\theta = 2\sin^2\theta - 1$ *instead of* $\cos 2\theta = 1 - 2\sin^2\theta$.

Q.13 Find the simplified form of

$$\cos^{-1}\left(\frac{3}{5}\cos x + \frac{4}{5}\sin x\right), \text{ where } x \in \left[\frac{-3\pi}{4}, \frac{\pi}{4}\right].$$

Ans. We have, $\cos^{-1}\left[\frac{3}{5}\cos x + \frac{4}{5}\sin x\right], x \in \left[\frac{-3\pi}{4}, \frac{\pi}{4}\right]$

Let $\cos y = \frac{3}{5}$

$\Rightarrow$ $\sin y = \frac{4}{5}$

$\Rightarrow$ $y = \cos^{-1}\frac{3}{5} = \sin^{-1}\frac{4}{5} = \tan^{-1}\left(\frac{4}{3}\right)$

$$\therefore \cos^{-1}\left(\frac{3}{5}\cos x + \frac{4}{5}\sin x\right) = \cos^{-1}[\cos y\cdot\cos x + \sin y\cdot\sin x]$$

$$= \cos^{-1}[\cos(y-x)] \qquad [\because \cos(A-B) = \cos A\cdot\cos B + \sin A\cdot\sin B]$$

$$= y - x = \tan^{-1}\frac{4}{3} - x \qquad \left[\because y = \tan^{-1}\frac{4}{3}\right]$$

Q.14 Prove that $\sin^{-1}\frac{8}{17} + \sin^{-1}\frac{3}{5} = \sin^{-1}\frac{77}{85}$.

Ans. To prove, $\sin^{-1}\frac{8}{17} + \sin^{-1}\frac{3}{5} = \sin^{-1}\frac{77}{85}$

Let $\sin^{-1}\frac{8}{17} = x$

$\Rightarrow$ $\sin x = \frac{8}{17}$

$\Rightarrow \qquad \cos x = \sqrt{1-\sin^2 x} = \sqrt{1-\left(\frac{8}{17}\right)^2}$

$= \sqrt{\frac{289-64}{289}} = \sqrt{\frac{225}{289}} = \frac{15}{17}$

Let $\qquad \sin^{-1}\frac{3}{5} = y$

$\Rightarrow \qquad \sin y = \frac{3}{5} \Rightarrow \sin^2 y = \frac{9}{25}$

$\therefore \qquad \cos^2 y = 1 - \frac{9}{25}$

$\Rightarrow \qquad \cos^2 y = \left(\frac{4}{5}\right)^2 \Rightarrow \cos y = \frac{4}{5}$

Now, $\qquad \sin(x+y) = \sin x \cdot \cos y + \cos x \cdot \sin y$

$= \frac{8}{17}\cdot\frac{4}{5} + \frac{15}{17}\cdot\frac{3}{5}$

$= \frac{32}{85} + \frac{45}{85} = \frac{77}{85}$

$\Rightarrow \qquad x + y = \sin^{-1}\left(\frac{77}{85}\right)$

$\Rightarrow \qquad \sin^{-1}\frac{8}{17} + \sin^{-1}\frac{3}{5} = \sin^{-1}\frac{77}{85}$ **Hence proved.**

Mistake Alert! *Students may incorrectly apply the identity* $\cos\theta = 1 - \sin^2\theta$ *instead of* $\cos\theta = \sqrt{1-\sin^2\theta}$.

Q.15 Show that $\sin^{-1}\frac{5}{13} + \cos^{-1}\frac{3}{5} = \tan^{-1}\frac{63}{16}$.

Ans. Let $\qquad \sin^{-1}\frac{5}{13} = x$

$\sin x = \frac{5}{13} \Rightarrow \tan x = \frac{5}{12}$

Also, let $\qquad \cos^{-1}\frac{3}{5} = y$

$\Rightarrow \qquad \cos y = \frac{3}{5} \Rightarrow \tan y = \frac{4}{3}$

Now, $\qquad \tan(x+y) = \frac{\tan x + \tan y}{1 - \tan x \tan y}$

$= \frac{\frac{5}{12} + \frac{4}{3}}{1 - \frac{5}{12} \times \frac{4}{3}} = \frac{\frac{5+16}{12}}{\frac{36-20}{36}} = \frac{21}{12} \times \frac{36}{16} = \frac{63}{16}$

$\Rightarrow \qquad x + y = \tan^{-1}\frac{63}{16}$

$\Rightarrow \qquad \sin^{-1}\frac{5}{13} + \cos^{-1}\frac{3}{5} = \tan^{-1}\frac{63}{16}$ **Hence proved.**

Q.16 Prove that $\tan^{-1}\frac{1}{4}+\tan^{-1}\frac{2}{9}=\sin^{-1}\frac{1}{\sqrt{5}}$.

- Concept Insight In this question, we will use the identity $\sin(A+B)=\sin A\cos B+\cos A\sin B$.

Ans. Let $\tan^{-1}\frac{1}{4}=x \Rightarrow \tan x=\frac{1}{4}$

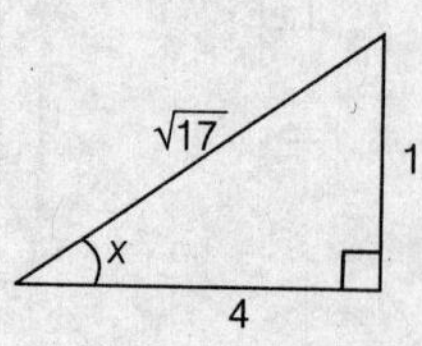

$\therefore$ $\sin x=\frac{1}{\sqrt{17}}$ and $\cos x=\frac{4}{\sqrt{17}}$

Also, let $\tan^{-1}\frac{2}{9}=y \Rightarrow \tan y=\frac{2}{9}$

$\therefore$ $\sin y=\frac{2}{\sqrt{85}}$ and $\cos y=\frac{9}{\sqrt{85}}$

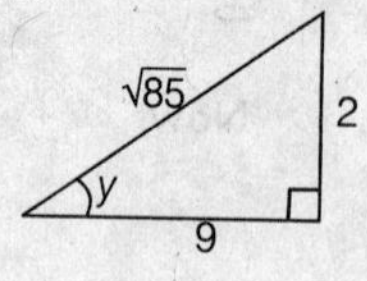

Now, $\sin(x+y)=\sin x\cos y+\cos x\sin y$

$$=\frac{1}{\sqrt{17}}\times\frac{9}{\sqrt{85}}+\frac{4}{\sqrt{17}}\times\frac{2}{\sqrt{85}}$$

$$=\frac{17}{\sqrt{17}\sqrt{85}}=\frac{\sqrt{17}}{\sqrt{85}}=\frac{1}{\sqrt{5}}$$

$\Rightarrow$ $x+y=\sin^{-1}\frac{1}{\sqrt{5}}$

$\Rightarrow$ $\tan^{-1}\frac{1}{4}+\tan^{-1}\frac{2}{9}=\sin^{-1}\frac{1}{\sqrt{5}}$ **Hence proved.**

*** Q.17** Find the value of $4\tan^{-1}\frac{1}{5}-\tan^{-1}\frac{1}{239}$.

Thinking Process Use the properties $2\tan^{-1}x=\tan^{-1}\left(\frac{2x}{1-x^2}\right)$ and $\tan^{-1}x-\tan^{-1}y=\tan^{-1}\left(\frac{x-y}{1+xy}\right)$ to get the desired value.

Ans. We have, $4\tan^{-1}\frac{1}{5}-\tan^{-1}\frac{1}{239}$

$$=2\cdot 2\tan^{-1}\frac{1}{5}-\tan^{-1}\frac{1}{239}$$

$$=2\cdot\left[\tan^{-1}\frac{\frac{2}{5}}{1-\left(\frac{1}{5}\right)^2}\right]-\tan^{-1}\frac{1}{239}\quad\left[\because 2\tan^{-1}x=\tan^{-1}\left(\frac{2x}{1-x^2}\right)\right]$$

$$=2\cdot\left[\tan^{-1}\left(\frac{\frac{2}{5}}{1-\frac{1}{25}}\right)\right]-\tan^{-1}\frac{1}{239}$$

$$=2\cdot\left[\tan^{-1}\left(\frac{2/5}{24/25}\right)\right]-\tan^{-1}\frac{1}{239}$$

$$=2\tan^{-1}\frac{5}{12}-\tan^{-1}\frac{1}{239}$$

Note (✱) Topics related to these question have been removed from latest NCERT.

$$= \tan^{-1}\left(\frac{2\cdot\frac{5}{12}}{1-\left(\frac{5}{12}\right)^2}\right) - \tan^{-1}\frac{1}{239} \quad \left[\because\ 2\tan^{-1}x = \tan^{-1}\left(\frac{2x}{1-x^2}\right)\right]$$

$$= \tan^{-1}\left(\frac{\frac{5}{6}}{1-\frac{25}{144}}\right) - \tan^{-1}\frac{1}{239}$$

$$= \tan^{-1}\left(\frac{144\times 5}{119\times 6}\right) - \tan^{-1}\frac{1}{239}$$

$$= \tan^{-1}\left(\frac{120}{119}\right) - \tan^{-1}\frac{1}{239}$$

$$= \tan^{-1}\left(\frac{\frac{120}{119}-\frac{1}{239}}{1+\frac{120}{119}\cdot\frac{1}{239}}\right) \quad \left[\because\ \tan^{-1}x - \tan^{-1}y = \tan^{-1}\left(\frac{x-y}{1+xy}\right)\right]$$

$$= \tan^{-1}\left(\frac{120\times 239-119}{119\times 239+120}\right)$$

$$= \tan^{-1}\left[\frac{28680-119}{28441+120}\right] = \tan^{-1}\left(\frac{28561}{28561}\right)$$

$$= \tan^{-1}(1) = \tan^{-1}\left(\tan\frac{\pi}{4}\right) = \frac{\pi}{4}$$

Q.18 Show that $\tan\left(\frac{1}{2}\sin^{-1}\frac{3}{4}\right) = \frac{4-\sqrt{7}}{3}$ and justify why the other value $\frac{4+\sqrt{7}}{3}$ is ignored?

Ans. We have to prove that, $\tan\left(\frac{1}{2}\sin^{-1}\frac{3}{4}\right) = \frac{4-\sqrt{7}}{3}$

Now, $\quad \text{LHS} = \tan\left[\frac{1}{2}\sin^{-1}\left(\frac{3}{4}\right)\right]$ ...(i)

Let $\quad \frac{1}{2}\sin^{-1}\frac{3}{4} = \theta \Rightarrow \sin^{-1}\frac{3}{4} = 2\theta$

$\Rightarrow \quad \sin 2\theta = \frac{3}{4} \Rightarrow \frac{2\tan\theta}{1+\tan^2\theta} = \frac{3}{4} \quad \left[\because\ \sin 2\theta = \frac{2\tan\theta}{1+\tan^2\theta}\right]$

$\Rightarrow \quad 3+3\tan^2\theta = 8\tan\theta$

$\Rightarrow \quad 3\tan^2\theta - 8\tan\theta + 3 = 0$

Let $\quad \tan\theta = y$

$\therefore \quad 3y^2 - 8y + 3 = 0$

$\Rightarrow \quad y = \frac{+8 \pm \sqrt{64-4\times 3\times 3}}{2\times 3} = \frac{8\pm\sqrt{28}}{6} = \frac{2[4\pm\sqrt{7}]}{2\times 3}$

$\Rightarrow \quad \tan\theta = \frac{4\pm\sqrt{7}}{3}$

$\Rightarrow \qquad \theta = \tan^{-1}\left[\frac{4 \pm \sqrt{7}}{3}\right]$

Since, $\qquad -\frac{\pi}{2} \le \sin^{-1}\frac{3}{4} \le \pi/2$

$\Rightarrow \qquad \frac{-\pi}{4} \le \frac{1}{2}\sin^{-1}\frac{3}{4} \le \pi/4$

$\therefore \qquad \tan\left(\frac{-\pi}{4}\right) \le \tan\frac{1}{2}\left(\sin^{-1}\frac{3}{4}\right) \le \tan\frac{\pi}{4}$

$\Rightarrow \qquad -1 \le \tan\left(\frac{1}{2}\sin^{-1}\frac{3}{4}\right) \le 1$

Thus, $\qquad \theta = \frac{4-\sqrt{7}}{3}$

$$\text{LHS} = \tan\left(\tan^{-1}\left(\frac{4-\sqrt{7}}{3}\right)\right)$$

$$= \frac{4-\sqrt{7}}{3} = \text{RHS}$$

Hence proved.

*** Q.19** If $a_1, a_2, a_3, \ldots, a_n$ is an arithmetic progression with common difference d, then evaluate the following expression.

$$\tan\left[\tan^{-1}\left(\frac{d}{1+a_1a_2}\right) + \tan^{-1}\left(\frac{d}{1+a_2a_3}\right) + \tan^{-1}\left(\frac{d}{1+a_3a_4}\right) + \ldots + \tan^{-1}\left(\frac{d}{1+a_{n-1}\,a_n}\right)\right]$$

Ans. In AP we have, $a_1 = a,\ a_2 = a + d,\ a_3 = a + 2d$

and $\qquad d = a_2 - a_1 = a_3 - a_2 = a_4 - a_3 = \ldots = a_n - a_{n-1}$

Given that, $\tan\left[\tan^{-1}\left(\frac{d}{1+a_1a_2}\right) + \tan^{-1}\left(\frac{d}{1+a_2a_3}\right) + \tan^{-1}\left(\frac{d}{1+a_3a_4}\right) + \ldots + \tan^{-1}\left(\frac{d}{1+a_{n-1}\cdot a_n}\right)\right]$

$$= \tan\left[\tan^{-1}\frac{a_2-a_1}{1+a_2\cdot a_1} + \tan^{-1}\frac{a_3-a_2}{1+a_3\cdot a_2} + \ldots + \tan^{-1}\frac{a_n-a_{n-1}}{1+a_n\cdot a_{n-1}}\right]$$

$$= \tan\left[(\tan^{-1}a_2 - \tan^{-1}a_1) + (\tan^{-1}a_3 - \tan^{-1}a_2) + \ldots + (\tan^{-1}a_n - \tan^{-1}a_{n-1})\right]$$

$$= \tan[\tan^{-1}a_n - \tan^{-1}a_1]$$

$$= \tan\left[\tan^{-1}\frac{a_n-a_1}{1+a_n\cdot a_1}\right] \qquad \left[\because \tan^{-1}x - \tan^{-1}y = \tan^{-1}\left(\frac{x-y}{1+xy}\right)\right]$$

$$= \frac{a_n-a_1}{1+a_n\cdot a_1} \qquad [\because \tan(\tan^{-1}x) = x]$$

Note (✱) Topics related to these question have been removed from latest NCERT.

Objective Type Questions

Choose the correct answers out of the given four options in each of the exercises from 20 to 37.

Q.20 Which of the following is the principal value branch of $\cos^{-1} x$?

(a) $\left[-\frac{\pi}{2}, \frac{\pi}{2}\right]$ (b) $(0, \pi)$ (c) $[0, \pi]$ (d) $(0, \pi) - \left\{\frac{\pi}{2}\right\}$

Ans. ***(c)*** We know that, the principal value branch of $\cos^{-1} x$ is $[0, \pi]$.

Q.21 Which of the following is the principal value branch of $\operatorname{cosec}^{-1} x$?

(a) $\left(\frac{-\pi}{2}, \frac{\pi}{2}\right)$ (b) $[0, \pi] - \left\{\frac{\pi}{2}\right\}$ (c) $\left[-\frac{\pi}{2}, \frac{\pi}{2}\right]$ (d) $\left[\frac{-\pi}{2}, \frac{\pi}{2}\right] - \{0\}$

Ans. ***(d)*** We know that, the principal value branch of $\operatorname{cosec}^{-1} x$ is $\left[\frac{-\pi}{2}, \frac{\pi}{2}\right] - \{0\}$.

Q.22 If $3\tan^{-1} x + \cot^{-1} x = \pi$, then x is equal to

(a) 0 (b) 1 (c) -1 (d) $\frac{1}{2}$

Ans. ***(b)*** We have, $3\tan^{-1} x + \cot^{-1} x = \pi$ ...(i)

Let $3\tan^{-1} x = \alpha \Rightarrow \tan^{-1} x = \frac{\alpha}{3} \Rightarrow x = \tan\frac{\alpha}{3}$ ...(ii)

Again, let $\cot^{-1} x = \beta \Rightarrow x = \cot\beta$ ...(iii)

From Eqs.(ii) and (iii), we get

$$\tan\frac{\alpha}{3} = \cot\beta \Rightarrow \tan\frac{\alpha}{3} = \tan\left(\frac{\pi}{2} - \beta\right)$$

$$\Rightarrow \frac{\alpha}{3} = \frac{\pi}{2} - \beta \Rightarrow \alpha = \frac{3\pi}{2} - 3\beta \quad \text{...(iv)}$$

From Eq.(i), we get

$$3\tan^{-1} x + \cot^{-1} x = \pi \Rightarrow \alpha + \beta = \pi$$

$$\Rightarrow \frac{3\pi}{2} - 3\beta + \beta = \pi \quad \text{[from Eq. (iv)]}$$

$$\Rightarrow 2\beta = \frac{3\pi}{2} - \pi = \frac{\pi}{2} \Rightarrow \beta = \frac{\pi}{4}$$

$$\therefore x = \cot\beta = \cot\frac{\pi}{4} = 1$$

Q.23 The value of $\sin^{-1}\left[\cos\left(\frac{33\pi}{5}\right)\right]$ is

(a) $\frac{3\pi}{5}$ (b) $\frac{-7\pi}{5}$ (c) $\frac{\pi}{10}$ (d) $\frac{-\pi}{10}$

Ans. ***(d)*** We have,

$$\sin^{-1}\left(\cos\frac{33\pi}{5}\right) = \sin^{-1}\left[\cos\left(6\pi + \frac{3\pi}{5}\right)\right] = \sin^{-1}\left[\cos\left(\frac{3\pi}{5}\right)\right] \quad [\because \cos(2n\pi + \theta) = \cos\theta]$$

$$= \sin^{-1}\left[\cos\left(\frac{\pi}{2} + \frac{\pi}{10}\right)\right] = \sin^{-1}\left(-\sin\frac{\pi}{10}\right) \quad \left[\because \cos\left(\frac{\pi}{2} + \theta\right) = -\sin\theta\right]$$

$$= \sin^{-1}\left(\sin\left(\frac{-\pi}{10}\right)\right) \quad [\because \sin(-x) = -\sin x]$$

$$= -\frac{\pi}{10} \quad \left[\because \sin^{-1}(\sin x) = x, x \in \left[\frac{-\pi}{2}, \frac{\pi}{2}\right]\right]$$

Q.24 The domain of the function $\cos^{-1}(2x-1)$ is

(a) [0, 1] (b) [−1, 1] (c) (−1, 1) (d) [0, π]

Ans. *(a)* We have, $f(x) = \cos^{-1}(2x-1)$

$\because \quad -1 \le 2x - 1 \le 1 \Rightarrow 0 \le 2x \le 2$

$\Rightarrow \quad 0 \le x \le 1$

$\therefore \quad x \in [0, 1]$

Q.25 The domain of the function defined by $f(x) = \sin^{-1}\sqrt{x-1}$ is

(a) [1, 2] (b) [−1, 1] (c) [0, 1] (d) None of these

Ans. *(a)* $\because \quad f(x) = \sin^{-1}\sqrt{x-1}$

$\therefore \quad 0 \le x - 1 \le 1 \qquad [\because \sqrt{x-1} \ge 0 \text{ and } -1 \le \sqrt{x-1} \le 1]$

$\Rightarrow \quad 1 \le x \le 2$

$\therefore \quad x \in [1, 2]$

Q.26 If $\cos\left(\sin^{-1}\frac{2}{5} + \cos^{-1}x\right) = 0$, then x is equal to

(a) $\frac{1}{5}$ (b) $\frac{2}{5}$ (c) 0 (d) 1

Thinking Process Use the identity $\cos(x+y) = \cos x \cos y - \sin x \sin y$ to get the required value.

Ans. *(b)* We have, $\cos\left(\sin^{-1}\frac{2}{5} + \cos^{-1}x\right) = 0$ ...(i)

Let $\quad \sin^{-1}\frac{2}{5} = \alpha \Rightarrow \sin\alpha = \frac{2}{5}$

$\therefore \quad \cos\alpha = \sqrt{1 - \left(\frac{2}{5}\right)^2} = \frac{\sqrt{21}}{5}$

Again, let $\quad \cos^{-1}x = \beta \Rightarrow \cos\beta = x$

$\therefore \quad \sin\beta = \sqrt{1-x^2}$

From Eq.(i), we get

$$\cos\left(\sin^{-1}\frac{2}{5} + \cos^{-1}x\right) = 0$$

$\Rightarrow \quad \cos(\alpha + \beta) = 0$

$\Rightarrow \quad \cos\alpha\cos\beta - \sin\alpha\sin\beta = 0$

$\Rightarrow \quad \frac{\sqrt{21}}{5} \times x - \frac{2}{5} \times \sqrt{1-x^2} = 0$

$\Rightarrow \quad x\sqrt{21} - 2\sqrt{1-x^2} = 0$

$\Rightarrow \quad x\sqrt{21} = 2\sqrt{1-x^2}$

On squaring both sides, we get

$21x^2 = 4(1-x^2)$

$\Rightarrow \quad 21x^2 = 4 - 4x^2$

$\Rightarrow \quad 25x^2 = 4 \Rightarrow x^2 = \frac{4}{25}$

$\therefore \quad x = \pm\frac{2}{5}$

Q.27 The value of $\sin[2\tan^{-1}(0.75)]$ is

(a) 0.75 (b) 1.5 (c) 0.96 (d) $\sin 1.5$

Ans. ***(c)*** We have, $\sin[2\tan^{-1}(0.75)] = \sin\left[2\tan^{-1}\frac{3}{4}\right]$ ...(i)

Let $\tan^{-1}\frac{3}{4} = x \Rightarrow \tan x = \frac{3}{4}$

$$\therefore \quad \sin\left[2\tan^{-1}\frac{3}{4}\right] = \sin[2x]$$

$$= \frac{2\tan x}{1+\tan^2 x}$$

$$= \frac{2\times\frac{3}{4}}{1+\left(\frac{3}{4}\right)^2} = \frac{\frac{3}{2}}{\frac{25}{16}}$$

$$= \frac{24}{25} = 0.96$$

Q.28 The value of $\cos^{-1}\left(\cos\frac{3\pi}{2}\right)$ is

(a) $\frac{\pi}{2}$ (b) $\frac{3\pi}{2}$ (c) $\frac{5\pi}{2}$ (d) $\frac{7\pi}{2}$

Ans. ***(a)*** We have, $\cos^{-1}\left(\cos\frac{3\pi}{2}\right) = \cos^{-1}\left[\cos\left(2\pi - \frac{\pi}{2}\right)\right]$

$$= \cos^{-1}\left(\cos\left(\frac{\pi}{2}\right)\right) \quad \left[\because \cos\left(2\pi - \frac{\pi}{2}\right) = \cos\frac{\pi}{2}\right]$$

$$= \frac{\pi}{2} \quad \{\because \cos^{-1}(\cos x) = x, x \in [0, \pi]\}$$

Note Remember that, $\cos^{-1}\left(\cos\frac{3\pi}{2}\right) \neq \frac{3\pi}{2}$ $\left[\because \frac{3\pi}{2} \notin [0, \pi]\right]$

Q.29 The value of the expression $2\sec^{-1} 2 + \sin^{-1}\left(\frac{1}{2}\right)$ is

(a) $\frac{\pi}{6}$ (b) $\frac{5\pi}{6}$ (c) $\frac{7\pi}{6}$ (d) 1

Ans. ***(b)*** We have, $2\sec^{-1}2 + \sin^{-1}\frac{1}{2} = 2\sec^{-1}\left(\sec\frac{\pi}{3}\right) + \sin^{-1}\left(\sin\frac{\pi}{6}\right)$

$$\left[\because \sec\frac{\pi}{3} = 2 \text{ and } \sin\frac{\pi}{6} = \frac{1}{2}\right]$$

$$= 2\cdot\frac{\pi}{3} + \frac{\pi}{6} \quad [\because \sec^{-1}(\sec x) = x \text{ and } \sin^{-1}(\sin x) = x]$$

$$= \frac{4\pi + \pi}{6} = \frac{5\pi}{6}$$

Q.30 If $\tan^{-1} x + \tan^{-1} y = \frac{4\pi}{5}$, then $\cot^{-1} x + \cot^{-1} y$ is equal to

(a) $\frac{\pi}{5}$ (b) $\frac{2\pi}{5}$ (c) $\frac{3\pi}{5}$ (d) π

Ans. *(a)* We have, $\tan^{-1} x + \tan^{-1} y = \frac{4\pi}{5}$ ...(i)

Let $\tan^{-1} x = \alpha \Rightarrow \tan\alpha = x$

Again, let $\tan^{-1} y = \beta \Rightarrow \tan\beta = y$

From Eq.(i), we get

$$\alpha + \beta = \frac{4\pi}{5} \quad ...(ii)$$

Now,

$$\cot^{-1} x + \cot^{-1} y = \cot^{-1}(\tan\alpha) + \cot^{-1}(\tan\beta)$$

$$= \cot^{-1}\left[\cot\left(\frac{\pi}{2} - \alpha\right)\right] + \cot^{-1}\left[\cot\left(\frac{\pi}{2} - \beta\right)\right]$$

$$= \frac{\pi}{2} - \alpha + \frac{\pi}{2} - \beta$$

$$= \pi - (\alpha + \beta)$$

$$= \pi - \frac{4\pi}{5} \quad \text{[from Eq. (ii)]}$$

$$= \frac{\pi}{5}$$

Q.31 If $\sin^{-1}\left(\frac{2a}{1+a^2}\right) + \cos^{-1}\left(\frac{1-a^2}{1+a^2}\right) = \tan^{-1}\left(\frac{2x}{1-x^2}\right)$, where $a, x \in]0, 1[$, then the value of x is

(a) 0 (b) $\frac{a}{2}$ (c) a (d) $\frac{2a}{1-a^2}$

Thinking Process Use the identities, $\tan 2\theta = \frac{2\tan\theta}{1-\tan^2\theta}$ and $\cos 2\theta = \frac{1-\tan^2\theta}{1+\tan^2\theta}$ to get the required value.

Ans. *(d)* We have, $\sin^{-1}\left(\frac{2a}{1+a^2}\right) + \cos^{-1}\left(\frac{1-a^2}{1+a^2}\right) = \tan^{-1}\left(\frac{2x}{1-x^2}\right)$

Let $a = \tan\theta \Rightarrow \theta = \tan^{-1} a$

$$\therefore \quad \sin^{-1}\left(\frac{2\tan\theta}{1+\tan^2\theta}\right) + \cos^{-1}\left(\frac{1-\tan^2\theta}{1+\tan^2\theta}\right) = \tan^{-1}\frac{2x}{1-x^2}$$

$$\Rightarrow \quad \sin^{-1}(\sin 2\theta) + \cos^{-1}(\cos 2\theta) = \tan^{-1}\frac{2x}{1-x^2}$$

$$\left[\because \sin 2\theta = \frac{2\tan\theta}{1+\tan^2\theta} \text{ and } \cos 2\theta = \frac{1-\tan^2\theta}{1+\tan^2\theta}\right]$$

$$\Rightarrow \quad 2\theta + 2\theta = \tan^{-1}\frac{2x}{1-x^2}$$

$$\Rightarrow \quad 4\tan^{-1} a = \tan^{-1}\frac{2x}{1-x^2}$$

$$\Rightarrow \quad 2 \times 2\tan^{-1} a = \tan^{-1}\frac{2x}{1-x^2} \quad ...(i)$$

Now, let $\quad 2\tan^{-1} a = 2\alpha \Rightarrow \tan^{-1} a = \alpha$

$\Rightarrow \quad \tan\alpha = a$

From Eq.(i), we get

$$2\cdot 2\alpha = \tan^{-1}\frac{2x}{1-x^2} \qquad \left[\because \tan 2\theta = \frac{2\tan\theta}{1-\tan^2\theta}\right]$$

$$\Rightarrow \quad \tan[2\cdot(2\alpha)] = \frac{2x}{1-x^2}$$

$$\Rightarrow \quad \frac{2\tan 2\alpha}{1-\tan^2 2\alpha} = \frac{2x}{1-x^2}$$

$$\Rightarrow \quad \frac{\dfrac{2\cdot 2\tan\alpha}{1-\tan^2\alpha}}{1-\left(\dfrac{2\tan\alpha}{1-\tan^2\alpha}\right)^2} = \frac{2x}{1-x^2}$$

$$\Rightarrow \quad \frac{\dfrac{2\cdot 2a}{1-a^2}}{1-\left(\dfrac{2a}{1-a^2}\right)^2} = \frac{2x}{1-x^2}$$

On comparing both sides, we get $x = \dfrac{2a}{1-a^2}$

Q.32 The value of $\cot\left[\cos^{-1}\left(\frac{7}{25}\right)\right]$ is

(a) $\frac{25}{24}$ (b) $\frac{25}{7}$ (c) $\frac{24}{25}$ (d) $\frac{7}{24}$

Ans. ***(d)*** We have, $\cot\left[\cos^{-1}\left(\frac{7}{25}\right)\right]$

Let $\quad \cos^{-1}\frac{7}{25} = x$

$$\Rightarrow \quad \cos x = \frac{7}{25}$$

$$\therefore \quad \sin x = \sqrt{1-\cos^2 x} = \sqrt{1-\left(\frac{7}{25}\right)^2}$$

$$= \sqrt{\frac{625-49}{625}} = \frac{24}{25}$$

and $$\cot x = \frac{\cos x}{\sin x} = \frac{\frac{7}{25}}{\frac{24}{25}} = \frac{7}{24} \qquad \text{...(i)}$$

$$\Rightarrow \quad x = \cot^{-1}\left(\frac{7}{24}\right) = \cos^{-1}\left(\frac{7}{25}\right)$$

$$\therefore \quad \cot\left(\cos^{-1}\frac{7}{25}\right) = \cot\left(\cot^{-1}\frac{7}{24}\right) = \frac{7}{24} \qquad \left[\because \cot^{-1}\frac{7}{24} = \cos^{-1}\frac{7}{25}\right]$$

Q.33 The value of the expression $\tan\left(\frac{1}{2}\cos^{-1}\frac{2}{\sqrt{5}}\right)$ is

(a) $2+\sqrt{5}$ (b) $\sqrt{5}-2$ (c) $\frac{\sqrt{5}+2}{2}$ (d) $5+\sqrt{2}$

Ans. ***(b)*** We have, $\tan\left(\frac{1}{2}\cos^{-1}\frac{2}{\sqrt{5}}\right)$

Let $\frac{1}{2}\cos^{-1}\frac{2}{\sqrt{5}}=\theta$ ...(i)

$\Rightarrow \quad \cos^{-1}\frac{2}{\sqrt{5}}=2\theta \Rightarrow \cos 2\theta=\frac{2}{\sqrt{5}}$

$\Rightarrow \quad (1-2\sin^2\theta)=\frac{2}{\sqrt{5}}$ $[\because \cos 2\theta = 1-2\sin^2\theta]$

$\Rightarrow \quad 2\sin^2\theta = 1-\frac{2}{\sqrt{5}}$

$\Rightarrow \quad \sin^2\theta=\frac{1}{2}-\frac{1}{\sqrt{5}}$

$\Rightarrow \quad \sin\theta=\sqrt{\frac{1}{2}-\frac{1}{\sqrt{5}}}$

$\because \quad \cos^2\theta = 1-\sin^2\theta = 1-\left\{\sqrt{\frac{1}{2}-\frac{1}{\sqrt{5}}}\right\}^2$

$=1-\frac{1}{2}+\frac{1}{\sqrt{5}}=\frac{1}{2}+\frac{1}{\sqrt{5}}$

$\Rightarrow \quad \cos\theta=\sqrt{\frac{1}{2}+\frac{1}{\sqrt{5}}}$

$\therefore \quad \tan\theta=\sqrt{\dfrac{\frac{1}{2}-\frac{1}{\sqrt{5}}}{\frac{1}{2}+\frac{1}{\sqrt{5}}}}=\sqrt{\frac{\sqrt{5}-2}{\sqrt{5}+2}}$ $\left[\because \tan\theta=\frac{\sin\theta}{\cos\theta}\right]$

$\Rightarrow \quad \theta=\tan^{-1}\sqrt{\frac{\sqrt{5}-2}{\sqrt{5}+2}}$

$\therefore \quad \tan\left(\frac{1}{2}\cos^{-1}\frac{2}{\sqrt{5}}\right)=\tan(\theta)=\tan\left[\tan^{-1}\sqrt{\frac{\sqrt{5}-2}{\sqrt{5}+2}}\right]$ [from Eq. (i)]

$=\sqrt{\frac{\sqrt{5}-2}{\sqrt{5}+2}\cdot\frac{\sqrt{5}-2}{\sqrt{5}-2}}$

$=\sqrt{\frac{(\sqrt{5}-2)^2}{5-4}}=\sqrt{5}-2$

Q.34 If $|x| \le 1$, then $2\tan^{-1} x + \sin^{-1}\left(\frac{2x}{1+x^2}\right)$ is equal to

(a) $4\tan^{-1} x$ (b) 0 (c) $\frac{\pi}{2}$ (d) π

Ans. **(a)** We have, $2\tan^{-1} x + \sin^{-1}\left(\frac{2x}{1+x^2}\right)$

Let $x = \tan\theta \Rightarrow \theta = \tan^{-1} x$

$$\therefore \quad 2\tan^{-1} x + \sin^{-1}\left(\frac{2x}{1+x^2}\right) = 2\tan^{-1}(\tan\theta) + \sin^{-1}\left(\frac{2\tan\theta}{1+\tan^2\theta}\right)$$
$$= 2\theta + \sin^{-1}(\sin 2\theta)$$
$$= 2\theta + 2\theta = 4\theta = 4\tan^{-1} x$$

Q.35 If $\cos^{-1}\alpha + \cos^{-1}\beta + \cos^{-1}\gamma = 3\pi$, then $\alpha(\beta+\gamma) + \beta(\gamma+\alpha) + \gamma(\alpha+\beta)$ is equal to

(a) 0 (b) 1 (c) 6 (d) 12

Ans. **(c)** We have, $\cos^{-1}\alpha + \cos^{-1}\beta + \cos^{-1}\gamma = 3\pi$

We know that, $0 \le \cos^{-1} x \le \pi$

$\Rightarrow$ $\cos^{-1}\alpha + \cos^{-1}\beta + \cos^{-1}\gamma = 3\pi$ is possible if and only if,

$$\cos^{-1}\alpha = \cos^{-1}\beta = \cos^{-1}\gamma = \pi$$
$$\Rightarrow \quad \cos\pi = \alpha = \beta = \gamma$$
$$\Rightarrow \quad \alpha = \beta = \gamma = -1$$
$$\therefore \alpha(\beta+\gamma) + \beta(\gamma+\alpha) + \gamma(\alpha+\beta) = -1(-1-1) - 1(-1-1) - 1(-1-1)$$
$$= 2 + 2 + 2 = 6$$

Q.36 The number of real solutions of the equation

$$\sqrt{1+\cos 2x} = \sqrt{2}\cos^{-1}(\cos x) \text{ in } \left[\frac{\pi}{2}, \pi\right] \text{ is}$$

(a) 0 (b) 1 (c) 2 (d) Infinite

Ans. **(a)** We have, $\sqrt{1+\cos 2x} = \sqrt{2}\cos^{-1}(\cos x), \left[\frac{\pi}{2}, \pi\right]$

$$\Rightarrow \quad \sqrt{1+2\cos^2 x - 1} = \sqrt{2}\cos^{-1}(\cos x) \qquad [\because \cos 2x = 2\cos^2 x - 1]$$
$$\Rightarrow \quad \sqrt{2}\cos x = \sqrt{2}\cos^{-1}(\cos x)$$
$$\Rightarrow \quad \cos x = \cos^{-1}(\cos x)$$
$$\Rightarrow \quad \cos x = x \qquad [\because \cos^{-1}(\cos x) = x]$$

which is not true for any real value of x.

Hence, there is no solution possible for the given equation.

Q.37 If $\cos^{-1} x > \sin^{-1} x$, then

(a) $\frac{1}{\sqrt{2}} < x \le 1$ (b) $0 \le x < \frac{1}{\sqrt{2}}$

(c) $-1 \le x < \frac{1}{\sqrt{2}}$ (d) $x > 0$

Ans. *(c)* We know that,

Range of $\cos^{-1} x = [0, \pi]$

and range of $\sin^{-1} x = \left[\frac{-\pi}{2}, \frac{\pi}{2}\right]$

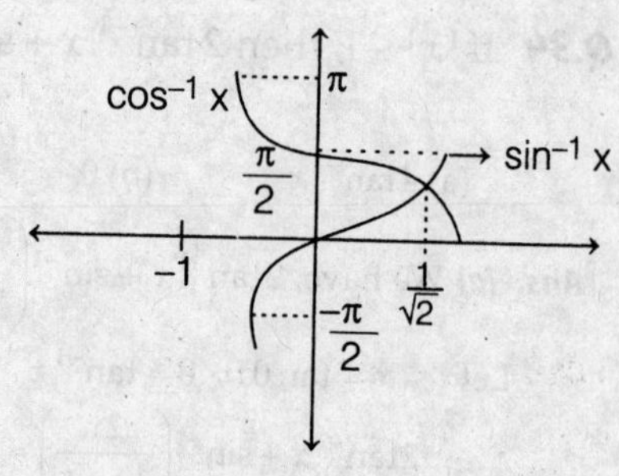

From the graph,

$$\cos^{-1} x > \sin^{-1} x$$

When $\quad x \in \left[-1, \frac{1}{\sqrt{2}}\right]$

$\Rightarrow \quad -1 \le x < \frac{1}{\sqrt{2}}$

Fill in the blanks in each of the exercises 38 to 48.

Q.38 The principal value of $\cos^{-1}\left(-\frac{1}{2}\right)$ is ....

Ans. $\because$

$$0 \le \cos^{-1} x \le \pi$$

$$\cos^{-1}\left(-\frac{1}{2}\right) = \cos^{-1}\left(-\cos\frac{\pi}{3}\right) = \cos^{-1}\left[\cos\left(\pi - \frac{\pi}{3}\right)\right]$$

$$= \cos^{-1}\left(\cos\frac{2\pi}{3}\right) = \frac{2\pi}{3}$$

Q.39 The value of $\sin^{-1}\left(\sin\frac{3\pi}{5}\right)$ is ....

Ans. $\because$

$$-\frac{\pi}{2} \le \sin^{-1} x \le \frac{\pi}{2}$$

$\therefore$

$$\sin^{-1}\left(\sin\frac{3\pi}{5}\right) = \sin^{-1}\sin\left(\pi - \frac{2\pi}{5}\right) = \sin^{-1}\left(\sin\frac{2\pi}{5}\right) = \frac{2\pi}{5}$$

Q.40 **If $\cos(\tan^{-1} x + \cot^{-1}\sqrt{3}) = 0$, then the value of x is ....**

Ans. We have, $\cos(\tan^{-1} x + \cot^{-1}\sqrt{3}) = 0 \quad$ ...(i)

Let $\quad \tan^{-1} x = \alpha$

$\Rightarrow \quad \tan\alpha = x$

$\therefore \quad \sin\alpha = \frac{1}{\sqrt{1+x^2}}$

and $\quad \cos\alpha = \frac{1}{\sqrt{1+x^2}}$

Now, let $\quad \cot^{-1}\sqrt{3} = \beta \Rightarrow \cot\beta = \sqrt{3}$

$\therefore \quad \sin\beta = \frac{1}{2}$ and $\cos\beta = \frac{\sqrt{3}}{2}$

From Eq.(i), we get

$$\cos(\alpha + \beta) = 0 \Rightarrow \cos\alpha\cos\beta - \sin\alpha\sin\beta = 0$$

$$\Rightarrow \quad \frac{1}{\sqrt{1+x^2}} \times \frac{\sqrt{3}}{2} - \frac{x}{\sqrt{1+x^2}} \times \frac{1}{2} = 0$$

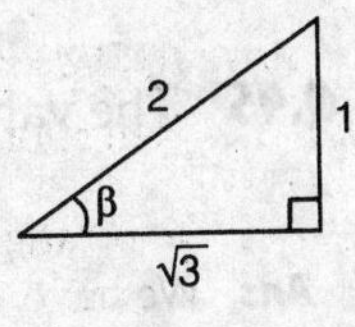

$$\Rightarrow \quad \frac{1}{2\sqrt{1+x^2}}(\sqrt{3} - x) = 0$$

$$\Rightarrow \quad \sqrt{3} - x = 0$$

$$\Rightarrow \quad x = \sqrt{3}$$

Q.41 The set of values of $\sec^{-1}\frac{1}{2}$ is ....

Ans. Since, domain of $\sec^{-1} x$ is $R - (-1, 1)$.

$$\Rightarrow \quad (-\infty, -1] \cup [1, \infty)$$

So, there is no set of values exist for $\sec^{-1}\frac{1}{2}$.

So, ϕ is the answer.

Q.42 The principal value of $\tan^{-1}\sqrt{3}$ is ....

Ans. $\because$

$$\tan^{-1}\sqrt{3} = \tan^{-1}\left[\tan\left(\frac{\pi}{3}\right)\right]$$

$$= \left(\frac{\pi}{3}\right) \qquad \left[\because \tan^{-1}(\tan x) = x, x \in \left(-\frac{\pi}{2}, \frac{\pi}{2}\right)\right]$$

Q.43 The value of $\cos^{-1}\left(\cos\frac{14\pi}{3}\right)$ is ....

Ans. We have,

$$\cos^{-1}\left(\cos\frac{14\pi}{3}\right) = \cos^{-1}\left[\cos\left(4\pi + \frac{2\pi}{3}\right)\right]$$

$$= \cos^{-1}\left(\cos\frac{2\pi}{3}\right) \qquad [\because \cos(2n\pi + \theta) = \cos\theta]$$

$$= \frac{2\pi}{3} \qquad \{\because \cos^{-1}(\cos x) = x, x \in [0, \pi]\}$$

Note Remember that, $\cos^{-1}\left(\cos\frac{14\pi}{3}\right) \neq \frac{14\pi}{3}$ $\qquad \left[\because \frac{14\pi}{3} \notin [0, \pi]\right]$

Q.44 The value of $\cos(\sin^{-1} x + \cos^{-1} x)$, where $|x| \le 1$, is ....

Ans. We have, $\cos(\sin^{-1} x + \cos^{-1} x)$

Let $\quad \sin^{-1} x = \alpha$

$$\Rightarrow \quad \sin\alpha = x$$

$$\therefore \quad \cos = \alpha\sqrt{1 - x^2} \qquad [\because \sin^2\theta + \cos^2\theta = 1]$$

and let $\quad \cos^{-1} x = \beta$

$$\Rightarrow \quad \cos\beta = x$$

$$\therefore \quad \sin\beta = \sqrt{1 - x^2} \qquad [\because \sin^2\theta + \cos^2\theta = 1]$$

$$\therefore \quad \cos(\sin^{-1} x + \cos^{-1} x) = \cos(\alpha + \beta) = \cos\alpha\cos\beta - \sin\alpha\sin\beta$$

$$= \sqrt{1 - x^2} \cdot x - x \cdot \sqrt{1 - x^2} = 0$$

Q.45 The value of expression $\tan\left(\dfrac{\sin^{-1}x+\cos^{-1}x}{2}\right)$, when $x=\dfrac{\sqrt{3}}{2}$, is ....

Ans. We have, $\tan\left(\dfrac{\sin^{-1}x+\cos^{-1}x}{2}\right)$ ...(i)

Let $\sin^{-1}x=\alpha\Rightarrow\sin\alpha=x$ ...(ii)

Again, let $\cos^{-1}x=\beta\Rightarrow\cos\beta=x$ ...(iii)

From Eqs.(ii) and (iii), we get

$$\sin\alpha=\cos\beta\Rightarrow\sin\alpha=\sin\left(\frac{\pi}{2}-\beta\right)$$

$$\Rightarrow\quad \alpha=\frac{\pi}{2}-\beta\Rightarrow\alpha+\beta=\frac{\pi}{2}$$

From Eq.(i), we get

$$\tan\left(\frac{\sin^{-1}x+\cos^{-1}x}{2}\right)=\tan\left(\frac{\pi/2}{2}\right)$$

$$=\tan\frac{\pi}{4}=1$$

Q.46 If $y=2\tan^{-1}x+\sin^{-1}\left(\dfrac{2x}{1+x^2}\right)$, then .... $< y <$ ....

Ans. We have, $y=2\tan^{-1}x+\sin^{-1}\left(\dfrac{2x}{1+x^2}\right)$

Let $x=\tan\theta\Rightarrow\theta=\tan^{-1}x$

$\therefore\quad y=2\tan^{-1}(\tan\theta)+\sin^{-1}\dfrac{2\tan\theta}{1+\tan^2\theta}$

$\Rightarrow\quad y=2\theta+\sin^{-1}(\sin 2\theta)$ $\left[\because \sin 2\theta=\dfrac{2\tan\theta}{1+\tan^2\theta}\right]$

$\Rightarrow\quad y=2\theta+2\theta=4\theta$

$\Rightarrow\quad y=4\tan^{-1}x$ $[\because \theta=\tan^{-1}x]$

$\because\quad -\pi/2<\tan^{-1}x<\pi/2$

$\therefore\quad -\dfrac{4\pi}{2}<4\tan^{-1}x<4\pi/2$

$\Rightarrow\quad -2\pi<4\tan^{-1}x<2\pi$

$\Rightarrow\quad -2\pi<y<2\pi$ $[\because y=4\tan^{-1}x]$

Q.47 The result $\tan^{-1}x-\tan^{-1}y=\tan^{-1}\left(\dfrac{x-y}{1+xy}\right)$ is true when the value of xy is ....

Ans. LHS $=\tan^{-1}x-\tan^{-1}y$

Let $\tan^{-1}x=\alpha\Rightarrow\tan\alpha=x$

and let $\tan^{-1}y=\beta\Rightarrow\tan\beta=y$

Now, $\tan(\alpha-\beta)=\dfrac{\tan\alpha-\tan\beta}{1+\tan\alpha\tan\beta}$

$$= \frac{x-y}{1+xy}, xy > -1$$

$$\Rightarrow \quad \alpha - \beta = \tan^{-1}\left(\frac{x-y}{1+xy}\right), xy > -1$$

$$\Rightarrow \quad \tan^{-1} x - \tan^{-1} \beta = \tan^{-1}\left(\frac{x-y}{1+xy}\right), xy > -1$$

So, $\tan^{-1} x - \tan^{-1} \beta = \tan^{-1}\left(\frac{x-y}{1+xy}\right)$, when $xy > -1$.

Q.48 The value of $\cot^{-1}(-x)\ x \in R$ in terms of $\cot^{-1} x$ is ....

Ans. Let $\cot^{-1}(-x) = \alpha$

$$\Rightarrow \quad \cot\alpha = -x$$

$$\Rightarrow \quad -\cot\alpha = x$$

$$\Rightarrow \quad \cot(\pi - \alpha) = x \qquad [\because \cot(\pi - \theta) = -\cot\theta]$$

$$\Rightarrow \quad \pi - \alpha = \cot^{-1} x$$

$$\Rightarrow \quad \alpha = \pi - \cot^{-1} x$$

$$\therefore \quad \cot^{-1}(-x) = \pi - \cot^{-1} x, x \in R$$

State True or False for the statement in each of the exercises 49 to 55.

Q.49 All trigonometric functions have inverse over their respective domains.

Ans. *False*

We know that, all trigonometric functions have inverse over their restricted domains.

***Q.50** The value of the expression $(\cos^{-1} x)^2$ is equal to $\sec^2 x$.

Ans. *False*

$$\because \quad [\cos^{-1} x]^2 = \left[\sec^{-1}\frac{1}{x}\right]^2 \neq \sec^2 x$$

Q.51 The domain of trigonometric functions can be restricted to any one of their branch (not necessarily principal value) in order to obtain their inverse functions.

Ans. *True*

We know that, the domain of trigonometric functions are restricted in their domain to obtain their inverse functions.

Q.52 The least numerical value, either positive or negative of angle θ is called principal value of the inverse trigonometric function.

Ans. *True*

We know that, the smallest numerical value, either positive or negative of θ is called the principal value of the function of the inverse trigonometric function.

Note (✱) Topics related to these question have been removed from latest NCERT.

Q.53 The graph of inverse trigonometric function can be obtained from the graph of their corresponding function by interchanging X and Y-axes.

Ans. *True*

We know that, the graph of an inverse function can be obtained from the corresponding graph of original function as a mirror image (*i.e.*, reflection) along the line $y = x$.

Q.54 The minimum value of n for which $\tan^{-1}\frac{n}{\pi} > \frac{\pi}{4}, n \in N$, is valid is 5.

Sol. *False*

$$\because \quad \tan^{-1}\frac{n}{\pi} > \frac{\pi}{4} \Rightarrow \frac{n}{\pi} > \tan\frac{\pi}{4}$$

$$\Rightarrow \quad \frac{n}{\pi} > 1 \qquad \left[\because \tan\frac{\pi}{4} = 1\right]$$

$$\Rightarrow \quad n > \pi$$

So, the minimum value of n is 4. $[\because n \in N \text{ and } \pi = 3.14...]$

Q.55 The principal value of $\sin^{-1}\left[\cos\left(\sin^{-1}\frac{1}{2}\right)\right]$ is $\frac{\pi}{3}$.

Ans. *True*

Given that, $$\sin^{-1}\left[\cos\left(\sin^{-1}\frac{1}{2}\right)\right] = \sin^{-1}\left[\cos \sin^{-1}\left(\sin\frac{\pi}{6}\right)\right]$$

$$= \sin^{-1}\left[\cos\frac{\pi}{6}\right] \qquad [\because \sin^{-1}(\sin x) = x]$$

$$= \sin^{-1}\frac{\sqrt{3}}{2}$$

$$= \sin^{-1}\left(\sin\frac{\pi}{3}\right) = \frac{\pi}{3}$$

Chapter 03

Matrices

Short Answer Type Questions

Q.1 If a matrix has 28 elements, what are the possible orders it can have? What if it has 13 elements?

Ans. We know that, if a matrix is of order $m \times n$, then it has mn elements, where m and n are natural numbers.

We have, $m \times n = 28$

$$\Rightarrow \quad (m, n) = \{(1, 28), (2, 14), (4, 7), (7, 4), (14, 2), (28, 1)\}$$

So, the possible orders are 1×28, 2×14, 4×7, 7×4, 14×2, 28×1.

Also, if it has 13 elements, then $m \times n = 13$

$$\Rightarrow \quad (m, n) = \{(1,13), (13,1)\}$$

Hence, the possible orders are 1×13, 13×1.

Note If number of elements is prime number, then the possible order of matrix will be two.

Q.2 In the matrix $A = \begin{bmatrix} a & 1 & x \\ 2 & \sqrt{3} & x^2 - y \\ 0 & 5 & \frac{-2}{5} \end{bmatrix}$, write

(i) the order of the matrix A.

(ii) the number of elements.

(iii) elements a_{23}, a_{31} and a_{12}.

Ans. We have,
$$A = \begin{bmatrix} a & 1 & x \\ 2 & \sqrt{3} & x^2 - y \\ 0 & 5 & \frac{-2}{5} \end{bmatrix}$$

(i) The order of matrix $A = 3 \times 3$

(ii) The number of elements $= 3 \times 3 = 9$

[since, the number of elements in an $m \times n$ matrix will be equal to $m \times n = mn$]

(iii) $a_{23} = x^2 - y$, $a_{31} = 0$, $a_{12} = 1$

[since, we know that a_{ij}, is a representation of element lying in the ith row and jth column]

Q.3 Construct $a_{2\times 2}$ matrix, where

(i) $a_{ij} = \dfrac{(i-2j)^2}{2}$ (ii) $a_{ij} = |-2i+3j|$

Ans. We know that, the notation, namely $A = [a_{ij}]_{m\times n}$ indicates that A is a matrix of order $m \times n$, also $1 \le i \le m$, $1 \le j \le n$; $i, j \in N$.

(i) Here, $A = [a_{ij}]_{2\times 2}$

$$\Rightarrow \quad A = \frac{(i-2j)^2}{2}, 1 \le i \le 2; 1 \le j \le 2 \quad ...(i)$$

$$\therefore \quad a_{11} = \frac{(1-2)^2}{2} = \frac{1}{2}$$

$$a_{12} = \frac{(1-2\times 2)^2}{2} = \frac{9}{2}$$

$$a_{21} = \frac{(2-2\times 1)^2}{2} = 0$$

$$a_{22} = \frac{(2-2\times 2)^2}{2} = 2$$

$$\text{Thus,} \quad A = \begin{bmatrix} \frac{1}{2} & \frac{9}{2} \\ 0 & 2 \end{bmatrix}_{2\times 2}$$

(ii) Here, $A = [a_{ij}]_{2\times 2} = |-2i + 3j|$, $1 \le i \le 2$; $1 \le j \le 2$

$$\therefore \quad a_{11} = |-2\times 1 + 3\times 1| = 1$$

$$a_{12} = |-2\times 1 + 3\times 2| = 4$$

$$a_{21} = |-2\times 2 + 3\times 1| = 1 \quad [\because |-1| = 1]$$

$$a_{22} = |-2\times 2 + 3\times 2| = 2$$

$$\therefore \quad A = \begin{bmatrix} 1 & 4 \\ 1 & 2 \end{bmatrix}_{2\times 2}$$

Q.4 Construct a 3×2 matrix whose elements are given by $a_{ij} = e^{i\cdot x} \cdot \sin jx$.

Ans. Since, $A = [a_{ij}]_{m\times n}$ $1 \le i \le m$ and $1 \le j \le n$, $i, j \in N$

$$\therefore \quad A = [e^{i\cdot x} \sin jx]_{3\times 2};\ 1 \le i \le 3;\ 1 \le j \le 2$$

$$\Rightarrow \quad a_{11} = e^{1\cdot x} \cdot \sin 1\cdot x = e^x \sin x$$

$$a_{12} = e^{1\cdot x} \cdot \sin 2\cdot x = e^x \sin 2x$$

$$a_{21} = e^{2\cdot x} \cdot \sin 1\cdot x = e^{2x} \sin x$$

$$a_{22} = e^{2\cdot x} \cdot \sin 2\cdot x = e^{2x} \sin 2x$$

$$a_{31} = e^{3\cdot x} \cdot \sin 1\cdot x = e^{3x} \sin x$$

$$a_{32} = e^{3\cdot x} \cdot \sin 2\cdot x = e^{3x} \sin 2x$$

$$\therefore \quad A = \begin{bmatrix} e^x \sin x & e^x \sin 2x \\ e^{2x} \sin x & e^{2x} \sin 2x \\ e^{3x} \sin x & e^{3x} \sin 2x \end{bmatrix}_{3\times 2}$$

Q.5 Find the values of a and b, if $A = B$, where

$$A = \begin{bmatrix} a+4 & 3b \\ 8 & -6 \end{bmatrix} \text{ and } B = \begin{bmatrix} 2a+2 & b^2+2 \\ 8 & b^2-5b \end{bmatrix}$$

Thinking Process Two matrices are said to be equal, if

(i) order of A = order of B

(ii) corresponding elements of A and B are equal

Ans. We have, $A = \begin{bmatrix} a+4 & 3b \\ 8 & -6 \end{bmatrix}_{2\times 2}$ and $B = \begin{bmatrix} 2a+2 & b^2+2 \\ 8 & b^2-5b \end{bmatrix}_{2\times 2}$

Also, $A = B$

For equality of matrices we know that each element of A is equal to the corresponding element of B, that is $a_{ij} = b_{ij}$ for all i and j.

$\therefore$ $a_{11} = b_{11} \Rightarrow a+4 = 2a+2 \Rightarrow a = 2$

$a_{12} = b_{12} \Rightarrow 3b = b^2+2 \Rightarrow b^2 = 3b-2$

and $a_{22} = b_{22} \Rightarrow -6 = b^2 - 5b$

$\Rightarrow$ $-6 = 3b - 2 - 5b$ $[\because b^2 = 3b-2]$

$\Rightarrow$ $2b = 4 \Rightarrow b = 2$

$\therefore$ $a = 2$ and $b = 2$

Q.6 If possible, find the sum of the matrices A and B, where

$$A = \begin{bmatrix} \sqrt{3} & 1 \\ 2 & 3 \end{bmatrix} \text{ and } B = \begin{bmatrix} x & y & z \\ a & b & 6 \end{bmatrix}.$$

Thinking Process We know that, two matrices can be added, if they have same order.

Ans. We have, $A = \begin{bmatrix} \sqrt{3} & 1 \\ 2 & 3 \end{bmatrix}_{2\times 2}$ and $B = \begin{bmatrix} x & y & z \\ a & b & 6 \end{bmatrix}_{2\times 3}$

Here, A and B are of different orders. Also, we know that the addition of two matrices A and B is possible only if order of both the matrices A and B should be same.

Hence, the sum of matrices A and B is not possible.

Q.7 If $X = \begin{bmatrix} 3 & 1 & -1 \\ 5 & -2 & -3 \end{bmatrix}$ and $Y = \begin{bmatrix} 2 & 1 & -1 \\ 7 & 2 & 4 \end{bmatrix}$, then find

(i) $X + Y$.

(ii) $2X - 3Y$.

(iii) a matrix Z such that $X + Y + Z$ is a zero matrix.

Ans. We have, $X = \begin{bmatrix} 3 & 1 & -1 \\ 5 & -2 & -3 \end{bmatrix}_{2\times 3}$ and $Y = \begin{bmatrix} 2 & 1 & -1 \\ 7 & 2 & 4 \end{bmatrix}_{2\times 3}$

(i) $X + Y = \begin{bmatrix} 3+2 & 1+1 & -1-1 \\ 5+7 & -2+2 & -3+4 \end{bmatrix} = \begin{bmatrix} 5 & 2 & -2 \\ 12 & 0 & 1 \end{bmatrix}$

(ii) $\because\ 2X = 2\begin{bmatrix} 3 & 1 & -1 \\ 5 & -2 & -3 \end{bmatrix} = \begin{bmatrix} 6 & 2 & -2 \\ 10 & -4 & -6 \end{bmatrix}$

and $3Y = 3\begin{bmatrix} 2 & 1 & -1 \\ 7 & 2 & 4 \end{bmatrix} = \begin{bmatrix} 6 & 3 & -3 \\ 21 & 6 & 12 \end{bmatrix}$

$\therefore\ 2X - 3Y = \begin{bmatrix} 6-6 & 2-3 & -2+3 \\ 10-21 & -4-6 & -6-12 \end{bmatrix} = \begin{bmatrix} 0 & -1 & 1 \\ -11 & -10 & -18 \end{bmatrix}$

(iii) $\because\ X + Y = \begin{bmatrix} 3+2 & 1+1 & -1-1 \\ 5+7 & -2+2 & -3+4 \end{bmatrix} = \begin{bmatrix} 5 & 2 & -2 \\ 12 & 0 & 1 \end{bmatrix}$

Also, $X + Y + Z = \begin{bmatrix} 0 & 0 & 0 \\ 0 & 0 & 0 \end{bmatrix}$

We see that Z is the additive inverse of $(X + Y)$ or negative of $(X+Y)$.

$\therefore\ Z = \begin{bmatrix} -5 & -2 & 2 \\ -12 & 0 & -1 \end{bmatrix}$ $[\because Z = -(X+Y)]$

Q.8 Find non-zero values of x satisfying the matrix equation

$$x\begin{bmatrix} 2x & 2 \\ 3 & x \end{bmatrix} + 2\begin{bmatrix} 8 & 5x \\ 4 & 4x \end{bmatrix} = 2\begin{bmatrix} (x^2+8) & 24 \\ (10) & 6x \end{bmatrix}.$$

Ans. Given that,

$$x\begin{bmatrix} 2x & 2 \\ 3 & x \end{bmatrix} + 2\begin{bmatrix} 8 & 5x \\ 4 & 4x \end{bmatrix} = 2\begin{bmatrix} (x^2+8) & 24 \\ 10 & 6x \end{bmatrix}$$

$$\Rightarrow \begin{bmatrix} 2x^2 & 2x \\ 3x & x^2 \end{bmatrix} + \begin{bmatrix} 16 & 10x \\ 8 & 8x \end{bmatrix} = \begin{bmatrix} 2x^2+16 & 48 \\ 20 & 12x \end{bmatrix}$$

$$\Rightarrow \begin{bmatrix} 2x^2+16 & 2x+10x \\ 3x+8 & x^2+8x \end{bmatrix} = \begin{bmatrix} 2x^2+16 & 48 \\ 20 & 12x \end{bmatrix}$$

On comparing the corresponding element of matrices, we get

[$\because$ both matrices have same order, so corresponding elements are equal]

$\Rightarrow\ 2x + 10x = 48$

$\Rightarrow\ 12x = 48$

$\therefore\ x = \frac{48}{12} = 4$

Q.9 If $A = \begin{bmatrix} 0 & 1 \\ 1 & 1 \end{bmatrix}$ and $B = \begin{bmatrix} 0 & -1 \\ 1 & 0 \end{bmatrix}$, then show that $(A+B)(A-B) \neq A^2 - B^2$.

Ans. We have, $A = \begin{bmatrix} 0 & 1 \\ 1 & 1 \end{bmatrix}$ and $B = \begin{bmatrix} 0 & -1 \\ 1 & 0 \end{bmatrix}$

$\therefore\ (A+B) = \begin{bmatrix} 0+0 & 1-1 \\ 1+1 & 1+0 \end{bmatrix} = \begin{bmatrix} 0 & 0 \\ 2 & 1 \end{bmatrix}_{2\times 2}$

and $(A-B) = \begin{bmatrix} 0-0 & 1+1 \\ 1-1 & 1-0 \end{bmatrix} = \begin{bmatrix} 0 & 2 \\ 0 & 1 \end{bmatrix}_{2\times 2}$

Since, $(A+B)\cdot(A-B)$ is defined, as the number of columns of $(A+B)$ is equal to the number of rows of $(A-B)$, so here multiplication of matrices $(A+B)\cdot(A-B)$ is possible.

Now, $(A+B)_{2\times2}\cdot(A-B)_{2\times2}=\begin{bmatrix}0+0 & 0+0\\ 0+0 & 4+1\end{bmatrix}=\begin{bmatrix}0 & 0\\ 0 & 5\end{bmatrix}$ ...(i)

Also,
$$A^2=A\cdot A$$
$$=\begin{bmatrix}0 & 1\\ 1 & 1\end{bmatrix}\cdot\begin{bmatrix}0 & 1\\ 1 & 1\end{bmatrix}$$
$$=\begin{bmatrix}0+1 & 0+1\\ 0+1 & 1+1\end{bmatrix}=\begin{bmatrix}1 & 1\\ 1 & 2\end{bmatrix}$$

and $B^2=B\cdot B$ $=\begin{bmatrix}0 & -1\\ 1 & 0\end{bmatrix}\cdot\begin{bmatrix}0 & -1\\ 1 & 0\end{bmatrix}=\begin{bmatrix}0-1 & 0+0\\ 0+0 & -1+0\end{bmatrix}=\begin{bmatrix}-1 & 0\\ 0 & -1\end{bmatrix}$

$\therefore$ $A^2-B^2=\begin{bmatrix}1 & 1\\ 1 & 2\end{bmatrix}-\begin{bmatrix}-1 & 0\\ 0 & -1\end{bmatrix}=\begin{bmatrix}2 & 1\\ 1 & 3\end{bmatrix}$ ...(ii)

Thus, we see that

$\begin{bmatrix}0 & 0\\ 0 & 5\end{bmatrix}\neq\begin{bmatrix}2 & 1\\ 1 & 3\end{bmatrix}$ [using Eqs. (i) and (ii)]

$\therefore$ $(A+B)\cdot(A-B)\neq A^2-B^2$ **Hence proved.**

Q.10 Find the value of x, if $[1\ x\ 1]\begin{bmatrix}1 & 3 & 2\\ 2 & 5 & 1\\ 15 & 3 & 2\end{bmatrix}\begin{bmatrix}1\\ 2\\ x\end{bmatrix}=O.$

Ans. We have, $[1\ \ x\ \ 1]_{1\times3}\begin{bmatrix}1 & 3 & 2\\ 2 & 5 & 1\\ 15 & 3 & 2\end{bmatrix}_{3\times3}\begin{bmatrix}1\\ 2\\ x\end{bmatrix}_{3\times1}=O$

$\Rightarrow$ $[1+2x+15\ \ 3+5x+3\ \ 2+x+2]_{1\times3}\begin{bmatrix}1\\ 2\\ x\end{bmatrix}_{3\times1}=O$

$\Rightarrow$ $[16+2x\ \ 5x+6\ \ x+4]_{1\times3}\begin{bmatrix}1\\ 2\\ x\end{bmatrix}_{3\times1}=O$

$\Rightarrow$ $[16+2x+(5x+6)\cdot2+(x+4)\cdot x]_{1\times1}=O$

$\Rightarrow$ $[16+2x+10x+12+x^2+4x]=O$

$\Rightarrow$ $[x^2+16x+28]=O$

$\Rightarrow$ $[x^2+2x+14x+28]=O$

On comparing left side with zero matrix, we get
$$x^2+2x+14x+28=0$$
$$x(x+2)+14(x+2)=0$$
$\Rightarrow$ $(x+2)(x+14)=0$

$\therefore$ $x=-2,-14$

Q.11 Show that $A=\begin{bmatrix}5 & 3\\ -1 & -2\end{bmatrix}$ satisfies the equation $A^2-3A-7I=O$ and hence find the value of A^{-1}.

Ans. We have, $$A=\begin{bmatrix}5 & 3\\ -1 & -2\end{bmatrix}$$

$\therefore$ $$A^2=A\cdot A=\begin{bmatrix}5 & 3\\ -1 & -2\end{bmatrix}\cdot\begin{bmatrix}5 & 3\\ -1 & -2\end{bmatrix}$$

$$=\begin{bmatrix}25-3 & 15-6\\ -5+2 & -3+4\end{bmatrix}=\begin{bmatrix}22 & 9\\ -3 & 1\end{bmatrix}$$

$$3A=3\begin{bmatrix}5 & 3\\ -1 & -2\end{bmatrix}=\begin{bmatrix}15 & 9\\ -3 & -6\end{bmatrix}$$

and $$7I=7\begin{bmatrix}1 & 0\\ 0 & 1\end{bmatrix}=\begin{bmatrix}7 & 0\\ 0 & 7\end{bmatrix}$$

$\therefore$ $$A^2-3A-7I=\begin{bmatrix}22 & 9\\ -3 & 1\end{bmatrix}-\begin{bmatrix}15 & 9\\ -3 & -6\end{bmatrix}-\begin{bmatrix}7 & 0\\ 0 & 7\end{bmatrix}$$

$$=\begin{bmatrix}22-15-7 & 9-9-0\\ -3+3-0 & 1+6-7\end{bmatrix}$$

$$=\begin{bmatrix}0 & 0\\ 0 & 0\end{bmatrix}$$

$$=O$$ **Hence proved.**

Since, $$A^2-3A-7I=O$$

$\Rightarrow$ $$A^{-1}[(A^2)-3A-7I]=A^{-1}O$$

$\Rightarrow$ $$A^{-1}A\cdot A-3A^{-1}A-7A^{-1}I=O \quad [\because A^{-1}O=O]$$

$\Rightarrow$ $$IA-3I-7A^{-1}=O \quad [\because A^{-1}A=I]$$

$\Rightarrow$ $$A-3I-7A^{-1}=O \quad [\because A^{-1}I=A^{-1}]$$

$\Rightarrow$ $$-7A^{-1}=-A+3I$$

$$=\begin{bmatrix}-5 & -3\\ 1 & 2\end{bmatrix}+\begin{bmatrix}3 & 0\\ 0 & 3\end{bmatrix}=\begin{bmatrix}-2 & -3\\ 1 & 5\end{bmatrix}$$

$\therefore$ $$A^{-1}=\frac{-1}{7}\begin{bmatrix}-2 & -3\\ 1 & 5\end{bmatrix}$$

Q.12 Find the matrix A satisfying the matrix equation

$$\begin{bmatrix}2 & 1\\ 3 & 2\end{bmatrix}A\begin{bmatrix}-3 & 2\\ 5 & -3\end{bmatrix}=\begin{bmatrix}1 & 0\\ 0 & 1\end{bmatrix}.$$

Thinking Process We know that, if two matrices A and B of order $m\times n$ and $p\times q$ respectively are multiplied, then necessary condition for multiplication of $A\cdot B$ is $n=p$. So, by taking a matrix of correct order we can get the desired elements of the required matrix.

Ans. We have, $$\begin{bmatrix}2 & 1\\ 3 & 2\end{bmatrix}_{2\times 2}A\cdot\begin{bmatrix}-3 & 2\\ 5 & -3\end{bmatrix}_{2\times 2}=\begin{bmatrix}1 & 0\\ 0 & 1\end{bmatrix}_{2\times 2}$$

Let $A=\begin{bmatrix}a & b\\ c & d\end{bmatrix}$

$$\therefore \quad \begin{bmatrix} 2 & 1 \\ 3 & 2 \end{bmatrix}\begin{bmatrix} a & b \\ c & d \end{bmatrix}\begin{bmatrix} -3 & 2 \\ 5 & -3 \end{bmatrix} = \begin{bmatrix} 1 & 0 \\ 0 & 1 \end{bmatrix}$$

$$\Rightarrow \quad \begin{bmatrix} 2a+c & 2b+d \\ 3a+2c & 3b+2d \end{bmatrix}\begin{bmatrix} -3 & 2 \\ 5 & -3 \end{bmatrix} = \begin{bmatrix} 1 & 0 \\ 0 & 1 \end{bmatrix}$$

$$\Rightarrow \quad \begin{bmatrix} -6a-3c+10b+5d & 4a+2c-6b-3d \\ -9a-6c+15b+10d & 6a+4c-9b-6d \end{bmatrix} = \begin{bmatrix} 1 & 0 \\ 0 & 1 \end{bmatrix}$$

On comparing the corresponding elements of matrices, we get

$$\Rightarrow \quad -6a-3c+10b+5d = 1 \quad \text{...(i)}$$

$$\Rightarrow \quad 4a+2c-6b-3d = 0 \quad \text{...(ii)}$$

$$\Rightarrow \quad -9a-6c+15b+10d = 0 \quad \text{...(iii)}$$

$$\Rightarrow \quad 6a+4c-9b-6d = 1 \quad \text{...(iv)}$$

On adding Eqs. (i) and (iv), we get

$$c+b-d = 2 \Rightarrow d = c+b-2 \quad \text{...(v)}$$

On adding Eqs. (ii) and (iii), we get

$$-5a-4c+9b+7d = 0 \quad \text{...(vi)}$$

On adding Eqs. (vi) and (iv), we get

$$a+0+0+d = 1 \Rightarrow d = 1-a \quad \text{...(vii)}$$

From Eqs. (v) and (vii), we get

$$c+b-2 = 1-a \Rightarrow a+b+c = 3 \quad \text{...(viii)}$$

$$\Rightarrow \quad a = 3-b-c$$

Now, putting the values of a and d in Eq. (iii), we get

$$-9(3-b-c)-6c+15b+10(-2+b+c) = 0$$

$$\Rightarrow \quad -27+9b+9c-6c+15b-20+10b+10c = 0$$

$$\Rightarrow \quad 34b+13c = 47 \quad \text{...(ix)}$$

Now, putting the values of a and d in Eq. (ii), we get

$$4(3-b-c)+2c-6b-3(b+c-2) = 0$$

$$\Rightarrow \quad 12-4b-4c+2c-6b-3b-3c+6 = 0$$

$$\Rightarrow \quad -13b-5c = -18 \quad \text{...(x)}$$

On multiplying Eq. (ix) by 5 and Eq. (x) by 13, then adding, we get

$$\begin{aligned} -169b-65c &= -234 \\ 170\,b+65c &= 235 \\ \hline b &= 1 \end{aligned}$$

$$\Rightarrow \quad -13\times 1-5c = -18 \quad \text{[from Eq. (x)]}$$

$$\Rightarrow \quad -5c = -18+13 = -5 \Rightarrow c = 1$$

$$\therefore \quad a = 3-1-1 = 1 \text{ and } d = 1-1 = 0$$

$$\therefore \quad A = \begin{bmatrix} 1 & 1 \\ 1 & 0 \end{bmatrix}$$

Q.13 Find A, if $\begin{bmatrix} 4 \\ 1 \\ 3 \end{bmatrix} A = \begin{bmatrix} -4 & 8 & 4 \\ -1 & 2 & 1 \\ -3 & 6 & 3 \end{bmatrix}$.

Ans. We have, $\begin{bmatrix} 4 \\ 1 \\ 3 \end{bmatrix}_{3\times 1} \cdot A = \begin{bmatrix} -4 & 8 & 4 \\ -1 & 2 & 1 \\ -3 & 6 & 3 \end{bmatrix}_{3\times 3}$

Let $A = [x\ y\ z]$

$$\therefore \quad \begin{bmatrix} 4 \\ 1 \\ 3 \end{bmatrix}_{3\times1} [x\ y\ z]_{1\times3} = \begin{bmatrix} -4 & 8 & 4 \\ -1 & 2 & 1 \\ -3 & 6 & 3 \end{bmatrix}_{3\times3}$$

$$\Rightarrow \quad \begin{bmatrix} 4x & 4y & 4z \\ x & y & z \\ 3x & 3y & 3z \end{bmatrix} = \begin{bmatrix} -4 & 8 & 4 \\ -1 & 2 & 1 \\ -3 & 6 & 3 \end{bmatrix}$$

Since, both matrices are equal.

On comparing the corresponding elements of matrices, we get

$$\therefore \quad 4x = -4 \Rightarrow x = -1,\ 4y = 8$$

$$\Rightarrow \quad y = 2 \text{ and } 4z = 4$$

$$\Rightarrow \quad z = 1$$

$$\therefore \quad A = [-1\ 2\ 1]$$

Q.14 If $A \begin{bmatrix} 3 & -4 \\ 1 & 1 \\ 2 & 0 \end{bmatrix}$ and $B = \begin{bmatrix} 2 & 1 & 2 \\ 1 & 2 & 4 \end{bmatrix}$, then verify $(BA)^2 \neq B^2 A^2$.

Ans. We have, $A = \begin{bmatrix} 3 & -4 \\ 1 & 1 \\ 2 & 0 \end{bmatrix}_{3\times2}$ and $B = \begin{bmatrix} 2 & 1 & 2 \\ 1 & 2 & 4 \end{bmatrix}_{2\times3}$

$$\therefore \quad BA = \begin{bmatrix} 2 & 1 & 2 \\ 1 & 2 & 4 \end{bmatrix}_{2\times3} \begin{bmatrix} 3 & -4 \\ 1 & 1 \\ 2 & 0 \end{bmatrix}_{3\times2}$$

$$= \begin{bmatrix} 6+1+4 & -8+1+0 \\ 3+2+8 & -4+2+0 \end{bmatrix} = \begin{bmatrix} 11 & -7 \\ 13 & -2 \end{bmatrix}$$

and $$(BA)\cdot(BA) = \begin{bmatrix} 11 & -7 \\ 13 & -2 \end{bmatrix} \begin{bmatrix} 11 & -7 \\ 13 & -2 \end{bmatrix}$$

$$\Rightarrow \quad (BA)^2 = \begin{bmatrix} 121-91 & -77+14 \\ 143-26 & -91+4 \end{bmatrix} = \begin{bmatrix} 30 & -63 \\ 117 & -87 \end{bmatrix} \quad \text{...(i)}$$

Also, $$B^2 = B\cdot B = \begin{bmatrix} 2 & 1 & 2 \\ 1 & 2 & 4 \end{bmatrix}_{2\times3} \begin{bmatrix} 2 & 1 & 2 \\ 1 & 2 & 4 \end{bmatrix}_{2\times3}$$

So, B^2 is not possible, since B is not a square matrix.

Hence, $(BA)^2 \neq B^2 A^2$.

Q.15 If possible, find BA and AB, where

$$A = \begin{bmatrix} 2 & 1 & 2 \\ 1 & 2 & 4 \end{bmatrix} \text{ and } B = \begin{bmatrix} 4 & 1 \\ 2 & 3 \\ 1 & 2 \end{bmatrix}.$$

Ans. We have, $A=\begin{bmatrix}2&1&2\\1&2&4\end{bmatrix}_{2\times3}$ and $B=\begin{bmatrix}4&1\\2&3\\1&2\end{bmatrix}_{3\times2}$

So, AB and BA both are possible.

[since, in both $A\cdot B$ and $B\cdot A$, the number of columns of first is equal to the number of rows of second.]

$$\therefore \quad AB=\begin{bmatrix}2&1&2\\1&2&4\end{bmatrix}_{2\times3}\cdot\begin{bmatrix}4&1\\2&3\\1&2\end{bmatrix}_{3\times2}$$

$$=\begin{bmatrix}8+2+2&2+3+4\\4+4+4&1+6+8\end{bmatrix}=\begin{bmatrix}12&9\\12&15\end{bmatrix}$$

and
$$BA=\begin{bmatrix}4&1\\2&3\\1&2\end{bmatrix}_{3\times2}\begin{bmatrix}2&1&2\\1&2&4\end{bmatrix}_{2\times3}$$

$$=\begin{bmatrix}8+1&4+2&8+4\\4+3&2+6&4+12\\2+2&1+4&2+8\end{bmatrix}=\begin{bmatrix}9&6&12\\7&8&16\\4&5&10\end{bmatrix}$$

Q.16 Show by an example that for $A\neq O,\ B\neq O$ and $AB=O$.

Ans. Let $A=\begin{bmatrix}0&-4\\0&2\end{bmatrix}\neq O$ and $B=\begin{bmatrix}3&5\\0&0\end{bmatrix}\neq O$

$$\therefore \quad AB=\begin{bmatrix}0&0\\0&0\end{bmatrix}=O$$ **Hence proved.**

Q.17 Given, $A=\begin{bmatrix}2&4&0\\3&9&6\end{bmatrix}$ and $B=\begin{bmatrix}1&4\\2&8\\1&3\end{bmatrix}$. Is $(AB)'=B'A'$?

Ans. We have, $A=\begin{bmatrix}2&4&0\\3&9&6\end{bmatrix}_{2\times3}$ and $B=\begin{bmatrix}1&4\\2&8\\1&3\end{bmatrix}_{3\times2}$

$$\therefore \quad AB=\begin{bmatrix}2+8+0&8+32+0\\3+18+6&12+72+18\end{bmatrix}=\begin{bmatrix}10&40\\27&102\end{bmatrix}$$

$$\therefore \quad (AB)'=\begin{bmatrix}10&27\\40&102\end{bmatrix} \quad ...(i)$$

Also, $B'=\begin{bmatrix}1&2&1\\4&8&3\end{bmatrix}_{2\times3}$ and $A'=\begin{bmatrix}2&3\\4&9\\0&6\end{bmatrix}_{3\times2}$

$$\therefore \quad B'A'=\begin{bmatrix}2+8+0&3+18+6\\8+32+0&12+72+18\end{bmatrix}=\begin{bmatrix}10&27\\40&102\end{bmatrix} \quad ...(ii)$$

Thus, we see that, $(AB)'=B'A'$ [using Eqs. (i) and (ii)]

Q.18 Solve for x and y, $x\begin{bmatrix}2\\1\end{bmatrix}+y\begin{bmatrix}3\\5\end{bmatrix}+\begin{bmatrix}-8\\-11\end{bmatrix}=O.$

Ans. We have, $$x\begin{bmatrix}2\\1\end{bmatrix}+y\begin{bmatrix}3\\5\end{bmatrix}+\begin{bmatrix}-8\\-11\end{bmatrix}=O$$

$$\Rightarrow \begin{bmatrix}2x\\x\end{bmatrix}+\begin{bmatrix}3\cdot y\\5\cdot y\end{bmatrix}+\begin{bmatrix}-8\\-11\end{bmatrix}=O$$

$$\Rightarrow \begin{bmatrix}2x+3y-8\\x+5y-11\end{bmatrix}=\begin{bmatrix}0\\0\end{bmatrix}$$

On comparing the corresponding elements of matrices, we get

$$2x+3y-8=0 \quad \text{...(i)}$$

On multiplying Eq. (i) by 2, we get

$$4x+6y=16 \quad \text{...(ii)}$$

and $$x+5y-11=0 \quad \text{...(iii)}$$

On multiplying Eq. (iii) by 4, we get

$$4x+20y=44 \quad \text{...(iv)}$$

On subtracting Eq. (ii) from Eq. (iv), we get

$$14y=28 \Rightarrow y=2$$

On putting the value of $y=2$ in Eq. (i), we get

$$2x+3\times 2-8=0$$

$$\Rightarrow 2x=2 \Rightarrow x=1$$

$$\therefore x=1 \text{ and } y=2$$

Q.19 If X and Y are 2×2 matrices, then solve the following matrix equations for X and Y

$$2X+3Y=\begin{bmatrix}2&3\\4&0\end{bmatrix}, 3X+2Y=\begin{bmatrix}-2&2\\1&-5\end{bmatrix}.$$

Ans. We have,

$$2X+3Y=\begin{bmatrix}2&3\\4&0\end{bmatrix} \quad \text{...(i)}$$

and $$3X+2Y=\begin{bmatrix}-2&2\\1&-5\end{bmatrix} \quad \text{...(ii)}$$

On subtracting Eq. (i) from Eq. (ii), we get

$$\therefore (3X+2Y)-(2X+3Y)=\begin{bmatrix}-2-2&2-3\\1-4&-5-0\end{bmatrix}$$

$$(X-Y)=\begin{bmatrix}-4&-1\\-3&-5\end{bmatrix} \quad \text{...(iii)}$$

On adding Eqs. (i) and (ii), we get

$$(5X+5Y)=\begin{bmatrix}0&5\\5&-5\end{bmatrix}$$

$$\Rightarrow (X+Y)=\frac{1}{5}\begin{bmatrix}0&5\\5&-5\end{bmatrix}=\begin{bmatrix}0&1\\1&-1\end{bmatrix} \quad \text{...(iv)}$$

On adding Eqs. (iii) and (iv), we get

$$(X-Y)+(X+Y)=\begin{bmatrix}-4 & 0\\ -2 & -6\end{bmatrix}$$

$$\Rightarrow \quad 2X=2\begin{bmatrix}-2 & 0\\ -1 & -3\end{bmatrix}$$

$$\therefore \quad X=\begin{bmatrix}-2 & 0\\ -1 & -3\end{bmatrix}$$

From Eq. (iv), we get

$$\begin{bmatrix}-2 & 0\\ -1 & -3\end{bmatrix}+Y=\begin{bmatrix}0 & 1\\ 1 & -1\end{bmatrix}$$

$$\Rightarrow \quad Y=\begin{bmatrix}0 & 1\\ 1 & -1\end{bmatrix}-\begin{bmatrix}-2 & 0\\ -1 & -3\end{bmatrix}=\begin{bmatrix}0+2 & 1-0\\ 1+1 & -1+3\end{bmatrix}=\begin{bmatrix}2 & 1\\ 2 & 2\end{bmatrix}$$

$$\therefore \quad Y=\begin{bmatrix}2 & 1\\ 2 & 2\end{bmatrix} \text{ and } X=\begin{bmatrix}-2 & 0\\ -1 & -3\end{bmatrix}$$

Q.20 If $A=[3\ \ 5]$ and $B=[7\ \ 3]$, then find a non-zero matrix C such that $AC=BC$.

Ans. We have, $A=[3\ 5]_{1\times 2}$ and $B=[7\ \ 3]_{1\times 2}$

As for multiplication AB. If order of $A=m\times n$ and order of $B=n\times p$, then order of $AB=m\times p$.

Let $C=\begin{bmatrix}x\\ y\end{bmatrix}_{2\times 1}$ is a non-zero matrix of order 2×1.

$$\therefore \quad AC=[3\ \ 5]\begin{bmatrix}x\\ y\end{bmatrix}=[3x+5y]$$

$$\text{and} \quad BC=[7\ \ 3]\begin{bmatrix}x\\ y\end{bmatrix}=[7x+3y]$$

Since, $AC=BC$

$\therefore \quad [3x+5y]=[7x+3y]$

On using equality of matrix, we get

$$3x+5y=7x+3y$$

$$\Rightarrow \quad 4x=2y$$

$$\Rightarrow \quad y=2x$$

$$\therefore \quad C=\begin{bmatrix}x\\ 2x\end{bmatrix}$$

We see that on taking C of order $2\times 1, 2\times 2, 2\times 3, \ldots$, we get

$$C=\begin{bmatrix}x\\ 2x\end{bmatrix}, \begin{bmatrix}x & x\\ 2x & 2x\end{bmatrix}, \begin{bmatrix}x & x & x\\ 2x & 2x & 2x\end{bmatrix}\ldots$$

In general,

$$C=\begin{bmatrix}k\\ 2k\end{bmatrix}, \begin{bmatrix}k & k\\ 2k & 2k\end{bmatrix} \text{ etc.}$$

where, k is any real number.

Q.21 Give an example of matrices A, B and C, such that $AB = AC$, where A is non-zero matrix but $B \neq C$.

Ans. Let $$A = \begin{bmatrix} 1 & 0 \\ 0 & 0 \end{bmatrix}, B = \begin{bmatrix} 2 & 3 \\ 4 & 0 \end{bmatrix} \text{ and } C = \begin{bmatrix} 2 & 3 \\ 4 & 4 \end{bmatrix} \quad [\because B \neq C]$$

$$\therefore \quad AB = \begin{bmatrix} 1 & 0 \\ 0 & 0 \end{bmatrix}\begin{bmatrix} 2 & 3 \\ 4 & 0 \end{bmatrix} = \begin{bmatrix} 2 & 3 \\ 0 & 0 \end{bmatrix} \quad \text{...(i)}$$

and $$AC = \begin{bmatrix} 1 & 0 \\ 0 & 0 \end{bmatrix} \cdot \begin{bmatrix} 2 & 3 \\ 4 & 4 \end{bmatrix} = \begin{bmatrix} 2 & 3 \\ 0 & 0 \end{bmatrix} \quad \text{...(ii)}$$

Thus, we see that $AB = AC$ [using Eqs. (i) and (ii)]
where, A is non-zero matrix but $B \neq C$.

Q.22 If $A = \begin{bmatrix} 1 & 2 \\ -2 & 1 \end{bmatrix}$, $B = \begin{bmatrix} 2 & 3 \\ 3 & -4 \end{bmatrix}$ and $C = \begin{bmatrix} 1 & 0 \\ -1 & 0 \end{bmatrix}$, verify

(i) $(AB)\,C = A(BC)$.

(ii) $A\,(B + C) = AB + AC$.

Ans. We have, $A = \begin{bmatrix} 1 & 2 \\ -2 & 1 \end{bmatrix}$, $B = \begin{bmatrix} 2 & 3 \\ 3 & -4 \end{bmatrix}$ and $C = \begin{bmatrix} 1 & 0 \\ -1 & 0 \end{bmatrix}$

(i) $$(AB) = \begin{bmatrix} 1 & 2 \\ -2 & 1 \end{bmatrix}\begin{bmatrix} 2 & 3 \\ 3 & -4 \end{bmatrix} = \begin{bmatrix} 2+6 & 3-8 \\ -4+3 & -6-4 \end{bmatrix} = \begin{bmatrix} 8 & -5 \\ -1 & -10 \end{bmatrix}$$

and $$(AB)C = \begin{bmatrix} 8 & -5 \\ -1 & -10 \end{bmatrix}\begin{bmatrix} 1 & 0 \\ -1 & 0 \end{bmatrix}$$

$$= \begin{bmatrix} 8+5 & 0 \\ -1+10 & 0 \end{bmatrix} = \begin{bmatrix} 13 & 0 \\ 9 & 0 \end{bmatrix} \quad \text{...(i)}$$

Again, $$(BC) = \begin{bmatrix} 2 & 3 \\ 3 & -4 \end{bmatrix}\begin{bmatrix} 1 & 0 \\ -1 & 0 \end{bmatrix}$$

$$= \begin{bmatrix} 2-3 & 0 \\ 3+4 & 0 \end{bmatrix} = \begin{bmatrix} -1 & 0 \\ 7 & 0 \end{bmatrix}$$

and $$A(BC) = \begin{bmatrix} 1 & 2 \\ -2 & 1 \end{bmatrix}\begin{bmatrix} -1 & 0 \\ 7 & 0 \end{bmatrix}$$

$$= \begin{bmatrix} -1+14 & 0 \\ 2+7 & 0 \end{bmatrix} = \begin{bmatrix} 13 & 0 \\ 9 & 0 \end{bmatrix} \quad \text{...(ii)}$$

$$\therefore \quad (AB)\,C = A(BC) \quad \text{[using Eqs. (i) and (ii)]}$$

(ii) $$(B+C) = \begin{bmatrix} 2 & 3 \\ 3 & -4 \end{bmatrix} + \begin{bmatrix} 1 & 0 \\ -1 & 0 \end{bmatrix} = \begin{bmatrix} 3 & 3 \\ 2 & -4 \end{bmatrix}$$

and $$A \cdot (B+C) = \begin{bmatrix} 1 & 2 \\ -2 & 1 \end{bmatrix}\begin{bmatrix} 3 & 3 \\ 2 & -4 \end{bmatrix}$$

$$= \begin{bmatrix} 3+4 & 3-8 \\ -6+2 & -6-4 \end{bmatrix} = \begin{bmatrix} 7 & -5 \\ -4 & -10 \end{bmatrix} \quad \text{...(iii)}$$

Also, $$AB = \begin{bmatrix} 1 & 2 \\ -2 & 1 \end{bmatrix} \cdot \begin{bmatrix} 2 & 3 \\ 3 & -4 \end{bmatrix}$$

$$= \begin{bmatrix} 2+6 & 3-8 \\ -4+3 & -6-4 \end{bmatrix} = \begin{bmatrix} 8 & -5 \\ -1 & -10 \end{bmatrix}$$

and $$AC = \begin{bmatrix} 1 & 2 \\ -2 & 1 \end{bmatrix} \begin{bmatrix} 1 & 0 \\ -1 & 0 \end{bmatrix} = \begin{bmatrix} 1-2 & 0 \\ -2-1 & 0 \end{bmatrix} = \begin{bmatrix} -1 & 0 \\ -3 & 0 \end{bmatrix}$$

$\therefore$ $$AB + AC = \begin{bmatrix} 8 & -5 \\ -1 & -10 \end{bmatrix} + \begin{bmatrix} -1 & 0 \\ -3 & 0 \end{bmatrix}$$

$\Rightarrow$ $$AB + AC = \begin{bmatrix} 7 & -5 \\ -4 & -10 \end{bmatrix} \quad \text{...(iv)}$$

From Eqs. (iii) and (iv), we get

$$A(B + C) = AB + AC$$

Q.23 If $P = \begin{bmatrix} x & 0 & 0 \\ 0 & y & 0 \\ 0 & 0 & z \end{bmatrix}$ and $Q = \begin{bmatrix} a & 0 & 0 \\ 0 & b & 0 \\ 0 & 0 & c \end{bmatrix}$, then prove that

$$PQ = \begin{bmatrix} xa & 0 & 0 \\ 0 & yb & 0 \\ 0 & 0 & zc \end{bmatrix} = QP.$$

Ans. $$PQ = \begin{bmatrix} x & 0 & 0 \\ 0 & y & 0 \\ 0 & 0 & z \end{bmatrix} \begin{bmatrix} a & 0 & 0 \\ 0 & b & 0 \\ 0 & 0 & c \end{bmatrix} = \begin{bmatrix} xa & 0 & 0 \\ 0 & yb & 0 \\ 0 & 0 & zc \end{bmatrix} \quad \text{...(i)}$$

and $$QP = \begin{bmatrix} a & 0 & 0 \\ 0 & b & 0 \\ 0 & 0 & c \end{bmatrix} \begin{bmatrix} x & 0 & 0 \\ 0 & y & 0 \\ 0 & 0 & z \end{bmatrix} = \begin{bmatrix} ax & 0 & 0 \\ 0 & by & 0 \\ 0 & 0 & zc \end{bmatrix} \quad \text{...(ii)}$$

Thus, we see that, $PQ = QP$ [using Eqs. (i) and (ii)]

Hence proved.

Q.24 If $[2 \ 1 \ 3] \begin{bmatrix} -1 & 0 & -1 \\ -1 & 1 & 0 \\ 0 & 1 & 1 \end{bmatrix} \begin{bmatrix} 1 \\ 0 \\ -1 \end{bmatrix} = A$, then find the value of A.

Ans. We have, $$[2 \ 1 \ 3] \begin{bmatrix} -1 & 0 & -1 \\ -1 & 1 & 0 \\ 0 & 1 & 1 \end{bmatrix} \begin{bmatrix} 1 \\ 0 \\ -1 \end{bmatrix} = A$$

$\therefore$ $$[2 \ 1 \ 3] \begin{bmatrix} -1 & 0 & -1 \\ -1 & 1 & 0 \\ 0 & 1 & 1 \end{bmatrix} = [-2-1+0 \quad 0+1+3 \quad -2+0+3]$$

$$= [-3 \quad 4 \quad 1]$$

Now, $$[-3 \ 4 \ 1] \begin{bmatrix} 1 \\ 0 \\ -1 \end{bmatrix} = A$$

$$A = [-3 + 0 - 1] = [-4]$$

Q.25 If $A=[2\ \ 1]$, $B=\begin{bmatrix}5 & 3 & 4\\ 8 & 7 & 6\end{bmatrix}$ and $C=\begin{bmatrix}-1 & 2 & 1\\ 1 & 0 & 2\end{bmatrix}$, then verify that $A(B+C)=(AB+AC)$.

Ans. We have to verify that, $A(B+C)=AB+AC$

We have, $A=[2\ \ 1]$, $B=\begin{bmatrix}5 & 3 & 4\\ 8 & 7 & 6\end{bmatrix}$ and $C=\begin{bmatrix}-1 & 2 & 1\\ 1 & 0 & 2\end{bmatrix}$

$$\therefore \quad A(B+C)=[2\ \ 1]\begin{bmatrix}5-1 & 3+2 & 4+1\\ 8+1 & 7+0 & 6+2\end{bmatrix}$$

$$=[2\ \ 1]\begin{bmatrix}4 & 5 & 5\\ 9 & 7 & 8\end{bmatrix}$$

$$=[8+9\ \ 10+7\ \ 10+8]$$

$$=[17\ \ 17\ \ 18] \quad \text{...(i)}$$

Also, $$AB=[2\ \ 1]\begin{bmatrix}5 & 3 & 4\\ 8 & 7 & 6\end{bmatrix}$$

$$=[10+8\ \ 6+7\ \ 8+6]=[18\ \ 13\ \ 14]$$

and $$AC=[2\ \ 1]\begin{bmatrix}-1 & 2 & 1\\ 1 & 0 & 2\end{bmatrix}$$

$$=[-2+1\ \ 4+0\ \ 2+2]=[-1\ \ 4\ \ 4]$$

$$\therefore \quad AB+AC=[18\ \ 13\ \ 14]+[-1\ \ 4\ \ 4]$$

$$=[17\ \ 17\ \ 18] \quad \text{...(ii)}$$

$$\therefore \quad A(B+C)=(AB+AC) \quad \text{[using Eqs. (i) and (ii)]}$$

Hence proved.

Q.26 If $A=\begin{bmatrix}1 & 0 & -1\\ 2 & 1 & 3\\ 0 & 1 & 1\end{bmatrix}$, then verify that $A^2+A=A(A+I)$, where I is 3×3 unit matrix.

Ans. We have, $$A=\begin{bmatrix}1 & 0 & -1\\ 2 & 1 & 3\\ 0 & 1 & 1\end{bmatrix}$$

$$\therefore \quad A^2=A\cdot A$$

$$=\begin{bmatrix}1 & 0 & -1\\ 2 & 1 & 3\\ 0 & 1 & 1\end{bmatrix}\begin{bmatrix}1 & 0 & -1\\ 2 & 1 & 3\\ 0 & 1 & 1\end{bmatrix}=\begin{bmatrix}1 & -1 & -2\\ 4 & 4 & 4\\ 2 & 2 & 4\end{bmatrix}$$

$$\therefore \quad A^2+A=\begin{bmatrix}1 & -1 & -2\\ 4 & 4 & 4\\ 2 & 2 & 4\end{bmatrix}+\begin{bmatrix}1 & 0 & -1\\ 2 & 1 & 3\\ 0 & 1 & 1\end{bmatrix}$$

$$=\begin{bmatrix}2 & -1 & -3\\ 6 & 5 & 7\\ 2 & 3 & 5\end{bmatrix} \quad \text{...(i)}$$

Now, $$A+I=\begin{bmatrix}1&0&-1\\2&1&3\\0&1&1\end{bmatrix}+\begin{bmatrix}1&0&0\\0&1&0\\0&0&1\end{bmatrix}=\begin{bmatrix}2&0&-1\\2&2&3\\0&1&2\end{bmatrix}$$

and $$A(A+I)=\begin{bmatrix}1&0&-1\\2&1&3\\0&1&1\end{bmatrix}\cdot\begin{bmatrix}2&0&-1\\2&2&3\\0&1&2\end{bmatrix}=\begin{bmatrix}2&-1&-3\\6&5&7\\2&3&5\end{bmatrix} \quad \text{...(ii)}$$

Thus, we see that $A^2 + A = A(A+I)$ [using Eqs. (i) and (ii)]

Q.27 If $A=\begin{bmatrix}0&-1&2\\4&3&-4\end{bmatrix}$ and $B=\begin{bmatrix}4&0\\1&3\\2&6\end{bmatrix}$, then verify that

(i) $(A')'=A$
(ii) $(AB)'=B'A'$
(iii) $(kA)'=(kA')$.

Ans. We have, $A=\begin{bmatrix}0&-1&2\\4&3&-4\end{bmatrix}$ and $B=\begin{bmatrix}4&0\\1&3\\2&6\end{bmatrix}$

(i) We have to verify that, $(A')'=A$

$$\therefore \quad A'=\begin{bmatrix}0&4\\-1&3\\2&-4\end{bmatrix}$$

and $$(A')'=\begin{bmatrix}0&-1&2\\4&3&-4\end{bmatrix}=A$$ **Hence proved.**

(ii) We have to verify that, $(AB')=B'A'$

Since $$AB=\begin{bmatrix}0&-1&2\\4&3&-4\end{bmatrix}\begin{bmatrix}4&0\\1&3\\2&6\end{bmatrix}$$

$$\therefore \quad AB=\begin{bmatrix}3&9\\11&-15\end{bmatrix}$$

$$\Rightarrow \quad (AB)'=\begin{bmatrix}3&11\\9&-15\end{bmatrix}$$

and $$B'A'=\begin{bmatrix}4&1&2\\0&3&6\end{bmatrix}\begin{bmatrix}0&4\\-1&3\\2&-4\end{bmatrix}=\begin{bmatrix}3&11\\9&-15\end{bmatrix}$$

$$=(AB)'$$ **Hence proved.**

(iii) We have to verify that, $(kA)'=(kA')$

Now, $$(kA)=\begin{bmatrix}0&-k&2k\\4k&3k&-4k\end{bmatrix}$$

and $$(kA)' = \begin{bmatrix} 0 & 4k \\ -k & 3k \\ 2k & -4k \end{bmatrix}$$

Also, $$kA' = \begin{bmatrix} 0 & 4k \\ -k & 3k \\ 2k & -4k \end{bmatrix}$$

$$= (kA)'$$ **Hence proved.**

Q.28 If $A = \begin{bmatrix} 1 & 2 \\ 4 & 1 \\ 5 & 6 \end{bmatrix}$ and $B = \begin{bmatrix} 1 & 2 \\ 6 & 4 \\ 7 & 3 \end{bmatrix}$, then verify that

(i) $(2A + B)' = 2A' + B'$.

(ii) $(A - B)' = A' - B'$.

Ans. We have, $$A = \begin{bmatrix} 1 & 2 \\ 4 & 1 \\ 5 & 6 \end{bmatrix} \text{ and } B = \begin{bmatrix} 1 & 2 \\ 6 & 4 \\ 7 & 3 \end{bmatrix}$$

(i) $\therefore$ $$(2A + B) = \begin{bmatrix} 2 & 4 \\ 8 & 2 \\ 10 & 12 \end{bmatrix} + \begin{bmatrix} 1 & 2 \\ 6 & 4 \\ 7 & 3 \end{bmatrix} = \begin{bmatrix} 3 & 6 \\ 14 & 6 \\ 17 & 15 \end{bmatrix}$$

and $$(2A + B)' = \begin{bmatrix} 3 & 14 & 17 \\ 6 & 6 & 15 \end{bmatrix}$$

Also, $$2A' + B' = 2\begin{bmatrix} 1 & 4 & 5 \\ 2 & 1 & 6 \end{bmatrix} + \begin{bmatrix} 1 & 6 & 7 \\ 2 & 4 & 3 \end{bmatrix}$$

$$= \begin{bmatrix} 2 & 8 & 10 \\ 4 & 2 & 12 \end{bmatrix} + \begin{bmatrix} 1 & 6 & 7 \\ 2 & 4 & 3 \end{bmatrix}$$

$$= \begin{bmatrix} 3 & 14 & 17 \\ 6 & 6 & 15 \end{bmatrix} = (2A + B)'$$ **Hence proved.**

(ii) $$(A - B) = \begin{bmatrix} 1 & 2 \\ 4 & 1 \\ 5 & 6 \end{bmatrix} - \begin{bmatrix} 1 & 2 \\ 6 & 4 \\ 7 & 3 \end{bmatrix} = \begin{bmatrix} 0 & 0 \\ -2 & -3 \\ -2 & 3 \end{bmatrix}$$

and $$(A - B)' = \begin{bmatrix} 0 & -2 & -2 \\ 0 & -3 & 3 \end{bmatrix}$$

Also, $$A' - B' = \begin{bmatrix} 1 & 4 & 5 \\ 2 & 1 & 6 \end{bmatrix} - \begin{bmatrix} 1 & 6 & 7 \\ 2 & 4 & 3 \end{bmatrix}$$

$$= \begin{bmatrix} 0 & -2 & -2 \\ 0 & -3 & 3 \end{bmatrix}$$

$$= (A - B)'$$ **Hence proved.**

Q.29 Show that $A'A$ and $A A'$ are both symmetric matrices for any matrix A.

Thinking Process We know that, for a matrix A to be symmetric matrix, $A' = A$. Also by using the result $(AB)' = B'A'$, we can prove that $A'A$ and AA' are both symmetric matrices for any matrix A.

Ans. Let $P = A'A$

$\therefore$
$$P' = (AA')'$$
$$= A'(A')' \quad [\because (AB)' = B'A']$$
$$= A'A = P \quad [\because (A')' = A]$$

So, $A'A$ is symmetric matrix for any matrix A.

Similarly, let $Q = A A'$

$\therefore$
$$Q' = (AA')' = (A')' A'$$
$$= A A' = Q$$

So, AA' is symmetric matrix for any matrix A.

Q.30 Let A and B be square matrices of the order 3×3. Is $(AB)^2 = A^2B^2$? Give reasons.

Ans. Since, A and B are square matrices of order 3×3.

$\therefore$
$$(AB)^2 = AB \cdot AB$$
$$= (AB)AB$$
$$= AABB \quad [\because AB = BA]$$
$$= A^2B^2$$

So, $AB^2 = A^2B^2$ is true only when $AB = BA$.

Q.31 Show that, if A and B are square matrices such that $AB = BA$, then $(A+B)^2 = A^2 + 2AB + B^2$.

Ans. Since, A and B are square matrices such that $AB = BA$.

$\therefore$
$$(A+B)^2 = (A+B)\cdot(A+B)$$
$$= A^2 + AB + BA + B^2$$
$$= A^2 + AB + AB + B^2 \quad [\because AB = BA]$$
$$= A^2 + 2AB + B^2$$

Hence proved.

Q.32 If $A = \begin{bmatrix} 1 & 2 \\ -1 & 3 \end{bmatrix}, B = \begin{bmatrix} 4 & 0 \\ 1 & 5 \end{bmatrix}, C = \begin{bmatrix} 2 & 0 \\ 1 & -2 \end{bmatrix}$, $a = 4$, and $b = -2$, then show that

(i) $A + (B + C) = (A + B) + C$

(ii) $A(BC) = (AB)C$

(iii) $(a + b)B = aB + bB$

(iv) $a(C - A) = aC - aA$

(v) $(A^T)^T = A$

(vi) $(bA)^T = b A^T$

(vii) $(AB)^T = B^T A^T$

(viii) $(A - B)C = AC - BC$

(ix) $(A - B)^T = A^T - B^T$

Ans. Given, $A=\begin{bmatrix}1 & 2\\-1 & 3\end{bmatrix}, B=\begin{bmatrix}4 & 0\\1 & 5\end{bmatrix}$

$C=\begin{bmatrix}2 & 0\\1 & -2\end{bmatrix}$ and $a=4, b=-2$

(i) $A+(B+C)=\begin{bmatrix}1 & 2\\-1 & 3\end{bmatrix}+\begin{bmatrix}6 & 0\\2 & 3\end{bmatrix}=\begin{bmatrix}7 & 2\\1 & 6\end{bmatrix}$

and $(A+B)+C=\begin{bmatrix}5 & 2\\0 & 8\end{bmatrix}+\begin{bmatrix}2 & 0\\1 & -2\end{bmatrix}$

$=\begin{bmatrix}7 & 2\\1 & 6\end{bmatrix}=A+(B+C)$ **Hence proved.**

(ii) $(BC)=\begin{bmatrix}4 & 0\\1 & 5\end{bmatrix}\begin{bmatrix}2 & 0\\1 & -2\end{bmatrix}=\begin{bmatrix}8 & 0\\7 & -10\end{bmatrix}$ $\left[\because \begin{matrix}4\times 2+0\times 1 & 4\times 0+0(-2)\\1\times 2+5\times 1 & 1\times 0+5(-2)\end{matrix}\right]$

and $A(BC)=\begin{bmatrix}1 & 2\\-1 & 3\end{bmatrix}\begin{bmatrix}8 & 0\\7 & -10\end{bmatrix}$

$=\begin{bmatrix}8+14 & 0-20\\-8+21 & 0-30\end{bmatrix}=\begin{bmatrix}22 & -20\\13 & -30\end{bmatrix}$

Also, $(AB)=\begin{bmatrix}1 & 2\\-1 & 3\end{bmatrix}\cdot\begin{bmatrix}4 & 0\\1 & 5\end{bmatrix}=\begin{bmatrix}6 & 10\\-1 & 15\end{bmatrix}$

$(AB)C=\begin{bmatrix}6 & 10\\-1 & 15\end{bmatrix}\begin{bmatrix}2 & 0\\1 & -2\end{bmatrix}$

$=\begin{bmatrix}22 & -20\\13 & -30\end{bmatrix}=A(BC)$ **Hence proved.**

(iii) $(a+b)B=(4-2)\begin{bmatrix}4 & 0\\1 & 5\end{bmatrix}$ $[\because a=4, b=-2]$

$=\begin{bmatrix}8 & 0\\2 & 10\end{bmatrix}$

and $aB+bB=4B-2B$

$=4\begin{bmatrix}4 & 0\\1 & 5\end{bmatrix}-2\begin{bmatrix}4 & 0\\1 & 5\end{bmatrix}$

$=\begin{bmatrix}16 & 0\\4 & 20\end{bmatrix}-\begin{bmatrix}8 & 0\\2 & 10\end{bmatrix}=\begin{bmatrix}8 & 0\\2 & 10\end{bmatrix}$

$=(a+b)B$ **Hence proved.**

(iv) $(C-A)=\begin{bmatrix}2 & 0\\1 & -2\end{bmatrix}-\begin{bmatrix}1 & 2\\-1 & 3\end{bmatrix}=\begin{bmatrix}2-1 & 0-2\\1+1 & -2-3\end{bmatrix}=\begin{bmatrix}1 & -2\\2 & -5\end{bmatrix}$

and $a(C-A)=4\begin{bmatrix}1 & -2\\2 & -5\end{bmatrix}=\begin{bmatrix}4 & -8\\8 & -20\end{bmatrix}$ $[\because a=4]$

Also, $aC-aA=4\begin{bmatrix}2 & 0\\1 & -2\end{bmatrix}-4\begin{bmatrix}1 & 2\\-1 & 3\end{bmatrix}=\begin{bmatrix}8 & 0\\4 & -8\end{bmatrix}-\begin{bmatrix}4 & 8\\-4 & 12\end{bmatrix}=\begin{bmatrix}4 & -8\\8 & -20\end{bmatrix}$

$=a(C-A)$ **Hence proved.**

(v) $$A^T = \begin{bmatrix} 1 & 2 \\ -1 & 3 \end{bmatrix}^T = \begin{bmatrix} 1 & -1 \\ 2 & 3 \end{bmatrix}$$

Now, $$(A^T)^T = \begin{bmatrix} 1 & 2 \\ -1 & 3 \end{bmatrix}$$

$$= A$$ **Hence proved.**

(vi) $$(bA)^T = \begin{bmatrix} -2 & -4 \\ 2 & -6 \end{bmatrix}^T$$ $[\because\ b = -2]$

$$\Rightarrow \quad (bA)^T = \begin{bmatrix} -2 & 2 \\ -4 & -6 \end{bmatrix}$$

and $$A^T = \begin{bmatrix} 1 & -1 \\ 2 & 3 \end{bmatrix}$$

$$\Rightarrow \quad bA^T = -2\begin{bmatrix} 1 & -1 \\ 2 & 3 \end{bmatrix}$$

$$\Rightarrow \quad bA^T = \begin{bmatrix} (-2)\times 1 & (-2)\times(-1) \\ (-2)\times 2 & (-2)\times 3 \end{bmatrix}$$

$$= \begin{bmatrix} -2 & 2 \\ -4 & -6 \end{bmatrix} = (bA)^T$$ **Hence proved.**

(vii) $$AB = \begin{bmatrix} 1 & 2 \\ -1 & 3 \end{bmatrix}\begin{bmatrix} 4 & 0 \\ 1 & 5 \end{bmatrix} = \begin{bmatrix} 4+2 & 0+10 \\ -4+3 & 0+15 \end{bmatrix} = \begin{bmatrix} 6 & 10 \\ -1 & 15 \end{bmatrix}$$

$$\therefore \quad (AB)^T = \begin{bmatrix} 6 & -1 \\ 10 & 15 \end{bmatrix}$$

Now, $$B^T A^T = \begin{bmatrix} 4 & 1 \\ 0 & 5 \end{bmatrix}\begin{bmatrix} 1 & -1 \\ 2 & 3 \end{bmatrix} = \begin{bmatrix} 6 & -1 \\ 10 & 15 \end{bmatrix}$$

$$= (AB)^T$$ **Hence proved.**

(viii) $$(A-B) = \begin{bmatrix} 1-4 & 2-0 \\ -1-1 & 3-5 \end{bmatrix} = \begin{bmatrix} -3 & 2 \\ -2 & -2 \end{bmatrix}$$

$$(A-B)\ C = \begin{bmatrix} -3 & 2 \\ -2 & -2 \end{bmatrix}\begin{bmatrix} 2 & 0 \\ 1 & -2 \end{bmatrix} = \begin{bmatrix} -4 & -4 \\ -6 & 4 \end{bmatrix}$$ ...(i)

Now, $$AC = \begin{bmatrix} 1 & 2 \\ -1 & 3 \end{bmatrix}\begin{bmatrix} 2 & 0 \\ 1 & -2 \end{bmatrix} = \begin{bmatrix} 4 & -4 \\ 1 & -6 \end{bmatrix}$$ ...(ii)

and $$BC = \begin{bmatrix} 4 & 0 \\ 1 & 5 \end{bmatrix}\begin{bmatrix} 2 & 0 \\ 1 & -2 \end{bmatrix} = \begin{bmatrix} 8 & 0 \\ 7 & -10 \end{bmatrix}$$ ...(iii)

$$\therefore \quad AC - BC = \begin{bmatrix} 4-8 & -4-0 \\ 1-7 & -6+10 \end{bmatrix}$$ [using Eqs. (ii) and (iii)]

$$= \begin{bmatrix} -4 & -4 \\ -6 & 4 \end{bmatrix}$$

$$= (A-B)C$$ [using Eq. (i)]

Hence proved.

(ix) $(A-B)^T = \begin{bmatrix} 1-4 & 2-0 \\ -1-1 & 3-5 \end{bmatrix}^T$

$$= \begin{bmatrix} -3 & 2 \\ -2 & -2 \end{bmatrix}^T = \begin{bmatrix} -3 & -2 \\ 2 & -2 \end{bmatrix}$$

$$A^T - B^T = \begin{bmatrix} 1 & -1 \\ 2 & 3 \end{bmatrix} - \begin{bmatrix} 4 & 1 \\ 0 & 5 \end{bmatrix}$$

$$= \begin{bmatrix} -3 & -2 \\ 2 & -2 \end{bmatrix} = (A-B)^T$$

Hence proved.

Q.33 If $A = \begin{bmatrix} \cos\theta & \sin\theta \\ -\sin\theta & \cos\theta \end{bmatrix}$, then show that $A^2 = \begin{bmatrix} \cos 2\theta & \sin 2\theta \\ -\sin 2\theta & \cos 2\theta \end{bmatrix}$.

Ans. We have, $A = \begin{bmatrix} \cos\theta & \sin\theta \\ -\sin\theta & \cos\theta \end{bmatrix}$

$\therefore$ $$A^2 = A \cdot A = \begin{bmatrix} \cos\theta & \sin\theta \\ -\sin\theta & \cos\theta \end{bmatrix} \cdot \begin{bmatrix} \cos\theta & \sin\theta \\ -\sin\theta & \cos\theta \end{bmatrix}$$

$$= \begin{bmatrix} \cos^2\theta - \sin^2\theta & \cos\theta \cdot \sin\theta + \sin\theta\cos\theta \\ -\sin\theta\cos\theta - \cos\theta\sin\theta & -\sin^2\theta + \cos^2\theta \end{bmatrix}$$

$$= \begin{bmatrix} \cos 2\theta & 2\sin\theta\cos\theta \\ -2\sin\theta\cos\theta & \cos 2\theta \end{bmatrix} \quad [\because \cos^2\theta - \sin^2\theta = \cos 2\theta]$$

$$= \begin{bmatrix} \cos 2\theta & \sin 2\theta \\ -\sin 2\theta & \cos 2\theta \end{bmatrix} \quad [\because \sin 2\theta = 2\sin\theta \cdot \cos\theta]$$

Hence proved.

Q.34 If $A = \begin{bmatrix} 0 & -x \\ x & 0 \end{bmatrix}$, $B = \begin{bmatrix} 0 & 1 \\ 1 & 0 \end{bmatrix}$ and $x^2 = -1$, then show that $(A+B)^2 = A^2 + B^2$.

Ans. We have, $A = \begin{bmatrix} 0 & -x \\ x & 0 \end{bmatrix}$, $B = \begin{bmatrix} 0 & 1 \\ 1 & 0 \end{bmatrix}$ and $x^2 = -1$

$\therefore$ $$(A+B) = \begin{bmatrix} 0 & -x+1 \\ x+1 & 0 \end{bmatrix}$$

and $$(A+B)^2 = \begin{bmatrix} 0 & -x+1 \\ x+1 & 0 \end{bmatrix} \begin{bmatrix} 0 & -x+1 \\ x+1 & 0 \end{bmatrix}$$

$$= \begin{bmatrix} 1-x^2 & 0 \\ 0 & 1-x^2 \end{bmatrix}$$

$$= \begin{bmatrix} 1+1 & 0 \\ 0 & 1+1 \end{bmatrix} \quad [\because x^2 = -1]$$

$$= \begin{bmatrix} 2 & 0 \\ 0 & 2 \end{bmatrix}$$

Also, $$A^2 = A \cdot A = \begin{bmatrix} 0 & -x \\ x & 0 \end{bmatrix} \begin{bmatrix} 0 & -x \\ x & 0 \end{bmatrix} = \begin{bmatrix} -x^2 & 0 \\ 0 & -x^2 \end{bmatrix} = \begin{bmatrix} 1 & 0 \\ 0 & 1 \end{bmatrix}$$

and $$B^2 = B \cdot B = \begin{bmatrix} 0 & 1 \\ 1 & 0 \end{bmatrix} \begin{bmatrix} 0 & 1 \\ 1 & 0 \end{bmatrix} = \begin{bmatrix} 1 & 0 \\ 0 & 1 \end{bmatrix}$$

Now, $$A^2 + B^2 = \begin{bmatrix} 1+1 & 0 \\ 0 & 1+1 \end{bmatrix} = \begin{bmatrix} 2 & 0 \\ 0 & 2 \end{bmatrix}$$

$$= (A+B)^2$$

Hence proved.

Q.35 Verify that $A^2 = I$, when $A = \begin{bmatrix} 0 & 1 & -1 \\ 4 & -3 & 4 \\ 3 & -3 & 4 \end{bmatrix}$.

Ans. We have, $A = \begin{bmatrix} 0 & 1 & -1 \\ 4 & -3 & 4 \\ 3 & -3 & 4 \end{bmatrix}$

$$\therefore \quad A^2 = \begin{bmatrix} 0 & 1 & -1 \\ 4 & -3 & 4 \\ 3 & -3 & 4 \end{bmatrix} \cdot \begin{bmatrix} 0 & 1 & -1 \\ 4 & -3 & 4 \\ 3 & -3 & 4 \end{bmatrix} \qquad [\because A^2 = A \cdot A]$$

$$= \begin{bmatrix} 0+4-3 & 0-3+3 & 0+4-4 \\ 0-12+12 & 4+9-12 & -4-12+16 \\ 0-12+12 & 3+9-12 & -3-12+16 \end{bmatrix}$$

$$= \begin{bmatrix} 1 & 0 & 0 \\ 0 & 1 & 0 \\ 0 & 0 & 1 \end{bmatrix} = I$$

Hence proved.

***Q.36** Prove by mathematical induction that $(A')^n = (A^n)'$ where $n \in N$ for any square matrix A.

Ans. Let $\quad P(n): (A')^n = (A^n)'$

$\therefore \quad P(1): (A')^1 = (A)'$

$\Rightarrow \quad A' = A' \Rightarrow P(1)$ is true.

Now, $\quad P(k): (A')^k = (A^k)'$,

where $k \in N$

and $\quad P(k+1): (A')^{k+1} = (A^{k+1})'$

where $P(k+1)$ is true whenever $P(k)$ is true.

$\therefore \quad P(k+1): (A')^k . (A')^1 = [A^{k+1}]'$

$\Rightarrow \quad (A^k)'.(A)' = [A^{k+1}]' \qquad [\because (A')^k = (A^k)']$

$\Rightarrow \quad (A \cdot A^k)' = [A^{k+1}]' \qquad [\because (AB)' = B'A']$

$\Rightarrow \quad (A^{k+1})' = [A^{k+1}]'$

Hence proved.

***Q.37** Find inverse, by elementary row operations (if possible), of the following matrices.

(i) $\begin{bmatrix} 1 & 3 \\ -5 & 7 \end{bmatrix}$ (ii) $\begin{bmatrix} 1 & -3 \\ -2 & 6 \end{bmatrix}$

Thinking Process To find the inverse of a matrix A, we know that $A = IA$ is used for elementary row operations. So, with the help of this method we can get the desired result.

Ans. (i) Let $A = \begin{bmatrix} 1 & 3 \\ -5 & 7 \end{bmatrix}$

In order to use elementary row operations we write $A = IA$.

Note (*) Topics related to these question have been removed from latest NCERT.

$\therefore \quad \begin{bmatrix} 1 & 3 \\ -5 & 7 \end{bmatrix} = \begin{bmatrix} 1 & 0 \\ 0 & 1 \end{bmatrix} A$

$\Rightarrow \quad \begin{bmatrix} 1 & 3 \\ 0 & 22 \end{bmatrix} = \begin{bmatrix} 1 & 0 \\ 5 & 1 \end{bmatrix} A \qquad [\because R_2 \to R_2 + 5R_1]$

$\Rightarrow \quad \begin{bmatrix} 1 & 3 \\ 0 & 1 \end{bmatrix} = \begin{bmatrix} 1 & 0 \\ 5/22 & 1/22 \end{bmatrix} A \qquad \left[\because R_2 \to \frac{1}{22} R_2\right]$

$\Rightarrow \quad \begin{bmatrix} 1 & 0 \\ 0 & 1 \end{bmatrix} = \begin{bmatrix} 7/22 & -3/22 \\ 5/22 & 1/22 \end{bmatrix} A \qquad [\because R_1 \to R_1 - 3R_2]$

$\Rightarrow \quad \begin{bmatrix} 1 & 0 \\ 0 & 1 \end{bmatrix} = \frac{1}{22} \begin{bmatrix} 7 & -3 \\ 5 & 1 \end{bmatrix} A$

$\Rightarrow I = B A$, where B is the inverse of A.

$\therefore \quad B = \frac{1}{22} \begin{bmatrix} 7 & -3 \\ 5 & -1 \end{bmatrix}$

(ii) Let $A = \begin{bmatrix} 1 & -3 \\ -2 & 6 \end{bmatrix}$

In order to use elementary row operations, we write $A = IA$

$\Rightarrow \quad \begin{bmatrix} 1 & -3 \\ -2 & 6 \end{bmatrix} = \begin{bmatrix} 1 & 0 \\ 0 & 1 \end{bmatrix} A$

$\Rightarrow \quad \begin{bmatrix} 1 & -3 \\ 0 & 0 \end{bmatrix} = \begin{bmatrix} 1 & 0 \\ 2 & 1 \end{bmatrix} A \qquad [\because R_2 \to R_2 + 2R_1]$

Since, we obtain all zeroes in a row of the matrix A on LHS, so A^{-1} does not exist.

Q.38 If $\begin{bmatrix} xy & 4 \\ z+6 & x+y \end{bmatrix} = \begin{bmatrix} 8 & w \\ 0 & 6 \end{bmatrix}$, then find the values of x, y, z and w.

Ans. We have, $\begin{bmatrix} xy & 4 \\ z+6 & x+y \end{bmatrix} = \begin{bmatrix} 8 & w \\ 0 & 6 \end{bmatrix}$

By equality of matrix, $x + y = 6$ ...(i)

$xy = 8$ ...(ii)

From Eq. (i), we get $x = 6 - y$ ...(iii)

On putting $x = 6 - y$ in Eq. (ii), we get

$(6 - y) \cdot y = 8$

$\Rightarrow \quad y^2 - 6y + 8 = 0$

$\Rightarrow \quad y^2 - 4y - 2y + 8 = 0$

$\Rightarrow \quad (y-2)(y-4) = 0$

$\Rightarrow \quad y = 2$ or $y = 4$

$\therefore \quad x = 6 - 2 = 4 \qquad [\because x = 6 - y]$

or $\quad x = 6 - 4 = 2$

Also, $\quad z + 6 = 0$

$\Rightarrow \quad z = -6$ and $w = 4$

$\therefore \quad x = 2, y = 4$ or $x = 4, y = 2, z = -6$ and $w = 4$

Q.39 If $A=\begin{bmatrix}1 & 5\\7 & 12\end{bmatrix}$ and $B=\begin{bmatrix}9 & 1\\7 & 8\end{bmatrix}$, then find a matrix C such that $3A+5B+2C$ is a null matrix.

Ans. We have, $A=\begin{bmatrix}1 & 5\\7 & 12\end{bmatrix}$ and $B=\begin{bmatrix}9 & 1\\7 & 8\end{bmatrix}$

Let $$C=\begin{bmatrix}a & b\\c & d\end{bmatrix}$$

$\therefore$ $$3A+5B+2C=O$$

$$\Rightarrow \begin{bmatrix}3 & 15\\21 & 36\end{bmatrix}+\begin{bmatrix}45 & 5\\35 & 40\end{bmatrix}+\begin{bmatrix}2a & 2b\\2c & 2d\end{bmatrix}=\begin{bmatrix}0 & 0\\0 & 0\end{bmatrix}$$

$$\Rightarrow \begin{bmatrix}48+2a & 20+2b\\56+2c & 76+2d\end{bmatrix}=\begin{bmatrix}0 & 0\\0 & 0\end{bmatrix}$$

On comparing the corresponding elements of matrices, we get

$$2a+48=0 \Rightarrow a=-24$$

Also, $$20+2b=0 \Rightarrow b=-10$$

$$56+2c=0 \Rightarrow c=-28$$

and $$76+2d=0 \Rightarrow d=-38$$

$\therefore$ $$C=\begin{bmatrix}-24 & -10\\-28 & -38\end{bmatrix}$$

Q.40 If $A=\begin{bmatrix}3 & -5\\-4 & 2\end{bmatrix}$, then find $A^2-5A-14I$. Hence, obtain A^3.

Ans. We have, $$A=\begin{bmatrix}3 & -5\\-4 & 2\end{bmatrix} \quad \text{...(i)}$$

$\therefore$ $$A^2=A\cdot A=\begin{bmatrix}3 & -5\\-4 & 2\end{bmatrix}\begin{bmatrix}3 & -5\\-4 & 2\end{bmatrix}$$

$$=\begin{bmatrix}9+20 & -15-10\\-12-8 & 20+4\end{bmatrix}$$

$$=\begin{bmatrix}29 & -25\\-20 & 24\end{bmatrix} \quad \text{...(ii)}$$

$\therefore$ $$A^2-5A-14I=\begin{bmatrix}29 & -25\\-20 & 24\end{bmatrix}-\begin{bmatrix}15 & -25\\-20 & 10\end{bmatrix}-\begin{bmatrix}14 & 0\\0 & 14\end{bmatrix}$$

$$=\begin{bmatrix}29-15-14 & -25+25-0\\-20+20-0 & 24-10-14\end{bmatrix}$$

$$=\begin{bmatrix}0 & 0\\0 & 0\end{bmatrix}=O$$

Now, $$A^2-5A-14I=O$$

$$\Rightarrow A\cdot A^2-5A\cdot A-14AI=O$$

$$\Rightarrow A^3-5A^2-14A=O \quad [\because AI=A]$$

$$\Rightarrow A^3=5A^2+14A$$

$$= 5\begin{bmatrix} 29 & -25 \\ -20 & 24 \end{bmatrix} + 14\begin{bmatrix} 3 & -5 \\ -4 & 2 \end{bmatrix} \quad \text{[using Eqs. (i) and (ii)]}$$

$$= \begin{bmatrix} 145 & -125 \\ -100 & 120 \end{bmatrix} + \begin{bmatrix} 42 & -70 \\ -56 & 28 \end{bmatrix}$$

$$= \begin{bmatrix} 145+42 & -125-70 \\ -100-56 & 120+28 \end{bmatrix}$$

$$= \begin{bmatrix} 187 & -195 \\ -156 & 148 \end{bmatrix}$$

Q.41 Find the values of a, b, c and d, if

$$3\begin{bmatrix} a & b \\ c & d \end{bmatrix} = \begin{bmatrix} a & 6 \\ -1 & 2d \end{bmatrix} + \begin{bmatrix} 4 & a+b \\ c+d & 3 \end{bmatrix}$$

Ans. We have, $$3\begin{bmatrix} a & b \\ c & d \end{bmatrix} = \begin{bmatrix} a & 6 \\ -1 & 2d \end{bmatrix} + \begin{bmatrix} 4 & a+b \\ c+d & 3 \end{bmatrix}$$

$$\Rightarrow \begin{bmatrix} 3a & 3b \\ 3c & 3d \end{bmatrix} = \begin{bmatrix} a+4 & 6+a+b \\ c+d-1 & 3+2d \end{bmatrix}$$

On comparing the corresponding elements of both sides, we get

$\Rightarrow \quad 3a = a + 4 \Rightarrow a = 2;$

$3b = 6 + a + b$

$\Rightarrow \quad 3b - b = 8 \Rightarrow b = 4; \qquad [\because a = 2]$

$3d = 3 + 2d \Rightarrow d = 3$

and $\quad 3c = c + d - 1$

$\Rightarrow \quad 2c = 3 - 1 \Rightarrow c = 1 \qquad [\because d = 3]$

$\therefore \quad a = 2,\ b = 4,\ c = 1 \text{ and } d = 3$

Q.42 Find the matrix A such that $\begin{bmatrix} 2 & -1 \\ 1 & 0 \\ -3 & 4 \end{bmatrix} A = \begin{bmatrix} -1 & -8 & -10 \\ 1 & -2 & -5 \\ 9 & 22 & 15 \end{bmatrix}$.

Ans. We have, $$\begin{bmatrix} 2 & -1 \\ 1 & 0 \\ -3 & 4 \end{bmatrix}_{3\times 2} A = \begin{bmatrix} -1 & -8 & -10 \\ 1 & -2 & -5 \\ 9 & 22 & 15 \end{bmatrix}_{3\times 3}$$

From the given equation, it is clear that order of A should be 2×3.

Let $$A = \begin{bmatrix} a & b & c \\ d & e & f \end{bmatrix}_{2\times 3}$$

$$\therefore \begin{bmatrix} 2 & -1 \\ 1 & 0 \\ -3 & 4 \end{bmatrix} \begin{bmatrix} a & b & c \\ d & e & f \end{bmatrix} = \begin{bmatrix} -1 & -8 & -10 \\ 1 & -2 & -5 \\ 9 & 22 & 15 \end{bmatrix}$$

$$\Rightarrow \begin{bmatrix} 2a-d & 2b-e & 2c-f \\ a+0d & b+0\cdot e & c+0\cdot f \\ -3a+4d & -3b+4e & -3c+4f \end{bmatrix} = \begin{bmatrix} -1 & -8 & -10 \\ 1 & -2 & -5 \\ 9 & 22 & 15 \end{bmatrix}$$

$$\Rightarrow \begin{bmatrix} 2a-d & 2b-e & 2c-f \\ a & b & c \\ -3a+4d & -3b+4e & -3c+4f \end{bmatrix} = \begin{bmatrix} -1 & -8 & -10 \\ 1 & -2 & -5 \\ 9 & 22 & 15 \end{bmatrix}$$

Since, both matrices are equal.

$\therefore$ On comparing corresponding elements of both sides, we get

$$a = 1, b = -2, c = -5$$

and $2a - d = -1 \Rightarrow d = 2a + 1 = 3;$ $[\because a = 1]$

$2b - e = -8 \Rightarrow e = 2b + 8 = 4$ $[\because b = -2]$

$2c - f = -10 \Rightarrow f = 2c + 10 = 0$ $[\because c = -5]$

$$\therefore \quad A = \begin{bmatrix} 1 & -2 & -5 \\ 3 & 4 & 0 \end{bmatrix}$$

Q.43 If $A = \begin{bmatrix} 1 & 2 \\ 4 & 1 \end{bmatrix}$, then find $A^2 + 2A + 7I$.

Ans. We have, $A = \begin{bmatrix} 1 & 2 \\ 4 & 1 \end{bmatrix}$

$$\therefore \quad A^2 = \begin{bmatrix} 1 & 2 \\ 4 & 1 \end{bmatrix}\begin{bmatrix} 1 & 2 \\ 4 & 1 \end{bmatrix} \quad [\because A^2 = A \cdot A]$$

$$= \begin{bmatrix} 1+8 & 2+2 \\ 4+4 & 8+1 \end{bmatrix} = \begin{bmatrix} 9 & 4 \\ 8 & 9 \end{bmatrix}$$

$$\therefore \quad A^2 + 2A + 7I = \begin{bmatrix} 9 & 4 \\ 8 & 9 \end{bmatrix} + 2\begin{bmatrix} 1 & 2 \\ 4 & 1 \end{bmatrix} + 7\begin{bmatrix} 1 & 0 \\ 0 & 1 \end{bmatrix}$$

$$= \begin{bmatrix} 9 & 4 \\ 8 & 9 \end{bmatrix} + \begin{bmatrix} 2 & 4 \\ 8 & 2 \end{bmatrix} + \begin{bmatrix} 7 & 0 \\ 0 & 7 \end{bmatrix}$$

$$= \begin{bmatrix} 9+2+7 & 4+4+0 \\ 8+8+0 & 9+2+7 \end{bmatrix} = \begin{bmatrix} 18 & 8 \\ 16 & 18 \end{bmatrix}$$

Q.44 If $A = \begin{bmatrix} \cos\alpha & \sin\alpha \\ -\sin\alpha & \cos\alpha \end{bmatrix}$ and $A^{-1} = A'$, then find the value of α.

Ans. We have,

$$A = \begin{bmatrix} \cos\alpha & \sin\alpha \\ -\sin\alpha & \cos\alpha \end{bmatrix} \text{ and } A' = \begin{bmatrix} \cos\alpha & -\sin\alpha \\ \sin\alpha & \cos\alpha \end{bmatrix}$$

Also, $A^{-1} = A'$

$\Rightarrow \quad AA^{-1} = AA'$

$$\Rightarrow \quad I = \begin{bmatrix} \cos\alpha & \sin\alpha \\ -\sin\alpha & \cos\alpha \end{bmatrix}\begin{bmatrix} \cos\alpha & -\sin\alpha \\ \sin\alpha & \cos\alpha \end{bmatrix}$$

$$I = \begin{bmatrix} \cos^2\alpha + \sin^2\alpha & -\cos\alpha \cdot \sin\alpha + \sin\alpha \cdot \cos\alpha \\ -\sin\alpha \cdot \cos\alpha + \cos\alpha \cdot \sin\alpha & \sin^2\alpha + \cos^2\alpha \end{bmatrix}$$

$$\Rightarrow \quad \begin{bmatrix} 1 & 0 \\ 0 & 1 \end{bmatrix} = \begin{bmatrix} \cos^2\alpha + \sin^2\alpha & 0 \\ 0 & \sin^2\alpha + \cos^2\alpha \end{bmatrix}$$

Since, both matrices are equal.

$\therefore$ On comparing the corresponding elements of both sides, we get

$$\cos^2\alpha + \sin^2\alpha = 1$$

which is true for all real values of α.

Q.45 If matrix $\begin{bmatrix} 0 & a & 3 \\ 2 & b & -1 \\ c & 1 & 0 \end{bmatrix}$ is a skew-symmetric matrix, then find the values of a, b and c.

Thinking Process We know that, a matrix A is skew-symmetric matrix, if $A' = -A$, so by using this we can get the values of a, b and c.

Ans. Let $A = \begin{bmatrix} 0 & a & 3 \\ 2 & b & -1 \\ c & 1 & 0 \end{bmatrix}$

Since, A is skew-symmetric matrix.

$$\therefore \quad A' = -A$$

$$\Rightarrow \quad \begin{bmatrix} 0 & 2 & c \\ a & b & 1 \\ 3 & -1 & 0 \end{bmatrix} = -\begin{bmatrix} 0 & a & 3 \\ 2 & b & -1 \\ c & 1 & 0 \end{bmatrix}$$

$$\Rightarrow \quad \begin{bmatrix} 0 & 2 & c \\ a & b & 1 \\ 3 & -1 & 0 \end{bmatrix} = \begin{bmatrix} 0 & -a & -3 \\ -2 & -b & 1 \\ -c & -1 & 0 \end{bmatrix}$$

Since, both matrices are equal.

$\therefore$ On comparing the corresponding elements of both sides, we get

$$a = -2,\ c = -3 \text{ and } b = -b \Rightarrow b = 0$$

$$\therefore \quad a = -2,\ b = 0 \text{ and } c = -3$$

Q.46 If $P(x) = \begin{bmatrix} \cos x & \sin x \\ -\sin x & \cos x \end{bmatrix}$, then show that

$P(x) \cdot P(y) = P(x + y) = P(y) \cdot P(x)$.

Ans. We have,

$$P(x) = \begin{bmatrix} \cos x & \sin x \\ -\sin x & \cos x \end{bmatrix}$$

$$\therefore \quad P(y) = \begin{bmatrix} \cos y & \sin y \\ -\sin y & \cos y \end{bmatrix}$$

Now, $$P(x) \cdot P(y) = \begin{bmatrix} \cos x & \sin x \\ -\sin x & \cos x \end{bmatrix} \begin{bmatrix} \cos y & \sin y \\ -\sin y & \cos y \end{bmatrix}$$

$$= \begin{bmatrix} \cos x \cdot \cos y - \sin x \cdot \sin y & \cos x \cdot \sin y + \sin x \cdot \cos y \\ -\sin x \cdot \cos y - \cos x \cdot \sin y & -\sin x \cdot \sin y + \cos x \cdot \cos y \end{bmatrix}$$

$$= \begin{bmatrix} \cos(x + y) & \sin(x + y) \\ -\sin(x + y) & \cos(x + y) \end{bmatrix} \quad \text{...(i)}$$

$$\begin{bmatrix} \because \cos(x + y) = \cos x \cdot \cos y - \sin x \cdot \sin y \\ \text{and } \sin(x + y) = \sin x \cdot \cos y + \cos x \cdot \sin y \end{bmatrix}$$

and $$P(x + y) = \begin{bmatrix} \cos(x + y) & \sin(x + y) \\ -\sin(x + y) & \cos(x + y) \end{bmatrix} \quad \text{...(ii)}$$

Also, $$P(y) \cdot P(x) = \begin{bmatrix} \cos y & \sin y \\ -\sin y & \cos y \end{bmatrix} \begin{bmatrix} \cos x & \sin x \\ -\sin x & \cos x \end{bmatrix}$$

$$= \begin{bmatrix} \cos y \cdot \cos x - \sin y \cdot \sin x & \cos y \cdot \sin x + \sin y \cdot \cos x \\ -\sin y \cdot \cos x - \sin x \cdot \cos y & -\sin y \cdot \sin x + \cos y \cdot \cos x \end{bmatrix}$$

$$= \begin{bmatrix} \cos(x+y) & \sin(x+y) \\ -\sin(x+y) & \cos(x+y) \end{bmatrix} \quad \text{...(iii)}$$

Thus, we see from the Eqs. (i), (ii) and (iii)

$$P(x) \cdot P(y) = P(x+y) = P(y) \cdot P(x)$$ **Hence proved.**

Q.47 If A is square matrix such that $A^2 = A$, then show that $(I + A)^3 = 7A + I$.

Ans. Since, $A^2 = A$ and $(I + A) \cdot (I + A) = I^2 + IA + AI + A^2$

$$= I^2 + 2AI + A^2 \qquad [\because AI = A]$$
$$= I + 2A + A = I + 3A$$

and $(I + A) \cdot (I + A)(I + A) = (I + A)(I + 3A)$

$$= I^2 + 3AI + AI + 3A^2$$
$$= I + 4AI + 3A \qquad [\because A^2 = A]$$
$$= I + 7A = 7A + I \qquad [\because AI = A]$$ **Hence proved.**

Q.48 If A, B are square matrices of same order and B is a skew-symmetric matrix, then show that $A'BA$ is skew-symmetric.

Ans. Since, A and B are square matrices of same order and B is a skew-symmetric matrix *i.e.*, $B' = -B$.

Now, we have to prove that $A'BA$ is a skew-symmetric matrix.

$$\therefore \quad (A'BA)' = (BA)'(A')' = A'B' \cdot A \qquad [\because (AB)' = B'A']$$
$$= A'(-B)A$$
$$= -A'BA \qquad [\because B' = -B]$$

Hence, $A'BA$ is a skew-symmetric matrix.

Long Answer Type Questions

*** Q.49** If $AB = BA$ for any two square matrices, then prove by mathematical induction that $(AB)^n = A^n B^n$.

Ans. Let $\quad P(n) : (AB)^n = A^n B^n$

Put $n = 1$, $\quad P(1) : (AB)^1 = A^1 B^1 \Rightarrow AB = AB$

So, $P(1)$ is true.

Now, Put $n = k$, $\quad P(k) : (AB)^k = A^k B^k$, $k \in N$

So, $P(k)$ is true, whenever $P(k+1)$ is true.

Now, put $n = k + 1$ $\quad P(K+1) : (AB)^{k+1} = A^{k+1} B^{k+1}$ ...(i)

Now, $\quad$ LHS $(AB)^{k+1} = (AB)^k \cdot AB$

$\Rightarrow \quad = A^k \cdot B^k \cdot B \cdot A \qquad [\because AB = BA]$

$\Rightarrow \quad = A^k \cdot A \cdot B^{k+1} = A^{k+1} \cdot B^{k+1} =$ RHS

So, $P(k + 1)$ is true for all $n \in N$, whenever $P(k)$ is true.

By mathematical induction $(AB)^n = A^n B^n$ is true for all $n \in N$.

Q.50 Find x, y and z, if $A = \begin{bmatrix} 0 & 2y & z \\ x & y & -z \\ x & -y & z \end{bmatrix}$ satisfies $A' = A^{-1}$.

Note (✱) Topics related to these question have been removed from latest NCERT.

Ans. We have,

$$A = \begin{bmatrix} 0 & 2y & z \\ x & y & -z \\ x & -y & z \end{bmatrix} \text{ and } A' = \begin{bmatrix} 0 & x & x \\ 2y & y & -y \\ z & -z & z \end{bmatrix}$$

Given, $A' = A^{-1}$

$\Rightarrow \quad AA' = AA^{-1}$ $\quad [\because AA^{-1} = I]$

$\Rightarrow \quad AA' = I$

$$\Rightarrow \begin{bmatrix} 0 & 2y & z \\ x & y & -z \\ x & -y & z \end{bmatrix}\begin{bmatrix} 0 & x & x \\ 2y & y & -y \\ z & -z & z \end{bmatrix} = \begin{bmatrix} 1 & 0 & 0 \\ 0 & 1 & 0 \\ 0 & 0 & 1 \end{bmatrix}$$

$$\Rightarrow \begin{bmatrix} 4y^2 + z^2 & 2y^2 - z^2 & -2y^2 + z^2 \\ 2y^2 - z^2 & x^2 + y^2 + z^2 & x^2 - y^2 - z^2 \\ -2y^2 + z^2 & x^2 - y^2 - z^2 & x^2 + y^2 + z^2 \end{bmatrix} = \begin{bmatrix} 1 & 0 & 0 \\ 0 & 1 & 0 \\ 0 & 0 & 1 \end{bmatrix}$$

Since, both the matrices are equal.

$\therefore$ On comparing corresponding elements of the both sides, we get

$\Rightarrow \quad 2y^2 - z^2 = 0 \Rightarrow 2y^2 = z^2$ ...(i)

$\Rightarrow \quad 4y^2 + z^2 = 1$ ...(ii)

$\Rightarrow \quad 2 \cdot z^2 + z^2 = 1$ $\quad [\because 2y^2 = z^2]$

$\Rightarrow \quad z = \pm \dfrac{1}{\sqrt{3}}$

On putting $z = \pm \dfrac{1}{\sqrt{3}}$ in Eq. (i), we get

$$y^2 = \frac{z^2}{2} \Rightarrow y = \pm \frac{1}{\sqrt{6}}$$

Also, $x^2 + y^2 + z^2 = 1$ ...(iii)

On putting values of $y = \pm \dfrac{1}{\sqrt{6}}, z = \pm \dfrac{1}{\sqrt{3}}$ in Eq. (iii), we get

$$x^2 = 1 - y^2 - z^2 = 1 - \frac{1}{6} - \frac{1}{3} = 1 - \frac{3}{6} = \frac{1}{2}$$

$\Rightarrow \quad x = \pm \dfrac{1}{\sqrt{2}}$

$\therefore \quad x = \pm, \dfrac{1}{\sqrt{2}}, y = \pm \dfrac{1}{\sqrt{6}}$

and $\quad z = \pm \dfrac{1}{\sqrt{3}}$

***Q.51** If possible, using elementary row transformations, find the inverse of the following matrices.**

$$\text{(i)} \begin{bmatrix} 2 & -1 & 3 \\ -5 & 3 & 1 \\ -3 & 2 & 3 \end{bmatrix} \quad \text{(ii)} \begin{bmatrix} 2 & 3 & -3 \\ -1 & -2 & 2 \\ 1 & 1 & -1 \end{bmatrix} \quad \text{(iii)} \begin{bmatrix} 2 & 0 & -1 \\ 5 & 1 & 0 \\ 0 & 1 & 3 \end{bmatrix}$$

Ans. For getting the inverse of the given matrix A by row elementary operations we may write the given matrix as

$$A = IA$$

Note (*) Topics related to these question have been removed from latest NCERT.

(i) Let $A=\begin{bmatrix}2 & -1 & 3\\ -5 & 3 & 1\\ -3 & 2 & 3\end{bmatrix}$

$$\because \quad \begin{bmatrix}2 & -1 & 3\\ -5 & 3 & 1\\ -3 & 2 & 3\end{bmatrix}=\begin{bmatrix}1 & 0 & 0\\ 0 & 1 & 0\\ 0 & 0 & 1\end{bmatrix}A$$

$$\Rightarrow \quad \begin{bmatrix}2 & -1 & 3\\ -3 & 2 & 4\\ -3 & 2 & 3\end{bmatrix}=\begin{bmatrix}1 & 0 & 0\\ 1 & 1 & 0\\ 0 & 0 & 1\end{bmatrix}A \qquad [\because R_2 \to R_2 + R_1]$$

$$\Rightarrow \quad \begin{bmatrix}2 & -1 & 3\\ -3 & 2 & 4\\ 0 & 0 & -1\end{bmatrix}=\begin{bmatrix}1 & 0 & 0\\ 1 & 1 & 0\\ -1 & -1 & 1\end{bmatrix}A \qquad [\because R_3 \to R_3 - R_2]$$

$$\Rightarrow \quad \begin{bmatrix}-1 & 1 & 7\\ -3 & 2 & 4\\ 0 & 0 & -1\end{bmatrix}=\begin{bmatrix}2 & 1 & 0\\ 1 & 1 & 0\\ -1 & -1 & 1\end{bmatrix}A \qquad [\because R_1 \to R_1 + R_2]$$

$$\Rightarrow \quad \begin{bmatrix}-1 & 1 & 7\\ 0 & -1 & -17\\ 0 & 0 & -1\end{bmatrix}=\begin{bmatrix}2 & 1 & 0\\ -5 & -2 & 0\\ -1 & -1 & 1\end{bmatrix}A \qquad [\because R_2 \to R_2 - 3R_1]$$

$$\Rightarrow \quad \begin{bmatrix}-1 & 0 & -10\\ 0 & -1 & -17\\ 0 & 0 & 1\end{bmatrix}=\begin{bmatrix}-3 & -1 & 0\\ -5 & -2 & 0\\ 1 & 1 & -1\end{bmatrix}A \qquad \begin{bmatrix}\because R_1 \to R_1 + R_2\\ \text{and } R_3 \to -1\cdot R_3\end{bmatrix}$$

$$\Rightarrow \quad \begin{bmatrix}-1 & 0 & 0\\ 0 & -1 & 0\\ 0 & 0 & 1\end{bmatrix}=\begin{bmatrix}7 & 9 & -10\\ 12 & 15 & -17\\ 1 & 1 & -1\end{bmatrix}A \qquad \begin{bmatrix}\because R_1 \to R_1 + 10R_3\\ \text{and } R_2 \to R_2 + 17R_3\end{bmatrix}$$

$$\Rightarrow \quad \begin{bmatrix}1 & 0 & 0\\ 0 & 1 & 0\\ 0 & 0 & 1\end{bmatrix}=\begin{bmatrix}-7 & -9 & 10\\ -12 & -15 & 17\\ 1 & 1 & -1\end{bmatrix}A \qquad \begin{bmatrix}\because R_1 \to -1R_1\\ \text{and } R_2 \to -1R_2\end{bmatrix}$$

So, the inverse of A is $\begin{bmatrix}-7 & -9 & 10\\ -12 & -15 & 17\\ 1 & 1 & -1\end{bmatrix}$.

(ii) Let, $A=\begin{bmatrix}2 & 3 & -3\\ -1 & -2 & 2\\ 1 & 1 & -1\end{bmatrix}$

$$\because \quad \begin{bmatrix}2 & 3 & -3\\ -1 & -2 & 2\\ 1 & 1 & -1\end{bmatrix}=\begin{bmatrix}1 & 0 & 0\\ 0 & 1 & 0\\ 0 & 0 & 1\end{bmatrix}A$$

$$\Rightarrow \quad \begin{bmatrix}0 & 1 & -1\\ 0 & -1 & 1\\ 1 & 1 & -1\end{bmatrix}=\begin{bmatrix}1 & 0 & -2\\ 0 & 1 & 1\\ 0 & 0 & 1\end{bmatrix}A \qquad \begin{bmatrix}\because R_2 \to R_2 + R_3\\ \text{and } R_1 \to R_1 - 2R_3\end{bmatrix}$$

$$\Rightarrow \begin{bmatrix} 0 & 1 & -1 \\ 0 & 0 & 0 \\ 1 & 1 & -1 \end{bmatrix} = \begin{bmatrix} 1 & 0 & -2 \\ 1 & 1 & -1 \\ 0 & 0 & 1 \end{bmatrix} A \quad [\because R_2 \to R_2 + R_1]$$

Since, second row of the matrix A on LHS is containing all zeroes, so we can say that inverse of matrix A does not exist.

(iii) Let, $A = \begin{bmatrix} 2 & 0 & -1 \\ 5 & 1 & 0 \\ 0 & 1 & 3 \end{bmatrix}$

$$\because \begin{bmatrix} 2 & 0 & -1 \\ 5 & 1 & 0 \\ 0 & 1 & 3 \end{bmatrix} = \begin{bmatrix} 1 & 0 & 0 \\ 0 & 1 & 0 \\ 0 & 0 & 1 \end{bmatrix} A$$

$$\Rightarrow \begin{bmatrix} 2 & 0 & -1 \\ 3 & 1 & 1 \\ 0 & 1 & 3 \end{bmatrix} = \begin{bmatrix} 1 & 0 & 0 \\ -1 & 1 & 0 \\ 0 & 0 & 1 \end{bmatrix} A \quad [\because R_2 \to R_2 - R_1]$$

$$\Rightarrow \begin{bmatrix} 2 & 0 & -1 \\ 1 & 1 & 2 \\ 2 & 1 & 2 \end{bmatrix} = \begin{bmatrix} 1 & 0 & 0 \\ -2 & 1 & 0 \\ 1 & 0 & 1 \end{bmatrix} A \quad \begin{bmatrix} \because R_2 \to R_2 - R_1 \\ \text{and } R_3 \to R_3 + R_1 \end{bmatrix}$$

$$\Rightarrow \begin{bmatrix} 2 & 0 & -1 \\ 0 & 1 & \frac{5}{2} \\ 4 & 1 & 1 \end{bmatrix} = \begin{bmatrix} 1 & 0 & 0 \\ \frac{-5}{2} & 1 & 0 \\ 2 & 0 & 1 \end{bmatrix} A \quad \begin{bmatrix} \because R_3 \to R_3 + R_1 \\ \text{and } R_2 \to R_2 - \frac{1}{2} R_1 \end{bmatrix}$$

$$\Rightarrow \begin{bmatrix} 2 & 0 & -1 \\ 0 & 1 & 5/2 \\ 0 & 1 & 3 \end{bmatrix} = \begin{bmatrix} 1 & 0 & 0 \\ -5/2 & 1 & 0 \\ 0 & 0 & 1 \end{bmatrix} A \quad [\because R_3 \to R_3 - 2R_1]$$

$$\Rightarrow \begin{bmatrix} 2 & 0 & -1 \\ 0 & 1 & 5/2 \\ 0 & 0 & 1/2 \end{bmatrix} = \begin{bmatrix} 1 & 0 & 0 \\ -5/2 & 1 & 0 \\ 5/2 & -1 & 1 \end{bmatrix} A \quad [\because R_3 \to R_3 - R_2]$$

$$\Rightarrow \begin{bmatrix} 1 & 0 & -1/2 \\ 0 & 1 & 5/2 \\ 0 & 0 & 1 \end{bmatrix} = \begin{bmatrix} 1/2 & 0 & 0 \\ -5/2 & 1 & 0 \\ 5 & -2 & 2 \end{bmatrix} A \quad \begin{bmatrix} \because R_1 \to \frac{1}{2} R_1 \\ \text{and } R_3 \to 2R_3 \end{bmatrix}$$

$$\Rightarrow \begin{bmatrix} 1 & 0 & 0 \\ 0 & 1 & 0 \\ 0 & 0 & 1 \end{bmatrix} = \begin{bmatrix} 3 & -1 & 1 \\ -15 & 6 & -5 \\ 5 & -2 & 2 \end{bmatrix} A \quad \begin{bmatrix} \because R_1 \to R_1 + \frac{1}{2} R_3 \\ \text{and } R_2 \to R_2 - \frac{5}{2} R_3 \end{bmatrix}$$

Hence, $\begin{bmatrix} 3 & -1 & 1 \\ -15 & 6 & -5 \\ 5 & -2 & 2 \end{bmatrix}$ is the inverse of given matrix A.

Q.52 Express the matrix $\begin{bmatrix} 2 & 3 & 1 \\ 1 & -1 & 2 \\ 4 & 1 & 2 \end{bmatrix}$ as the sum of a symmetric and a skew-symmetric matrix.

Thinking Process We know that, any square matrix A can be expressed as the sum of a symmetric matrix and skew-symmetric matrix, *i.e.*, $A = \frac{A+A'}{2} + \frac{A-A'}{2}$, where $A + A'$ and $A - A'$ are a symmetric matrix and a skew-symmetric matrix, respectively.

Ans. We have,
$$A = \begin{bmatrix} 2 & 3 & 1 \\ 1 & -1 & 2 \\ 4 & 1 & 2 \end{bmatrix}$$

$\therefore$
$$A' = \begin{bmatrix} 2 & 1 & 4 \\ 3 & -1 & 1 \\ 1 & 2 & 2 \end{bmatrix}$$

Now,
$$\frac{A+A'}{2} = \frac{1}{2}\left[\begin{bmatrix} 2 & 3 & 1 \\ 1 & -1 & 2 \\ 4 & 1 & 2 \end{bmatrix} + \begin{bmatrix} 2 & 1 & 4 \\ 3 & -1 & 1 \\ 1 & 2 & 2 \end{bmatrix}\right]$$

$$= \frac{1}{2}\begin{bmatrix} 2+2 & 3+1 & 1+4 \\ 1+3 & -1-1 & 2+1 \\ 4+1 & 1+2 & 2+2 \end{bmatrix}$$

$$= \frac{1}{2}\begin{bmatrix} 4 & 4 & 5 \\ 4 & -2 & 3 \\ 5 & 3 & 4 \end{bmatrix} = \begin{bmatrix} 2 & 2 & \frac{5}{2} \\ 2 & -1 & \frac{3}{2} \\ \frac{5}{2} & \frac{3}{2} & 2 \end{bmatrix}$$

and
$$\frac{A-A'}{2} = \frac{1}{2}\left[\begin{bmatrix} 2 & 3 & 1 \\ 1 & -1 & 2 \\ 4 & 1 & 2 \end{bmatrix} - \begin{bmatrix} 2 & 1 & 4 \\ 3 & -1 & 1 \\ 1 & 2 & 2 \end{bmatrix}\right]$$

$$= \frac{1}{2}\begin{bmatrix} 2-2 & 3-1 & 1-4 \\ 1-3 & -1+1 & 2-1 \\ 4-1 & 1-2 & 2-2 \end{bmatrix}$$

$$= \frac{1}{2}\begin{bmatrix} 0 & 2 & -3 \\ -2 & 0 & 1 \\ 3 & -1 & 0 \end{bmatrix} = \begin{bmatrix} 0 & 1 & \frac{-3}{2} \\ -1 & 0 & \frac{1}{2} \\ \frac{3}{2} & \frac{-1}{2} & 0 \end{bmatrix}$$

$\therefore$
$$\frac{A+A'}{2} + \frac{A-A'}{2} = \begin{bmatrix} 2 & 2 & \frac{5}{2} \\ 2 & -1 & \frac{3}{2} \\ \frac{5}{2} & \frac{3}{2} & 2 \end{bmatrix} + \begin{bmatrix} 0 & 1 & \frac{-3}{2} \\ -1 & 0 & \frac{1}{2} \\ \frac{3}{2} & \frac{-1}{2} & 0 \end{bmatrix}$$

which is the required expression.

Objective Type Questions

Choose the correct answer out of the given four options of the exercises from 53 to 67.

Q.53 The matrix $P=\begin{bmatrix}0 & 0 & 4\\0 & 4 & 0\\4 & 0 & 0\end{bmatrix}$ is a

(a) square matrix (b) diagonal matrix (c) unit matrix (d) None of these

Ans. ***(a)*** We know that, in a square matrix number of rows is equal to the number of columns, so the matrix $P=\begin{bmatrix}0 & 0 & 4\\0 & 4 & 0\\4 & 0 & 0\end{bmatrix}$ is a square matrix.

Q.54 Total number of possible matrices of order 3×3 with each entry 2 or 0 is

(a) 9 (b) 27 (c) 81 (d) 512

Ans. ***(d)*** Total number of possible matrices of order 3×3 with each entry 2 or 0 is 2^9 *i.e.,* 512.

Q.55 If $\begin{bmatrix}2x+y & 4x\\5x-7 & 4x\end{bmatrix}=\begin{bmatrix}7 & 7y-13\\y & x+6\end{bmatrix}$, then the value of $x+y$ is

(a) 2 (b) 5 (c) 6 (d) 4

Ans. ***(b)*** Since, both the matrices are equal.

$\therefore$ On comparing the corresponding elements of both sides, we get

$$4x = x+6 \Rightarrow x=2 \quad \text{...(i)}$$

and $$4x = 7y-13 \Rightarrow 8 = 7y-13 \quad [\because x=2, \text{ from Eq. (i)}]$$

$\Rightarrow$ $$7y=21 \Rightarrow y=3$$

$\therefore$ $$x+y=2+3=5$$

Q.56 If $A=\frac{1}{\pi}\begin{bmatrix}\sin^{-1}(x\pi) & \tan^{-1}\left(\frac{x}{\pi}\right)\\ \sin^{-1}\left(\frac{x}{\pi}\right) & \cot^{-1}(\pi x)\end{bmatrix}$ and $B=\frac{1}{\pi}\begin{bmatrix}-\cos^{-1}(x\pi) & \tan^{-1}\left(\frac{x}{\pi}\right)\\ \sin^{-1}\left(\frac{x}{\pi}\right) & -\tan^{-1}(\pi x)\end{bmatrix}$,

then $A-B$ is equal to

(a) I (b) O (c) $2I$ (d) $\frac{1}{2}I$

Ans. ***(d)*** We have, $A=\begin{bmatrix}\frac{1}{\pi}\sin^{-1}x\pi & \frac{1}{\pi}\tan^{-1}\frac{x}{\pi}\\ \frac{1}{\pi}\sin^{-1}\frac{x}{\pi} & \frac{1}{\pi}\cot^{-1}\pi x\end{bmatrix}$

and $$B=\begin{bmatrix}\frac{-1}{\pi}\cos^{-1}x\pi & \frac{1}{\pi}\tan^{-1}\frac{x}{\pi}\\ \frac{1}{\pi}\sin^{-1}\frac{x}{\pi} & \frac{-1}{\pi}\tan^{-1}\pi x\end{bmatrix}$$

$$\therefore \quad A-B=\begin{bmatrix} \frac{1}{\pi}(\sin^{-1} x\pi+\cos^{-1} x\pi) & \frac{1}{\pi}\left(\tan^{-1}\frac{x}{\pi}-\tan^{-1}\frac{x}{\pi}\right) \\ \frac{1}{\pi}\left(\sin^{-1}\frac{x}{\pi}-\sin^{-1}\frac{x}{\pi}\right) & \frac{1}{\pi}(\cot^{-1}\pi x+\tan^{-1}\pi x) \end{bmatrix}$$

$$=\begin{bmatrix} \frac{1}{\pi}\cdot\frac{\pi}{2} & 0 \\ 0 & \frac{1}{\pi}\cdot\frac{\pi}{2} \end{bmatrix} \qquad \left[\because \sin^{-1} x+\cos^{-1} x=\frac{\pi}{2} \text{ and } \tan^{-1} x+\cot^{-1} x=\frac{\pi}{2}\right]$$

$$=\frac{1}{2}\begin{bmatrix} 1 & 0 \\ 0 & 1 \end{bmatrix}=\frac{1}{2}I$$

Q.57 If A and B are two matrices of the order $3\times m$ and $3\times n$, respectively and $m=n$, then order of matrix $(5A-2B)$ is

(a) $m\times 3$ (b) 3×3 (c) $m\times n$ (d) $3\times n$

Ans. *(d)* $A_{3\times m}$ and $B_{3\times n}$ are two matrices. If $m=n$, then A and B have same orders as $3\times n$ each, so the order of $(5A-2B)$ should be same as $3\times n$.

Q.58 If $A=\begin{bmatrix} 0 & 1 \\ 1 & 0 \end{bmatrix}$, then A^2 is equal to

(a) $\begin{bmatrix} 0 & 1 \\ 1 & 0 \end{bmatrix}$ (b) $\begin{bmatrix} 1 & 0 \\ 1 & 0 \end{bmatrix}$ (c) $\begin{bmatrix} 0 & 1 \\ 0 & 1 \end{bmatrix}$ (d) $\begin{bmatrix} 1 & 0 \\ 0 & 1 \end{bmatrix}$

Ans. *(d)* $\because A^2=A\cdot A=\begin{bmatrix} 0 & 1 \\ 1 & 0 \end{bmatrix}\cdot\begin{bmatrix} 0 & 1 \\ 1 & 0 \end{bmatrix}=\begin{bmatrix} 0+1 & 0+0 \\ 0+0 & 1+0 \end{bmatrix}=\begin{bmatrix} 1 & 0 \\ 0 & 1 \end{bmatrix}$

Q.59 If matrix $A=[a_{ij}]_{2\times 2}$, where $a_{ij}=1$, if $i\neq j$ and $a_{ij}=0$ if $i=j$, then A^2 is equal to

(a) I (b) A (c) O (d) None of these

Ans. *(a)* We have, $A=[a_{ij}]_{2\times 2}$, where $a_{ij}=1$, if $i\neq j$ and $a_{ij}=0$ if $i=j$

Let $A=\begin{bmatrix} a_{11} & a_{12} \\ a_{21} & a_{22} \end{bmatrix}$

So, $a_{11}=0$ [since, $i=j$]

$a_{12}=1$ [since, $i\neq j$]

$a_{21}=1$ [since, $i\neq j$]

$a_{22}=0$ [since, $i=j$]

$\therefore \quad A=\begin{bmatrix} 0 & 1 \\ 1 & 0 \end{bmatrix}$

and $\quad A^2=\begin{bmatrix} 0 & 1 \\ 1 & 0 \end{bmatrix}\begin{bmatrix} 0 & 1 \\ 1 & 0 \end{bmatrix}=\begin{bmatrix} 1 & 0 \\ 0 & 1 \end{bmatrix}=I$

Q.60 The matrix $\begin{bmatrix} 1 & 0 & 0 \\ 0 & 2 & 0 \\ 0 & 0 & 4 \end{bmatrix}$ is a

(a) identity matrix (b) symmetric matrix

(c) skew-symmetric matrix (d) None of these

Ans. ***(b)*** Let $$A=\begin{bmatrix}1 & 0 & 0\\0 & 2 & 0\\0 & 0 & 4\end{bmatrix}$$

$\therefore$ $$A'=\begin{bmatrix}1 & 0 & 0\\0 & 2 & 0\\0 & 0 & 4\end{bmatrix}=A$$

So, the given matrix is a symmetric matrix.

[since, for a square matrix A, if $A'=A$, then A is called symmetric matrix]

Q.61 The matrix $\begin{bmatrix}0 & -5 & 8\\5 & 0 & 12\\-8 & -12 & 0\end{bmatrix}$ is a

(a) diagonal matrix
(b) symmetric matrix
(c) skew-symmetric matrix
(d) scalar matrix

Ans. ***(c)*** We know that, in a square matrix, if $b_{ij}=0$, when $i \neq j$, then it is said to be a diagonal matrix. Here, $b_{12}, b_{13}, \ldots \neq 0$, so the given matrix is not a diagonal matrix.

Let $$B=\begin{bmatrix}0 & -5 & 8\\5 & 0 & 12\\-8 & -12 & 0\end{bmatrix}$$

$\therefore$ $$B'=\begin{bmatrix}0 & 5 & -8\\-5 & 0 & -12\\8 & 12 & 0\end{bmatrix}=-\begin{bmatrix}0 & -5 & 8\\5 & 0 & 12\\-8 & -12 & 0\end{bmatrix}=-B$$

Since, $B'=-B$

So, the given matrix is a skew-symmetric matrix.

Q.62 If A is matrix of order $m \times n$ and B is a matrix such that AB' and $B'A$ are both defined, then order of matrix B is

(a) $m \times m$
(b) $n \times n$
(c) $n \times m$
(d) $m \times n$

Ans. ***(d)*** Let $A=[a_{ij}]_{m\times n}$ and $B=[b_{ij}]_{p\times q}$

$\therefore$ $B'=[b_{ji}]_{q\times p}$

Now, since AB' is defined, so $n=q$

and $B'A$ is also defined, so $p=m$

$\therefore$ Order of $B'=[b_{ji}]_{n\times m}$

So, Order of $B=[b_{ij}]_{m\times n}$

Q.63 If A and B are matrices of same order, then $(AB'-BA')$ is a

(a) skew-symmetric matrix
(b) null matrix
(c) symmetric matrix
(d) unit matrix

Ans. ***(a)*** We have matrices A and B of same order.

Let $P = (AB' - BA')$

Then, $P' = (AB' - BA')' = (AB')' - (BA')'$

$= (B')'(A)' - (A')' B' = BA' - AB'$

Since, $P' = -P$ $= -(AB' - BA') = -P$

Hence, $(AB' - BA')$ is a skew-symmetric matrix.

Q.64 If A is a square matrix such that $A^2 = I$, then $(A - I)^3 + (A + I)^3 - 7A$ is equal to

(a) A (b) $I - A$ (c) $I + A$ (d) $3A$

Ans. ***(a)*** We have, $A^2 = I$

$$\therefore (A - I)^3 + (A + I)^3 - 7A = [[(A - I) + (A + I)]\{(A - I)^2 + (A + I)^2 - (A - I)(A + I)\}] - 7A$$

$[\because a^3 + b^3 = (a + b)(a^2 + b^2 - ab)]$

$$= [(2A)\{A^2 + I^2 - 2AI + A^2 + I^2 + 2AI - (A^2 - I^2)\}] - 7A$$

$$= 2A[I + I^2 + I + I^2 - A^2 + I^2] - 7A \quad [\because A^2 = I]$$

$$= 2A[5I - I] - 7A$$

$$= 8AI - 7AI$$

$$= AI = A \quad [\because A = AI]$$

Q.65 For any two matrices A and B, we have

(a) $AB = BA$ (b) $AB \neq BA$ (c) $AB = O$ (d) None of these

Ans. ***(d)*** For any two matrices A and B, we may have $AB = BA = I$, $AB \neq BA$ and $AB = O$ but it is not always true.

****Q.66*** On using elementary column operations $C_2 \to C_2 - 2C_1$ in the following matrix equation $\begin{bmatrix} 1 & -3 \\ 2 & 4 \end{bmatrix} = \begin{bmatrix} 1 & -1 \\ 0 & 1 \end{bmatrix}\begin{bmatrix} 3 & 1 \\ 2 & 4 \end{bmatrix}$, we have

(a) $\begin{bmatrix} 1 & -5 \\ 0 & 4 \end{bmatrix} = \begin{bmatrix} 1 & -1 \\ -2 & 2 \end{bmatrix}\begin{bmatrix} 3 & -5 \\ 2 & 0 \end{bmatrix}$ (b) $\begin{bmatrix} 1 & -5 \\ 0 & 4 \end{bmatrix} = \begin{bmatrix} 1 & -1 \\ 0 & 1 \end{bmatrix}\begin{bmatrix} 3 & -5 \\ -0 & 2 \end{bmatrix}$

(c) $\begin{bmatrix} 1 & -5 \\ 2 & 0 \end{bmatrix} = \begin{bmatrix} 1 & -3 \\ 0 & 1 \end{bmatrix}\begin{bmatrix} 3 & 1 \\ -2 & 4 \end{bmatrix}$ (d) $\begin{bmatrix} 1 & -5 \\ 2 & 0 \end{bmatrix} = \begin{bmatrix} 1 & -1 \\ 0 & 1 \end{bmatrix}\begin{bmatrix} 3 & -5 \\ 2 & 0 \end{bmatrix}$

Ans. ***(d)*** Given that, $\begin{bmatrix} 1 & -3 \\ 2 & 4 \end{bmatrix} = \begin{bmatrix} 1 & -1 \\ 0 & 1 \end{bmatrix}\begin{bmatrix} 3 & 1 \\ 2 & 4 \end{bmatrix}$

On using $C_2 \to C_2 - 2C_1$, we get $\begin{bmatrix} 1 & -5 \\ 2 & 0 \end{bmatrix} = \begin{bmatrix} 1 & -1 \\ 0 & 1 \end{bmatrix}\begin{bmatrix} 3 & -5 \\ 2 & 0 \end{bmatrix}$

Since, by using elementary column operation on $X = AB$, we apply these operations simultaneously on X and on the second matrix B of the product AB on RHS.

Note (✱) Topics related to these question have been removed from latest NCERT.

***Q.67** On using elementary row operation $R_1 \to R_1 - 3R_2$ in the following matrix equation $\begin{bmatrix} 4 & 2 \\ 3 & 3 \end{bmatrix} = \begin{bmatrix} 1 & 2 \\ 0 & 3 \end{bmatrix} \begin{bmatrix} 2 & 0 \\ 1 & 1 \end{bmatrix}$, we have

(a) $\begin{bmatrix} -5 & -7 \\ 3 & 3 \end{bmatrix} = \begin{bmatrix} 1 & -7 \\ 0 & 3 \end{bmatrix} \begin{bmatrix} 2 & 0 \\ 1 & 1 \end{bmatrix}$

(b) $\begin{bmatrix} -5 & -7 \\ 3 & 3 \end{bmatrix} = \begin{bmatrix} 1 & 2 \\ 0 & 3 \end{bmatrix} \begin{bmatrix} -1 & -3 \\ 1 & 1 \end{bmatrix}$

(c) $\begin{bmatrix} -5 & -7 \\ 3 & 3 \end{bmatrix} = \begin{bmatrix} 1 & 2 \\ 1 & -7 \end{bmatrix} \begin{bmatrix} 2 & 0 \\ 1 & 1 \end{bmatrix}$

(d) $\begin{bmatrix} 4 & 2 \\ -5 & -7 \end{bmatrix} = \begin{bmatrix} 1 & 2 \\ -3 & -3 \end{bmatrix} \begin{bmatrix} 2 & 0 \\ 1 & 1 \end{bmatrix}$

Ans. *(a)* We have, $\begin{bmatrix} 4 & 2 \\ 3 & 3 \end{bmatrix} = \begin{bmatrix} 1 & 2 \\ 0 & 3 \end{bmatrix} \begin{bmatrix} 2 & 0 \\ 1 & 1 \end{bmatrix}$

On using elementary row operation $R_1 \to R_1 - 3R_2$, we get

$$\begin{bmatrix} -5 & -7 \\ 3 & 3 \end{bmatrix} = \begin{bmatrix} 1 & -7 \\ 0 & 3 \end{bmatrix} \begin{bmatrix} 2 & 0 \\ 1 & 1 \end{bmatrix}$$

Since, on using elementary row operation on $X = AB$, we apply these operation simultaneously on X and on the first matrix A of the product AB on RHS.

Fill in the blanks in each of the exercises 68 to 81.

Q.68 .... matrix is both symmetric and skew-symmetric matrix.

Ans. Null matrix is both symmetric and skew-symmetric matrix.

Q.69 Sum of two skew-symmetric matrices is always .... matrix.

Ans. Let A and B be two skew-symmetric matrices.

$\therefore$ $A' = -A$ and $B' = -B$

Now, $(A + B)' = A' + B' = -A - B = -(A + B)$

So, $A + B$ is a skew-symmetric matrix.

Q.70 The negative of a matrix is obtained by multiplying it by ....

Ans. Let A is a given matrix.

$\therefore$ $$-A = -1 \times A$$

So, the negative of a matrix is obtained by multiplying it by – 1.

Q.71 The product of any matrix by the scalar .... is the null matrix.

Ans. The product of any matrix by the scalar 0 is the null matrix. *i.e.,* $0 \cdot A = O$.

[where, A is any matrix]

Q.72 A matrix which is not a square matrix is called a .... matrix.

Ans. A matrix which is not a square matrix is called a rectangular matrix. For example a rectangular matrix is $A = [a_{ij}]_{m \times n}$, where $m \neq n$.

Q.73 Matrix multiplication is ...... over addition.

Ans. Matrix multiplication is distributive over addition.

e.g., For three matrices A, B and C,

(i) $A(B + C) = AB + AC$

(ii) $(A + B)C = AC + BC$

Note (✱) Topics related to these question have been removed from latest NCERT.

Q.74 If A is a symmetric matrix, then A^3 is a .... matrix.

Ans. Since, A is a symmetric matrix,

$$\because \quad A' = A$$

$$\therefore \quad (A^3)' = A'^3 = A^3 \qquad [\because (A')^n = (A^n)']$$

So, A^3 is a symmetric matrix.

Q.75 If A is a skew-symmetric matrix, then A^2 is a ....

Ans. Since, A is a skew-symmetric matrix,

$$\therefore \quad A' = -A$$

$$\therefore \quad (A^2)' = (A')^2 \qquad [\because (A^n)' = (A')^n]$$

$$= (-A)^2 \qquad [\because A' = -A]$$

$$= A^2$$

So, A^2 is a symmetric matrix.

Q.76 If A and B are square matrices of the same order, then

(i) $(AB)' =$ ....

(ii) $(kA)' =$ .... (where, k is any scalar)

(iii) $[k(A-B)]' =$ ....

Ans. (i) $(AB)' = B'A'$

(ii) $(kA)' = kA'$

(iii) $[k(A-B)]' = k(A' - B')$

Q.77 If A is a skew-symmetric, then kA is a .... (where, k is any scalar).

Ans. If A is a skew-symmetric, then kA is a skew-symmetric matrix (where, k is any scalar).

$$[\because A' = -A \Rightarrow (kA)' = k(A)' = -(kA)]$$

Q.78 If A and B are symmetric matrices, then

(i) $AB - BA$ is a ....

(ii) $BA - 2AB$ is a ....

Ans. (i) Since, A and B are symmetric matrices,

$$\therefore \quad A' = A \text{ and } B' = B$$

Now,
$$[AB - BA]' = (AB)' - (BA)'$$

$$= B'A' - A'B' \qquad [\because (AB)' = B'A']$$

$$= BA - AB$$

$$= -[AB - BA]$$

So, $[AB - BA]$ is a skew-symmetric matrix.

(ii) Here,

$$\therefore \quad (BA - 2AB)' = (BA)' - 2(AB)'$$

$$= A'B' - 2B'A'$$

$$= AB - 2BA$$

$$= -(2BA - AB)$$

So, $[BA - 2AB]$ is neither symmetric nor skew-symmetric matrix.

Q.79 If A is symmetric matrix, then $B'AB$ is ....

Ans. Since, A is a symmetric matrix,

$\therefore A' = A$

Now,
$$[B'AB]' = [B'(AB)]' = (AB)'(B')' \quad [\because (AB)' = B'A']$$
$$= B'A'B$$
$$= B'AB \quad [\because A' = A]$$

So, $B'AB$ is a symmetric matrix.

Q.80 If A and B are symmetric matrices of same order, then AB is symmetric if and only if......... .

Ans. If A and B are symmetric matrices of same order, then AB is symmetric if and only if $AB = BA$.

$\because$
$$(AB)' = B'A' = BA \quad [\because B' = B, A' = A]$$
$$= AB \quad [\because AB = BA]$$

Q.81 In applying one or more row operations while finding A^{-1} by elementary row operations, we obtain all zeroes in one or more, then A^{-1} ....

Ans. In applying one or more row operations while finding A^{-1} by elementary row operations, we obtain all zeroes in one or more, then A^{-1} does not exist.

State True or False for the statements in each of the exercises 82 to 101.

Q.82 A matrix denotes a number.

Ans. *False*

A matrix is an ordered rectangular array of numbers or functions.

Q.83 Matrices of any order can be added.

Ans. *False*

Two matrices can added, if they are of the same order.

Q.84 Two matrices are equal, if they have same number of rows and same number of columns.

Ans. *False*

If two matrices have same number of rows and same number of columns, then they are said to be square matrices and if two square matrices have same elements in both the matrices only, then they are called equal.

Q.85 Matrices of different order cannot be subtracted.

Ans. *True*

Two matrices of same order can be subtracted

Q.86 Matrix addition is associative as well as commutative.

Ans. *True*

Matrix addition is associative as well as commutative *i.e.*,

$(A + B) + C = A + (B + C)$ and $A + B = B + A$, where A, B and C are matrices of same order.

Q.87 Matrix multiplication is commutative.

Ans. *False*

If AB and BA both are defined, then matrix multiplication may or may not be commutative.

Q.88 A square matrix where every element is unity is called an identity matrix.

Ans. *False*

Since, in an identity matrix, the diagonal elements are all one and rest are all zero.

Q.89 If A and B are two square matrices of the same order, then $A+B=B+A$.

Ans. *True*

Since, matrix addition is commutative *i.e.,* $A+B=B+A$, where A and B are two square matrices.

Q.90 If A and B are two matrices of the same order, then $A-B=B-A$.

Ans. *False*

Since, the addition of two matrices of same order are commutative.

$$\therefore \quad A+(-B)=A-B=-[B-A]\neq B-A$$

Q.91 If matrix $AB=O$, then $A=O$ or $B=O$ or both A and B are null matrices.

Ans. *False*

Since, for two non-zero matrices A and B of same order, it can be possible that $A\cdot B=O=$ null matrix

e.g., Let $A=\begin{bmatrix}0 & 1\\ 0 & 2\end{bmatrix}$, $B=\begin{bmatrix}3 & 4\\ 0 & 0\end{bmatrix}$

$$\text{Here } A\cdot B=\begin{bmatrix}0 & 1\\ 0 & 2\end{bmatrix}\cdot\begin{bmatrix}3 & 4\\ 0 & 0\end{bmatrix}$$

$$=\begin{bmatrix}0+0 & 0+0\\ 0+0 & 0+0\end{bmatrix}=\begin{bmatrix}0 & 0\\ 0 & 0\end{bmatrix}$$

Q.92 Transpose of a column matrix is a column matrix.

Ans. *False*

Transpose of a column matrix is a row matrix.

Q.93 If A and B are two square matrices of the same order, then $AB=BA$.

Ans. *False*

For two square matrices of same order it is not always true that $AB=BA$.

Q.94 If each of the three matrices of the same order are symmetric, then their sum is a symmetric matrix.

Ans. *True*

Let A, B and C are three matrices of same order, then

$$\therefore \quad A'=A,\ B'=B \text{ and } C'=C \qquad \text{[given]}$$

$$\therefore \quad (A+B+C)'=A'+B'+C'$$

$$=(A+B+C)$$

Q.95 If A and B are any two matrices of the same order, then $(AB)' = A'B'$.

Ans. *False*

$\because \quad (AB)' = B'A'$

Q.96 If $(AB)' = B'A'$, where A and B are not square matrices, then number of rows in A is equal to number of columns in B and number of columns in A is equal to number of rows in B.

Ans. *True*

Let A is of order $m \times n$ and B is of order $p \times q$.

Since, $\quad (AB)' = B'A'$

$\therefore \quad A_{(m \times n)} B_{(p \times q)}$ is defined $\Rightarrow n = p$ ...(i)

and AB is of order $m \times q$.

$\Rightarrow \quad (AB)'$ is of order $q \times m$ ...(ii)

Also, B' is of order $q \times p$ and A' is of order $n \times m$

$\therefore \quad B'A'$ is defined $\Rightarrow p = n$

and $\quad B'A'$ is of order $q \times m$. ...(iii)

Also, equality of matrices $(AB)' = B'A'$, we get the given statement as true.

e.g., If A is of order (3×1) and B is of order (1×3), we get

Order of $(AB)'$ = Order of $(B'A') = 3 \times 3$

Q.97 If A, B and C are square matrices of same order, then $AB = AC$ always implies that $B = C$.

Ans. *False*

If $AB = AC = O$, then it can be possible that B and C are two non-zero matrices such that $B \neq C$.

Let $\quad A = \begin{bmatrix} 1 & 0 \\ 0 & 0 \end{bmatrix}, B = \begin{bmatrix} 0 & 0 \\ 1 & 3 \end{bmatrix}$ and $C = \begin{bmatrix} 0 & 0 \\ 3 & 1 \end{bmatrix}$

$\therefore \quad AB = \begin{bmatrix} 1 & 0 \\ 0 & 0 \end{bmatrix} \begin{bmatrix} 0 & 0 \\ 1 & 3 \end{bmatrix} = \begin{bmatrix} 0 & 0 \\ 0 & 0 \end{bmatrix}$

and $\quad AC = \begin{bmatrix} 1 & 0 \\ 0 & 0 \end{bmatrix} \cdot \begin{bmatrix} 0 & 0 \\ 3 & 1 \end{bmatrix} = \begin{bmatrix} 0 & 0 \\ 0 & 0 \end{bmatrix}$

$\Rightarrow \quad AB = AC$ but $B \neq C$

Q.98 AA' is always a symmetric matrix for any matrix A.

Ans. *True*

$\because \quad [AA']' = (A')'A' = [AA']$

Q.99 If $A = \begin{bmatrix} 2 & 3 & -1 \\ 1 & 4 & 2 \end{bmatrix}$ and $B = \begin{bmatrix} 2 & 3 \\ 4 & 5 \\ 2 & 1 \end{bmatrix}$, then AB and BA are defined and equal.

Ans. *False*

Since, AB is defined.

$$\therefore \quad AB = \begin{bmatrix} 2 & 3 & -1 \\ 1 & 4 & 2 \end{bmatrix} \begin{bmatrix} 2 & 3 \\ 4 & 5 \\ 2 & 1 \end{bmatrix} = \begin{bmatrix} 14 & 20 \\ 22 & 25 \end{bmatrix}$$

Also, BA is defined.

$$\therefore \quad BA = \begin{bmatrix} 2 & 3 \\ 4 & 5 \\ 2 & 1 \end{bmatrix} \begin{bmatrix} 2 & 3 & -1 \\ 1 & 4 & 2 \end{bmatrix}$$

$$= \begin{bmatrix} 7 & 18 & 4 \\ 13 & 32 & 6 \\ 5 & 10 & 0 \end{bmatrix}$$

$$\therefore \quad AB \neq BA$$

Alternate Method

False,

Given, $A = \begin{bmatrix} 2 & 3 & -1 \\ 1 & 4 & 2 \end{bmatrix}_{2 \times 3}$ $B = \begin{bmatrix} 2 & 3 \\ 4 & 5 \\ 2 & 1 \end{bmatrix}_{3 \times 2}$

Since, number columns of A = number of rows of B

$\therefore$ AB is defined and order of $A \cdot B = 2 \times 2$

Since, number of columns of B = number of rows of A

$\therefore$ BA is defined and order of $BA = 3 \times 3$

Since, order of $AB \neq$ order of BA

$\therefore$ $AB \neq BA$

Q.100 If A is skew-symmetric matrix, then A^2 is a symmetric matrix.

Ans. *True*

$$\because \quad [A^2]' = [A']^2 \quad [\because (A^n)' = (A')^n]$$
$$= [-A]^2 \quad [\because A' = -A]$$
$$= A^2$$

Hence, A^2 is symmetric matrix.

Q.101 $(AB)^{-1} = A^{-1} \cdot B^{-1}$, where A and B are invertible matrices satisfying commutative property with respect to multiplication.

Ans. *True*

We know that, if A and B are invertible matrices of the same order, then

$$(AB)^{-1} = A^{-1}B^{-1} \quad [\because AB = BA]$$
$$B^{-1}A^{-1} = A^{-1}B^{-1}$$

[since, A and B are satisfying commutative property with respect to multiplications].

Chapter

04 Determinants

Short Answer Type Questions

Using the properties of determinants in exercises 1 to 6, evaluate.

***Q.1** $\begin{vmatrix} x^2 - x + 1 & x - 1 \\ x + 1 & x + 1 \end{vmatrix}$

Ans. We have, $\begin{vmatrix} x^2 - x + 1 & x - 1 \\ x + 1 & x + 1 \end{vmatrix} = \begin{vmatrix} x^2 - 2x + 2 & x - 1 \\ 0 & x + 1 \end{vmatrix}$ $[\because C_1 \to C_1 - C_2]$

$$= (x^2 - 2x + 2)\cdot(x + 1) - (x - 1)\cdot 0$$
$$= x^3 - 2x^2 + 2x + x^2 - 2x + 2$$
$$= x^3 - x^2 + 2$$

***Q.2** $\begin{vmatrix} a + x & y & z \\ x & a + y & z \\ x & y & a + z \end{vmatrix}$

Ans. We have, $\begin{vmatrix} a + x & y & z \\ x & a + y & z \\ x & y & a + z \end{vmatrix} = \begin{vmatrix} a & -a & 0 \\ 0 & a & -a \\ x & y & a + z \end{vmatrix}$ $\begin{bmatrix} \because R_1 \to R_1 - R_2 \\ \text{and } R_2 \to R_2 - R_3 \end{bmatrix}$

$$= \begin{vmatrix} a & 0 & 0 \\ 0 & a & -a \\ x & x + y & a + z \end{vmatrix} \quad [\because C_2 \to C_2 + C_1]$$

On expanding along R_1, we get

$$= a\,(a^2 + az + ax + ay)$$
$$= a^2\,(a + z + x + y)$$

***Q.3** $\begin{vmatrix} 0 & xy^2 & xz^2 \\ x^2y & 0 & yz^2 \\ x^2z & zy^2 & 0 \end{vmatrix}$

Ans. We have, $\begin{vmatrix} 0 & xy^2 & xz^2 \\ x^2y & 0 & yz^2 \\ x^2z & zy^2 & 0 \end{vmatrix} = x^2y^2z^2 \begin{vmatrix} 0 & x & x \\ y & 0 & y \\ z & z & 0 \end{vmatrix}$

[taking x^2, y^2 and z^2 common from C_1, C_2 and C_3, respectively]

Note (*) Topics related to these question have been removed from latest NCERT.

$$= x^2y^2z^2\begin{vmatrix} 0 & 0 & x \\ y & -y & y \\ z & z & 0 \end{vmatrix} \qquad [\because C_2 \to C_2 - C_3]$$

On expanding along R_1, we get

$$= x^2y^2z^2\,[x\,(yz + yz)]$$
$$= x^2y^2z^2 \cdot 2xyz = 2x^3y^3z^3$$

***Q.4** $\begin{vmatrix} 3x & -x+y & -x+z \\ x-y & 3y & z-y \\ x-z & y-z & 3z \end{vmatrix}$

Ans. We have, $\begin{vmatrix} 3x & -x+y & -x+z \\ x-y & 3y & z-y \\ x-z & y-z & 3z \end{vmatrix}$

On applying, $C_1 \to C_1 + C_2 + C_3$, we get

$$= \begin{vmatrix} x+y+z & -x+y & -x+z \\ x+y+z & 3y & z-y \\ x+y+z & y-z & 3z \end{vmatrix}$$

$$= (x+y+z)\begin{vmatrix} 1 & -x+y & -x+z \\ 1 & 3y & z-y \\ 1 & y-z & 3z \end{vmatrix}$$

[taking $(x+y+z)$ common from column C_1]

$$= (x+y+z)\begin{vmatrix} 1 & -x+y & -x+z \\ 0 & 2y+x & x-y \\ 0 & x-z & 2z+x \end{vmatrix}$$

$[\because R_2 \to R_2 - R_1 \text{ and } R_3 \to R_3 - R_1]$

Now, on expanding along C_1, we get

$$(x+y+z) \cdot 1\,[(2y+x)(2z+x) - (x-y)(x-z)]$$
$$= (x+y+z)(4yz + 2yx + 2xz + x^2 - x^2 + xz + yx - yz)$$
$$= (x+y+z)(3yz + 3yx + 3xz)$$
$$= 3\,(x+y+z)(yz + yx + xz)$$

***Q.5** $\begin{vmatrix} x+4 & x & x \\ x & x+4 & x \\ x & x & x+4 \end{vmatrix}$

Ans. We have, $\begin{vmatrix} x+4 & x & x \\ x & x+4 & x \\ x & x & x+4 \end{vmatrix}$

On applying, $R_1 \to R_1 + R_2 + R_3$, we get

$$\begin{vmatrix} 3x+4 & 3x+4 & 3x+4 \\ x & x+4 & x \\ x & x & x+4 \end{vmatrix}$$

Note (✱) Topics related to these question have been removed from latest NCERT.

On taking $(3x + 4)$ common from First row, we get

$$(3x+4)\begin{vmatrix} 1 & 1 & 1 \\ x & x+4 & x \\ x & x & x+4 \end{vmatrix}$$

Now, on applying $C_2 \to C_2 - C_1$, $C_3 \to C_3 - C_1$, we get

$$(3x+4)\begin{vmatrix} 1 & 0 & 0 \\ x & 4 & 0 \\ x & 0 & 4 \end{vmatrix}$$

On expanding along R_1, we get

$$(3x+4)\,[1(4 \times 4 - 0 \times 0)]$$
$$= 16 \cdot (3x+4)$$

***Q.6** $\begin{vmatrix} a-b-c & 2a & 2a \\ 2b & b-c-a & 2b \\ 2c & 2c & c-a-b \end{vmatrix}$

Ans. We have,

$$\begin{vmatrix} a-b-c & 2a & 2a \\ 2b & b-c-a & 2b \\ 2c & 2c & c-a-b \end{vmatrix}$$

$$= \begin{vmatrix} a+b+c & a+b+c & a+b+c \\ 2b & b-c-a & 2b \\ 2c & 2c & c-a-b \end{vmatrix} \quad [\because R_1 \to R_1 + R_2 + R_3]$$

$$= (a+b+c) \cdot \begin{vmatrix} 1 & 1 & 1 \\ 2b & b-c-a & 2b \\ 2c & 2c & c-a-b \end{vmatrix}$$

[taking $(a + b + c)$ common from the first row]

$$= (a+b+c)\begin{vmatrix} 0 & 0 & 1 \\ 0 & -(a+b+c) & 2b \\ (a+b+c) & (a+b+c) & (c-a-b) \end{vmatrix}$$

$[\because C_1 \to C_1 - C_3 \text{ and } C_2 \to C_2 - C_3]$

On taking $(a + b + c)$ common from C_1 and C_2, we get

$$(a+b+c)^3\begin{vmatrix} 0 & 0 & 1 \\ 0 & -1 & 2b \\ 1 & 1 & c-a-b \end{vmatrix}$$

On expanding along R_1, we get

$$(a+b+c)^3 \cdot 1[0-(-1)]$$
$$= (a+b+c)^3$$

Note (✱) Topics related to these question have been removed from latest NCERT.

Using the properties of determinants in questions 7 to 9 prove that.

*** Q.7** $\begin{vmatrix} y^2z^2 & yz & y+z \\ z^2x^2 & zx & z+x \\ x^2y^2 & xy & x+y \end{vmatrix} = 0$

Ans. We have to prove,

$$\begin{vmatrix} y^2z^2 & yz & y+z \\ z^2x^2 & zx & z+x \\ x^2y^2 & xy & x+y \end{vmatrix} = 0$$

$$\therefore \quad \text{LHS} = \begin{vmatrix} y^2z^2 & yz & y+z \\ z^2x^2 & zx & z+x \\ x^2y^2 & xy & x+y \end{vmatrix} = \frac{1}{xyz}\begin{vmatrix} xy^2z^2 & xyz & xy+xz \\ x^2yz^2 & xyz & yz+xy \\ x^2y^2z & xyz & xz+yz \end{vmatrix}$$

[$\because R_1 \to x R_1$, $R_2 \to y R_2$, $R_3 \to z R_3$ and divide determinant by xyz]

$$= \frac{1}{xyz}(xyz)^2 \begin{vmatrix} yz & 1 & xy+xz \\ xz & 1 & yz+xy \\ xy & 1 & xz+yz \end{vmatrix}$$

[taking (xyz) common from C_1 and C_2]

$$= xyz \begin{vmatrix} yz & 1 & xy+yz+zx \\ xz & 1 & xy+yz+zx \\ xy & 1 & xy+yz+zx \end{vmatrix} \quad [\because C_3 \to C_3 + C_1]$$

$$= xyz\,(xy+yz+zx) \begin{vmatrix} yz & 1 & 1 \\ xz & 1 & 1 \\ xy & 1 & 1 \end{vmatrix}$$

[taking $(xy+yz+zx)$ common from C_3]

$= 0$ [since, C_2 and C_3 are identicals]

$=$ RHS **Hence proved.**

Note Using the property of determinants. If two columns or rows are identical, then the value of determinant will be zero.

*** Q.8** $\begin{vmatrix} y+z & z & y \\ z & z+x & x \\ y & x & x+y \end{vmatrix} = 4xyz$

Thinking Process First in LHS use $C_1 \to C_1 + C_2 + C_3$ and then by using $C_1 \to C_1 - C_2$ and $R_1 \to R_1 - R_3$, we can get two zeroes in column 1 and then by simplification we will get the desired result.

Ans. We have to prove,

$$\begin{vmatrix} y+z & z & y \\ z & z+x & x \\ y & x & x+y \end{vmatrix} = 4xyz$$

$$\therefore \quad \text{LHS} = \begin{vmatrix} y+z & z & y \\ z & z+x & x \\ y & x & x+y \end{vmatrix}$$

Note (✱) Topics related to these question have been removed from latest NCERT.

$$= \begin{vmatrix} y+z+z+y & z & y \\ z+z+x+x & z+x & x \\ y+x+x+y & x & x+y \end{vmatrix} \quad [\because C_1 \to C_1 + C_2 + C_3]$$

$$= 2\begin{vmatrix} (y+z) & z & y \\ (z+x) & z+x & x \\ (x+y) & x & x+y \end{vmatrix} \quad \text{[taking 2 common from } C_1]$$

$$= 2\begin{vmatrix} y & z & y \\ 0 & z+x & x \\ y & x & x+y \end{vmatrix} \quad [\because C_1 \to C_1 - C_2]$$

$$= 2\begin{vmatrix} 0 & z-x & -x \\ 0 & z+x & x \\ y & x & x+y \end{vmatrix} \quad [\because R_1 \to R_1 - R_3]$$

Now, on expanding determinant along with C_1, we get

$$2\,[y(xz - x^2 + xz + x^2)] = 4xyz = \text{RHS}$$ **Hence proved.**

***Q.9** $\begin{vmatrix} a^2+2a & 2a+1 & 1 \\ 2a+1 & a+2 & 1 \\ 3 & 3 & 1 \end{vmatrix} = (a-1)^3$

Thinking Process Here, by using $R_1 \to R_1 - R_2$ and $R_2 \to R_2 - R_3$ in LHS, we can easily get the desired result.

Ans. We have to prove,

$$\begin{vmatrix} a^2+2a & 2a+1 & 1 \\ 2a+1 & a+2 & 1 \\ 3 & 3 & 1 \end{vmatrix} = (a-1)^3$$

$$\therefore \quad \text{LHS} = \begin{vmatrix} a^2+2a & 2a+1 & 1 \\ 2a+1 & a+2 & 1 \\ 3 & 3 & 1 \end{vmatrix}$$

$$= \begin{vmatrix} a^2+2a-2a-1 & 2a+1-a-2 & 0 \\ 2a+1-3 & a+2-3 & 0 \\ 3 & 3 & 1 \end{vmatrix}$$

$$[\because R_1 \to R_1 - R_2 \text{ and } R_2 \to R_2 - R_3]$$

$$= \begin{vmatrix} (a-1)(a+1) & (a-1) & 0 \\ 2(a-1) & (a-1) & 0 \\ 3 & 3 & 1 \end{vmatrix} = (a-1)^2 \begin{vmatrix} (a+1) & 1 & 0 \\ 2 & 1 & 0 \\ 3 & 3 & 1 \end{vmatrix}$$

[taking $(a-1)$ common from R_1 and R_2 each]

Now, on expanding along the C_3, we get

$$(a-1)^2[1(a+1) - 2] = (a-1)^2(a+1-2)$$

$$= (a-1)^2\,(a-1) = (a-1)^3 = \text{RHS}.$$ **Hence proved.**

Note (✱) Topics related to these question have been removed from latest NCERT.

Q.10 If $A+B+C=0$, then prove that $\begin{vmatrix} 1 & \cos C & \cos B \\ \cos C & 1 & \cos A \\ \cos B & \cos A & 1 \end{vmatrix} = 0.$

Thinking Process We are, given $A+B+C=0$, so on solving the determinant by expansion, we can use $\cos(A+B)=\cos(-C)$ *i.e.* $\cos(A+B)=\cos C$ as $\cos(-\theta)=\cos\theta$ and similarly after simplification this expression we will get the desired result.

Ans. We have to prove, $\begin{vmatrix} 1 & \cos C & \cos B \\ \cos C & 1 & \cos A \\ \cos B & \cos A & 1 \end{vmatrix} = 0$

$$\therefore \quad \text{LHS} = \begin{vmatrix} 1 & \cos C & \cos B \\ \cos C & 1 & \cos A \\ \cos B & \cos A & 1 \end{vmatrix}$$

On expanding along R_1, we get

$$= 1(1-\cos^2 A) - \cos C(\cos C - \cos A \cdot \cos B) + \cos B(\cos C \cdot \cos A - \cos B)$$
$$= \sin^2 A - \cos^2 C + \cos A \cdot \cos B \cdot \cos C + \cos A \cdot \cos B \cdot \cos C - \cos^2 B$$
$$= \sin^2 A - \cos^2 B + 2\cos A \cdot \cos B \cdot \cos C - \cos^2 C$$
$$= -\cos(A+B) \cdot \cos(A-B) + 2\cos A \cdot \cos B \cdot \cos C - \cos^2 C$$
$$[\because \cos^2 B - \sin^2 A = \cos(A+B) \cdot \cos(A-B)]$$
$$= -\cos(-C) \cdot \cos(A-B) + \cos C(2\cos A \cdot \cos B - \cos C)$$
$$= -\cos C(\cos A \cdot \cos B + \sin A \cdot \sin B - 2\cos A \cdot \cos B + \cos C)$$
$$[\because \cos(-\theta) = \cos\theta \text{ and } \cos(A-B) = \cos A \cos B + \sin A \sin B]$$
$$= \cos C(\cos A \cdot \cos B - \sin A \cdot \sin B - \cos C)$$
$$= \cos C[\cos(A+B) - \cos C] \qquad [\because \cos(A+B) = \cos A \cos B - \sin A \sin B]$$
$$= \cos C[\cos(-C) - \cos C]$$
$$= \cos C(\cos C - \cos C) = 0 = \text{RHS} \qquad [\because \cos(A+B) = \cos(-C)]$$

Hence proved.

Q.11 If the coordinates of the vertices of an equilateral triangle with sides of length 'a' are $(x_1, y_1), (x_2, y_2)$ and (x_3, y_3), then

$$\begin{vmatrix} x_1 & y_1 & 1 \\ x_2 & y_2 & 1 \\ x_3 & y_3 & 1 \end{vmatrix}^2 = \frac{3a^4}{4}.$$

Ans. Since, we know that area of a triangle with vertices $(x_1, y_1), (x_2, y_2)$ and (x_3, y_3), is given by

$$\Delta = \frac{1}{2}\begin{vmatrix} x_1 & y_1 & 1 \\ x_2 & y_2 & 1 \\ x_3 & y_3 & 1 \end{vmatrix}$$

$$\Rightarrow \quad \Delta^2 = \frac{1}{4}\begin{vmatrix} x_1 & y_1 & 1 \\ x_2 & y_2 & 1 \\ x_3 & y_3 & 1 \end{vmatrix}^2 \qquad \text{...(i)}$$

We know that, area of an equilateral triangle with side a,

$$\Delta = \frac{\sqrt{3}}{4}a^2 \Rightarrow \Delta^2 = \frac{3}{16}a^4 \quad \text{...(ii)}$$

From Eqs. (i) and (ii), we get $\frac{3}{16}a^4 = \frac{1}{4}\begin{vmatrix} x_1 & y_1 & 1 \\ x_2 & y_2 & 1 \\ x_3 & y_3 & 1 \end{vmatrix}^2$

$$\Rightarrow \begin{vmatrix} x_1 & y_1 & 1 \\ x_2 & y_2 & 1 \\ x_3 & y_3 & 1 \end{vmatrix}^2 = \frac{3}{4}a^4$$

Hence proved.

Q.12 Find the value of θ satisfying $\begin{bmatrix} 1 & 1 & \sin 3\theta \\ -4 & 3 & \cos 2\theta \\ 7 & -7 & -2 \end{bmatrix} = 0$

Ans. We have, $\begin{vmatrix} 1 & 1 & \sin 3\theta \\ -4 & 3 & \cos 2\theta \\ 7 & -7 & -2 \end{vmatrix} = 0$

On expanding along R_1, we get

$$1(-6 + 7\cos 2\theta) - 1(8 - 7\cos 2\theta) + \sin 3\theta (28 - 21) = 0$$

$$\Rightarrow -6 + 7\cos 2\theta - 8 + 7\cos 2\theta + 7\sin 3\theta = 0$$

$$\Rightarrow -14 + 7\cos 2\theta + 7\cos 2\theta + 7\sin 3\theta = 0$$

$$\Rightarrow -2 + \cos 2\theta + \cos 2\theta + \sin 3\theta = 0$$

$$\Rightarrow 2\cos 2\theta + \sin 3\theta = 2$$

$$\Rightarrow 2(1 - 2\sin^2\theta) + 3\sin\theta - 4\sin^3\theta = 2$$

$$[\because \cos 2\theta = 1 - 2\sin^2\theta \text{ and } \sin 3\theta = 3\sin\theta - 4\sin^3\theta]$$

$$\Rightarrow 2 - 4\sin^2\theta + 3\sin\theta - 4\sin^3\theta = 2$$

$$\Rightarrow -4\sin^2\theta + 3\sin\theta - 4\sin^3\theta = 0$$

$$\Rightarrow 4\sin^3\theta - 3\sin\theta + 4\sin^2\theta = 0$$

$$\Rightarrow \sin\theta(4\sin^2\theta + 4\sin\theta - 3) = 0$$

Now, either $\sin\theta = 0$

$$\Rightarrow \theta = n\pi$$

or $4\sin^2\theta + 4\sin\theta - 3 = 0$

$$\Rightarrow 4\sin^2\theta + 6\sin\theta - 2\sin\theta - 3 = 0$$

$$\Rightarrow 2\sin\theta(2\sin\theta + 3) - 1(2\sin\theta + 3) = 0$$

$$\Rightarrow (2\sin\theta + 3)(2\sin\theta - 1) = 0$$

$$\Rightarrow \sin\theta = \frac{1}{2}, \text{ or } \sin\theta = -\frac{3}{2}$$

(not possible as $-1 \le \sin\theta \le 1$)

$$\therefore \sin\theta = \frac{1}{2} \Rightarrow \sin\theta = \sin\frac{\pi}{6}$$

$$\Rightarrow \theta = n\pi + (-1)^n \frac{\pi}{6} \quad [\because \sin\theta = \sin\alpha \Rightarrow \theta = n\pi + (-1)^n \alpha]$$

Hence, required values of θ is $n\pi$ or $n\pi + (-1)^n \frac{\pi}{6}$.

Q.13 If $\begin{bmatrix} 4-x & 4+x & 4+x \\ 4+x & 4-x & 4+x \\ 4+x & 4+x & 4-x \end{bmatrix} = 0$, then find the values of x.

Ans. Given, $\begin{vmatrix} 4-x & 4+x & 4+x \\ 4+x & 4-x & 4+x \\ 4+x & 4+x & 4-x \end{vmatrix} = 0$

On expanding along R_1, we get

$$(4-x)[(4-x)^2-(4+x)^2]-(4+x)\ [(4+x)(4-x)-(4+x)^2]+(4+x)$$
$$[(4+x)^2-(4+x)(4-x)]=0$$
$$\Rightarrow (4-x)[16+x^2-8x-16-x^2-8x]-(4+x)^2[4-x-4-x]+(4+x)^2$$
$$[(4+x)-4+x]\ =0$$
$$\Rightarrow (4-x)(-16x)-(4+x)^2(-2x)+(4+x)^2(2x)=0$$
$$\Rightarrow \quad (4-x)(-16x)+(4+x)^2(2x+2x)=0$$
$$\Rightarrow \quad -64x+16x^2+64x+4x^3+32x^2=0$$
$$\Rightarrow \quad 4x^3+48x^2=0$$
$$\Rightarrow \quad 4x^2(x+12)=0 \Rightarrow x^2=0 \text{ or } x+12=0$$
$$\Rightarrow \quad x=0 \text{ or } x=-12$$

Hence, the values of x are 0 and -12

***Q.14** If $a_1, a_2, a_3, \ldots, a_r$ are in GP, then prove that the determinant $\begin{vmatrix} a_{r+1} & a_{r+5} & a_{r+9} \\ a_{r+7} & a_{r+11} & a_{r+15} \\ a_{r+11} & a_{r+17} & a_{r+21} \end{vmatrix}$ is independent of r.

Thinking Process We know that, nth term of a GP has value ar^{n-1}, where a = first term and r = common ratio. So, by using this result, we can prove the given determinant as independent of r.

Ans. We know that, $a_{r+1} = AR^{(r+1)-1} = AR^r$

where $r = r$ th term of a GP, A = First term of a GP and R = Common ratio of GP

We have, $\begin{vmatrix} a_{r+1} & a_{r+5} & a_{r+9} \\ a_{r+7} & a_{r+11} & a_{r+15} \\ a_{r+11} & a_{r+17} & a_{r+21} \end{vmatrix}$

$$= \begin{vmatrix} AR^r & AR^{r+4} & AR^{r+8} \\ AR^{r+6} & AR^{r+10} & AR^{r+14} \\ AR^{r+10} & AR^{r+16} & AR^{r+20} \end{vmatrix}$$

$$= AR^r \cdot AR^{r+6} \cdot AR^{r+10} \begin{vmatrix} 1 & AR^4 & AR^8 \\ 1 & AR^4 & AR^8 \\ 1 & AR^6 & AR^{10} \end{vmatrix} = 0$$

[taking AR^r, AR^{r+6} and AR^{r+10} common from R_1, R_2 and R_3, respectively]

[since, R_1 and R_2 are identicals then according to property determinant will be zero].

Note (✱) Topics related to these question have been removed from latest NCERT.

Q.15 Show that the points $(a+5, a-4)$, $(a-2, a+3)$ and (a, a) do not lie on a straight line for any value of a.

Thinking Process The area of triangle is given by

$$\Delta = \frac{1}{2}\begin{vmatrix} x_1 & y_1 & 1 \\ x_2 & y_2 & 1 \\ x_3 & y_3 & 1 \end{vmatrix}$$

We know that, if three points lie on a straight line, then area formed by these points will be equal to zero. So, by showing area formed by these points other than zero, we can prove the result.

Ans. We are given points $(a+5, a-4)$, $(a-2, a+3)$ and (a, a).

$$\Delta = \frac{1}{2}\begin{vmatrix} a+5 & a-4 & 1 \\ a-2 & a+3 & 1 \\ a & a & 1 \end{vmatrix}$$

On expanding along R_1, we get

$$\Delta = \frac{1}{2}[(a+5)(a+3-a) - (a-4)(a-2-a) + 1(a^2 - 2a - a^2 - 3a)]$$

$$= \frac{1}{2}[(a+5)(3) - (a-4)(-2) + (-5a)]$$

$$= \frac{1}{2}[3a + 15 + 2a - 8 - 5a] = \frac{7}{2} \neq 0$$

Since, $\Delta \neq 0$, therefore, given points form a triangle.
Hence, points do not lie on a straight line.

***Q.16** Show that ΔABC is an isosceles triangle, if the determinant

$$\Delta = \begin{vmatrix} 1 & 1 & 1 \\ 1+\cos A & 1+\cos B & 1+\cos C \\ \cos^2 A + \cos A & \cos^2 B + \cos B & \cos^2 C + \cos C \end{vmatrix} = 0.$$

Ans. We have, $\begin{vmatrix} 1 & 1 & 1 \\ 1+\cos A & 1+\cos B & 1+\cos C \\ \cos^2 A + \cos A & \cos^2 B + \cos B & \cos^2 C + \cos C \end{vmatrix} = 0$

$$\Rightarrow \begin{vmatrix} 0 & 0 & 1 \\ \cos A - \cos C & \cos B - \cos C & 1+\cos C \\ \cos^2 A + \cos A - \cos^2 C - \cos C & \cos^2 B + \cos B - \cos^2 C - \cos C & \cos^2 C + \cos C \end{vmatrix} = 0$$

$[\because C_1 \to C_1 - C_3 \text{ and } C_2 \to C_2 - C_3]$

$$\Rightarrow \begin{vmatrix} 0 & 0 & 1 \\ \cos A - \cos C & \cos B - \cos C & 1+\cos C \\ \cos^2 A - \cos^2 C + \cos A - \cos C & \cos^2 B - \cos^2 C + \cos B - \cos C & \cos^2 C + \cos C \end{vmatrix} = 0$$

$$\Rightarrow \begin{vmatrix} 0 & 0 & 1 \\ \cos A - \cos C & \cos B - \cos C & 1+\cos C \\ (\cos A - \cos C)(\cos A + \cos C + 1) & (\cos B - \cos C)(\cos B + \cos C + 1) & \cos^2 C + \cos C \end{vmatrix}$$

Note (✱) Topics related to these question have been removed from latest NCERT.

$$\Rightarrow (\cos A - \cos C)\cdot(\cos B - \cos C)\begin{vmatrix} 0 & 0 & 1 \\ 1 & 1 & 1+\cos C \\ \cos A + \cos C + 1 & \cos B + \cos C + 1 & \cos^2 C + \cos C \end{vmatrix} = 0$$

[taking $(\cos A - \cos C)$ common from C_1 and $(\cos B - \cos C)$ common from C_2]

On expanding along R_1, we get

$\Rightarrow$ $(\cos A - \cos C)\cdot(\cos B - \cos C)[(\cos B + \cos C + 1) - (\cos A + \cos C + 1)] = 0$

$\Rightarrow$ $(\cos A - \cos C)\cdot(\cos B - \cos C)\ (\cos B + \cos C + 1 - \cos A - \cos C - 1) = 0$

$\Rightarrow$ $(\cos A - \cos C)\cdot(\cos B - \cos C)(\cos B - \cos A) = 0$

i.e., $\cos A = \cos C$ or $\cos B = \cos C$ or $\cos B = \cos A$

$\Rightarrow$ $\angle A = \angle C$ or $\angle B = \angle C$ or $\angle B = \angle A$

Hence, ABC is an isosceles triangle.

Q.17 Find A^{-1}, if $A = \begin{bmatrix} 0 & 1 & 1 \\ 1 & 0 & 1 \\ 1 & 1 & 0 \end{bmatrix}$ and show that $A^{-1} = \dfrac{A^2 - 3I}{2}$.

Ans. We have, $A = \begin{bmatrix} 0 & 1 & 1 \\ 1 & 0 & 1 \\ 1 & 1 & 0 \end{bmatrix}$

$$|A| = 0\begin{vmatrix} 0 & 1 \\ 1 & 0 \end{vmatrix} - 1\begin{vmatrix} 1 & 1 \\ 1 & 0 \end{vmatrix} + 1\begin{vmatrix} 1 & 0 \\ 1 & 1 \end{vmatrix} = 0 - 1(-1) + 1 = 2 \neq 0$$

$\therefore$ A^{-1} exists.

Now, we will find co-factors

$$A_{11} = \begin{vmatrix} 0 & 1 \\ 1 & 0 \end{vmatrix} = (0-1) = -1,\ A_{21} = -\begin{vmatrix} 1 & 1 \\ 1 & 0 \end{vmatrix} = -(0-1) = 1$$

$$A_{12} = -\begin{vmatrix} 1 & 1 \\ 1 & 0 \end{vmatrix} = -(0-1) = 1,\ A_{22} = \begin{vmatrix} 0 & 1 \\ 1 & 0 \end{vmatrix} = (0-1) = -1$$

$$A_{13} = \begin{vmatrix} 1 & 0 \\ 1 & 1 \end{vmatrix} = (1-0) = 1,\ A_{23} = -\begin{vmatrix} 0 & 1 \\ 1 & 1 \end{vmatrix} = -(0-1) = 1$$

$$A_{31} = \begin{vmatrix} 1 & 1 \\ 0 & 1 \end{vmatrix} = (1-0) = 1,\ A_{32} = -\begin{vmatrix} 0 & 1 \\ 1 & 1 \end{vmatrix} = -(0-1) = 1$$

$$A_{33} = \begin{vmatrix} 0 & 1 \\ 1 & 0 \end{vmatrix} = (0-1) = -1$$

$\therefore$ $A_{11} = -1, A_{12} = 1, A_{13} = 1, A_{21} = 1, A_{22} = -1, A_{23} = 1, A_{31} = 1, A_{32} = 1$

and $A_{33} = -1$

$$\therefore\quad \text{adj } A = \begin{bmatrix} -1 & 1 & 1 \\ 1 & -1 & 1 \\ 1 & 1 & -1 \end{bmatrix}^T = \begin{bmatrix} -1 & 1 & 1 \\ 1 & -1 & 1 \\ 1 & 1 & -1 \end{bmatrix}$$

$$\therefore\quad A^{-1} = \frac{\text{adj } A}{|A|} = \frac{1}{2}\begin{bmatrix} -1 & 1 & 1 \\ 1 & -1 & 1 \\ 1 & 1 & -1 \end{bmatrix} \qquad [\because |A| = 2] \ldots\text{(i)}$$

and $$A^2 = \begin{bmatrix} 0 & 1 & 1 \\ 1 & 0 & 1 \\ 1 & 1 & 0 \end{bmatrix} \cdot \begin{bmatrix} 0 & 1 & 1 \\ 1 & 0 & 1 \\ 1 & 1 & 0 \end{bmatrix} = \begin{bmatrix} 2 & 1 & 1 \\ 1 & 2 & 1 \\ 1 & 1 & 2 \end{bmatrix} \quad \text{...(ii)}$$

$$\therefore \quad \frac{A^2 - 3I}{2} = \frac{1}{2}\left\{\begin{bmatrix} 2 & 1 & 1 \\ 1 & 2 & 1 \\ 1 & 1 & 2 \end{bmatrix} - \begin{bmatrix} 3 & 0 & 0 \\ 0 & 3 & 0 \\ 0 & 0 & 3 \end{bmatrix}\right\} = \frac{1}{2}\begin{bmatrix} -1 & 1 & 1 \\ 1 & -1 & 1 \\ 1 & 1 & -1 \end{bmatrix} \quad \text{[using Eq. (i)]}$$

$$= A^{-1}$$

Hence proved.

Long Answer Type Questions

Q.18 If $A = \begin{bmatrix} 1 & 2 & 0 \\ -2 & -1 & -2 \\ 0 & -1 & 1 \end{bmatrix}$, then find the value of A^{-1}.

Using A^{-1}, solve the system of linear equations $x - 2y = 10$, $2x - y - z = 8$ and $-2y + z = 7$.

Ans. We have, $$A = \begin{bmatrix} 1 & 2 & 0 \\ -2 & -1 & -2 \\ 0 & -1 & 1 \end{bmatrix} \quad \text{...(i)}$$

Expanding along R_1

$$|A| = 1(-3) - 2(-2) + 0 = 1 \neq 0$$

$\therefore$ A^{-1} exists

Now, $$A_{11} = -3,\ A_{12} = 2,\ A_{13} = 2,\ A_{21} = -2,\ A_{22} = 1,\ A_{23} = 1,$$
$$A_{31} = -4,\ A_{32} = 2 \text{ and } A_{33} = 3$$

$$\therefore \quad \text{adj}(A) = \begin{bmatrix} -3 & 2 & 2 \\ -2 & 1 & 1 \\ -4 & 2 & 3 \end{bmatrix}^T = \begin{bmatrix} -3 & -2 & -4 \\ 2 & 1 & 2 \\ 2 & 1 & 3 \end{bmatrix}$$

$$\therefore \quad A^{-1} = \frac{\text{adj } A}{|A|}$$

$$= \frac{1}{1}\begin{bmatrix} -3 & -2 & -4 \\ 2 & 1 & 2 \\ 2 & 1 & 3 \end{bmatrix}$$

$$\Rightarrow \quad A^{-1} = \begin{bmatrix} -3 & -2 & -4 \\ 2 & 1 & 2 \\ 2 & 1 & 3 \end{bmatrix} \quad \text{...(ii)}$$

Also, we have the system of linear equations as

$$x - 2y = 10,$$
$$2x - y - z = 8$$
and $$-2y + z = 7$$

In the form of $CX = D$,

$$\begin{bmatrix} 1 & -2 & 0 \\ 2 & -1 & -1 \\ 0 & -2 & 1 \end{bmatrix} \begin{bmatrix} x \\ y \\ z \end{bmatrix} = \begin{bmatrix} 10 \\ 8 \\ 7 \end{bmatrix}$$

where, $C = \begin{bmatrix} 1 & -2 & 0 \\ 2 & -1 & -1 \\ 0 & -2 & 1 \end{bmatrix}, X = \begin{bmatrix} x \\ y \\ z \end{bmatrix}$ and $D = \begin{bmatrix} 10 \\ 8 \\ 7 \end{bmatrix}$

We know that, $(A^T)^{-1} = (A^{-1})^T$

$\therefore$ $C^T = \begin{bmatrix} 1 & 2 & 0 \\ -2 & -1 & -2 \\ 0 & -1 & 1 \end{bmatrix} = A$ [using Eq. (i)]

$\therefore$ $X = C^{-1} D$

$\Rightarrow$ $\begin{bmatrix} x \\ y \\ z \end{bmatrix} = \begin{bmatrix} -3 & 2 & 2 \\ -2 & 1 & 1 \\ -4 & 2 & 3 \end{bmatrix} \begin{bmatrix} 10 \\ 8 \\ 7 \end{bmatrix}$

$$= \begin{bmatrix} -30+16+14 \\ -20+8+7 \\ -40+16+21 \end{bmatrix} = \begin{bmatrix} 0 \\ -5 \\ -3 \end{bmatrix}$$

$\therefore$ $x = 0, y = -5$ and $z = -3$

Q.19 Using matrix method, solve the system of equations $3x + 2y - 2z = 3$, $x + 2y + 3z = 6$ and $2x - y + z = 2$.

Thinking Process We know that, for given system of equations in the matrix form, we get $AX = B \Rightarrow X = A^{-1} B$, where $A^{-1} = \dfrac{\text{adj}(A)}{|A|}$ and then by getting inverse of A and determinant of A, we can get the desired result.

Ans. Given, system of equations is

$$3x + 2y - 2z = 3,$$
$$x + 2y + 3z = 6$$

and $$2x - y + z = 2$$

In the form of $AX = B$,

$$\begin{bmatrix} 3 & 2 & -2 \\ 1 & 2 & 3 \\ 2 & -1 & 1 \end{bmatrix} \begin{bmatrix} x \\ y \\ z \end{bmatrix} = \begin{bmatrix} 3 \\ 6 \\ 2 \end{bmatrix}$$

where, $A = \begin{bmatrix} 3 & 2 & -2 \\ 1 & 2 & 3 \\ 2 & -1 & 1 \end{bmatrix}, X = \begin{bmatrix} x \\ y \\ z \end{bmatrix}, B = \begin{bmatrix} 3 \\ 6 \\ 2 \end{bmatrix}$

For A^{-1}, expanding along R_1, $|A| = 3(5) - 2(1-6) + (-2)(-5)$

$$= 15 + 10 + 10 = 35 \neq 0$$

Since, $|A| \neq 0$

$\therefore$ A^{-1} exists

$\therefore A_{11} = 5, A_{12} = 5, A_{13} = -5, A_{21} = 0, A_{22} = 7, A_{23} = 7, A_{31} = 10, A_{32} = -11$ and $A_{33} = 4$

$$\therefore \quad \text{adj } A = \begin{bmatrix} 5 & 5 & -5 \\ 0 & 7 & 7 \\ 10 & -11 & 4 \end{bmatrix}^T = \begin{bmatrix} 5 & 0 & 10 \\ 5 & 7 & -11 \\ -5 & 7 & 4 \end{bmatrix}$$

Now,
$$A^{-1} = \frac{\text{adj } A}{|A|} = \frac{1}{35}\begin{bmatrix} 5 & 0 & 10 \\ 5 & 7 & -11 \\ -5 & 7 & 4 \end{bmatrix}$$

For $X = A^{-1}B$,

$$\begin{bmatrix} x \\ y \\ z \end{bmatrix} = \frac{1}{35}\begin{bmatrix} 5 & 0 & 10 \\ 5 & 7 & -11 \\ -5 & 7 & 4 \end{bmatrix}\begin{bmatrix} 3 \\ 6 \\ 2 \end{bmatrix}$$

$$= \frac{1}{35}\begin{bmatrix} 15 + 20 \\ 15 + 42 - 22 \\ -15 + 42 + 8 \end{bmatrix} = \frac{1}{35}\begin{bmatrix} 35 \\ 35 \\ 35 \end{bmatrix} = \begin{bmatrix} 1 \\ 1 \\ 1 \end{bmatrix}$$

$\therefore$ $x = 1, y = 1$ and $z = 1$

Q.20 Given $A = \begin{bmatrix} 2 & 2 & -4 \\ -4 & 2 & -4 \\ 2 & -1 & 5 \end{bmatrix}, B = \begin{bmatrix} 1 & -1 & 0 \\ 2 & 3 & 4 \\ 0 & 1 & 2 \end{bmatrix}$, find BA and use this to solve the system of equations $y + 2z = 7, x - y = 3, 2x + 3y + 4z = 17$.

Ans. We have,
$$A = \begin{bmatrix} 2 & 2 & -4 \\ -4 & 2 & -4 \\ 2 & -1 & 5 \end{bmatrix} \text{ and } B = \begin{bmatrix} 1 & -1 & 0 \\ 2 & 3 & 4 \\ 0 & 1 & 2 \end{bmatrix}$$

$$\therefore \quad BA = \begin{bmatrix} 1 & -1 & 0 \\ 2 & 3 & 4 \\ 0 & 1 & 2 \end{bmatrix}\begin{bmatrix} 2 & 2 & -4 \\ -4 & 2 & -4 \\ 2 & -1 & 5 \end{bmatrix} = \begin{bmatrix} 6 & 0 & 0 \\ 0 & 6 & 0 \\ 0 & 0 & 6 \end{bmatrix} = 6I$$

$$\therefore \quad B^{-1} = \frac{A}{6} = \frac{1}{6}A = \frac{1}{6}\begin{bmatrix} 2 & 2 & -4 \\ -4 & 2 & -4 \\ 2 & -1 & 5 \end{bmatrix} \quad \text{...(i)}$$

Also, $\quad x - y = 3, 2x + 3y + 4z = 17$ and $y + 2z = 7$

$$\Rightarrow \quad \begin{bmatrix} 1 & -1 & 0 \\ 2 & 3 & 4 \\ 0 & 1 & 2 \end{bmatrix}\begin{bmatrix} x \\ y \\ z \end{bmatrix} = \begin{bmatrix} 3 \\ 17 \\ 7 \end{bmatrix}$$

$$\therefore \quad \begin{bmatrix} x \\ y \\ z \end{bmatrix} = \begin{bmatrix} 1 & -1 & 0 \\ 2 & 3 & 4 \\ 0 & 1 & 2 \end{bmatrix}^{-1}\begin{bmatrix} 3 \\ 17 \\ 7 \end{bmatrix}$$

$$= \frac{1}{6}\begin{bmatrix} 2 & 2 & -4 \\ -4 & 2 & -4 \\ 2 & -1 & 5 \end{bmatrix}\begin{bmatrix} 3 \\ 17 \\ 7 \end{bmatrix} \quad \text{[using Eq. (i)]}$$

$$= \frac{1}{6}\begin{bmatrix} 6+34-28 \\ -12+34-28 \\ 6-17+35 \end{bmatrix} = \frac{1}{6}\begin{bmatrix} 12 \\ -6 \\ 24 \end{bmatrix} = \begin{bmatrix} 2 \\ -1 \\ 4 \end{bmatrix}$$

$\therefore$ $x = 2, y = -1$ and $z = 4$

Q.21 If $a + b + c \neq 0$ and $\begin{vmatrix} a & b & c \\ b & c & a \\ c & a & b \end{vmatrix} = 0$, then prove that $a = b = c$.

Ans. Let $A = \begin{vmatrix} a & b & c \\ b & c & a \\ c & a & b \end{vmatrix} = 0$

Given, $a + b + c \neq 0$

Now, on expanding A along R_1, we get

$$a(bc - a^2) - b(b^2 - ac) + c(ab - c^2) = 0$$

$\Rightarrow$ $abc - a^3 - b^3 + abc + abc - c^3 = 0$

$\Rightarrow$ $a^3 + b^3 + c^3 - 3abc = 0$

$\Rightarrow (a + b + c)(a^2 + b^2 + c^2 - ab - bc - ca) = 0$

On multiplying and dividing by 2, we get

$\Rightarrow \frac{1}{2}(a + b + c)(a^2 + b^2 - 2ab + b^2 + c^2 - 2bc + a^2 + c^2 - 2ac) = 0$

$\Rightarrow \frac{1}{2}(a + b + c)[(a - b)^2 + (b - c)^2 + (a - c)^2] = 0$

$\Rightarrow$ $(a + b + c)[(a - b)^2 + (b - c)^2 + (a - c)^2] = 0$

Since, $a + b + c \neq 0$

$\therefore$ $(a - b)^2 + (b - c)^2 + (a - c)^2 = 0$ ...(ii)

The Eq. (ii) is only possible, when

$$(a - b)^2 = (b - c)^2 = (a - c)^2 = 0$$

$\Rightarrow$ $(a - b) = (b - c)(a - c) = 0$

$\Rightarrow$ $a = b = c$ **Hence proved.**

***Q.22** Prove that $\begin{vmatrix} bc - a^2 & ca - b^2 & ab - c^2 \\ ca - b^2 & ab - c^2 & bc - a^2 \\ ab - c^2 & bc - a^2 & ca - b^2 \end{vmatrix}$ is divisible by $(a + b + c)$ and find the quotient.

Ans. Let $\Delta = \begin{vmatrix} bc - a^2 & ca - b^2 & ab - c^2 \\ ca - b^2 & ab - c^2 & bc - a^2 \\ ab - c^2 & bc - a^2 & ca - b^2 \end{vmatrix}$

$$= \begin{vmatrix} bc - a^2 - ca + b^2 & ca - b^2 - ab + c^2 & ab - c^2 \\ ca - b^2 - ab + c^2 & ab - c^2 - bc + a^2 & bc - a^2 \\ ab - c^2 - bc + a^2 & bc - a^2 - ca + b^2 & ca - b^2 \end{vmatrix}$$ $[\because C_1 \to C_1 - C_2$ and $C_2 \to C_2 - C_3]$

Note (*) Topics related to these question have been removed from latest NCERT.

$$= \begin{vmatrix} bc - ca + b^2 - a^2 & ca - ab + c^2 - b^2 & ab - c^2 \\ ca - ab + c^2 - b^2 & ab - bc + a^2 - c^2 & bc - a^2 \\ ab - bc + a^2 - c^2 & bc - ca + b^2 - a^2 & ca - b^2 \end{vmatrix}$$

$$= \begin{vmatrix} c(b-a) + (b-a)(b+a) & a(c-b) + (c-b)(c+b) & ab - c^2 \\ a(c-b) + (c-b)(c+b) & b(a-c) + (a-c)(a+c) & bc - a^2 \\ b(a-c) + (a-c)(a+c) & c(b-a) + (b-a)(b+a) & ca - b^2 \end{vmatrix}$$

$$= \begin{vmatrix} (b-a)(a+b+c) & (c-b)(a+b+c) & ab - c^2 \\ (c-b)(a+b+c) & (a-c)(a+b+c) & bc - a^2 \\ (a-c)(a+b+c) & (b-a)(a+b+c) & ca - b^2 \end{vmatrix} = (a+b+c)^2 \begin{vmatrix} b-a & c-b & ab-c^2 \\ c-b & a-c & bc-a^2 \\ a-c & b-a & ca-b^2 \end{vmatrix}$$

[taking $(a+b+c)$ common from C_1 and C_2 each]

$$= (a+b+c)^2 \begin{vmatrix} 0 & 0 & ab+bc+ca-(a^2+b^2+c^2) \\ c-b & a-c & bc-a^2 \\ a-c & b-a & ca-b^2 \end{vmatrix} \quad [\because R_1 \to R_1 + R_2 + R_3]$$

Now, on expanding along R_1, we get

$= (a+b+c)^2\,[ab+bc+ca-(a^2+b^2+c^2)][(c-b)(b-a)-(a-c)^2]$

$= (a+b+c)^2\,(ab+bc+ca-a^2-b^2-c^2)(cb-ac-b^2+ab-a^2-c^2+2ac)$

$= (a+b+c)^2\,(a^2+b^2+c^2-ab-bc-ca)(a^2+b^2+c^2-ac-ab-bc)$

On multiplying and dividing by 2, we get

$= \dfrac{1}{2}(a+b+c)\,[(a+b+c)(a^2+b^2+c^2-ab-bc-ca)][(a-b)^2+(b-c)^2+(c-a)^2]$

$= \dfrac{1}{2}(a+b+c)(a^3+b^3+c^3-3abc)\,[(a-b)^2+(b-c)^2+(c-a)^2]$

$[\because a^3+b^3+c^3-3abc = [(a+b+c)(a^2+b^2+c^2-ab-bc-ca)]$

Hence, given determinant is divisible by $(a+b+c)$ and quotient is

$= \dfrac{1}{2}(a^3+b^3+c^3-3abc)[(a-b)^2+(b-c)^2+(c-a)^2]$. **Hence proved.**

***Q.23** If $x+y+z=0$, then prove that $\begin{vmatrix} xa & yb & zc \\ yc & za & xb \\ zb & xc & ya \end{vmatrix} = xyz \begin{vmatrix} a & b & c \\ c & a & b \\ b & c & a \end{vmatrix}$.

Thinking Process We have, given $x+y+z=0 \Rightarrow x^3+y^3+z^3=3xyz$. So, by using this in solving the given determinant from both the sides, we can equate the obtained result from both the sides to desired result.

Ans. We have, $x+y+z=0$, also we have to prove

$$\begin{vmatrix} xa & yb & zc \\ yc & za & xb \\ zb & xc & ya \end{vmatrix} = xyz \begin{vmatrix} a & b & c \\ c & a & b \\ b & c & a \end{vmatrix}$$

$$\therefore \quad \text{LHS} = \begin{vmatrix} xa & yb & zc \\ yc & za & xb \\ zb & xc & ya \end{vmatrix}$$

Note (✱) Topics related to these question have been removed from latest NCERT.

On expanding along R_1, we get

$$= xa\,(za\cdot ya - xb\cdot xc) - yb\,(yc\cdot ya - xb\cdot zb) + zc\,(yc\cdot xc - za\cdot zb)$$
$$= xa\,(a^2yz - x^2bc) - yb\,(y^2ac - b^2xz) + zc\,(c^2xy - z^2ab)$$
$$= xyza^3 - x^3abc - y^3abc + b^3xyz + c^3xyz - z^3abc$$
$$= xyz\,(a^3 + b^3 + c^3) - abc\,(x^3 + y^3 + z^3)$$
$$= xyz\,(a^3 + b^3 + c^3) - abc\,(3\,xyz)$$
$$[\because x + y + z = 0 \Rightarrow x^3 + y^3 + z^3 = 3xyz]$$
$$= xyz\,(a^3 + b^3 + c^3 - 3abc) \qquad \text{...(i)}$$

Now, $$\text{RHS} = xyz\begin{vmatrix} a & b & c \\ c & a & b \\ b & c & a \end{vmatrix} = xyz\begin{vmatrix} a+b+c & b & c \\ a+b+c & a & b \\ a+b+c & c & a \end{vmatrix} \quad [\because C_1 \to C_1 + C_2 + C_3]$$

$$= xyz\,(a+b+c)\begin{vmatrix} 1 & b & c \\ 1 & a & b \\ 1 & c & a \end{vmatrix} \quad [\text{taking } (a+b+c) \text{ common from } C_1]$$

$$= xyz\,(a+b+c)\begin{vmatrix} 0 & b-c & c-a \\ 0 & a-c & b-a \\ 1 & c & a \end{vmatrix}$$
$$[\because R_1 \to R_1 - R_3 \text{ and } R_2 \to R_2 - R_3]$$

On expanding along C_1, we get

$$= xyz\,(a+b+c)\,[(b-c)(b-a) - (a-c)(c-a)]$$
$$= xyz\,(a+b+c)\,(b^2 - ab - bc + ac + a^2 + c^2 - 2ac)$$
$$= xyz\,(a+b+c)\,(a^2 + b^2 + c^2 - ab - bc - ca)$$
$$= xyz\,(a^3 + b^3 + c^3 - 3abc) \qquad \text{...(ii)}$$
$$[\because a^3 + b^3 + c^3 - 3abc = (a+b+c)(a^2 + b^2 + c^2 - ab - bc - ca)]$$

From Eqs. (i) and (ii), we get

$$\text{LHS} = \text{RHS}$$

$$\Rightarrow \begin{vmatrix} xa & yb & zc \\ yc & za & xb \\ zb & xc & ya \end{vmatrix} = xyz\begin{vmatrix} a & b & c \\ c & a & b \\ b & c & a \end{vmatrix}$$

Hence proved.

Objective Type Questions

Choose the correct answer out of the given four options in each of the exercises from 24 to 37.

Q.24 If $\begin{vmatrix} 2x & 5 \\ 8 & x \end{vmatrix} = \begin{vmatrix} 6 & -2 \\ 7 & 3 \end{vmatrix}$, then the value of x is

(a) 3 (b) ± 3 (c) ± 6 (d) 6

Ans. ***(c)*** We have,
$$\begin{vmatrix} 2x & 5 \\ 8 & x \end{vmatrix} = \begin{vmatrix} 6 & -2 \\ 7 & 3 \end{vmatrix}$$

$$\Rightarrow \quad 2x^2 - 40 = 18 + 14$$
$$\Rightarrow \quad 2x^2 = 32 + 40$$
$$\Rightarrow \quad x^2 = \frac{72}{2} = 36$$
$$\therefore \quad x = \pm 6$$

Q.25 The value of $\begin{vmatrix} a-b & b+c & a \\ b-a & c+a & b \\ c-a & a+b & c \end{vmatrix}$ is

(a) $a^3 + b^3 + c^3$ (b) $3bc$
(c) $a^3 + b^3 + c^3 - 3abc$ (d) None of these

Ans. ***(d)*** We have, $\begin{vmatrix} a-b & b+c & a \\ b-a & c+a & b \\ c-a & a+b & c \end{vmatrix}$

On expanding along R_1, we get

$(a-b)[c(c+a) - b(a+b)] - (b+c)\ [c(b-a) - b(c-a)] + a[(a+b)(b-a) - (c+a)(c-a)]$

$= (a-b)[c^2 + ac - ba - b^2] - (b+c)[cb - ac - bc + ab] + a[ab - a^2 + b^2 - ab - c^2 + a^2 + ca - ac]$

$= (a-b)[a(c-b) + (c^2 - b^2)] - (b+c)\ [ab - ac] + a[b^2 - c^2]$

$= (a-b)[(c-b)(a+c+b)] - a(b+c)(b-c) + a(b+c)\cdot(b-c)(c-b)[(a-b)(a+b+c) + a(b+c) - a(b+c)]$

$= (c-b)(a-b)(a+b+c)$

Q.26 The area of a triangle with vertices $(-3, 0)$, $(3, 0)$ and $(0, k)$ is 9 sq. units. The value of k will be

(a) 9 (b) 3 (c) -9 (d) 6

Ans. ***(b)*** We know that, area of a triangle with vertices (x_1, y_1), (x_2, y_2) and (x_3, y_3) is given by

$$\Delta = \frac{1}{2}\begin{vmatrix} x_1 & y_1 & 1 \\ x_2 & y_2 & 1 \\ x_3 & y_3 & 1 \end{vmatrix}$$

$$\therefore \quad \Delta = \frac{1}{2}\begin{vmatrix} -3 & 0 & 1 \\ 3 & 0 & 1 \\ 0 & k & 1 \end{vmatrix}$$

On expanding along R_1, we get

$$\Delta = \frac{1}{2}[-3(-k) - 0 + 1(3k) = \frac{1}{2}[6k] = 3k$$

Since, area of triangle = 9 sq. units.

$$\pm 9 = 3k$$

$$\therefore \quad k = -\frac{+9}{3} = \pm 3$$

Q.27 The determinant $\begin{vmatrix} b^2 - ab & b - c & bc - ac \\ ab - a^2 & a - b & b^2 - ab \\ bc - ac & c - a & ab - a^2 \end{vmatrix}$ is equal to

(a) $abc(b-c)(c-a)(a-b)$ (b) $(b-c)(c-a)(a-b)$

(c) $(a+b+c)(b-c)(c-a)(a-b)$ (d) None of these

Ans. *(d)* We have, $\begin{vmatrix} b^2 - ab & b - c & bc - ac \\ ab - a^2 & a - b & b^2 - ab \\ bc - ac & c - a & ab - a^2 \end{vmatrix}$

On expanding along R_1, we get

$(b^2 - ab)[(a-b)(ab-a^2) - (c-a)(b^2-ab)] - (b-c)[(ab-a^2)^2 -$
$(bc-ac)\cdot(b^2-ab)] + (bc-ac)[(c-a)(ab-a^2) - (a-b)(bc-ac)]$
$= (b^2-ab)[(a^2b - a^3 - ab^2 + a^2b) - (cb^2 - abc - ab^2 + a^2b)] - (b-c)$
$[a^2b^2 + a^4 - 2ab\cdot a^2 - (b^3c - ab^2c - ab^2c + a^2bc)] + (bc-ac)[(abc - a^2c$
$- a^2b + a^3) - (abc - a^2c - b^2c + abc)]$
$= (b^2-ab)[a^2b - a^3 - ab^2 + a^2b - cb^2 + abc + ab^2 - a^2b] - (b-c)[a^2b^2 + a^4$
$- 2a^3b - b^3c + ab^2c + ab^2c - a^2bc] + (bc-ac)[abc - a^2c - a^2b + a^3 - abc$
$+ a^2c + b^2c - abc]$
$= (b^2-ab)[-a^3 - cb^2 + abc + a^2b] - (b-c)[a^2b^2 + a^4 - 2a^3b - b^3c + 2ab^2c$
$- a^2bc] + (bc-ac)[-abc - a^2b + a^3 + b^2c]$
$= -a^3b^2 - cb^4 + ab^3c + a^2b^3 + a^4b + acb^3 - a^2b^2c - a^3b^2 - a^2b^3 - a^4b$
$+ 2a^3b^2 + b^4c - 2ab^3c + a^2b^2c + a^2b^2c + a^4c - 2a^3bc - b^3c^2 + 2ab^2c^2 - a^2bc^2$
$- ab^2c^2 - a^2cb^2 + a^3bc + b^3c^2 + a^2bc^2 + a^3bc - a^4c - ab^2c^2 = 0$

Q.28 The number of distinct real roots of $\begin{vmatrix} \sin x & \cos x & \cos x \\ \cos x & \sin x & \cos x \\ \cos x & \cos x & \sin x \end{vmatrix} = 0$ in the interval $-\frac{\pi}{4} \le x \le \frac{\pi}{4}$ is

(a) 0 (b) 2 (c) 1 (d) 3

Ans. *(c)* We have,

$$\begin{vmatrix} \sin x & \cos x & \cos x \\ \cos x & \sin x & \cos x \\ \cos x & \cos x & \sin x \end{vmatrix} = 0$$

On expanding along R_1, we get

$$\sin x\,[\sin^2 x - \cos^2 x] - \cos x[\sin x \cos x - \cos^2 x] + \cos x[\cos^2 x - \cos x \sin x] = 0$$

$$\Rightarrow \sin x\,(\sin x + \cos x)\,(\sin x - \cos x) - \cos^2 x(\sin x - \cos x)$$
$$- \cos^2 x(\sin x - \cos x) = 0$$

$$\Rightarrow (\sin x - \cos x)\,[\sin^2 x + \sin x \cos x - 2\cos^2 x] = 0$$

$$\Rightarrow (\sin x - \cos x)\,[\sin^2 x + 2\sin x \cos x - \sin x \cos x - 2\cos^2 x] = 0$$

$$\Rightarrow (\sin x - \cos x\,[\sin x(\sin x + 2\cos x) - \cos x(\sin x + 2\cos x)] = 0$$

$$\Rightarrow (\sin x - \cos x)^2\,(\sin x + 2\cos x) = 0$$

Either $\quad 2\cos x = -\sin x$

$$\Rightarrow \cos x = -\frac{1}{2}\sin x$$

$$\Rightarrow \tan x = -2 \quad \text{...(i)}$$

But here for $-\frac{\pi}{4} \le x \le \frac{\pi}{4}$, we get $-1 \le \tan x \le 1$ so, no solution possible for $\tan x = -2$

or $\quad (\sin x - \cos x)^2 = 0,$

$$\Rightarrow \sin x - \cos x = 0$$

$$\Rightarrow \sin x = \cos x$$

$$\Rightarrow \tan x = 1 = \tan\frac{\pi}{4}$$

$$\therefore x = \frac{\pi}{4}$$

So, only one distinct real root exist.

***Q.29** If A, B and C are angles of a triangle, then the determinant $\begin{vmatrix} -1 & \cos C & \cos B \\ \cos C & -1 & \cos A \\ \cos B & \cos A & -1 \end{vmatrix}$ is equal to

(a) 0 (b) –1 (c) 1 (d) None of these

Ans. *(a)* We have, $\begin{vmatrix} -1 & \cos C & \cos B \\ \cos C & -1 & \cos A \\ \cos B & \cos A & -1 \end{vmatrix}$

On applying $C_1 \to a\,C_1 + b\,C_2 + c\,C_3$, we get

$$\begin{vmatrix} -a + b\cos C + c\cos B & \cos C & \cos B \\ a\cos C - b + c\cos A & -1 & \cos A \\ a\cos B + b\cos A - c & \cos A & -1 \end{vmatrix}$$

Also, by projection rule in a triangle, we know that

$$a = b\cos C + c\cos B,\ b = c\cos A + a\cos C \text{ and } c = a\cos B + b\cos A$$

Note (✱) Topics related to these question have been removed from latest NCERT.

Using above equation in column first, we get

$$\begin{vmatrix} -a+a & \cos C & \cos B \\ b-b & -1 & \cos A \\ c-c & \cos A & -1 \end{vmatrix} = \begin{vmatrix} 0 & \cos C & \cos B \\ 0 & -1 & \cos A \\ 0 & \cos A & -1 \end{vmatrix} = 0$$

[since, all elements of column C_1 are zero]

Q.30 If $f(t) = \begin{bmatrix} \cos t & t & 1 \\ 2\sin t & t & 2t \\ \sin t & t & t \end{bmatrix}$, then $\lim_{t \to 0} \frac{f(t)}{t^2}$ is equal to

(a) 0 (b) −1 (c) 2 (d) 3

Ans. *(a)* We are given, $f(t) = \begin{vmatrix} \cos t & t & 1 \\ 2\sin t & t & 2t \\ \sin t & t & t \end{vmatrix}$

On expanding along R_1, we get

$$f(t) = \cos t[t^2 - 2t^2] - t\,[2t\sin t - 2t\sin t] + 1[2t\sin t - t\sin t]$$
$$= \cos t\,[-t^2] - 0 + [2t\sin t - t\sin t]$$
$$= -t^2\cos t + t\sin t \quad \text{...(i)}$$

On dividing Eq. (i) by t^2, we get

$$\frac{f(t)}{t^2} = -\cos t + \frac{\sin t}{t}$$

Now, $\lim_{t \to 0} \frac{f(t)}{t^2} = \lim_{t \to 0} (-\cos t) + \lim_{t \to 0} \frac{\sin t}{t} = -1 + 1 = 0$ $\left[\because \lim_{x \to 0} \frac{\sin x}{x} = 1, \cos 0 = 1\right]$

Q.31 The maximum value of

$$\Delta = \begin{vmatrix} 1 & 1 & 1 \\ 1 & 1+\sin\theta & 1 \\ 1+\cos\theta & 1 & 1 \end{vmatrix} \text{ is (where, } \theta \text{ is real number)}$$

(a) $\frac{1}{2}$ (b) $\frac{\sqrt{3}}{2}$ (c) $\sqrt{2}$ (d) $\frac{2\sqrt{3}}{4}$

Ans. *(a)* Since, $\Delta = \begin{vmatrix} 1 & 1 & 1 \\ 1 & 1+\sin\theta & 1 \\ 1+\cos\theta & 1 & 1 \end{vmatrix}$

On expanding along R_1, we get

$$\Delta = 1[(1+\sin\theta) - 1] - 1[1-(1+\cos\theta)] + 1[1-(1+\sin\theta)(1+\cos\theta)]$$
$$= [\sin\theta] - [-\cos\theta] + [1 - 1 - \cos\theta - \sin\theta - \sin\theta \cdot \cos\theta]$$
$$= \sin\theta + \cos\theta - \cos\theta - \sin\theta - \sin\theta \cdot \cos\theta$$
$$= -\sin\theta \cdot \cos\theta$$
$$= -\frac{1}{2}\, 2\sin\theta \cdot \cos\theta \quad \text{[divide and multiply by 2]}$$
$$= -\frac{1}{2}\sin 2\theta \quad [\because \sin 2\theta = 2\sin\theta \cdot \cos\theta]$$

Δ is maximum when $\sin 2\theta$ is maximum.

Since, $-1 \le \sin 2\theta \le 1$

$\therefore \quad \Delta_{max} = -\frac{1}{2}(-1) = \frac{1}{2}$

Q.32 If $f(x)=\begin{vmatrix} 0 & x-a & x-b \\ x+a & 0 & x-c \\ x+b & x+c & 0 \end{vmatrix}$, then

(a) $f(a)=0$ (b) $f(b)=0$ (c) $f(0)=0$ (d) $f(1)=0$

Ans. *(c)* We have,

$$f(x)=\begin{vmatrix} 0 & x-a & x-b \\ x+a & 0 & x-c \\ x+b & x+c & 0 \end{vmatrix} \quad ...(i)$$

On substituting $x = a$ in Eq. (i) we get

$$\Rightarrow \quad f(a)=\begin{vmatrix} 0 & 0 & a-b \\ 2a & 0 & a-c \\ a+b & a+c & 0 \end{vmatrix}$$

On expanding along R_1, we get

$$=[(a-b)\{2a\cdot(a+c)\}]\neq 0$$

On substituting $x = b$ in Eq. (i) we get

$$\therefore \quad f(b)=\begin{vmatrix} 0 & b-a & 0 \\ b+a & 0 & b-c \\ 2b & b+c & 0 \end{vmatrix}$$

On expanding along R_1, we get

$$=-(b-a)[-2b(b-c)]$$
$$=2b(b-a)(b-c)\neq 0$$

On substituting $x = 0$ in Eq. (i) we get

$$\therefore \quad f(0)=\begin{vmatrix} 0 & -a & -b \\ a & 0 & -c \\ b & c & 0 \end{vmatrix}$$

On expanding along R_1, we get

$$=a(bc)-b(ac)$$
$$=abc-abc=0$$

Q.33 If $A=\begin{bmatrix} 2 & \lambda & -3 \\ 0 & 2 & 5 \\ 1 & 1 & 3 \end{bmatrix}$, then A^{-1} exists, if

(a) $\lambda = 2$ (b) $\lambda \neq 2$ (c) $\lambda \neq -2$ (d) None of these

Ans. *(d)* We have,

$$A=\begin{bmatrix} 2 & \lambda & -3 \\ 0 & 2 & 5 \\ 1 & 1 & 3 \end{bmatrix}$$

On expanding along R_1, we get

$$|A|=2(6-5)-\lambda(-5)-3(-2)=2+5\lambda+6$$

We know that, A^{-1} exists, if A is non-singular matrix *i.e.*, $|A|\neq 0$.

$$\therefore \quad 2+5\lambda+6\neq 0$$

$\Rightarrow$ $$5\lambda \neq -8$$

$\therefore$ $$\lambda \neq \frac{-8}{5}$$

So, A^{-1} exists if and only if $\lambda \neq \frac{-8}{5}$.

Q.34 If A and B are invertible matrices, then which of the following is not correct?

(a) adj $A = |A| \cdot A^{-1}$
(b) $\det (A)^{-1} = [\det (A)]^{-1}$
(c) $(AB)^{-1} = B^{-1} A^{-1}$
(d) $(A + B)^{-1} = B^{-1} + A^{-1}$

Ans. ***(d)*** Since, A and B are invertible matrices. So, we can say that

$$(AB)^{-1} = B^{-1} A^{-1} \quad \text{...(i)}$$

Also, $$A^{-1} = \frac{1}{|A|} (\text{adj } A)$$

$\Rightarrow$ $$\text{adj } A = |A| \cdot A^{-1} \quad \text{...(ii)}$$

Also, $$\det (A)^{-1} = [\det (A)]^{-1}$$

$\Rightarrow$ $$\det (A)^{-1} = \frac{1}{[\det (A)]}$$

$\Rightarrow$ $$\det (A) \cdot \det (A)^{-1} = 1 \quad \text{...(iii)}$$

which is true.

Again, $$(A + B)^{-1} = \frac{1}{|(A + B)|} \text{adj } (A + B)$$

$\Rightarrow$ $$(A + B)^{-1} \neq B^{-1} + A^{-1} \quad \text{...(iv)}$$

Hence, (i), (ii), (iii) represent option (*a*), (*b*), (*c*) are true and (iv) represent (*d*) is not true.

Q.35 If x, y and z are all different from zero and $\begin{vmatrix} 1+x & 1 & 1 \\ 1 & 1+y & 1 \\ 1 & 1 & 1+z \end{vmatrix} = 0$, then the value of $x^{-1} + y^{-1} + z^{-1}$ is

(a) xyz
(b) $x^{-1}y^{-1}z^{-1}$
(c) $-x - y - z$
(d) -1

Ans. ***(d)*** We have, $\begin{vmatrix} 1+x & 1 & 1 \\ 1 & 1+y & 1 \\ 1 & 1 & 1+z \end{vmatrix} = 0,$

where $x \neq 0, y \neq 0, z \neq 0$.

Now, on expanding along R_1, we get

$(1 + x) [(1 + y) (1 + z) - 1] - 1 [(1 + z) - 1] + 1 [1 - (1 + y)] = 0$

$\Rightarrow$ $$(1 + x) [1 + z + y + yz - 1] - [z] + [-y] = 0$$

$\Rightarrow$ $$(1 + x) [z + y + yz] - z - y = 0$$

$\Rightarrow$ $$z + y + yz + xz + xy + xyz - z - y = 0$$

$\Rightarrow$ $$yz + xz + xy + xyz = 0 \quad \text{...(i)}$$

On dividing eq. (i) by xyz, we get

$$\frac{yz}{xyz} + \frac{xz}{xyz} + \frac{xy}{xyz} + \frac{xyz}{xyz} = 0$$

$$\Rightarrow \quad \frac{1}{x} + \frac{1}{y} + \frac{1}{z} + 1 = 0$$

$$\Rightarrow \quad x^{-1} + {}^{-1} + z^{-1} = -1$$

Q.36 The value of $\begin{vmatrix} x & x+y & x+2y \\ x+2y & x & x+y \\ x+y & x+2y & x \end{vmatrix}$ is

(a) $9x^2 (x + y)$ (b) $9y^2 (x + y)$

(c) $3y^2 (x + y)$ (d) $7x^2 (x + y)$

Ans. ***(b)*** We have, $\begin{vmatrix} x & x+y & x+2y \\ x+2y & x & x+y \\ x+y & x+2y & x \end{vmatrix}$

On expanding along R_1, we get

$$= x[x^2 - (x+2y)(x+y)] - (x+y)[x(x+2y) - (x+y)^2] + x + 2y[(x+2y)^2 - x(x+y)]$$

$$= x[x^2 - x^2 - xy - 2xy - 2y^2] - (x+y)[x^2 + 2xy - x^2 - y^2 - 2xy] + (x+2y)[x^2 + 4y^2 + 4xy - x^2 - xy]$$

$$= x[-3xy - 2y^2] - (x+y)[-y^2] + (x+2y)[4y^2 + 3xy]$$

$$= -3x^2y - 2xy^2 + xy^2 + y^3 + 4xy^2 + 3x^2y + 8y^3 + 6xy^2$$

$$= 9y^3 + 9xy^2 = 9y^2(x+y)$$

Q.37 If there are two values of a which makes determinant,

$\Delta = \begin{vmatrix} 1 & -2 & 5 \\ 2 & a & -1 \\ 0 & 4 & 2a \end{vmatrix} = 86$, then the sum of these number is

(a) 4 (b) 5 (c) -4 (d) 9

Ans. ***(c)*** We have,

$$\Delta = \begin{vmatrix} 1 & -2 & 5 \\ 2 & a & -1 \\ 0 & 4 & 2a \end{vmatrix} = 86$$

$$\Rightarrow \quad 1(2a^2 + 4) - 2(-4a - 20) + 0 = 86$$

[expanding along first column *i.e.*, C_1]

$$\Rightarrow \quad 2a^2 + 4 + 8a + 40 = 86$$

$$\Rightarrow \quad 2a^2 + 8a + 44 - 86 = 0$$

$$\Rightarrow \quad 2a^2 + 8a - 42 = 0$$

$$\Rightarrow \quad a^2 + 4a - 21 = 0$$

$$\Rightarrow \quad a^2 + 7a - 3a - 21 = 0$$

$$\Rightarrow \quad (a+7)(a-3) = 0$$

$$a = -7 \text{ and } 3$$

$$\therefore \quad \text{Required sum} = -7 + 3 = -4$$

Fill in the blanks in each of the exercises 38 to 47.

Q.38 If A is a matrix of order 3×3, then $|3A| =$ ....

Thinking Process If A is a square matrix of order n, then $|KA| = K^n \cdot |A|$.

Ans. If A is a matrix of order 3×3, then $|3A| = 3 \times 3 \times 3\,|A| = 27\,|A|$

Q.39 If A is invertible matrix of order 3×3, then $|A^{-1}|$ is equal to ....

Ans. If A is invertible matrix of order 3×3, then $|A^{-1}| = \dfrac{1}{|A|}$. [since, $|A| \cdot |A^{-1}| = 1$]

***Q.40** If $x, y, z \in R$, then the value of $\begin{vmatrix} (2^x+2^{-x})^2 & (2^x-2^{-x})^2 & 1 \\ (3^x+3^{-x})^2 & (3^x-3^{-x})^2 & 1 \\ (4^x+4^{-x})^2 & (4^x-4^{-x})^2 & 1 \end{vmatrix}$

is equal to ....

Ans. We have,

$$\begin{vmatrix} (2^x+2^{-x})^2 & (2^x-2^{-x})^2 & 1 \\ (3^x+3^{-x})^2 & (3^x-3^{-x})^2 & 1 \\ (4^x+4^{-x})^2 & (4^x-4^{-x})^2 & 1 \end{vmatrix}$$

$$= \begin{vmatrix} (2\cdot 2^x)(2\cdot 2^{-x}) & (2^x-2^{-x})^2 & 1 \\ (2\cdot 3^x)(2\cdot 3^{-x}) & (3^x-3^{-x})^2 & 1 \\ (2\cdot 4^x)(2\cdot 4^{-x}) & (4^x-4^{-x})^2 & 1 \end{vmatrix} \quad \begin{matrix}[\because C_1 \to C_1 - C_2] \\ [\because (a+b)^2-(a-b)^2 = 4ab]\end{matrix}$$

$$= \begin{vmatrix} 4 & (2^x-2^{-x})^2 & 1 \\ 4 & (3^x-3^{-x})^2 & 1 \\ 4 & (4^x-4^{-x})^2 & 1 \end{vmatrix}$$

$$= 4\begin{vmatrix} 1 & (2^x-2^{-x})^2 & 1 \\ 1 & (3^x-3^{-x})^2 & 1 \\ 1 & (4^x-4^{-x})^2 & 1 \end{vmatrix} = 0$$

As, C_1 and C_2 are identical, therefore determinant will be zero.

Q.41 If $\cos 2\theta = 0$, then $\begin{vmatrix} 0 & \cos\theta & \sin\theta \\ \cos\theta & \sin\theta & 0 \\ \sin\theta & 0 & \cos\theta \end{vmatrix}^2 =$ ....

Ans. Given, $\cos 2\theta = 0 = \cos\dfrac{\pi}{2} \Rightarrow 2\theta = \dfrac{\pi}{2} \Rightarrow \theta = \dfrac{\pi}{4}$

On putting $\theta = \dfrac{\pi}{4}$ in given determinant, we get

$$\begin{vmatrix} 0 & \cos\frac{\pi}{4} & \sin\frac{\pi}{4} \\ \cos\frac{\pi}{4} & \sin\frac{\pi}{4} & 0 \\ \sin\frac{\pi}{4} & 0 & \cos\frac{\pi}{4} \end{vmatrix}^2$$

Note (*) Topics related to these question have been removed from latest NCERT.

$$=\begin{vmatrix} 0 & \frac{1}{\sqrt{2}} & \frac{1}{\sqrt{2}} \\ \frac{1}{\sqrt{2}} & \frac{1}{\sqrt{2}} & 0 \\ \frac{1}{\sqrt{2}} & 0 & \frac{1}{\sqrt{2}} \end{vmatrix}^2 \qquad \left[\because \cos\frac{\pi}{4}=\sin\frac{\pi}{4}=\frac{1}{\sqrt{2}}\right]$$

On expanding the determinant along R_1, we get

$$\left[-\frac{1}{\sqrt{2}}\left(\frac{1}{2}\right)+\frac{1}{\sqrt{2}}\left(-\frac{1}{2}\right)\right]^2=\left[\frac{-2}{2\sqrt{2}}\right]^2-\left(\frac{1}{\sqrt{2}}\right)^2=\frac{1}{2}$$

Q.42 If A is a matrix of order 3×3, then $(A^2)^{-1}=$ ....

Ans. If A is a matrix of order 3×3, then $(A^2)^{-1}=(A^{-1})^2$.

Q.43 If A is a matrix of order 3×3, then the number of minors in determinant of A are ....

Ans. If A is a matrix of order 3×3, then the number of minors in determinant of A is 9.

[since, in a 3×3 matrix, there are 9 elements]

Q.44 The sum of products of elements of any row with the cofactors of corresponding elements is equal to ....

Ans. Let $\Delta=\begin{vmatrix} a_{11} & a_{12} & a_{13} \\ a_{21} & a_{22} & a_{23} \\ a_{31} & a_{32} & a_{33} \end{vmatrix}$

On expanding along R_1, we get

$$\Delta = a_{11}A_{11}+a_{12}A_{12}+a_{13}A_{13}$$

= Sum of products of elements of R_1 with their corresponding cofactors

$\therefore$ The sum of products of elements of any row with the cofactors of corresponding elements is equal to value of the determinant.

Q.45 If $x=-9$ is a root of $\begin{vmatrix} x & 3 & 7 \\ 2 & x & 2 \\ 7 & 6 & x \end{vmatrix}=0$, then other two roots are ....

Ans. Since, $\begin{vmatrix} x & 3 & 7 \\ 2 & x & 2 \\ 7 & 6 & x \end{vmatrix}=0$

On expanding along R_1, we get

$$x(x^2-12)-3(2x-14)+7(12-7x)=0$$

$$\Rightarrow \quad x^3-12x-6x+42+84-49x=0$$

$$\Rightarrow \quad x^3-67x+126=0 \qquad \text{...(i)}$$

Now, find the roots with the help of remainder theorem.

i.e., The expression $x^3-67x+126=0$ can be written as

$$x^2(x+9)-9x(x+9)+14(x+9)=0$$

$$\Rightarrow \quad (x+9)(x^2-9x+14)=0$$

$$\Rightarrow \quad (x+9)(x-7)(x-2)=0$$

Hence, roots will be $-9, 7$ and 2.

(-9) is the given root, therefore other roots will be 7 and 2.

Q.46 $\begin{vmatrix} 0 & xyz & x-z \\ y-x & 0 & y-z \\ z-x & z-y & 0 \end{vmatrix}$ is equal to ....

Ans. We have, $\begin{vmatrix} 0 & xyz & x-z \\ y-x & 0 & y-z \\ z-x & z-y & 0 \end{vmatrix}$

On expanding along R_3, we get

$(z-x)[(y-z)(xyz)-0]-(z-y)[0-(x-z)(y-x)]=0$
$=(z-x)(y-z)(xyz)+(y-z)(z-x)(y-x)$
$=(z-x)(y-z)[xyz+y-x]$

*** Q.47** If $f(x)=\begin{vmatrix} (1+x)^{17} & (1+x)^{19} & (1+x)^{23} \\ (1+x)^{23} & (1+x)^{29} & (1+x)^{34} \\ (1+x)^{41} & (1+x)^{43} & (1+x)^{47} \end{vmatrix}=A+Bx+Cx^2+\ldots$, **then** $A=\ldots.$

Ans. Since, $f(x)=(1+x)^{17}(1+x)^{23}(1+x)^{41}\begin{vmatrix} 1 & (1+x)^2 & (1+x)^6 \\ 1 & (1+x)^6 & (1+x)^{11} \\ 1 & (1+x)^2 & (1+x)^6 \end{vmatrix}=0$

[taking $(1+x)^{17}(1+x)^{23}(1+x)^{41}$ common from R_1, R_2 and R_3, respectively]
[since, R_1 and R_3 are identical
Then, value of determinant will be zero]

$\therefore \quad A=0$

State True or False for statements of the following exercises 48 to 58.

Q.48 $(A^3)^{-1}=(A^{-1})^3$, where A is a square matrix and $|A|\neq 0$.

Ans. *True*
Since, $(A^n)^{-1}=(A^{-1})^n$, where $n\in N$ and $|A|\neq 0$.

Q.49 $(aA)^{-1}=\frac{1}{a}A^{-1}$, where a is any real number and A is a square matrix.

Ans. *False*
Since, we know that, if A is a non-singular square matrix, then for any scalar a (non-zero), aA is invertible such that

$$(aA)\left(\frac{1}{a}A^{-1}\right)=\left(a\cdot\frac{1}{a}\right)(A\cdot A^{-1})$$
$$=I$$

i.e., (aA) is inverse of $\left(\frac{1}{a}A^{-1}\right)$ or $(aA)^{-1}=\frac{1}{a}A^{-1}$, where a is any non-zero scalar.

In the above statement a is any real number. So, we can conclude that above statement is false.

Note (*) Topics related to these question have been removed from latest NCERT.

Q.50 $|A^{-1}| \neq |A|^{-1}$, where A is a non-singular matrix.

Ans. *False*

$|A^{-1}| = |A|^{-1}$, where A is a non-singular matrix.

Q.51 If A and B are matrices of order 3 and $|A| = 5$, $|B| = 3$, then $|3AB| = 27 \times 5 \times 3 = 405$.

Ans. *True*

We know that, $|AB| = |A| \cdot |B|$ $\quad [\because |AB| = |A| \cdot |B|]$

$\therefore \quad |3AB| = 27\,|AB|$ $\quad [\because |kA| = k^n |A|]$

$= 27\,|A| \cdot |B|$

$= 27 \times 5 \times 3 = 405$

Q.52 If the value of a third order determinant is 12, then the value of the determinant formed by replacing each element by its cofactor will be 144.

Ans. *True*

Let A be the third order determinant.

$\therefore \quad |A| = 12$

Also, we know that, if A is a square matrix of order n, then $|\text{adj } A| = |A|^{n-1}$

For $n = 3$, $|\text{adj } A| = |A|^{3-1} = |A|^2 = (12)^2 = 144$

***Q.53** $\begin{vmatrix} x+1 & x+2 & x+a \\ x+2 & x+3 & x+b \\ x+3 & x+4 & x+c \end{vmatrix} = 0$, where a, b and c are in AP.

Ans. *True*

Since, a, b and c are in AP, then $2b = a + c$

$$\therefore \quad \begin{vmatrix} x+1 & x+2 & x+a \\ x+2 & x+3 & x+b \\ x+3 & x+4 & x+c \end{vmatrix} = 0$$

$$\Rightarrow \quad \begin{vmatrix} 2x+4 & 2x+6 & 2x+a+c \\ x+2 & x+3 & x+b \\ x+3 & x+4 & x+c \end{vmatrix} = 0 \qquad [\because R_1 \to R_1 + R_3]$$

$$\Rightarrow \quad \begin{vmatrix} 2(x+2) & 2(x+3) & 2(x+b) \\ x+2 & x+3 & x+b \\ x+3 & x+4 & x+c \end{vmatrix} = 0 \qquad [\because 2b = a + c]$$

$\Rightarrow \quad 0 = 0$ [since, R_1 and R_2 are in proportional to each other]

Hence, statement is true.

Q.54 $|\text{adj } A| = |A|^2$, where A is a square matrix of order two.

Ans. *False,*

If A is a square matrix of order n, then

$|\text{adj } A| = |A|^{n-1}$

$\Rightarrow \quad |\text{adj } A| = |A|^{2-1} = |A|$ $\quad [\because n = 2]$

Hence, given statement is false.

Note (✱) Topics related to these question have been removed from latest NCERT.

***Q.55** The determinant $\begin{vmatrix} \sin A & \cos A & \sin A + \cos B \\ \sin B & \cos A & \sin B + \cos B \\ \sin C & \cos A & \sin C + \cos B \end{vmatrix}$ is equal to zero.

Ans. *True*

We have, $\begin{vmatrix} \sin A & \cos A & \sin A + \cos B \\ \sin B & \cos A & \sin B + \cos B \\ \sin C & \cos A & \sin C + \cos B \end{vmatrix}$

$$= \begin{vmatrix} \sin A & \cos A & \sin A \\ \sin B & \cos A & \sin B \\ \sin C & \cos A & \sin C \end{vmatrix} + \begin{vmatrix} \sin A & \cos A & \cos B \\ \sin B & \cos A & \cos B \\ \sin C & \cos A & \cos B \end{vmatrix}$$

$$= \begin{vmatrix} \sin A & \cos A & \sin A \\ \sin B & \cos A & \sin B \\ \sin C & \cos A & \sin C \end{vmatrix} + \cos A \cdot \cos B \begin{vmatrix} \sin A & 1 & 1 \\ \sin B & 1 & 1 \\ \sin C & 1 & 1 \end{vmatrix}$$

$$= 0 + 0 = 0$$

[since, in first determinant C_1 and C_3 are identical and in second determinant C_2 and C_3 are identical therefore the value of both determinant will be zero]

***Q.56** If the determinant $\begin{vmatrix} x+a & p+u & l+f \\ y+b & q+v & m+g \\ z+c & r+w & n+h \end{vmatrix}$ splits into exactly k determinants of order 3, each element of which contains only one term, then the value of K is 8.

Ans. *True*

We have, $\begin{vmatrix} x+a & p+u & l+f \\ y+b & q+v & m+g \\ z+c & r+w & n+h \end{vmatrix}$

$$= \begin{vmatrix} x & p & l \\ y+b & q+v & m+g \\ z+c & r+w & n+h \end{vmatrix} + \begin{vmatrix} a & u & f \\ y+b & q+v & m+g \\ z+c & r+w & n+h \end{vmatrix}$$ [splitting first row]

$$= \begin{vmatrix} x & p & l \\ y & q & m \\ z+c & r+w & n+h \end{vmatrix} + \begin{vmatrix} x & p & l \\ b & v & g \\ z+c & r+w & n+h \end{vmatrix}$$

$$+ \begin{vmatrix} a & u & f \\ y & q & m \\ z+c & r+w & n+h \end{vmatrix} + \begin{vmatrix} a & u & f \\ b & v & g \\ z+c & r+w & n+h \end{vmatrix}$$ [splitting second row]

Similarly, we can split these 4 determinants in 8 determinants by splitting each one in two determinants further. So, given statement is true.

Note (✱) Topics related to these question have been removed from latest NCERT.

***Q.57** If $\Delta = \begin{vmatrix} a & p & x \\ b & q & y \\ c & r & z \end{vmatrix} = 16$, then $\Delta_1 = \begin{vmatrix} p+x & a+x & a+p \\ q+y & b+y & b+q \\ r+z & c+z & c+r \end{vmatrix} = 32$.

Ans. *True*

We have, $$\Delta = \begin{vmatrix} a & p & x \\ b & q & y \\ c & r & z \end{vmatrix} = 16$$

and we have to prove, $$\Delta_1 = \begin{vmatrix} p+x & a+x & a+p \\ q+y & b+y & b+q \\ r+z & c+z & c+r \end{vmatrix} = 32$$

Taking $$\Delta_1 = \begin{vmatrix} p+x & a+x & a+p \\ q+y & b+y & b+q \\ r+z & c+z & c+r \end{vmatrix}$$

$$\Delta_1 = \begin{vmatrix} 2p+2x+2a & a+x & a+p \\ 2q+2y+2b & b+y & b+q \\ 2r+2z+2c & c+z & c+r \end{vmatrix} \quad [\because C_1 \to C_1 + C_2 + C_3]$$

$$= 2\begin{vmatrix} p & x-p & a+p \\ q & y-q & b+q \\ r & z-r & c+r \end{vmatrix}$$

[taking 2 common from C_1 and then $C_1 \to C_1 - C_2$, $C_2 \to C_2 - C_3$]

$$= 2\left[\begin{vmatrix} p & x & a+p \\ q & y & b+q \\ r & z & c+r \end{vmatrix} - \begin{vmatrix} p & p & a+p \\ q & q & b+q \\ r & r & c+r \end{vmatrix}\right]$$

$$= 2\begin{vmatrix} p & x & a+p \\ q & y & b+q \\ r & z & c+r \end{vmatrix} - 0$$

[since, two columns C_1 and C_2 are identicals in 2nd determinant]

$$= 2\begin{vmatrix} p & x & a \\ q & y & b \\ r & z & c \end{vmatrix} + 2\begin{vmatrix} p & x & p \\ q & y & q \\ r & z & r \end{vmatrix}$$

$$= 2\begin{vmatrix} a & p & x \\ b & q & y \\ c & r & z \end{vmatrix} + 0$$

[since, C_1 and C_3 are identical in second determinant and in first determinant, $C_1 \leftrightarrow C_2$ and then $C_1 \leftrightarrow C_3$]

$$= 2 \times 16 \quad [\because \Delta = 16]$$
$$= 32$$

Hence proved.

Note (✱) Topics related to these question have been removed from latest NCERT.

Q.58 The maximum value of $\begin{vmatrix} 1 & 1 & 1 \\ 1 & 1+\sin\theta & 1 \\ 1 & 1 & 1+\cos\theta \end{vmatrix}$ is $\frac{1}{2}$.

Ans. *True*

Let $\Delta = \begin{vmatrix} 1 & 1 & 1 \\ 1 & 1+\sin\theta & 1 \\ 1 & 1 & 1+\cos\theta \end{vmatrix}$

$\therefore$ On expanding along R_1, we get

$$\Delta = 1[(1+\sin\theta)(1+\cos\theta) - 1] - 1[(1+\cos\theta) - 1] + 1[1-(1+\sin\theta)]$$

$$= [1+\cos\theta + \sin\theta + \sin\theta\cdot\cos\theta - 1] - [\cos\theta] + [-\sin\theta]$$

$$= [\cos\theta + \sin\theta - \cos\theta - \sin\theta + \sin\theta\cdot\cos\theta]$$

$$= \sin\theta\cdot\cos\theta$$

$$= \frac{1}{2}\times 2\sin\theta\cdot\cos\theta \qquad [\because \text{multiply and divide by 2}]$$

$$= \frac{1}{2}\sin 2\theta \qquad [\because \sin 2\theta = 2\sin\theta\cos\theta]$$

$$\therefore \Delta_{max} = \frac{1}{2}\times 1 \qquad [\because -1 \le \sin 2\theta \le 1]$$

$$= \frac{1}{2}$$

Chapter

05 Continuity and Differentiability

Short Answer Type Questions

Q.1 Examine the continuity of the function $f(x) = x^3 + 2x^2 - 1$ at $x = 1$.

Thinking Process We know that, function f will be continuous at $x = a$, if $\lim_{x \to a^-} f(x) = \lim_{x \to a^+} f(x) = f(a)$.

Ans. We have, $f(x) = x^3 + 2x^2 - 1$ at $x = 1$.

Put $x = 1 + h$ $\quad \lim_{x \to 1^+} f(x) = \lim_{h \to 0} (1+h)^3 + 2(1+h)^2 - 1 = 2$

Put $x = 1 - h$ $\quad \lim_{x \to 1^-} f(x) = \lim_{h \to 0} (1-h)^3 + 2(1-h)^2 - 1 = 2$

$\therefore \quad \lim_{x \to 1^+} f(x) = \lim_{x \to 1^-} f(x) = f(1)$ ...(i)

and $\quad f(1) = 1 + 2 - 1 = 2$

So, $f(x)$ is continuous at $x = 1$.

Note Every polynomial function is continuous at any real point.

Find which of the functions from exercise 2 to 10 is continuous or discontinuous at the indicated points.

Q.2 $f(x) = \begin{cases} 3x + 5, & \text{if } x \geq 2 \\ x^2, & \text{if } x < 2 \end{cases}$ at $x = 2$.

Ans. We have, $\quad f(x) = \begin{cases} 3x + 5, & \text{if } x \geq 2 \\ x^2, & \text{if } x < 2 \end{cases}$ at $x = 2$.

At $x = 2$, $\quad \text{LHL} = \lim_{x \to 2^-} (x)^2$

Put $x = 2 - h$, $\quad = \lim_{h \to 0} (2-h)^2 = \lim_{h \to 0} (4 + h^2 - 4h) = 4$

and $\quad \text{RHL} = \lim_{x \to 2^+} (3x + 5)$ $\quad [\because (a-b)^2 = a^2 + b^2 - 2ab]$

Put $x = 2 + h$, $\quad = \lim_{h \to 0} [3(2+h) + 5] = 11$

Since, $\quad \text{LHL} \neq \text{RHL}$ at $x = 2$

So, $\quad f(x)$ is discontinuous at $x = 2$.

Q.3 $f(x)=\begin{cases}\dfrac{1-\cos 2x}{x^2}, & \text{if } x\neq 0\\ 5, & \text{if } x=0\end{cases}$ at $x=0$.

Ans. We have, $f(x)=\begin{cases}\dfrac{1-\cos 2x}{x^2}, & \text{if } x\neq 0\\ 5, & \text{if } x=0\end{cases}$ at $x=0$.

At $x=0$, LHL $= \lim\limits_{x\to 0^-} \dfrac{1-\cos 2x}{x^2}$

Put $x=0-h$, $= \lim\limits_{h\to 0} \dfrac{1-\cos 2(0-h)}{(0-h)^2}$

$= \lim\limits_{h\to 0} \dfrac{1-\cos 2h}{h^2}$ [$\because \cos(-\theta)=\cos\theta$]

$= \lim\limits_{h\to 0} \dfrac{1-1+2\sin^2 h}{h^2}$ [$\because \cos 2\theta = 1-2\sin^2\theta$]

$= \lim\limits_{h\to 0} \dfrac{2(\sin h)^2}{(h)^2}$ $\left[\because \lim\limits_{h\to 0}\dfrac{\sin h}{h}=1\right]$

$=2$

RHL $= \lim\limits_{x\to 0^+} \dfrac{1-\cos 2x}{x^2}$

Put $x=0+h$, $= \lim\limits_{h\to 0} \dfrac{1-\cos 2(0+h)}{(0+h)^2}$

$= \lim\limits_{h\to 0} \dfrac{2\sin^2 h}{h^2} = 2$ $\left[\because \lim\limits_{h\to 0}\dfrac{\sin h}{h}=1\right]$

and $f(0)=5$

Since, LHL = RHL $\neq f(0)$

Hence, $f(x)$ is not continuous at $x=0$.

Q.4 $f(x)=\begin{cases}\dfrac{2x^2-3x-2}{x-2}, & \text{if } x\neq 2\\ 5, & \text{if } x=2\end{cases}$ at $x=2$.

Ans. We have, $f(x)=\begin{cases}\dfrac{2x^2-3x-2}{x-2}, & \text{if } x\neq 2\\ 5, & \text{if } x=2\end{cases}$ at $x=2$

At $x=2$, LHL $= \lim\limits_{x\to 2^-} \dfrac{2x^2-3x-2}{x-2}$

Put $x=2-h$, $= \lim\limits_{h\to 0} \dfrac{2(2-h)^2-3(2-h)-2}{(2-h)-2}$

$= \lim\limits_{h\to 0} \dfrac{8+2h^2-8h-6+3h-2}{-h}$

$= \lim\limits_{h\to 0} \dfrac{2h^2-5h}{-h} = \lim\limits_{h\to 0} \dfrac{h(2h-5)}{-h} = 5$

$$\text{RHL} = \lim_{x \to 2^+} \frac{2x^2 - 3x - 2}{x - 2}$$

Put $x = 2 + h$,
$$= \lim_{h \to 0} \frac{2(2+h)^2 - 3(2+h) - 2}{(2+h) - 2}$$
$$= \lim_{h \to 0} \frac{8 + 2h^2 + 8h - 6 - 3h - 2}{h}$$
$$= \lim_{h \to 0} \frac{2h^2 + 5h}{h} = \lim_{h \to 0} \frac{h(2h+5)}{h} = 5$$

and $f(2) = 5$

$\therefore$ LHL = RHL = $f(2)$

So, $f(x)$ is continuous at $x = 2$.

Alternate Method

The function can be simplified as
$$\frac{2x^2 - 3x - 2}{(x-2)} = \frac{2x^2 - 4x + x - 2}{x - 2} = \frac{(2x+1)\ (x-2)}{x-2} = 2x + 1$$
Then, we can find LHL and RHL to prove $f(x)$ is continuous at $x = 2$, as same as above.

Q.5 $f(x) = \begin{cases} \dfrac{|x-4|}{2(x-4)}, & \text{if } x \neq 4 \\ 0, & \text{if } x = 4 \end{cases}$ at $x = 4$.

Ans. We have, $f(x) = \begin{cases} \dfrac{|x-4|}{2(x-4)}, & \text{if } x \neq 4 \\ 0, & \text{if } x = 4 \end{cases}$ at $x = 4$

At $x = 4$,
$$\text{LHL} = \lim_{x \to 4^-} \frac{|x-4|}{2(x-4)}$$

Put $x = 4 - h$,
$$= \lim_{h \to 0} \frac{|4 - h - 4|}{2[(4-h) - 4]} = \lim_{h \to 0} \frac{|0 - h|}{(8 - 2h - 8)}$$
$$= \lim_{h \to 0} \frac{h}{-2h} = \frac{-1}{2} \quad \text{and} \quad f(4) = 0 \neq \text{LHL} \quad \text{[given]}$$

So, $f(x)$ is discontinuous at $x = 4$.

Q.6 $f(x) = \begin{cases} |x| \cos \dfrac{1}{x}, & \text{if } x \neq 0 \\ 0, & \text{if } x = 0 \end{cases}$ at $x = 0$.

Ans. We have, $f(x) = \begin{cases} |x| \cos \dfrac{1}{x}, & \text{if } x \neq 0 \\ 0, & \text{if } x = 0 \end{cases}$ at $x = 0$

At $x = 0$,
$$\text{LHL} = \lim_{x \to 0^-} |x| \cos \frac{1}{x} = \lim_{h \to 0} |0 - h| \cos \frac{1}{0 - h}$$

Put $x = 0 - h$,
$$= \lim_{h \to 0} h \cos \left(\frac{-1}{h}\right)$$
$$= 0 \times [\text{an oscillating number between } -1 \text{ and } 1] = 0$$

$$\text{RHL} = \lim_{x \to 0^+} |x| \cos \frac{1}{x}$$

$$= \lim_{h \to 0} |0 + h| \cos \frac{1}{(0+h)}$$

$$= \lim_{h \to 0} h \cos \frac{1}{h}$$

$= 0 \times$ [an oscillating number between -1 and 1] $= 0$

and $f(0) = 0$ [given]

Since, LHL = RHL = $f(0)$

So, $f(x)$ is continuous at $x = 0$.

Note In Mathematics, the oscillation number is quantifies how much function varies between its extreme values.

Q.7 $f(x) = \begin{cases} |x-a| \sin \dfrac{1}{x-a}, & \text{if } x \neq 0 \\ 0, & \text{if } x = a \end{cases}$ at $x = a$.

Ans. We have, $f(x) = \begin{cases} |x-a| \sin \dfrac{1}{x-a}, & \text{if } x \neq 0 \\ 0, & \text{if } x = a \end{cases}$ at $x = a$

At $x = a$, $\text{LHL} = \lim_{x \to a^-} |x-a| \sin \dfrac{1}{x-a}$

Put $x = a - h$, $= \lim_{h \to 0} |a-h-a| \sin \left(\dfrac{1}{a-h-a}\right)$

$= \lim_{h \to 0} -h \sin \left(\dfrac{1}{h}\right)$ $[\because \sin(-\theta) = -\sin\theta]$

$= 0 \times$ [an oscillating number between -1 and 1]

$= 0$

$\text{RHL} = \lim_{x \to a^+} |x-a| \sin \left(\dfrac{1}{x-a}\right)$

Put $x = a + h$, $= \lim_{h \to 0} |a+h-a| \sin \left(\dfrac{1}{a+h-a}\right)$

$= \lim_{h \to 0} h \sin \dfrac{1}{h}$

$= 0 \times$ [an oscillating number between -1 and 1]

$= 0$

and $f(a) = 0$ [given]

$\therefore$ LHL = RHL = $f(a)$

So, $f(x)$ is continuous at $x = a$.

Q.8 $f(x) = \begin{cases} \dfrac{e^{1/x}}{1+e^{1/x}}, & \text{if } x \neq 0 \\ 0, & \text{if } x = 0 \end{cases}$ at $x = 0$.

Ans. We have, $f(x) = \begin{cases} \dfrac{e^{1/x}}{1+e^{1/x}}, & \text{if } x \neq 0 \\ 0, & \text{if } x = 0 \end{cases}$ at $x = 0$

At $x = 0$, $\text{LHL} = \lim_{x \to 0^-} \dfrac{e^{1/x}}{1+e^{1/x}} = \lim_{h \to 0} \dfrac{e^{1/(0-h)}}{1+e^{1/(0-h)}}$

Put $x = 0 - h$, $= \lim_{h \to 0} \dfrac{e^{-1/h}}{1+e^{-1/h}} = \lim_{h \to 0} \dfrac{1}{e^{1/h}(1+e^{-1/h})}$

$= \lim_{h \to 0} \dfrac{1}{e^{1/h}+1} = \dfrac{1}{e^{\infty}+1} = \dfrac{1}{\infty+1} = \dfrac{1}{\infty} = 0$ $[\because e^{\infty} = \infty]$

$\text{RHL} = \lim_{x \to 0^+} \dfrac{e^{1/x}}{1+e^{1/x}}$

Put $x = 0 + h$, $= \lim_{h \to 0} \dfrac{e^{1/(0+h)}}{1+e^{1/(0+h)}} = \lim_{h \to 0} \dfrac{e^{1/h}}{1+e^{1/h}} = \lim_{h \to 0} \dfrac{1}{e^{-1/h}+1} = \dfrac{1}{e^{-\infty}+1}$

$= \dfrac{1}{0+1} = 1$ $[\because e^{-\infty} = 0]$

Hence, LHL $\neq$ RHL at $x = 0$.
So, $f(x)$ is discontinuous at $x = 0$.

Q.9 $f(x) = \begin{cases} \dfrac{x^2}{2}, & \text{if } 0 \le x \le 1 \\ 2x^2 - 3x + \dfrac{3}{2}, & \text{if } 1 < x \le 2 \end{cases}$ at $x = 1$.

Ans. We have, $f(x) = \begin{cases} \dfrac{x^2}{2}, & \text{if } 0 \le x \le 1 \\ 2x^2 - 3x + \dfrac{3}{2}, & \text{if } 1 < x \le 2 \end{cases}$ at $x = 1$

At $x = 1$, $\text{LHL} = \lim_{x \to 1^-} \dfrac{x^2}{2} = \lim_{h \to 0} \dfrac{(1-h)^2}{2}$

Put $x = 1 - h$, $= \lim_{h \to 0} \dfrac{1+h^2-2h}{2} = \dfrac{1}{2}$ $[\because (a-b)^2 = a^2 + b^2 - 2ab]$

$\text{RHL} = \lim_{x \to 1^+} \left(2x^2 - 3x + \dfrac{3}{2}\right)$

Put $x = 1 + h$, $= \lim_{h \to 0} \left[2(1+h)^2 - 3(1+h) + \dfrac{3}{2}\right] = \lim_{h \to 0} \left(2 + 2h^2 + 4h - 3 - 3h + \dfrac{3}{2}\right)$

$= -1 + \dfrac{3}{2} = \dfrac{1}{2}$ $[\because (a+b)^2 = a^2 + b^2 + 2ab]$

and $f(1) = \dfrac{1^2}{2} = \dfrac{1}{2}$

$\therefore$ LHL = RHL = $f(1)$

Hence, $f(x)$ is continuous at $x = 1$.

Q.10 $f(x) = |x| + |x-1|$ at $x = 1$.

Ans. We have, $f(x) = |x| + |x-1|$ at $x = 1$

At $x = 1$, LHL $= \lim_{x \to 1^-} [|x| + |x-1|]$

Put $x = 1 - h$, $= \lim_{h \to 0} [|1-h| + |1-h-1|] = 1 + 0 = 1$

and RHL $= \lim_{x \to 1^-} [|x| + |x-1|]$

Put $x = 1 + h$, $= \lim_{h \to 0} [|1+h| + |1+h-1|] = 1 + 0 = 1$ and $f(1) = |1| + |0| = 1$

$\therefore$ LHL = RHL = $f(1)$

Hence, $f(x)$ is continuous at $x = 1$.

Note Every modulus function is a continuous function at any real point.

Find the value of k in each of the exercises 11 to 14 so that the function f is continuous at the indicated point :

Q.11 $f(x) = \begin{cases} 3x - 8, & \text{if } x \le 5 \\ 2k, & \text{if } x > 5 \end{cases}$ at $x = 5$.

Ans. We have, $f(x) = \begin{cases} 3x - 8, & \text{if } x \le 5 \\ 2k, & \text{if } x > 5 \end{cases}$ at $x = 5$

Now at $x = 5$ LHL $= \lim_{x \to 5^-} (3x - 8) = \lim_{h \to 0} [3(5-h) - 8]$

Put $x = 5 - h$ $= \lim_{h \to 0} [15 - 3h - 8] = 7$

RHL $= \lim_{x \to 5^+} 2k = 2k$ and $f(5) = 3 \times 5 - 8 = 7$

Since, $f(x)$ is continuous at $x = 5$.

$\therefore$ LHL = RHL = $f(5)$

$\therefore$ $2k = 7 \Rightarrow k = \frac{7}{2}$

Q.12 $f(x) = \begin{cases} \dfrac{2^{x+2} - 16}{4^x - 16}, & \text{if } x \ne 2 \\ k, & \text{if } x = 2 \end{cases}$ at $x = 2$.

Ans. We have, $f(x) = \begin{cases} \dfrac{2^{x+2} - 16}{4^x - 16}, & \text{if } x \ne 2 \\ k, & \text{if } x = 2 \end{cases}$ at $x = 2$

At $x = 2$, $\lim_{x \to 2} \dfrac{2^x \cdot 2^2 - 2^4}{4^x - 4^2} = \lim_{x \to 2} \dfrac{4 \cdot (2^x - 4)}{(2^x)^2 - (4)^2}$

$= \lim_{x \to 2} \dfrac{4 \cdot (2^x - 4)}{(2^x - 4)(2^x + 4)}$ $[\because a^2 - b^2 = (a+b)(a-b)]$

$= \lim_{x \to 2} \dfrac{4}{2^x + 4} = \dfrac{4}{8} = \dfrac{1}{2}$

Since, $f(x)$ is continuous at $x = 2$.

$\therefore$ $$\text{LHL} = \text{RHL} = f(2)$$

But $$f(2) = k$$

$\therefore$ $$k = \frac{1}{2} \quad [\because f(x) \text{ is continuous at } x = 2]$$

Q.13 $f(x) = \begin{cases} \dfrac{\sqrt{1+kx} - \sqrt{1-kx}}{x}, & \text{if } -1 \le x < 0 \\ \dfrac{2x+1}{x-1}, & \text{if } \ 0 \le x \le 1 \end{cases}$ at $x = 0$.

Ans. We have, $f(x) = \begin{cases} \dfrac{\sqrt{1+kx} - \sqrt{1-kx}}{x}, & \text{if } -1 \le x < 0 \\ \dfrac{2x+1}{x-1}, & \text{if } \ 0 \le x \le 1 \end{cases}$ at $x = 0$.

$\therefore$ $$\text{LHL} = \lim_{x \to 0^-} \frac{\sqrt{1+kx} - \sqrt{1-kx}}{x}$$

$$= \lim_{x \to 0^-} \left(\frac{\sqrt{1+kx} - \sqrt{1-kx}}{x}\right) \cdot \left(\frac{\sqrt{1+kx} + \sqrt{1-kx}}{\sqrt{1+kx} + \sqrt{1-kx}}\right)$$

$[\because$ multiply and divided by $\sqrt{1+kx} + \sqrt{1-kx}]$

$$= \lim_{x \to 0^-} \frac{1 + kx - 1 + kx}{x[\sqrt{1+kx} + \sqrt{1-kx}]} \quad [\because a^2 - b^2 = (a+b)(a-b)]$$

$$= \lim_{x \to 0^-} \frac{2kx}{x\sqrt{1+kx} + \sqrt{1-kx}}$$

Put $x = 0 + h$, $$= \lim_{h \to 0} \frac{2k}{\sqrt{1 + k(0-h)} + \sqrt{1 - k(0-h)}}$$

$$= \lim_{h \to 0} \frac{2k}{\sqrt{1-kh} + \sqrt{1+kh}} = \frac{2k}{2} = k$$

and $$f(0) = \frac{2 \times 0 + 1}{0 - 1} = -1$$

$\Rightarrow$ $$k = -1 \quad [\because \text{LHL} = \text{RHL} = f(0)]$$

Alternate Method

It can be solved as LHL $= \lim_{x \to 0^-} \dfrac{\sqrt{1+kx} - \sqrt{1-kx}}{x}$

$$= \lim_{x \to 0^-} \frac{(1+kx)^{1/2} - (1-kx)^{1/2}}{x}$$

As $(1+x)^n = 1 + nx$, then

$$= \lim_{x \to 0^-} \frac{\left(1 + \frac{1}{2}kx\right) - \left(1 - \frac{kx}{2}\right)}{x}$$

$$= \lim_{x \to 0^-} \frac{kx}{x} = k$$

Q.14 $f(x) = \begin{cases} \dfrac{1-\cos kx}{x \sin x}, & \text{if } x \neq 0 \\ \dfrac{1}{2}, & \text{if } x = 0 \end{cases}$ at $x = 0$.

Ans. We have, $f(x) = \begin{cases} \dfrac{1-\cos kx}{x \sin x}, & \text{if} x \neq 0 \\ \dfrac{1}{2}, & \text{if} x = 0 \end{cases}$ at $x = 0$

At $x = 0$, $\text{LHL} = \lim_{x \to 0^-} \dfrac{1-\cos kx}{x \sin x}$

Put $x = 0 - h$, $= \lim_{h \to 0} \dfrac{1-\cos k(0-h)}{(0-h)\sin(0-h)} = \lim_{h \to 0} \dfrac{1-\cos(-kh)}{-h\sin(-h)}$

$= \lim_{h \to 0} \dfrac{1-\cos kh}{h \sin h}$ [$\because \cos(-\theta) = \cos\theta, \sin(-\theta) = -\sin\theta$]

$= \lim_{h \to 0} \dfrac{1-1+2\sin^2 \dfrac{kh}{2}}{h \sin h}$ $\left[\because \cos\theta = 1 - 2\sin^2\dfrac{\theta}{2}\right]$

$= \lim_{h \to 0} \dfrac{2\sin^2 \dfrac{kh}{2}}{h \sin h} = \lim_{h \to 0} \dfrac{2\sin\dfrac{kh}{2}}{\dfrac{kh}{2}} \cdot \dfrac{\sin\dfrac{kh}{2}}{\dfrac{kh}{2}} \cdot \dfrac{1}{\dfrac{\sin h}{h}} \cdot \dfrac{k^2h/4}{h}$

$= \dfrac{2k^2}{4} = \dfrac{k^2}{2}$ $\left[\because \lim_{h \to 0} \dfrac{\sin h}{h} = 1\right]$

Also, $f(0) = \dfrac{1}{2} \Rightarrow \dfrac{k^2}{2} = \dfrac{1}{2} \Rightarrow k = \pm 1$ [$\because \text{LHL} = \text{RHL} = f(0)$]

Q.15 Prove that the function f defined by $f(x) = \begin{cases} \dfrac{x}{|x| + 2x^2}, & \text{if } x \neq 0 \\ k, & \text{if } x = 0 \end{cases}$ remains discontinuous at $x = 0$, regardless the choice of k.

Ans. We have, $f(x) = \begin{cases} \dfrac{x}{|x| + 2x^2}, & \text{if } x \neq 0 \\ k, & \text{if } x = 0 \end{cases}$

At $x = 0$, $\text{LHL} = \lim_{x \to 0^-} \dfrac{x}{|x| + 2x^2}$

Put $x = 0 - h$, $= \lim_{h \to 0} \dfrac{(0-h)}{|0-h| + 2(0-h)^2} = \lim_{h \to 0} \dfrac{-h}{h + 2h^2} = \lim_{h \to 0} \dfrac{-h}{h(1+2h)} = -1$

$\text{RHL} = \lim_{x \to 0^+} \dfrac{x}{|x| + 2x^2}$

Put $x = 0 - h$, $= \lim_{h \to 0} \dfrac{0+h}{|0+h| + 2(0+h)^2} = \lim_{h \to 0} \dfrac{h}{h + 2h^2} = \lim_{h \to 0} \dfrac{h}{h(1+2h)} = 1$

and $f(0) = k$

Since, $\text{LHL} \neq \text{RHL}$ for any value of k.

Hence, $f(x)$ is discontinuous at $x = 0$ regardless the choice of k.

Q.16 Find the values of a and b such that the function f defined by

$$f(x)=\begin{cases}\dfrac{x-4}{|x-4|}+a, & \text{if } x<4\\ a+b, & \text{if } x=4\\ \dfrac{x-4}{|x-4|}+b, & \text{if } x>4\end{cases}$$

is a continuous function at $x=4$.

Ans. We have,
$$f(x)=\begin{cases}\dfrac{x-4}{|x-4|}+a, & \text{if } x<4\\ a+b, & \text{if } x=4\\ \dfrac{x-4}{|x-4|}+b, & \text{if } x>4\end{cases}$$

At $x=4$, $\quad \text{LHL}=\lim\limits_{x\to 4^-}\dfrac{x-4}{|x-4|}+a$

Put $x=4-h$, $\quad =\lim\limits_{h\to 0}\dfrac{4-h-4}{|4-h-4|}+a=\lim\limits_{h\to 0}\dfrac{-h}{h}+a=-1+a$

$\text{RHL}=\lim\limits_{x\to 4^+}\dfrac{x-4}{|x-4|}+b$

Put $x=4+h$, $\quad =\lim\limits_{h\to 0}\dfrac{4+h-4}{|4+h-4|}+b=\lim\limits_{h\to 0}\dfrac{h}{h}+b=1+b$

$\therefore \quad \text{LHL}=\text{RHL}=f(4)$

$\Rightarrow \quad -1+a=1+b=a+b$

Now, $\quad -1+a=a+b$ and $1+b=a+b$

Hence, $b=-1$ and $a=1$

Q.17 If the function $f(x)=\dfrac{1}{x+2}$, then find the points of discontinuity of the composite function $y=f\{f(x)\}$.

Ans. We have, $\quad f(x)=\dfrac{1}{x+2}$

$\therefore \quad y=f\{f(x)\}$

$$=f\left(\frac{1}{x+2}\right)=\frac{1}{\dfrac{1}{x+2}+2}=\frac{1}{\dfrac{1+2(x+2)}{x+2}}=\frac{x+2}{1+2x+4}$$

$$=\frac{1}{1+2x+4}\cdot(x+2)=\frac{(x+2)}{(2x+5)}$$

So, the function y will not be continuous at those points, because it is not defined as it is a rational function.

Therefore, $y=\dfrac{x+2}{(2x+5)}$ is not defined, when $2x+5=0$

$\therefore \quad x=\dfrac{-5}{2}$

Hence, y is discontinuous at $x=\dfrac{-5}{2}$.

Q.18 Find all points of discontinuity of the function $f(t)=\frac{1}{t^2+t-2}$, where $t=\frac{1}{x-1}$.

Ans. We have, $f(t)=\frac{1}{t^2+t-2}$ and $t=\frac{1}{x-1}$

On putting $t=\frac{1}{x-1}$ in $f(t)$, we get

$$f(t)=\frac{1}{\left(\frac{1}{x-1}\right)^2+\left(\frac{1}{x-1}\right)-\frac{2}{1}}$$

$$=\frac{1}{\frac{1}{(x-1)^2}+\frac{1}{(x-1)}-\frac{2}{1}}$$

$$=\frac{1}{\frac{1+x-1-2(x-1)^2}{(x-1)^2}}$$

$$=\frac{(x-1)^2}{1+x-1-2(x^2+1-2x)} \quad [\because (a-b)^2=a^2+b^2-2ab]$$

$$=\frac{(x-1)^2}{1+x-1-2x^2-2+4x}=\frac{(x-1)^2}{-2x^2+5x-2}$$

$$=\frac{(x-1)^2}{-(2x^2-5x+2)}=\frac{(x-1)^2}{-(2x-1)(x-2)}$$

$$=\frac{(x-1)^2}{(2x-1)(2-x)}$$

So, $f(t)$ is discontinuous at $2x-1=0 \Rightarrow x=1/2$

and $2-x=0 \Rightarrow x=2$.

Q.19 Show that the function $f(x)=|\sin x+\cos x|$ is continuous at $x=\pi$.

Ans. We have, $f(x)=|\sin x+\cos x|$ at $x=\pi$

Let $g(x)=\sin x+\cos x$

and $h(x)=|x|$

$\therefore$ $hog(x)=h[g(x)]$

$=h(\sin x+\cos x)$

$=|\sin x+\cos x|$

Since, $g(x)=\sin x+\cos x$ is a continuous function as it is forming with addition of two continuous functions $\sin x$ and $\cos x$.

Also, $h(x)=|x|$ is also a continuous function. Since, we know that composite functions of two continuous functions is also a continuous function.

Hence, $f(x)=|\sin x+\cos x|$ is a continuous function everywhere.

So, $f(x)$ is continuous at $x=\pi$.

Note Modulus function is always continuous.

Q.20 Examine the differentiability of f, where f is defined by

$$f(x)=\begin{cases} x[x], & \text{if } 0\le x<2 \\ (x-1)x, & \text{if } 2\le x<3 \end{cases} \text{ at } x=2.$$

Thinking Process We know that, a function f is differentiable at a point a in its domain, if both $Lf'(a)$ and $Rf'(a)$ are finite and equal, where $Lf'(c)=\lim_{h\to 0}\frac{f(a-h)-f(a)}{-h}$ and $Rf'(c)=\lim_{h\to 0}\frac{f(a+h)-f(a)}{h}$.

Ans. We have, $f(x)=\begin{cases} x[x], & \text{if } 0\le x<2 \\ (x-1)x & \text{if } 2\le x<3 \end{cases}$ at $x=2$

At $x=2$, $Lf'(2)=\lim_{h\to 0}\frac{f(2-h)-f(2)}{-h}$

$$=\lim_{h\to 0}\frac{(2-h)[2-h]-(2-1)2}{-h}$$

$\{\because [a-h]=a-1$, where a is any positive number$\}$

$$=\lim_{h\to 0}\frac{(2-h)(1)-2}{-h}$$

$$=\lim_{h\to 0}\frac{2-h-2}{-h}=\lim_{h\to 0}\frac{-h}{-h}=1$$

$$Rf'(2)=\lim_{h\to 0}\frac{f(2+h)-f(2)}{h}$$

$$=\lim_{h\to 0}\frac{(2+h-1)(2+h)-(2-1)\cdot 2}{h}$$

$$=\lim_{h\to 0}\frac{(1+h)(2+h)-2}{h}=\lim_{h\to 0}\frac{2+h+2h+h^2-2}{h}$$

$$=\lim_{h\to 0}\frac{h^2+3h}{h}=\lim_{h\to 0}\frac{h(h+3)}{h}=3$$

$\therefore$ $Lf'(2)\ne Rf'(2)$

So, $f(x)$ is not differentiable at $x=2$.

Q.21 $f(x)=\begin{cases} x^2\sin\frac{1}{x}, & \text{if } x\ne 0 \\ 0, & \text{if } x=0 \end{cases}$ at $x=0$.

Ans. We have, $f(x)=\begin{cases} x^2\sin\frac{1}{x}, & \text{if } x\ne 0 \\ 0, & \text{if } x=0 \end{cases}$ at $x=0$

For differentiability at $x=0$,

$$Lf'(0)=\lim_{x\to 0^-}\frac{f(x)-f(0)}{x-0}$$

$$=\lim_{x\to 0^-}\frac{x^2\sin\frac{1}{x}-0}{x-0}$$

Put $x = 0 - h$, $$= \lim_{h \to 0} \frac{(0-h)^2 \sin\left(\frac{1}{0-h}\right)}{0-h-0} = \lim_{h \to 0} \frac{h^2 \sin\left(\frac{-1}{h}\right)}{-h}$$

$$= \lim_{h \to 0} h \sin\left(\frac{1}{h}\right) \qquad [\because \sin(-\theta) = -\sin\theta]$$

$$= 0 \times [\text{an oscillating number between } -1 \text{and} 1] = 0$$

$$Rf'(0) = \lim_{x \to 0^+} \frac{f(x) - f(0)}{x - 0} = \lim_{x \to 0^+} \frac{x^2 \sin \frac{1}{x} - 0}{x - 0}$$

Put $x = 0 + h$, $$= \lim_{h \to 0} \frac{(0+h)^2 \sin\left(\frac{1}{0+h}\right)}{0+h-0} = \lim_{h \to 0} \frac{h^2 \sin(1/h)}{h} = \lim_{h \to 0} h \sin(1/h)$$

$$= 0 \times [\text{an oscillating number between } -1 \text{ and } 1] = 0$$

$\because$ $Lf'(0) = Rf'(0)$

So, $f(x)$ is differentiable at $x = 0$.

Q.22 $f(x) = \begin{cases} 1 + x, \text{ if } x \le 2 \\ 5 - x, \text{ if } x > 2 \end{cases}$ at $x = 2$.

Ans. We have, $f(x) = \begin{cases} 1 + x, \text{if } x \le 2 \\ 5 - x, \text{if } x > 2 \end{cases}$ at $x = 2$.

For differentiability at $x = 2$,

$$Lf'(2) = \lim_{x \to 2^-} \frac{f(x) - f(2)}{x - 2} = \lim_{x \to 2^-} \frac{(1 + x) - (1 + 2)}{x - 2}$$

Put $x = 2 - h$ $$= \lim_{h \to 0} \frac{(1 + 2 - h) - 3}{2 - h - 2} = \lim_{h \to 0} \frac{-h}{-h} = 1$$

$$Rf'(2) = \lim_{x \to 2^+} \frac{f(x) - f(2)}{x - 2} = \lim_{x \to 2^+} \frac{(5 - x) - 3}{x - 2}$$

Put $x = 2 + h$ $$= \lim_{h \to 0} \frac{5 - (2 + h) - 3}{2 + h - 2} = \lim_{h \to 0} \frac{5 - 2 - h - 3}{h} = \lim_{h \to 0} \frac{-h}{h}$$

$$= -1$$

$\because$ $Lf'(2) \ne Rf'(2)$

So, $f(x)$ is not differentiable at $x = 2$.

Q.23 Show that $f(x) = |x - 5|$ is continuous but not differentiable at $x = 5$.

Ans. We have, $f(x) = |x - 5|$

$\therefore$ $f(x) = \begin{cases} -(x - 5), \text{if } x < 5 \\ x - 5, \quad \text{if } x \ge 5 \end{cases}$

For continuity at $x = 5$,

$$\text{LHL} = \lim_{x \to 5^-} (-x + 5)$$

[when $x \to 5^-$, then it will be $-(x - 5) = -x + 5$]

Put $x = 5 - h$, $$= \lim_{h \to 0} [-(5 - h) + 5] = \lim_{h \to 0} h = 0$$

$$\text{RHL} = \lim_{x \to 5^+} (x - 5)$$

Put $x = 5 + h$, $\quad = \lim_{h \to 0} (5 + h - 5) = \lim_{h \to 0} h = 0$

$\therefore \quad f(5) = 5 - 5 = 0$

$\Rightarrow \quad \text{LHL} = \text{RHL} = f(5)$

Hence, $f(x)$ is continuous at $x = 5$.

Now,
$$Lf'(5) = \lim_{x \to 5^-} \frac{f(x) - f(5)}{x - 5}$$
$$= \lim_{x \to 5^-} \frac{-x + 5 - 0}{x - 5} = \lim_{x \to 5^-} \frac{-(x - 5)}{x - 5} = -1$$
$$Rf'(5) = \lim_{x \to 5^+} \frac{f(x) - f(5)}{x - 5} = \lim_{x \to 5^+} \frac{x - 5 - 0}{x - 5} = 1$$

$\therefore \quad Lf'(5) \neq Rf'(5)$

So, $f(x) = |x - 5|$ is not differentiable at $x = 5$.

Q.24 A function $f : R \to R$ satisfies the equation $f(x + y) = f(x) \cdot f(y)$ for all $x, y \in R$, $f(x) \neq 0$. Suppose that the function is differentiable at $x = 0$ and $f'(0) = 2$, then prove that $f'(x) = 2 f(x)$.

Ans. Let $f : R \to R$ satisfies the equation $f(x + y) = f(x) \cdot f(y)$, $\forall\, x, y \in R$, $f(x) \neq 0$.

On substituting $x = 0$, $y = h$, we get

$f(0 + h) = f(0) \cdot f(h)$

$\Rightarrow \; f(h) = f(0) \cdot f(h) \Rightarrow \; f(0) = 1$

Let $f(x)$ is differentiable at $x = 0$ and $f'(0) = 2$, then

$$f'(0) = \lim_{x \to 0} \frac{f(x) - f(0)}{x - 0}$$
$$\Rightarrow \quad 2 = \lim_{x \to 0} \frac{f(x) - f(0)}{x}$$
$$\Rightarrow \quad 2 = \lim_{h \to 0} \frac{f(0 + h) - f(0)}{0 + h}$$
$$\Rightarrow \quad 2 = \lim_{h \to 0} \frac{f(0) \cdot f(h) - f(0)}{h}$$
$$\Rightarrow \quad 2 = \lim_{h \to 0} \frac{f(0)\,[f(h) - 1]}{h}$$
$$\Rightarrow \quad 2 = \lim_{h \to 0} \frac{f(h) - 1}{h} \qquad [\because f(0) = 1] \quad \text{...(i)}$$

Also,
$$f'(x) = \lim_{h \to 0} \frac{f(x + h) - f(x)}{h}$$
$$= \lim_{h \to 0} \frac{f(x) \cdot f(h) - f(x)}{h} \qquad [\because f(x + y) = f(x) \cdot f(y)]$$
$$= \lim_{h \to 0} \frac{f(x)\,[f(h) - 1]}{h} = 2 f(x) \qquad \text{[using Eq. (i)]}$$

$\therefore \quad f'(x) = 2 f(x)$

Differentiate each of the following w.r.t. x exercises 25 to 43.

Q.25 $2^{\cos^2 x}$

Ans. Let $y = 2^{\cos^2 x}$

Taking log on both sides, we get

$$\log y = \log 2^{\cos^2 x} = \cos^2 x \cdot \log 2 \quad [\because \log m^n = n \log m]$$

On differentiating w.r.t. x, we get

$$\frac{d}{dy}\log y.\frac{dy}{dx} = \frac{d}{dx}(\log 2 \cdot \cos^2 x)$$

$$\Rightarrow \frac{1}{y}\cdot\frac{dy}{dx} = \log 2\frac{d}{dx}(\cos x)^2 \Rightarrow \frac{1}{y}\cdot\frac{dy}{dx} = \log 2\cdot[2\cos x]\cdot\frac{d}{dx}\cos x$$

$$= \log 2\cdot 2\cos x\cdot(-\sin x) \quad \left[\because \frac{d}{dx}\cos x = -\sin x\right]$$

$$= \log 2\cdot[-(\sin 2x)] \quad [\because \sin 2x = 2\sin x\cdot\cos x]$$

$$\Rightarrow \frac{dy}{dx} = -y\cdot\log 2(\sin 2x)$$

$$\therefore \frac{dy}{dx} = -2^{\cos^2 x}\cdot\log 2(\sin 2x) \quad [\because y = 2^{\cos^2 x}]$$

Q.26 $\dfrac{8^x}{x^8}$

Ans. Let $y = \dfrac{8^x}{x^8}$

Taking log on both sides, we get $\log y = \log\dfrac{8^x}{x^8}$

On differentiating w.r.t. x, we get

$$\frac{d}{dy}\log y\cdot\frac{dy}{dx} = \frac{d}{dx}[\log 8^x - \log x^8] \quad \left[\because \log\frac{m}{n} = \log m - \log n\right]$$

$$\Rightarrow \frac{1}{y}\cdot\frac{dy}{dx} = \frac{d}{dx}[x\cdot\log 8 - 8\cdot\log x] \quad [\because \log m^n = n\log m]$$

$$\Rightarrow \frac{1}{y}\cdot\frac{dy}{dx} = \log 8\cdot 1 - 8\cdot\frac{1}{x} \quad \left[\because \frac{d}{dx}\log x = \frac{1}{x}\right]$$

$$\Rightarrow \frac{1}{y}\cdot\frac{dy}{dx} = \log 8 - \frac{8}{x}$$

$$\therefore \frac{dy}{dx} = y\left(\log 8 - \frac{8}{x}\right) = \frac{8^x}{x^8}\left(\log 8 - \frac{8}{x}\right) \quad \left[\because y = \frac{8^x}{x^8}\right]$$

Q.27 $\log(x + \sqrt{x^2 + a})$

Ans. Let $y = \log(x + \sqrt{x^2 + a})$

On differentiating w.r.t. x, we get

$$\frac{dy}{dx} = \frac{d}{dx}\log(x + \sqrt{x^2 + a})$$

$$= \frac{1}{(x + \sqrt{x^2 + a})}\cdot\frac{d}{dx}[x + \sqrt{x^2 + a}] \quad \left[\because \frac{d}{dx}\log x = \frac{1}{x}\right]$$

[by chain rule]

$$= \frac{1}{(x + \sqrt{x^2 + a})} \left[1 + \frac{1}{2} (x^2 + a)^{-1/2} \cdot 2x \right]$$

$$= \frac{1}{(x + \sqrt{x^2 + a})} \cdot \left(1 + \frac{x}{\sqrt{x^2 + a}} \right)$$

$$= \frac{(\sqrt{x^2 + a} + x)}{(x + \sqrt{x^2 + a})(\sqrt{x^2 + a})} = \frac{1}{(\sqrt{x^2 + a})}$$

Q.28 $\log [\log (\log x^5)]$

Ans. Let $\quad y = \log [\log (\log x^5)]$

On differentiating w.r.t. to x, we get

$$\therefore \quad \frac{dy}{dx} = \frac{d}{dx} [\log [\log (\log x^5]] = \frac{1}{\log (\log x^5)} \cdot \frac{d}{dx} (\log \cdot \log x^5)$$

$$= \frac{1}{\log (\log x^5)} \cdot \left(\frac{1}{\log x^5} \right) \cdot \frac{d}{dx} \log x^5$$

$$= \frac{1}{\log (\log x^5)} \cdot \frac{1}{\log x^5} \cdot \frac{d}{dx} (5 \log x) = \frac{5}{x \cdot \log (\log x^5) \cdot \log (x^5)}$$

Q.29 $\sin \sqrt{x} + \cos^2 \sqrt{x}$

Ans. Let $\quad y = \sin \sqrt{x} + (\cos \sqrt{x})^2$

On differentiating w.r.t. x, we get

$$\frac{dy}{dx} = \frac{d}{dx} \sin(x^{1/2}) + \frac{d}{dx} [\cos (x^{1/2})]^2$$

$$= \cos x^{1/2} \cdot \frac{d}{dx} x^{1/2} + 2\cos (x^{1/2}) \frac{d}{dx} [\cos (x^{1/2})] \quad \left[\because \frac{d}{dx} \sin x = \cos x \right]$$

$$= \cos (x^{1/2}) \frac{1}{2} x^{-1/2} + 2 \cdot \cos (x^{1/2}) \cdot \left[- \sin (x^{1/2}) . \frac{d}{dx} x^{1/2} \right] \quad \left[\because \frac{d}{dx} x^n = nx^{n-1} \right]$$

$$= \cos \sqrt{x} \cdot \frac{1}{2\sqrt{x}} - 2\cos (x^{1/2}) \cdot \sin x^{1/2} \cdot \frac{1}{2\sqrt{x}}$$

$$= \frac{1}{2\sqrt{x}} [\cos (\sqrt{x}) - \sin (2\sqrt{x})] \quad [\because \sin 2x = 2 \sin x \cos x]$$

Q.30 $\sin^n (ax^2 + bx + c)$

Ans. Let $\quad y = \sin^n (ax^2 + bx + c)$

On differentiating w.r.t. x, we get

$$\frac{dy}{dx} = \frac{d}{dx} [\sin (ax^2 + bx + c)]^n$$

$$= n \cdot [\sin (ax^2 + bx + c)]^{n-1} \cdot \frac{d}{dx} \sin (ax^2 + bx + c)$$

$$= n \cdot \sin^{n-1} (ax^2 + bx + c) \cdot \cos (ax^2 + bx + c) . \frac{d}{dx} (ax^2 + bx + c)$$

$$= n \cdot \sin^{n-1} (ax^2 + bx + c) \cdot \cos (ax^2 + bx + c) \cdot (2ax + b)$$

$$= n \cdot (2ax + b) \cdot \sin^{n-1} (ax^2 + bx + c) \cdot \cos (ax^2 + bx + c)$$

Q.31 $\cos(\tan\sqrt{x+1})$

Ans. Let $y = \cos(\tan\sqrt{x+1})$

On differentiating w.r.t. x, we get

$$\therefore \quad \frac{dy}{dx} = \frac{d}{dx}\cos(\tan\sqrt{x+1}) = -\sin(\tan\sqrt{x+1})\cdot\frac{d}{dx}(\tan\sqrt{x+1})$$

$$= -\sin(\tan\sqrt{x+1})\cdot\sec^2\sqrt{x+1}\cdot\frac{d}{dx}(x+1)^{1/2} \qquad \left[\because \frac{d}{dx}(\tan x) = \sec^2 x\right]$$

$$= -\sin(\tan\sqrt{x+1})\cdot(\sec\sqrt{x+1})^2\cdot\frac{1}{2}(x+1)^{-1/2}\cdot\frac{d}{dx}(x+1)$$

$$= \frac{-1}{2\sqrt{x+1}}\cdot\sin(\tan\sqrt{x+1})\cdot\sec^2(\sqrt{x+1})$$

Q.32 $\sin x^2 + \sin^2 x + \sin^2(x^2)$

Ans. Let $y = \sin x^2 + \sin^2 x + \sin^2(x^2)$

On differentiating w.r.t. x, we get

$$\frac{dy}{dx} = \frac{d}{dx}\sin(x^2) + \frac{d}{dx}(\sin x)^2 + \frac{d}{dx}(\sin x^2)^2$$

$$= \cos(x^2)\cdot\frac{d}{dx}(x^2) + 2\sin x\cdot\frac{d}{dx}\sin x + 2\sin x^2\cdot\frac{d}{dx}\sin x^2$$

$$\left[\because \frac{d}{dx}\sin x = \cos x\right]$$

$$= \cos x^2\cdot 2x + 2\cdot\sin x\cdot\cos x + 2\sin x^2\cos x^2\cdot\frac{d}{dx}x^2$$

$$= 2x\cos(x)^2 + 2\cdot\sin x\cdot\cos x + 2\sin x^2\cdot\cos x^2\cdot 2x$$

$$= 2x\cos(x)^2 + \sin 2x + \sin 2(x)^2\cdot 2x \qquad [\because \sin 2x = 2\sin x\cos x]$$

$$= 2x\cos(x^2) + 2x\cdot\sin 2(x^2) + \sin 2x$$

Q.33 $\sin^{-1}\frac{1}{\sqrt{x+1}}$

Ans. Let $y = \sin^{-1}\frac{1}{\sqrt{x+1}}$

On differentiating w.r.t. x, we get

$$\frac{dy}{dx} = \frac{d}{dx}\sin^{-1}\frac{1}{\sqrt{x+1}}$$

$$= \frac{1}{\sqrt{1-\left(\frac{1}{\sqrt{x+1}}\right)^2}}\cdot\frac{d}{dx}\frac{1}{(x+1)^{1/2}} \qquad \left[\because \frac{d}{dx}(\sin^{-1}x) = \frac{1}{\sqrt{1-x^2}}\right]$$

$$= \frac{1}{\sqrt{\frac{x+1-1}{x+1}}}\cdot\frac{d}{dx}\cdot(x+1)^{-1/2} = \sqrt{\frac{x+1}{x}}\cdot\frac{-1}{2}(x+1)^{-\frac{1}{2}-1}\cdot\frac{d}{dx}(x+1)$$

$$= \frac{(x+1)^{1/2}}{x^{1/2}}\cdot\left(-\frac{1}{2}\right)(x+1)^{-3/2} = \frac{-1}{2\sqrt{x}}\cdot\left(\frac{1}{x+1}\right)$$

Q.34 $(\sin x)^{\cos x}$

Ans. Let $y = (\sin x)^{\cos x}$

Taking log on both sides, we get

$$\log y = \log(\sin x)^{\cos x} = \cos x \log \sin x \quad [\because \log m^n = n \log m]$$

On differentiating w.r.t. x, we get

$$\frac{d}{dy} \log y \cdot \frac{dy}{dx} = \frac{d}{dx}(\cos x \cdot \log \sin x)$$

$$\Rightarrow \frac{1}{y} \cdot \frac{dy}{dx} = \cos x \cdot \frac{d}{dx} \log \sin x + \log \sin x \cdot \frac{d}{dx} \cos x$$

$$\left[\because \frac{d}{dx}[f(x) \cdot g(x) = f(x)\frac{d}{dx}g(x) + g(x)\frac{d}{dx} f(x)\right]$$

$$= \cos x \cdot \frac{1}{\sin x} \cdot \frac{d}{dx} \sin x + \log \sin x \cdot (-\sin x)$$

$$= \cos x \cdot \frac{1}{\sin x} \cdot \cos x + \log \sin x \cdot (-\sin x)$$

$$\therefore \frac{dy}{dx} = y\left[\frac{\cos^2 x}{\sin x} - \sin x \cdot \log(\sin x)\right]$$

$$= \sin x^{\cos x}\left[\frac{\cos^2 x}{\sin x} - \sin x \cdot \log(\sin x)\right] \quad [\because y = (\sin x)^{\cos x}]$$

Q.35 $\sin^m x \cdot \cos^n x$

Ans. Let $y = \sin^m x \cdot \cos^n x$

On differentiating w.r.t. x, and applying product formula, we get

$$\frac{dy}{dx} = \frac{d}{dx}[(\sin x)^m \cdot (\cos x)^n]$$

$$= (\sin x)^m \cdot \frac{d}{dx}(\cos x)^n + (\cos x)^n \cdot \frac{d}{dx}(\sin x)^m$$

$$= (\sin x)^m \cdot n(\cos x)^{n-1} \cdot \frac{d}{dx} \cos x + (\cos x)^n\, m(\sin x)^{m-1} \cdot \frac{d}{dx} \sin x$$

$$= (\sin x)^m \cdot n(\cos x)^{n-1}\ (-\sin x) + (\cos x)^n \cdot m(\sin x)^{m-1} \cos x$$

$$= -n \sin^m x \cdot \cos^{n-1} x \cdot (\sin x) + m \cos^n x \cdot \sin^{m-1} x \cdot \cos x$$

$$= -n \cdot \sin^m x \cdot \sin x \cdot \cos^n x \cdot \frac{1}{\cos x} + m \cdot \sin^m x. \frac{1}{\sin x} \cdot \cos^n x \cdot \cos x$$

$$= -n \cdot \sin^m x \cdot \cos^n x \cdot \tan x + m \sin^m x \cdot \cos^n x \cdot \cot x$$

$$\left[\because \tan x = \frac{\sin x}{\cos x}, \cot x = \frac{\cos x}{\sin x}\right]$$

$$= \sin^m x \cdot \cos^n x\,[-n \tan x + m \cot x]$$

Q.36 $(x+1)^2(x+2)^3(x+3)^4$

Ans. Let $y = (x+1)^2\,(x+2)^3\,(x+3)^4$

Taking log on both sides, we get

$$\log y = \log\{(x+1)^2 \cdot (x+2)^3 (x+3)^4\}$$

$$= \log(x+1)^2 + \log(x+2)^3 + \log(x+3)^4 \quad [\because \log mn = \log m + \log n]$$

On differentiating w.r.t. to x, we get

$$\frac{d}{dy}\log y\cdot\frac{dy}{dx}=\frac{d}{dx}[2\log(x+1)]+\frac{d}{dx}[3\log(x+2)]+\frac{d}{dx}[4\log(x+3)]$$

$[\because \log m^n = n\log m]$

$$\Rightarrow \quad \frac{1}{y}\cdot\frac{dy}{dx}=\frac{2}{(x+1)}\cdot\frac{d}{dx}(x+1)+3\cdot\frac{1}{(x+2)}\cdot\frac{d}{dx}(x+2)$$

$$+4\cdot\frac{1}{(x+3)}\cdot\frac{d}{dx}(x+3)\left[\because \frac{d}{dx}(\log x)=\frac{1}{x}\right]$$

$$=\left[\frac{2}{x+1}+\frac{3}{x+2}+\frac{4}{x+3}\right]$$

$$\therefore \quad \frac{dy}{dx}=y\left[\frac{2}{(x+1)}+\frac{3}{(x+2)}+\frac{4}{(x+3)}\right]$$

$$=(x+1)^2\cdot(x+2)^3\cdot(x+3)^4\left[\frac{2}{(x+1)}+\frac{3}{(x+2)}+\frac{4}{(x+3)}\right]$$

$$=(x+1)^2\cdot(x+2)^3\cdot(x+3)^4\left[\frac{2(x+2)(x+3)+3(x+1)(x+3)+4(x+1)(x+2)}{(x+1)(x+2)(x+3)}\right]$$

$$=\frac{(x+1)^2(x+2)^3(x+3)^4}{(x+1)(x+2)(x+3)}[2(x^2+5x+6)+3(x^2+4x+3)+4(x^2+3x+2)]$$

$$=(x+1)(x+2)^2(x+3)^3[2x^2+10x+12+3x^2+12x+9+4x^2+12x+8]$$

$$=(x+1)(x+2)^2(x+3)^3[9x^2+34x+29]$$

Q.37 $\cos^{-1}\left(\frac{\sin x+\cos x}{\sqrt{2}}\right), -\frac{\pi}{4}<x<\frac{\pi}{4}$

Ans. Let $\quad y=\cos^{-1}\left(\frac{\sin x+\cos x}{\sqrt{2}}\right)$

$$\Rightarrow \quad y=\cos^{-1}\left(\frac{1}{\sqrt{2}}\sin x+\frac{1}{\sqrt{2}}\cos x\right)$$

$$\Rightarrow \quad y=\cos^{-1}\left(\frac{1}{\sqrt{2}}\cos x+\frac{1}{\sqrt{2}}\sin x\right)$$

$$\Rightarrow \quad y=\cos^{-1}\left(\cos\frac{\pi}{4}\cos x+\sin\frac{\pi}{4}\sin x\right) \quad \left[\because \cos\frac{\pi}{4}=\frac{1}{\sqrt{2}}=\sin\frac{\pi}{4}\right]$$

$$\Rightarrow \quad y=\cos^{-1}\left(\cos\left(\frac{\pi}{4}-x\right)\right) \quad [\because \cos(a-b)=\cos a\cos b+\sin a\sin b]$$

$$\therefore \quad y=\frac{\pi}{4}-x$$

On differentiating w.r.t. to x, we get

$$\frac{dy}{dx}=\frac{d}{dx}\left(\frac{\pi}{4}-x\right) \quad \Rightarrow \quad \frac{dy}{dx}=-1$$

Q.38 $\tan^{-1}\sqrt{\dfrac{1-\cos x}{1+\cos x}}, -\dfrac{\pi}{4}<x<\dfrac{\pi}{4}$

Ans. Let $y=\tan^{-1}\left(\sqrt{\dfrac{1-\cos x}{1+\cos x}}\right)=\tan^{-1}\left(\sqrt{\dfrac{1-1+2\sin^2\frac{x}{2}}{1+2\cos^2\frac{x}{2}-1}}\right)=\tan^{-1}\left(\tan\dfrac{x}{2}\right)=\dfrac{x}{2}$

$$\left[\because \cos x = 1-2\sin^2\frac{x}{2}=2\cos^2\frac{x}{2}-1\right]$$

On differentiating w.r.t. x, we get , $\dfrac{dy}{dx}=\dfrac{1}{2}$

Q.39 $\tan^{-1}(\sec x+\tan x), \dfrac{-\pi}{2}<x<\dfrac{\pi}{2}$

Ans. Let $y=\tan^{-1}(\sec x+\tan x)$

$$\therefore \quad \frac{dy}{dx}=\frac{d}{dx}\tan^{-1}(\sec x+\tan x)$$

$$=\frac{1}{1+(\sec x+\tan x)^2}\cdot\frac{d}{dx}(\sec x+\tan x) \qquad \left[\because \frac{d}{dx}(\tan^{-1}x)=\frac{1}{1+x^2}\right]$$

$$=\frac{1}{1+\sec^2 x+\tan^2 x+2\sec x\cdot\tan x}\cdot[\sec x\cdot\tan x+\sec^2 x]$$

$$\left[\begin{array}{l}\because \dfrac{d}{dx}\sec x=\sec x\cdot\tan x\\ \dfrac{d}{dx}\tan x=\sec^2 x\end{array}\right]$$

$$=\frac{1}{(\sec^2 x+\sec^2 x+2\sec x\cdot\tan x)}\cdot\sec x\cdot(\sec x+\tan x)$$

$$[\because 1+\tan^2 x=\sec^2 x]$$

$$=\frac{1}{2\sec x(\tan x+\sec x)}\cdot\sec x(\sec x+\tan x) \quad =\frac{1}{2}$$

Q.40 $\tan^{-1}\left(\dfrac{a\cos x-b\sin x}{b\cos x+a\sin x}\right), \dfrac{-\pi}{2}<x<\dfrac{\pi}{2}$ and $\dfrac{a}{b}\tan x>-1.$

Ans. Let $y=\tan^{-1}\left(\dfrac{a\cos x-b\sin x}{b\cos x+a\sin x}\right)$

On dividing by $b\cos x$ in both numerator and denominator, we get

$$y=\tan^{-1}\left[\frac{\dfrac{a\cos x}{b\cos x}-\dfrac{b\sin x}{b\cos x}}{\dfrac{b\cos x}{b\cos x}+\dfrac{a\sin x}{b\cos x}}\right]=\tan^{-1}\left[\frac{\dfrac{a}{b}-\tan x}{1+\dfrac{a}{b}\tan x}\right]$$

$$=\tan^{-1}\frac{a}{b}-\tan^{-1}(\tan x) \qquad \left[\because \tan^{-1}x-\tan^{-1}y=\tan^{-1}\left(\frac{x-y}{1+xy}\right)\right]$$

$$=\tan^{-1}\frac{a}{b}-x$$

On differentiating w.r.t. x, we get

$$\frac{dy}{dx}=\frac{d}{dx}\left(\tan^{-1}\frac{a}{b}\right)-\frac{d}{dx}(x)=0-1=-1$$

Q.41 $\sec^{-1}\left(\frac{1}{4x^3-3x}\right), 0< x< \frac{1}{\sqrt{2}}$

Ans. Let $$y=\sec^{-1}\left(\frac{1}{4x^3-3x}\right) \quad ...(i)$$

On putting $x=\cos\theta$ in Eq. (i), we get

$$y=\sec^{-1}\left(\frac{1}{4\cos^3\theta-3\cos\theta}\right)$$

$$=\sec^{-1}\frac{1}{\cos 3\theta} \quad [\because \cos 3\theta = 4\cos^3\theta - 3\cos\theta]$$

$$=\sec^{-1}(\sec 3\theta)=3\theta \quad \left[\because \sec\theta=\frac{1}{\cos\theta}\right]$$

$$=3\cos^{-1}x \quad [\because 0=\cos^{-1}x]$$

On differentiating w.r.t. x, we get

$$\frac{dy}{dx}=\frac{d}{dx}(3\cos^{-1}x)$$

$$=3\cdot\frac{-1}{\sqrt{1-x^2}} \quad \left[\because \frac{d}{dx}(\cos^{-1}x)=\frac{-1}{\sqrt{1-x^2}}\right]$$

Q.42 $\tan^{-1}\left(\frac{3a^2x-x^3}{a^3-3ax^2}\right), \frac{-1}{\sqrt{3}}<\frac{x}{a}<\frac{1}{\sqrt{3}}$

Ans. Let $$y=\tan^{-1}\left(\frac{3a^2x-x^3}{a^3-3ax^2}\right) \quad ...(i)$$

On putting $x=a\tan\theta \Rightarrow \theta=\tan^{-1}\frac{x}{a}$ in Eq. (i), we get

$$y=\tan^{-1}\left[\frac{3\tan\theta-\tan^3\theta}{1-3\tan^2\theta}\right]$$

$$=\tan^{-1}(\tan 3\theta)=3\theta \quad \left[\because \tan 3\theta=\frac{3\tan\theta-\tan^3\theta}{1-3\tan^2\theta}\right]$$

$$=3\tan^{-1}\frac{x}{a} \quad \left[\because \theta=\tan^{-1}\frac{x}{a}\right]$$

$$\therefore \quad \frac{dy}{dx}=3\cdot\frac{d}{dx}\tan^{-1}\frac{x}{a}=3\cdot\left[\frac{1}{1+\frac{x^2}{a^2}}\right]\cdot\frac{d}{dx}\cdot\left(\frac{x}{a}\right)$$

$$=3\cdot\frac{a^2}{a^2+x^2}\cdot\frac{1}{a}=\frac{3a}{a^2+x^2}$$

Q.43 $\tan^{-1}\left[\frac{\sqrt{1+x^2}+\sqrt{1-x^2}}{\sqrt{1+x^2}-\sqrt{1-x^2}}\right], -1< x<1, x\neq 0$

Ans. Let $$y=\tan^{-1}\left[\frac{\sqrt{1+x^2}+\sqrt{1-x^2}}{\sqrt{1+x^2}-\sqrt{1-x^2}}\right] \quad ...(i)$$

On putting $x^2 = \cos 2\theta$ in Eq. (i), we get

$$\therefore \quad y = \tan^{-1}\left(\frac{\sqrt{1+\cos 2\theta} + \sqrt{1-\cos 2\theta}}{\sqrt{1+\cos 2\theta} - \sqrt{1-\cos 2\theta}}\right)$$

$$= \tan^{-1}\left(\frac{\sqrt{1+2\cos^2\theta - 1} + \sqrt{1-1+2\sin^2\theta}}{\sqrt{1+2\cos^2\theta - 1} - \sqrt{1-1+2\sin^2\theta}}\right)$$

$$[\because \cos 2\theta = 2\cos^2\theta - 1 = 1 - 2\sin^2\theta]$$

$$= \tan^{-1}\left(\frac{\sqrt{2}\cos\theta + \sqrt{2}\sin\theta}{\sqrt{2}\cos\theta - \sqrt{2}\sin\theta}\right) = \tan^{-1}\left[\frac{\sqrt{2}(\cos\theta + \sin\theta)}{\sqrt{2}(\cos\theta - \sin\theta)}\right]$$

$$= \tan^{-1}\left(\frac{\cos\theta + \sin\theta}{\cos\theta - \sin\theta}\right)$$

On dividing by $\cos\theta$ in both numerator and denominator, we get

$$= \tan^{-1}\left(\frac{\frac{\cos\theta + \sin\theta}{\cos\theta}}{\frac{\cos\theta - \sin\theta}{\cos\theta}}\right) = \tan^{-1}\left(\frac{1+\tan\theta}{1-\tan\theta}\right) = \tan^{-1}\left(\frac{\tan\frac{\pi}{4} + \tan\theta}{1 - \tan\frac{\pi}{4}\tan\theta}\right)$$

$$= \tan^{-1}\tan\left(\frac{\pi}{4} + \theta\right) \qquad \left[\because \tan(a+b) = \frac{\tan a + \tan b}{1 - \tan a \cdot \tan b}, \tan\frac{\pi}{4} = 1\right]$$

$$= \frac{\pi}{4} + \theta = \frac{\pi}{4} + \frac{1}{2}\cos^{-1} x^2 \qquad \left[\because 2\theta = \cos^{-1} x^2 \Rightarrow \theta = \frac{1}{2}\cos^{-1} x^2\right]$$

On differentiating w.r.t. x, we get

$$\therefore \frac{dy}{dx} = \frac{d}{dx}\left(\frac{\pi}{4}\right) + \frac{d}{dx}\left(\frac{1}{2}\cos^{-1} x^2\right) = 0 + \frac{1}{2} \cdot \frac{-1}{\sqrt{1-x^4}} \cdot \frac{d}{dx} x^2 = \frac{1}{2} \cdot \frac{-2x}{\sqrt{1-x^4}} = \frac{-x}{\sqrt{1-x^4}}$$

Find $\frac{dy}{dx}$ of each of the functions expressed in parametric form in exercises from 44 to 48.

Q.44 $x = t + \frac{1}{t}, y = t - \frac{1}{t}$

Ans. We have, $x = t + \frac{1}{t}$ and $y = t - \frac{1}{t}$

Taking derivative of both equation w.r.t. t, we get

$$\therefore \quad \frac{dx}{dt} = \frac{d}{dt}\left(t + \frac{1}{t}\right) \quad \text{and} \quad \frac{dy}{dt} = \frac{d}{dt}\left(t - \frac{1}{t}\right)$$

$$\Rightarrow \quad \frac{dx}{dt} = 1 + (-1)t^{-2} \quad \text{and} \quad \frac{dy}{dt} = 1 - (-1)t^{-2}$$

$$\Rightarrow \quad \frac{dx}{dt} = 1 - \frac{1}{t^2} \quad \text{and} \quad \frac{dy}{dt} = 1 + \frac{1}{t^2}$$

$$\Rightarrow \quad \frac{dx}{dt} = \frac{t^2 - 1}{t^2} \quad \text{and} \quad \frac{dy}{dt} = \frac{t^2 + 1}{t^2}$$

$$\therefore \quad \frac{dy}{dx} = \frac{dy/dt}{dx/dt} = \frac{t^2 + 1/t^2}{t^2 - 1/t^2} = \frac{t^2 + 1}{t^2 - 1}$$

Q.45 $x = e^{\theta}\left(\theta + \frac{1}{\theta}\right), y = e^{-\theta}\left(\theta - \frac{1}{\theta}\right)$

Ans. We have, $x = e^{\theta}\left(\theta + \frac{1}{\theta}\right)$ and $y = e^{-\theta}\left(\theta - \frac{1}{\theta}\right)$

Taking derivative of both equations w.r.t. θ, we get

$$\frac{dx}{d\theta} = \frac{d}{d\theta}\left[e^{\theta}\cdot\left(\theta + \frac{1}{\theta}\right)\right]$$

$$= e^{\theta}\cdot\frac{d}{d\theta}\left(\theta + \frac{1}{\theta}\right) + \left(\theta + \frac{1}{\theta}\right)\cdot\frac{d}{d\theta}e^{\theta} \quad \text{[by product rule]}$$

$$= e^{\theta}\left(1 - \frac{1}{\theta^2}\right) + \left(\theta + \frac{1}{\theta}\right)e^{\theta} = e^{\theta}\left(1 - \frac{1}{\theta^2} + \theta + \frac{1}{\theta}\right)$$

$$= e^{\theta}\left(\frac{\theta^2 - 1 + \theta^3 + \theta}{\theta^2}\right) \quad \text{...(i)}$$

and $$\frac{dy}{d\theta} = \frac{d}{d\theta}\left[e^{-\theta}\cdot\left(\theta - \frac{1}{\theta}\right)\right]$$

$$= e^{-\theta}\cdot\frac{d}{d\theta}\left(\theta - \frac{1}{\theta}\right) + \left(\theta - \frac{1}{\theta}\right)\frac{d}{d\theta}e^{-\theta}$$

$$= e^{-\theta}\left(1 + \frac{1}{\theta^2}\right) + \left(\theta - \frac{1}{\theta}\right)e^{-\theta}\cdot\frac{d}{d\theta}(-\theta) \quad \text{[by product rule]}$$

$$= e^{-\theta}\left[\frac{\theta^2 + 1}{\theta^2} - \frac{\theta^2 - 1}{\theta}\right] = e^{-\theta}\left[\frac{\theta^2 + 1 - \theta^3 + \theta}{\theta^2}\right] \quad \text{...(ii)}$$

On dividing Eq. (ii) by Eq. (i), we get

$$\frac{dy}{dx} = \frac{dy/d\theta}{dx/d\theta} = \frac{e^{-\theta}\left(\frac{\theta^2 + 1 - \theta^3 + \theta}{\theta^2}\right)}{e^{\theta}\left(\frac{\theta^2 - 1 + \theta^3 + \theta}{\theta^2}\right)} = e^{-2\theta}\left(\frac{-\theta^3 + \theta^2 + \theta + 1}{\theta^3 + \theta^2 + \theta - 1}\right)$$

Q.46 $x = 3\cos\theta - 2\cos^3\theta, y = 3\sin\theta - 2\sin^3\theta$

Ans. We have, $x = 3\cos\theta - 2\cos^3\theta$ and $y = 3\sin\theta - 2\sin^3\theta$

Taking derivative of both equations w.r.t. θ, we get

$$\therefore \quad \frac{dx}{d\theta} = \frac{d}{d\theta}(3\cos\theta) - \frac{d}{d\theta}(2\cos^3\theta)$$

$$= 3\cdot(-\sin\theta) - 2\cdot 3\cos^2\theta\cdot\frac{d}{d\theta}\cdot\cos\theta$$

$$= -3\sin\theta + 6\cos^2\theta\sin\theta \quad \left[\because \frac{d}{dx}\cos x = -\sin x\right] \text{...(i)}$$

and $$\frac{dy}{d\theta} = \frac{d}{d\theta}(3\sin\theta) - \frac{d}{d\theta}(2\sin^3\theta)$$

$$\Rightarrow \quad \frac{dy}{d\theta} = 3\cos\theta - 2\cdot 3\sin^2\theta\cdot\frac{d}{d\theta}\cdot\sin\theta$$

$$= 3\cos\theta - 6\sin^2\theta\cdot\cos\theta \quad \text{...(ii)}$$

On dividing Eqs. (ii) by (i), we get

$$\frac{dy}{dx} = \frac{dy/d\theta}{dx/d\theta} = \frac{3\cos\theta - 6\sin^2\theta\cos\theta}{-3\sin\theta + 6\cos^2\theta\sin\theta}$$

$$= \frac{3\cos\theta(1-2\sin^2\theta)}{3\sin\theta(-1+2\cos^2\theta)} = \cot\theta\cdot\frac{\cos 2\theta}{\cos 2\theta} = \cot\theta$$

$[\because \cos 2\theta = 1 - 2\sin^2\theta = 2\cos^2\theta - 1]$

Q.47 $\sin x = \frac{2t}{1+t^2}$, $\tan y = \frac{2t}{1-t^2}$

Ans. We have, $\sin x = \frac{2t}{1+t^2}$ ...(i)

and $\tan y = \frac{2t}{1-t^2}$ ...(ii)

Taking derivative of Eqs. (i) and (ii) wr.t. t, we get

$$\frac{d}{dx}\sin x\cdot\frac{dx}{dt} = \frac{d}{dt}\left(\frac{2t}{1+t^2}\right)$$

$$\Rightarrow \cos x\frac{dx}{dt} = \frac{(1+t^2)\cdot\frac{d}{dt}(2t) - (2t)\cdot\frac{d}{dt}(1+t^2)}{(1+t^2)^2}$$ [$\because$ by quotient Rule]

$$= \frac{2(1+t^2) - 2t\cdot 2t}{(1+t^2)^2} = \frac{2+2t^2-4t^2}{(1+t^2)^2}$$

$$\Rightarrow \frac{dx}{dt} = \frac{2(1-t^2)}{(1+t^2)^2}\cdot\frac{1}{\cos x}$$

$$\Rightarrow \frac{dx}{dt} = \frac{2(1-t^2)}{(1+t^2)^2}\cdot\frac{1}{\sqrt{1-\sin^2 x}} = \frac{2(1-t^2)}{(1+t^2)^2}\cdot\frac{1}{\sqrt{1-\left(\frac{2t}{1+t^2}\right)^2}}$$

$[\because \cos^2 x = 1 - \sin^2 x]$

$$\Rightarrow \frac{dx}{dt} = \frac{2(1-t^2)}{(1+t^2)^2}\cdot\frac{(1+t^2)}{(1-t^2)} = \frac{2}{1+t^2}$$ $\left[\because \sin x = \frac{2t}{1+t^2}\right]$ ...(iii)

and $$\frac{d}{dy}\tan y\cdot\frac{dy}{dt} = \frac{d}{dt}\left(\frac{2t}{1-t^2}\right)$$

$$\sec^2 y\frac{dy}{dt} = \frac{(1-t^2)\frac{d}{dt}\cdot(2t) - 2t\cdot\frac{d}{dt}(1-t^2)}{(1-t^2)^2}$$

$$\frac{dy}{dt} = \frac{2-2t^2+4t^2}{(1-t^2)^2}\cdot\frac{1}{\sec^2 y}$$

$$= \frac{2(1+t^2)}{(1-t^2)^2}\cdot\frac{1}{(1+\tan^2 y)} = \frac{2(1+t^2)}{(1-t^2)^2}\cdot\frac{1}{1+\frac{4t^2}{(1-t^2)^2}}$$

$[\because \sec^2 x = 1 + \tan^2 x]$

$\left[\because \tan y = \frac{2t}{1-t^2}\right]$

$$= \frac{2(1+t^2)}{(1-t^2)^2} \cdot \frac{(1-t^2)^2}{(1+t^2)^2} = \frac{2}{1+t^2} \quad \text{...(iv)}$$

On dividing Eq. (iv) by Eq. (iii), we get

$$\therefore \quad \frac{dy}{dx} = \frac{dy/dt}{dx/dt} = \frac{2/1+t^2}{2/1+t^2} = 1$$

Q.48 $x = \dfrac{1+\log t}{t^2}, y = \dfrac{3+2\log t}{t}$

Ans. We have, $x = \dfrac{1+\log t}{t^2}$ and $y = \dfrac{3+2\log t}{t}$

Taking derivative of both equation w.r.t. t, we get

$$\therefore \quad \frac{dx}{dt} = \frac{t^2 . \frac{d}{dt}(1+\log t) - (1+\log t). \frac{d}{dt} t^2}{(t^2)^2} \quad \text{[by quotient rule]}$$

$$= \frac{t^2 \cdot \frac{1}{t} - (1+\log t)\cdot 2t}{t^4} = \frac{t - (1+\log t)\cdot 2t}{t^4}$$

$$= \frac{t}{t^4}[1 - 2(1+\log t)] = \frac{-1-2\log t}{t^3} \quad \text{... (i)}$$

and

$$\frac{dy}{dt} = \frac{t. \frac{d}{dt}(3+2\log t) - (3+2\log t). \frac{d}{dt} t}{t^2}$$

$$= \frac{t\cdot 2 \cdot \frac{1}{t} - (3+2\log t)\cdot 1}{t^2}$$

$$= \frac{2-3-2\log t}{t^2} = \frac{-1-2\log t}{t^2} \quad \text{...(ii)}$$

On dividing Eq. (ii) By Eq. (i), we get

$$\therefore \quad \frac{dy}{dx} = \frac{dy/dt}{dx/dt} = \frac{-1-2\log t / t^2}{-1-2\log t / t^3} = t$$

Q.49 If $x = e^{\cos 2t}$ and $y = e^{\sin 2t}$, then prove that $\dfrac{dy}{dx} = -\dfrac{y \log x}{x \log y}$.

Ans. Given, $x = e^{\cos 2t}$ and $y = e^{\sin 2t}$

Taking derivative of both equation w.r.t. t, we get

$$\therefore \quad \frac{dx}{dt} = \frac{d}{dt} e^{\cos 2t} = e^{\cos 2t} . \frac{d}{dt} \cos 2t \quad \left[\because \frac{d}{dx} e^x = e^x\right]$$

$$= e^{\cos 2t} \cdot (-\sin 2t) \cdot \frac{d}{dt}(2t)$$

$$\frac{dx}{dt} = -2e^{\cos 2t} \cdot \sin 2t \quad \text{...(i)}$$

and

$$\frac{dy}{dt} = \frac{d}{dt} e^{\sin 2t} = e^{\sin 2t} . \frac{d}{dt} \sin 2t$$

$$= e^{\sin 2t} \cos 2t \cdot \frac{d}{dt} 2t = 2e^{\sin 2t} . \cos 2t \quad \text{...(ii)}$$

On dividing Eq. (ii) and Eq. (i), we get

$$\therefore \quad \frac{dy}{dx} = \frac{dy/dt}{dx/dt} = \frac{2e^{\sin 2t} \cdot \cos 2t}{-2e^{\cos 2t} \cdot \sin 2t} = \frac{-e^{\sin 2t} \cdot \cos 2t}{e^{\cos 2t} \cdot \sin 2t} \quad \text{...(iii)}$$

Now, $x = e^{\cos 2t}$

Taking log on both sides, we get

$$\log x = \cos 2t \cdot \log e = \cos 2t \qquad [\because \log e = 1] \quad \text{...(iv)}$$

and
$$\log y = \sin 2t \cdot \log e = \sin 2t \qquad \text{...(v)}$$

$$\therefore \quad \frac{dy}{dx} = \frac{-y \log x}{x \log y}$$

Hence proved.

Q.50 If $x = a \sin 2t (1 + \cos 2t)$ and $y = b \cos 2t (1 - \cos 2t)$, then show that $\left(\frac{dy}{dx}\right)_{t = \pi/4} = \frac{b}{a}$.

Ans. Given, $x = a \sin 2t \ (1 + \cos 2t)$ and $y = b \cos 2t(1 - \cos 2t)$

Taking derivative of both equations w.r.t. t, we get

$$\frac{dx}{dt} = a\left[\sin 2t \cdot \frac{d}{dt}(1 + \cos 2t) + (1 + \cos 2t) \cdot \frac{d}{dt} \sin 2t\right]$$

$$= a\left[\sin 2t \cdot (-\sin 2t) \cdot \frac{d}{dt} 2t + (1 + \cos 2t) \cdot \cos 2t \cdot \frac{d}{dt} 2t\right]$$

$$= -2a \sin^2 2t + 2a \cos 2t \ (1 + \cos 2t)$$

$$\Rightarrow \quad \frac{dx}{dt} = -2a \left[\sin^2 2t - \cos 2t \ (1 + \cos 2t)\right] \qquad \text{...(i)}$$

and
$$\frac{dy}{dt} = b\left[\cos 2t \cdot \frac{d}{dt}(1 - \cos 2t) + (1 - \cos 2t) \cdot \frac{d}{dt} \cos 2t\right]$$

$$= b\left[\cos 2t \cdot (\sin 2t) \frac{d}{dt} 2t + (1 - \cos 2t)(-\sin 2t) \cdot \frac{d}{dt} 2t\right]$$

$$= b\left[2 \sin 2t \cdot \cos 2t + 2(1 - \cos 2t)(-\sin 2t)\right]$$

$$= 2b\left[\sin 2t \cdot \cos 2t - (1 - \cos 2t) \sin 2t\right] \qquad \text{...(ii)}$$

On dividing Eq. (ii) by Eq. (i), we get

$$\frac{dy}{dx} = \frac{dy/dt}{dx/dt} = \frac{-2b\left[-\sin 2t \cdot \cos 2t + (1 - \cos 2t) \sin 2t\right]}{-2a\left[\sin^2 2t - \cos 2t \ (1 + \cos 2t)\right]}$$

$$\Rightarrow \quad \left(\frac{dy}{dx}\right)_{t = \pi/4} = \frac{b}{a} \frac{\left[-\sin \frac{\pi}{2} \cos \frac{\pi}{2} + \left(1 - \cos \frac{\pi}{2}\right) \sin \frac{\pi}{2}\right]}{\left[\sin^2 \frac{\pi}{2} - \cos \frac{\pi}{2}\left(1 + \cos \frac{\pi}{2}\right)\right]}$$

$$= \frac{b}{a} \cdot \frac{(0 + 1)}{(1 - 0)} \qquad \left[\because \sin \frac{\pi}{2} = 1 \text{ and } \cos \frac{\pi}{2} = 0\right]$$

$$= \frac{b}{a}$$

Hence proved.

Q.51 If $x = 3\sin t - \sin 3t$, $y = 3\cos t - \cos 3t$, then find $\frac{dy}{dx}$ at $t = \frac{\pi}{3}$.

Ans. Given, $x = 3\sin t - \sin 3t$ and $y = 3\cos t - \cos 3t$

Taking derivative of both equations w.r.t. t, we get

$$\frac{dx}{dt} = 3\cdot\frac{d}{dt}\sin t - \frac{d}{dt}\sin 3t$$

$$= 3\cos t - \cos 3t\cdot\frac{d}{dt}3t = 3\cos t - 3\cos 3t \quad \text{...(i)}$$

and $$\frac{dy}{dt} = 3\cdot\frac{d}{dt}\cos t - \frac{d}{dt}\cos 3t$$

$$= -3\sin t + \sin 3t\cdot\frac{d}{dt}3t$$

$$\therefore \quad \frac{dy}{dt} = 3\sin 3t - 3\sin t \quad \text{...(ii)}$$

On dividing Eq. (ii) by Eq. (i), we get

$$\frac{dy}{dx} = \frac{dy/dt}{dx/dt} = \frac{3(\sin 3t - \sin t)}{3(\cos t - \cos 3t)}$$

Now, $$\left(\frac{dy}{dx}\right)_{t=\pi/3} = \frac{\sin\frac{3\pi}{3} - \sin\frac{\pi}{3}}{\left(\cos\frac{\pi}{3} - \cos 3\frac{\pi}{3}\right)} = \frac{0 - \sqrt{3}/2}{\frac{1}{2} - (-1)}$$

$$\left[\because \sin\pi = 0, \sin\frac{\pi}{3} = \frac{\sqrt{3}}{2}, \cos\pi = -1 \text{ and } \cos\frac{\pi}{3} = \frac{1}{2}\right]$$

$$= \frac{-\sqrt{3}/2}{3/2} = \frac{-\sqrt{3}}{3} = \frac{-1}{\sqrt{3}}$$

Q.52 Differentiate $\frac{x}{\sin x}$ w.r.t. $\sin x$.

Ans. Let $u = \frac{x}{\sin x}$ and $v = \sin x$

$$\therefore \quad \frac{du}{dx} = \frac{\sin x\cdot\frac{d}{dx}x - x\cdot\frac{d}{dx}\sin x}{(\sin x)^2} \quad \text{[by quotient rule]}$$

$$= \frac{\sin x - x\cos x}{\sin^2 x} \quad \text{...(i)}$$

and $$\frac{dv}{dx} = \frac{d}{dx}\sin x = \cos x \quad \text{...(ii)}$$

On dividing Eq. (i) by Eq. (ii), we get

$$\frac{du}{dv} = \frac{du/dx}{dv/dx} = \frac{\sin x - x\cos x/\sin^2 x}{\cos x}$$

$$= \frac{\sin x - x\cos x}{\sin^2 x\cos x} = \frac{\frac{\sin x - x\cos x}{\cos x}}{\frac{\sin^2 x\cos x}{\cos x}}$$

[dividing by $\cos x$ in both numerator and denominator]

$$= \frac{\tan x - x}{\sin^2 x}$$

Q.53 Differentiate $\tan^{-1}\dfrac{\sqrt{1+x^2}-1}{x}$ w.r.t. $\tan^{-1}x$, when $x \neq 0$.

Ans. Let $$u = \tan^{-1}\frac{\sqrt{1+x^2}-1}{x} \text{ and } v = \tan^{-1}x$$

Put $x = \tan\theta$

$$\Rightarrow \quad u = \tan^{-1}\frac{\sqrt{1+\tan^2\theta}-1}{\tan\theta} \quad \left[\because 1+\tan^2\theta = \sec^2\theta,\ \tan\theta = \frac{\sin\theta}{\cos\theta}\right]$$

$$= \tan^{-1}\frac{(\sec\theta-1)\cos\theta}{\sin\theta} \quad \left[\because \sec\theta = \frac{1}{\cos\theta}\right]$$

$$= \tan^{-1}\left(\frac{1-\cos\theta}{\sin\theta}\right)$$

$$= \tan^{-1}\left[\frac{1-1+2\sin^2\theta/2}{2\sin\theta/2\cdot\cos\theta/2}\right]$$

$$[\because \cos 2\theta = 1-2\sin^2\theta,\ \sin 2\theta = 2\sin\theta\cos\theta]$$

$$= \tan^{-1}\left[\tan\frac{\theta}{2}\right]$$

$$= \frac{\theta}{2} = \frac{1}{2}\tan^{-1}x \quad [\because \theta = \tan^{-1}x]$$

$$\therefore \quad \frac{du}{dx} = \frac{1}{2}\frac{d}{dx}\tan^{-1}x = \frac{1}{2}\cdot\frac{1}{1+x^2} \quad \text{...(i)}$$

and $$\frac{dv}{dx} = \frac{d}{dx}\tan^{-1}x = \frac{1}{1+x^2} \quad \text{...(ii)}$$

On dividing Eqs. (i) and (ii), we get

$$\therefore \quad \frac{du}{dv} = \frac{du/dx}{dv/dx} = \frac{1/2(1+x^2)}{1/(1+x^2)} = \frac{(1+x^2)}{2(1+x^2)} = \frac{1}{2}$$

Find $\dfrac{dy}{dx}$ when x and y are connected by the relation given in each of the exercise from 54 to 57.

Q.54 $\sin(xy) + \dfrac{x}{y} = x^2 - y$

Ans. We have, $$\sin(xy) + \frac{x}{y} = x^2 - y$$

On differentiating both sides w.r.t. x, we get

$$\frac{d}{dx}(\sin xy) + \frac{d}{dx}\left(\frac{x}{y}\right) = \frac{d}{dx}x^2 - \frac{d}{dx}y$$

$$\Rightarrow \quad \cos xy\cdot\frac{d}{dx}(xy) + \frac{y\dfrac{d}{dx}x - x\cdot\dfrac{d}{dx}y}{y^2} = 2x - \frac{dy}{dx} \quad [\because \text{by quotient rule}]$$

$$\Rightarrow \quad \cos xy\cdot\left[x\cdot\frac{d}{dx}y + y\cdot\frac{d}{dx}\cdot x\right] + \frac{y - x\dfrac{dy}{dx}}{y^2} = 2x - \frac{dy}{dx} \quad [\because \text{by product rule}]$$

$$\Rightarrow \quad x\cos xy\cdot\frac{dy}{dx}+y\cos xy+\frac{y}{y^2}-\frac{x}{y^2}\frac{dy}{dx}=2x-\frac{dy}{dx}$$

$$\Rightarrow \quad \frac{dy}{dx}\left[x\cos xy-\frac{x}{y^2}+1\right]=2x-y\cos xy-\frac{1}{y}$$

$$\therefore \quad \frac{dy}{dx}=\left[\frac{2xy-y^2\cos xy-1}{y}\right]\left[\frac{y^2}{xy^2\cos xy-x+y^2}\right]=\frac{(2xy-y^2\cos xy-1)\,y}{(xy^2\cos xy-x+y^2)}$$

Q.55 $\sec(x+y)=xy$

Ans. We have, $\sec(x+y)=xy$

On differentiating both sides w.r.t. x, we get

$$\frac{d}{dx}\sec(x+y)=\frac{d}{dx}(xy)$$

$$\Rightarrow \quad \sec(x+y)\cdot\tan(x+y)\cdot\frac{d}{dx}(x+y)=x\cdot\frac{d}{dx}y+y\cdot\frac{d}{dx}x$$

[∵ By product rule]

$$\Rightarrow \quad \sec(x+y)\cdot\tan(x+y)\cdot\left(1+\frac{dy}{dx}\right)=x\frac{dy}{dx}+y$$

$$\Rightarrow \quad \sec(x+y)\tan(x+y)+\sec(x+y)\cdot\tan(x+y).\frac{dy}{dx}=x\frac{dy}{dx}+y$$

$$\Rightarrow \quad \frac{dy}{dx}[\sec(x+y)\cdot\tan(x+y)-x]=y-\sec(x+y)\cdot\tan(x+y)$$

$$\therefore \quad \frac{dy}{dx}=\frac{y-\sec(x+y)\cdot\tan(x+y)}{\sec(x+y)\cdot\tan(x+y)-x}$$

Q.56 $\tan^{-1}(x^2+y^2)=a$

Ans. We have, $\tan^{-1}(x^2+y^2)=a,\Rightarrow x^2+y^2=\tan a$

On differentiating both sides w.r.t. x, we get

$$2x+2y\frac{dy}{dx}=0\Rightarrow 2y\frac{dy}{dx}=-2x\Rightarrow\frac{dy}{dx}=-\frac{x}{y}$$

Q.57 $(x^2+y^2)^2=xy$

Ans. We have, $(x^2+y^2)^2=xy$

On differentiating both sides w.r.t. x, we get

$$\frac{d}{dx}(x^2+y^2)^2=\frac{d}{dx}(xy)$$

$$\Rightarrow \quad 2(x^2+y^2)\cdot\frac{d}{dx}(x^2+y^2)=x\cdot\frac{d}{dx}y+y\cdot\frac{d}{dx}x \quad [\because \text{by product rule}]$$

$$\Rightarrow \quad 2(x^2+y^2)\cdot\left(2x+2y\frac{dy}{dx}\right)=x\frac{dy}{dx}+y$$

$$\Rightarrow 2x^2\cdot 2x+2x^2\cdot 2y\frac{dy}{dx}+2y^2\cdot 2x+2y^2\cdot 2y\frac{dy}{dx}=x\frac{dy}{dx}+y$$

$$\Rightarrow \quad \frac{dy}{dx}[4x^2y+4y^3-x]=y-4x^3-4xy^2$$

$$\therefore \quad \frac{dy}{dx}=\frac{(y-4x^3-4xy^2)}{(4x^2y+4y^3-x)}$$

Q.58 If $ax^2+2hxy+by^2+2gx+2fy+c=0$, then show that $\frac{dy}{dx}.\frac{dx}{dy}=1$.

Ans. We have, $ax^2 + 2hxy + by^2 + 2gx + 2fy + c = 0$ ...(i)

On differentiating both sides w.r.t. x, we get

$$\frac{d}{dx}(ax^2)+\frac{d}{dx}(2hxy)+\frac{d}{dx}(by^2)+\frac{d}{dx}(2gx)+\frac{d}{dx}(2fy)+\frac{d}{dx}(c)=0$$

$$\Rightarrow \quad 2ax+2h\left(x\cdot\frac{dy}{dx}+y\cdot 1\right)+b\cdot 2y\frac{dy}{dx}+2g+2f\frac{dy}{dx}+0=0$$

[∵ by product rule]

$$\Rightarrow \quad \frac{dy}{dx}[2hx+2by+2f]=-2ax-2hy-2g$$

$$\Rightarrow \quad \frac{dy}{dx}=\frac{-2(ax+hy+g)}{2(hx+by+f)}$$

$$=\frac{-(ax+hy+g)}{(hx+by+f)} \quad \text{...(ii)}$$

On differentiating Eq. (i) w.r.t. y, we get

$$\frac{d}{dy}(ax^2)+\frac{d}{dy}(2hxy)+\frac{d}{dy}(by^2)+\frac{d}{dy}(2gx)+\frac{d}{dy}(2fy)+\frac{d}{dy}(c)=0$$

$$\Rightarrow \quad a\cdot 2x\cdot\frac{dx}{dy}+2h\cdot\left(x\cdot\frac{d}{dy}y+y\cdot\frac{d}{dy}x\right)+b\cdot 2y+2g\cdot\frac{dx}{dy}+2f+0=0$$

$$\Rightarrow \quad \frac{dx}{dy}[2ax+2hy+2g]=-2hx-2by-2f$$

$$\Rightarrow \quad \frac{dx}{dy}=\frac{-2(hx+by+f)}{2(ax+hy+g)}=\frac{-(hx+by+f)}{(ax+hy+g)} \quad \text{...(iii)}$$

On multiplying Eqs. (ii) and (iii), we get

$$\therefore \quad \frac{dy}{dx}\cdot\frac{dx}{dy}=\frac{-(ax+hy+g)}{(hx+by+f)}\cdot\frac{-(hx+by+f)}{(ax+hy+g)}$$

$$=1=\text{RHS}$$

Hence proved.

Q.59 If $x=e^{x/y}$, then prove that $\frac{dy}{dx}=\frac{x-y}{x\log x}$.

Ans. We have, $x=e^{x/y}$

Taking log on both sides, we get

$$\log x=\frac{x}{y}\log e=\frac{x}{y} \quad [\because \log m^n = n\log m]$$

$$\Rightarrow \quad y\cdot\log x = x \quad [\because \log e = 1]$$

On differentiating w.r.t. x, we get

$$\frac{dy}{dx}\cdot\log x+\frac{1}{x}\cdot y=1 \quad [\because \text{by product rule}]$$

$$\Rightarrow \quad \frac{dy}{dx}\log x=1-\frac{y}{x}=\frac{x-y}{x}$$

$$\Rightarrow \quad \frac{dy}{dx}=\frac{x-y}{x\log x}$$

Q.60 If $y^x = e^{y-x}$, then prove that $\frac{dy}{dx} = \frac{(1+\log y)^2}{\log y}$.

Ans. We have, $y^x = e^{y-x}$

Taking log on both sides, we get

$$\log y^x = \log e^{y-x} \quad [\because \log m^n = n \log m]$$

$$\Rightarrow \quad x \log y = y - x \cdot \log_e = (y - x) \quad [\because \log_e = 1]$$

$$\Rightarrow \quad \log y = \frac{(y-x)}{x} \quad ...(i)$$

On differentiating w.r.t. x, we get

$$\frac{d}{dy}\log y \cdot \frac{dy}{dx} = \frac{d}{dx}\frac{(y-x)}{x}$$

$$\Rightarrow \quad \frac{1}{y}\cdot\frac{dy}{dx} = \frac{x \cdot \frac{d}{dx}(y-x) - (y-x)\cdot\frac{d}{dx}\cdot x}{x^2} \quad \text{[using quotient rule]}$$

$$\Rightarrow \quad \frac{1}{y}\frac{dy}{dx} = \frac{x\left(\frac{dy}{dx} - 1\right) - (y-x)}{x^2}$$

$$\Rightarrow \quad \frac{x^2}{y}\cdot\frac{dy}{dx} = x\frac{dy}{dx} - x - y + x$$

$$\Rightarrow \quad \frac{dy}{dx}\left(\frac{x^2}{y} - x\right) = -y$$

$$\therefore \quad \frac{dy}{dx} = \frac{-y^2}{x^2 - xy} = \frac{-y^2}{x(x-y)}$$

$$= \frac{y^2}{x(y-x)}\cdot\frac{x}{x} = \frac{y^2}{x^2}\cdot\frac{1}{\frac{(y-x)}{x}}$$

$$= \frac{(1+\log y)^2}{\log y}$$

$$\left[\because \log y = \frac{y-x}{x} \Rightarrow \log y = \frac{y}{x} - 1 \Rightarrow 1 + \log y = \frac{y}{x}\right]$$

Hence proved.

Q.61 If $y = (\cos x)^{(\cos x)^{(\cos x)^{\cdots\infty}}}$, then show that $\frac{dy}{dx} = \frac{y^2 \tan x}{y \log \cos x - 1}$.

Ans. We have, $y = (\cos x)^{(\cos x)^{(\cos x)^{\cdots\infty}}} \Rightarrow y = (\cos x)^y$

Taking log on both sides, we get

$$\therefore \quad \log y = \log(\cos x)^y \quad [\because \log m^n = n \log m]$$

$$\Rightarrow \quad \log y = y \log \cos x$$

On differentiating w.r.t. x, we get

$$\frac{1}{y}\cdot\frac{dy}{dx} = y\cdot\frac{d}{dx}\log\cos x + \log\cos x\cdot\frac{dy}{dx}$$

$$\Rightarrow \quad \frac{1}{y}\cdot\frac{dy}{dx} = \frac{y}{\cos x}\cdot\frac{d}{dx}\cos x + \log\cos x\cdot\frac{dy}{dx}$$

$$\Rightarrow \quad \frac{dy}{dx}\left[\frac{1}{y} - \log\cos x\right] = \frac{-y\sin x}{\cos x} = -y\tan x$$

$$\therefore \quad \frac{dy}{dx} = \frac{-y^2\tan x}{(1 - y\log\cos x)}$$

$$= \frac{y^2\tan x}{y\log\cos x - 1}$$

Hence proved.

Q.62 If $x\sin(a+y) + \sin a\cdot\cos(a+y) = 0$, then prove that

$$\frac{dy}{dx} = \frac{\sin^2(a+y)}{\sin a}.$$

Ans. We have, $x\sin(a+y) + \sin a\cdot\cos(a+y) = 0$

$$\Rightarrow \quad x\sin(a+y) = -\sin a\cdot\cos(a+y)$$

$$\Rightarrow \quad x = \frac{-\sin a\cdot\cos(a+y)}{\sin(a+y)}$$

$$\Rightarrow \quad x = -\sin a\cdot\cot(a+y)$$

$$\therefore \quad \frac{dx}{dy} = -\sin a\cdot[-\text{cosec}^2(a+y)]\cdot\frac{d}{dy}(a+y)$$

$$\left[\because \frac{d}{dx}\cot x = -\text{cosec}^2 x\right]$$

$$= \sin a\cdot\frac{1}{\sin^2(a+y)}\cdot 1$$

$$\Rightarrow \quad \frac{dy}{dx} = \frac{\sin^2(a+y)}{\sin a}$$

Hence proved.

Q.63 If $\sqrt{1-x^2} + \sqrt{1-y^2} = a(x-y)$, then prove that $\dfrac{dy}{dx} = \sqrt{\dfrac{1-y^2}{1-x^2}}.$

Ans. We have, $\sqrt{1-x^2} + \sqrt{1-y^2} = a(x-y)$

On putting $x = \sin\alpha$ and $y = \sin\beta$, we get

$$\sqrt{1-\sin^2\alpha} + \sqrt{1-\sin^2\beta} = a(\sin\alpha - \sin\beta)$$

$$\Rightarrow \quad \cos\alpha + \cos\beta = a(\sin\alpha - \sin\beta) \qquad [\because 1-\sin^2 x = \cos^2 x]$$

$$\Rightarrow \quad 2\cos\frac{\alpha+\beta}{2}.\cos\frac{\alpha-\beta}{2} = a\left(2\cos\frac{\alpha+\beta}{2}.\sin\frac{\alpha-\beta}{2}\right)$$

$$\left[\because \cos a + \cos b = 2\cos\frac{a+b}{2}\cos\frac{a-b}{2},\ \sin a - \sin b = 2\cos\frac{a+b}{2}\sin\frac{a-b}{2}\right]$$

$$\Rightarrow \quad \cos\frac{\alpha-\beta}{2} = a\sin\frac{\alpha-\beta}{2}$$

$$\Rightarrow \qquad \cot\frac{\alpha-\beta}{2} = a \Rightarrow \frac{\alpha-\beta}{2} = \cot^{-1} a$$

$$\Rightarrow \qquad \alpha - \beta = 2\cot^{-1} a$$

$$\Rightarrow \qquad \sin^{-1} x - \sin^{-1} y = 2\cot^{-1} a \qquad [\because x = \sin\alpha \text{ and } y = \sin\beta]$$

On differentiating both sides w.r.t. x, we get

$$\frac{1}{\sqrt{1-x^2}} - \frac{1}{\sqrt{1-y^2}}\frac{dy}{dx} = 0$$

$$\therefore \qquad \frac{dy}{dx} = \frac{\sqrt{1-y^2}}{\sqrt{1-x^2}} = \sqrt{\frac{1-y^2}{1-x^2}}$$

Hence proved.

Q.64 If $y = \tan^{-1} x$, then find $\dfrac{d^2y}{dx^2}$ in terms of y alone.

Ans. We have, $y = \tan^{-1} x$

$$\therefore \qquad \frac{dy}{dx} = \frac{1}{1+x^2} \qquad \text{[on differentiating w.r.t. } x]$$

Now,

$$\frac{d^2y}{dx^2} = \frac{d}{dx}(1+x^2)^{-1} \qquad \text{[again differentiating w.r.t. } x]$$

$$= -1(1+x^2)^{-2}.\frac{d}{dx}(1+x^2) = -\frac{1}{(1+x^2)^2}\cdot 2x$$

$$= \frac{-2\tan y}{(1+\tan^2 y)^2} \qquad [\because y = \tan^{-1} x \Rightarrow \tan y = x]$$

$$= \frac{-2\tan y}{(\sec^2 y)^2} = -2\frac{\sin y}{\cos y}\cdot\cos^2 y\cdot\cos^2 y$$

$$= 2\sin y\cos y\cdot\cos^2 y$$

$$= -\sin 2y\cdot\cos^2 y \qquad [\because \sin 2x = 2\sin x\cos x]$$

Verify the Rolle's theorem for each of the functions in exercises from 65 to 69.

****Q.65*** $f(x) = x(x-1)^2$ in $[0,1]$

Thinking Process We know that, Rolle's theorem states that, if f be a real valued function, defined in the closed interval $[a, b]$, such that (i) f is continuous on $[a, b]$. (ii) f is differentiable on $]a, b[$.(iii) $f(a) = f(b)$.
Then, there exists a real number c in the open interval $]a, b[$, such that $f'(c) = 0$'. Here, we shall verify the Rolle's theorem for the given function.

Ans. We have, $f(x) = x(x-1)^2$ in $[0, 1]$.

(i) Since, $f(x) = x(x-1)^2$ is a polynomial function.
So, it is continuous in $[0, 1]$.

(ii)

$$f'(x) = x\cdot\frac{d}{dx}(x-1)^2 + (x-1)^2\frac{d}{dx}x \qquad \text{[by product rule]}$$

$$= x\cdot 2(x-1)\cdot 1 + (x-1)^2$$

$$= 2x^2 - 2x + x^2 + 1 - 2x \qquad [\because (a-b)^2 = a^2 + b^2 - 2ab]$$

$$= 3x^2 - 4x + 1, \text{ which exists in } (0, 1).$$

So, $f(x)$ is differentiable in $(0, 1)$.

Note (*) *Topics related to these question have been removed from latest NCERT.*

(iii) $f(0) = 0$ and $f(1) = 0 \Rightarrow f(0) = f(1)$

f satisfies the above conditions of Rolle's theorem.

By Rolle's theorem $\exists\, c \in (0, 1)$ such that

$$f'(c) = 0$$
$$\Rightarrow \quad 3c^2 - 4c + 1 = 0$$
$$\Rightarrow \quad 3c^2 - 3c - c + 1 = 0$$
$$\Rightarrow \quad 3c(c-1) - 1(c-1) = 0$$
$$\Rightarrow \quad (3c-1)(c-1) = 0$$
$$\Rightarrow \quad c = \frac{1}{3}, 1 \Rightarrow \frac{1}{3} \in (0, 1)$$

Thus, we see that there exists a real number c in the open interval (0, 1).

Hence, Rolle's theorem has been verified.

***Q.66** $f(x) = \sin^4 x + \cos^4 x$ in $\left[0, \frac{\pi}{2}\right]$

Ans. We have, $f(x) = \sin^4 x + \cos^4 x$ in $\left[0, \frac{\pi}{2}\right]$ ...(i)

(i) $f(x)$ is continuous in $\left[0, \frac{\pi}{2}\right]$

[since, $\sin^4 x$ and $\cos^4 x$ are continuous functions and we know that, if g and h be continuous functions, then $(g + h)$ is a continuous function.]

(ii) $$f'(x) = 4(\sin x)^3 \cdot \cos x + 4(\cos x)^3 \cdot (-\sin x)$$
$$= 4\sin^3 x \cdot \cos x - 4\sin x \cdot \cos^3 x$$
$$= 4\sin x \cos x\,(\sin^2 x - \cos^2 x) \text{ which exists in } \left(0, \frac{\pi}{2}\right) \quad \text{...(ii)}$$

Hence, $f(x)$ is differentiable in $\left(0, \frac{\pi}{2}\right)$.

(iii) Also, $f(0) = 0 + 1 = 1$ and $f'\left(\frac{\pi}{2}\right) = 1 + 0 = 1 \Rightarrow f(0) = f\left(\frac{\pi}{2}\right)$

Since, all conditions of Rolle's theorem are satisfied.

Hence, there exists atleast one $c \in \left(0, \frac{\pi}{2}\right)$ such that $f'(c) = 0$.

$$\therefore \quad 4\sin c \cos c\,(\sin^2 c - \cos^2 c) = 0$$
$$\Rightarrow \quad 4\sin c \cos c\,(-\cos 2c) = 0 \qquad [\cos 2C = \cos^2 C - \sin^2 C]$$
$$\Rightarrow \quad -2\sin 2c \cdot \cos 2c = 0 \qquad [\because \sin 2x = 2\sin x \cos x]$$
$$\Rightarrow \quad -\sin 4c = 0$$
$$\Rightarrow \quad \sin 4c = 0$$
$$\Rightarrow \quad \sin 4c = \sin \pi \qquad [\because \sin \pi = 0]$$
$$\Rightarrow \quad 4c = \pi$$
$$\Rightarrow \quad c = \frac{\pi}{4} \text{ and } \frac{\pi}{4} \in \left(0, \frac{\pi}{2}\right)$$

Hence, Rolle's theorem has been verified.

Note (✱) *Topics related to these question have been removed from latest NCERT.*

***Q.67** $f(x) = \log(x^2 + 2) - \log 3$ in $[-1, 1]$

Ans. We have, $f(x) = \log(x^2 + 2) - \log 3.$

(i) Logarithmic functions are continuous in their domain.

Hence, $f(x) = \log(x^2 + 2) - \log 3$ is continuous in $[-1, 1]$.

(ii) $f'(x) = \dfrac{1}{x^2 + 2} \cdot 2x - 0$

$= \dfrac{2x}{x^2 + 2}$, which exists in $(-1, 1)$.

Hence, $f(x)$ is differentiable in $(-1, 1)$.

(iii) $f(-1) = \log[(-1)^2 + 2] - \log 3 = \log 3 - \log 3 = 0$ and

$f(1) = \log(1^2 + 2) - \log 3 = \log 3 - \log 3 = 0$

$\Rightarrow \quad f(-1) = f(1)$

Since, all conditions of Rolle's theorem are satisfied.

Hence, there exists a real number c such that

$$f'(c) = 0.$$

$$\Rightarrow \quad \frac{2c}{c^2 + 2} = 0 \Rightarrow c = 0 \in (-1, 1)$$

Hence, Rolle's theorem has been verified.

***Q.68** $f(x) = x(x + 3)e^{-x/2}$ in $[-3, 0]$

Ans. We have, $f(x) = x(x + 3)e^{-x/2}$

(i) $f(x)$ is a continuous function. [since, it is a combination of polynomial functions $x(x + 3)$ and an exponential function $e^{-x/2}$ which are continuous functions or product of continuous functions is continuous]

So, $f(x) = x(x + 3)e^{-x/2}$ is continuous in $[-3, 0]$.

(ii) $\therefore$ $f'(x) = (x^2 + 3x) \cdot \dfrac{d}{dx} e^{-x/2} + e^{-x/2} \cdot \dfrac{d}{dx}(x^2 + 3x)$

$= (x^2 + 3x) \cdot e^{-x/2} \cdot \left(-\dfrac{1}{2}\right) + e^{-x/2} \cdot (2x + 3)$

$= e^{-x/2}\left[2x + 3 - \dfrac{1}{2} \cdot (x^2 + 3x)\right]$

$= e^{-x/2}\left[\dfrac{4x + 6 - x^2 - 3x}{2}\right] = e^{-x/2} \cdot \dfrac{1}{2}[-x^2 + x + 6]$

$= \dfrac{-1}{2} e^{-x/2}[x^2 - x - 6] = \dfrac{-1}{2} e^{-x/2}[x^2 - 3x + 2x - 6]$

$= \dfrac{-1}{2} e^{-x/2}[(x + 2)(x - 3)]$, which exists in $(-3, 0)$.

Hence, $f(x)$ is differentiable in $(-3, 0)$.

(iii) $\therefore$ $f(-3) = -3(-3 + 3)e^{-3/2} = 0$

and $f(0) = 0(0 + 3)e^{-0/2} = 0$

$\Rightarrow \quad f(-3) = f(0)$

Note (*) *Topics related to these question have been removed from latest NCERT.*

Since, conditions of Rolle's theorem are satisfied.

Therefore, there exists a real number c such that $f'(c) = 0$

$\Rightarrow -\frac{1}{2} e^{-c/2} (c+2)(c-3) = 0 \Rightarrow c = -2, 3$, where $-2 \in (-3, 0)$

Hence, Rolle's theorem has been verified.

***Q.69** $f(x) = \sqrt{4 - x^2}$ in $[-2, 2]$

Ans. We have, $f(x) = \sqrt{4 - x^2} = (4 - x^2)^{1/2}$

(i) $f(x) = \sqrt{4 - x^2}$ is a continuous function.

[since every polynomial function is a continuous function]

Hence, $f(x)$ is continuous in $[-2, 2]$.

(ii) $f'(x) = \frac{1}{2}(4 - x^2)^{-1/2} \cdot (-2x)$ $\quad$ [$f'(x)$ = differentiation of $f(x)$]

$= -x. \frac{1}{\sqrt{4 - x^2}}$, which exists everywhere except at $x = \pm 2$

Hence, $f(x)$ is differentiable in $(-2, 2)$.

(iii) $f(-2) = \sqrt{(4-4)} = 0$ and $f(2) = \sqrt{(4-4)} = 0$

$\Rightarrow \quad f(-2) = f(2)$

Hence, all conditions of Rolle's theorem are satisfied.

Therefore, hence, there exists a real number c such that $f'(c) = 0$.

$\Rightarrow \quad -c\frac{1}{\sqrt{4 - c^2}} = 0 \Rightarrow c = 0 \in (-2, 2)$

Hence, Rolle's theorem has been verified.

***Q.70** Discuss the applicability of Rolle's theorem on the function given by

$$f(x) = \begin{cases} x^2 + 1, & \text{if } 0 \le x \le 1 \\ 3 - x, & \text{if } 1 \le x \le 2 \end{cases}$$

Ans. We have, $f(x) = \begin{cases} x^2 + 1, & \text{if } 0 \le x \le 1 \\ 3 - x, & \text{if } 1 \le x \le 2 \end{cases}$

We know that, polynomial function is everywhere continuous and differentiability.

So, $f(x)$ is continuous and differentiable at all points except possibly at $x = 1$.

Now, check the differentiability at $x = 1$,

At $x = 1$,

Left hand derivative LHD $= \lim_{x \to 1^-} \frac{f(x) - f(1)}{x - 1}$

$= \lim_{x \to 1} \frac{(x^2+1) - (1+1)}{x - 1}$ $\quad$ [$\because f(x) = x^2 + 1, \forall\, 0 \le x \le 1$]

$= \lim_{x \to 1} \frac{x^2 - 1}{x - 1}$

$= \lim_{x \to 1} \frac{(x+1)(x-1)}{x - 1}$ $\quad$ [$\because a^2 - b^2 = (a+b)(a-b)$]

$= 2$

Note (*) *Topics related to these question have been removed from latest NCERT.*

and Right hand derivative RHD = $\lim_{x\to 1^+} \frac{f(x)-f(1)}{x-1} = \lim_{x\to 1} \frac{(3-x)-(1+1)}{(x-1)}$

$$= \lim_{x\to 1} \frac{3-x-2}{x-1} = \lim_{x\to 1} \frac{-(x-1)}{x-1} = -1$$

$\therefore$ LHD $\neq$ RHD

So, $f(x)$ is not differentiable at $x = 1$.

Hence, Rolle's theorem is not applicable on the interval [0, 2].

Q.71 Find the points on the curve $y = (\cos x - 1)$ in $[0, 2\pi]$, where the tangent is parallel to X-axis.

Thinking Process For tangent to be parallel to X-axis, first find $\frac{dy}{dx}$ and set it equal to zero.

Ans. We have, $y = \cos x - 1$

$\therefore \quad \frac{dy}{dx} = -\sin x$

For tangent to be parallel to x-axis, we must have $\frac{dy}{dx} = 0$

$\therefore \quad -\sin x = 0$

$\therefore \quad x = \pi \in [0, 2\pi]$

$y(\pi) = \cos \pi - 1 = -2$ $\quad [\because \cos \pi = -1]$

Hence, the required point on the curve, where the tangent drawn is parallel to the x-axis is $(\pi, -2)$.

*** Q.72** Using Rolle's theorem, find the point on the curve $y = x(x-4)$, $x \in [0, 4]$, where the tangent is parallel to X-axis.

Ans. We have, $y = x(x-4)$, $x \in [0,4]$

(i) y is a continuous function since $x(x-4)$ is a polynomial function.
Hence, $y = x(x-4)$ is continuous in [0, 4].

(ii) $y' = (x-4)\cdot 1 + x \cdot 1 = 2x - 4$ which exists in (0,4).
Hence, y is differentiable in (0,4).

(iii) $y(0) = 0(0-4) = 0$ and $y(4) = 4(4-4) = 0 \Rightarrow y(0) = y(4)$
Since, conditions of Rolle's theorem are satisfied.
Hence, there exists a point c such that

$f'(c) = 0$ in (0,4) $\quad [\because f'(x) = y']$

$\Rightarrow 2c - 4 = 0 \Rightarrow 2c = 4 \Rightarrow c = 2 \Rightarrow x = 2; y = 2(2-4) = -4$

Thus, (2, − 4) is the point on the curve at which the tangent drawn is parallel to X-axis.

Verify mean value theorem for each of the functions given exercises from 73 to 76.

***Q.73** $f(x) = \frac{1}{4x-1}$ in [1, 4]

Thinking Process We know that, mean value theorem states that, if f be a real function such that

(i) $f(x)$ is continuous on $[a,b]$ (ii) $f(x)$ is differentiable on $]a,b[$

Then, there exists a real number $c \in]a,b[$ such that $f'(c) = \frac{f(b)-f(a)}{b-a}$, thus we can verify it for given function.

Note (*) *Topics related to these question have been removed from latest NCERT.*

Ans. We have, $f(x)=\dfrac{1}{4x-1}$ in [1, 4]

(i) $f(x)$ is continuous in [1, 4].

Also, at $x=\dfrac{1}{4}$, $f(x)$ is discontinuous.

Hence, $f(x)$ is continuous in [1, 4].

(ii) $f'(x)=-\dfrac{4}{(4x-1)^2}$, which exists in (1, 4).

Since, conditions of mean value theorem are satisfied.

Hence, there exists a real number $c\in]1,4[$ such that

$$f'(c)=\frac{f(4)-f(1)}{4-1}$$

$$\Rightarrow \quad \frac{-4}{(4c-1)^2}=\frac{\dfrac{1}{16-1}-\dfrac{1}{4-1}}{4-1}=\frac{\dfrac{1}{15}-\dfrac{1}{3}}{3}$$

$$\Rightarrow \quad \frac{-4}{(4c-1)^2}=\frac{1-5}{45}=\frac{-4}{45}$$

$$\Rightarrow \quad (4c-1)^2=45$$

$$\Rightarrow \quad 4c-1=\pm 3\sqrt{5}$$

$$\Rightarrow \quad c=\frac{3\sqrt{5}+1}{4}\in(1,4) \quad \text{[neglecting (– ve) value]}$$

Hence, mean value theorem has been verified.

***Q.74** $f(x)=x^3-2x^2-x+3$ in [0, 1]

Ans. We have, $f(x)=x^3-2x^2-x+3$ in [0, 1]

(i) Since, $f(x)$ is a polynomial function.

Hence, $f(x)$ is continuous in [0, 1].

(ii) $f'(x)=3x^2-4x-1$, which exists in (0,1).

Hence, $f(x)$ is differentiable in (0,1).

Since, conditions of mean value theorem are satisfied.

Therefore, by mean value theorem $\exists\, c\in(0,1)$, such that

$$f'(c)=\frac{f(1)-f(0)}{1-0}$$

$$\Rightarrow \quad 3c^2-4c-1=\frac{[1-2-1+3]-[0+3]}{1-0}$$

$$\Rightarrow \quad 3c^2-4c-1=\frac{-2}{1}$$

$$\Rightarrow \quad 3c^2-4c+1=0 \Rightarrow 3c^2-3c-c+1=0$$

$$\Rightarrow \quad 3c(c-1)-1(c-1)=0$$

$$\Rightarrow \quad (3c-1)(c-1)=0$$

$$\Rightarrow \quad c=\frac{1}{3}, 1, \text{ where } \frac{1}{3}\in(0,1)$$

Hence, the mean value theorem has been verified.

Note (*) *Topics related to these question have been removed from latest NCERT.*

***Q.75** $f(x) = \sin x - \sin 2x$ in $[0, \pi]$

Ans. We have, $f(x) = \sin x - \sin 2x$ in $[0,\pi]$

(i) Since, we know that sine functions are continuous functions hence $f(x) = \sin x - \sin 2x$ is a continuous function in $[0,\pi]$.

(ii) $f'(x) = \cos x - \cos 2x \cdot 2 = \cos x - 2\cos 2x$, which exists in $(0, \pi)$.

So, $f(x)$ is differentiable in $(0, \pi)$. Conditions of mean value theorem are satisfied.

Hence, $\exists\, c \in (0, \pi)$ such that, $f'(c) = \dfrac{f(\pi) - f(0)}{\pi - 0}$

$$\Rightarrow \quad \cos c - 2\cos 2c = \frac{\sin \pi - \sin 2\pi - \sin 0 + \sin 2 \cdot 0}{\pi - 0}$$

$$\Rightarrow \quad 2\cos 2c - \cos c = \frac{0}{\pi} \qquad [\because \cos 2c = 2\cos^2 c - 1]$$

$$\Rightarrow \quad 2 \cdot (2\cos^2 c - 1) - \cos c = 0$$

$$\Rightarrow \quad 4\cos^2 c - 2 - \cos c = 0$$

$$\Rightarrow \quad 4\cos^2 c - \cos c - 2 = 0 \Rightarrow \cos c = \frac{1 \pm \sqrt{1 + 32}}{8} = \frac{1 \pm \sqrt{33}}{8}$$

$$\therefore \quad c = \cos^{-1}\left(\frac{1 \pm \sqrt{33}}{8}\right)$$

Also, $\cos^{-1}\left(\dfrac{1 \pm \sqrt{33}}{8}\right) \in (0, \pi)$

Hence, mean value theorem has been verified.

***Q.76** $f(x) = \sqrt{25 - x^2}$ in $[1, 5]$

Ans. We have, $f(x) = \sqrt{25 - x^2}$ in $[1, 5]$

(i) Since, $f(x) = (25 - x^2)^{1/2}$, where $25 - x^2 \geq 0$

$\Rightarrow \quad x^2 \leq \pm 5 \Rightarrow -5 \leq x \leq 5$

Hence, $f(x)$ is continuous in $[1, 5]$.

(ii) $f'(x) = \dfrac{1}{2}(25 - x^2)^{-1/2} - 2x = \dfrac{-x}{\sqrt{25 - x^2}}$, which exists in $(1, 5)$.

Hence, $f'(x)$ is differentiable in $(1, 5)$.

Since, conditions of mean value theorem are satisfied.

By mean value theorem $\exists\, c \in (1, 5)$ such that

$$f'(c) = \frac{f(5) - f(1)}{5 - 1}$$

$$\Rightarrow \quad \frac{-c}{\sqrt{25 - c^2}} = \frac{0 - \sqrt{24}}{4} \Rightarrow \frac{c^2}{25 - c^2} = \frac{24}{16}$$

$$\Rightarrow \quad 16c^2 = 600 - 24c^2 \Rightarrow c^2 = \frac{600}{40} = 15$$

$$\therefore \quad c = \pm \sqrt{15}$$

Also, $\quad c = \sqrt{15} \in (1, 5)$

Hence, the mean value theorem has been verified.

Note (✱) *Topics related to these question have been removed from latest NCERT.*

Q.77 Find a point on the curve $y=(x-3)^2$, where the tangent is parallel to the chord joining the points (3, 0) and (4, 1).

Ans. First find the slope of a line joining two points (3, 0) and (4, 1) is

$$m=\frac{1-0}{4-3}=1 \qquad \left[\because m=\frac{y_2-y_1}{x_2-x_1}\right]$$

The slope of the tangent to the curve $y=(x-3)^2$ at any point is the derivative of y w.r.t. x

$$\frac{dy}{dx}=2(x-3)$$

For the tangent to be parallel to the chord, its slope must equal 1.

$$2(x-3)=1$$
$$x-3=\frac{1}{2}$$
$$x=\frac{7}{2}$$

Substitute $x=\frac{7}{2}$ into the equation of the curve $y=(x-3)^2$

$$y=\left(\frac{7}{2}-3\right)^2=\left(\frac{1}{2}\right)^2=\frac{1}{4}$$

Hence, the point on the curve is $\left(\frac{7}{2},\frac{1}{4}\right)$.

***Q.78** Using mean value theorem, prove that there is a point on the curve $y=2x^2-5x+3$ between the points $A(1, 0)$ and $B(2, 1)$, where tangent is parallel to the chord AB. Also, find that point.

Ans. We have, $y=2x^2-5x+3$, which is continuous in [1, 2] as it is a polynomial function.

Also, $y'=4x-5$, which exists in (1, 2).

By mean value theorem, $\exists\, c\in(1,2)$ at which drawn tangent is parallel to the chord AB, where A and B are (1, 0) and (2,1), respectively.

$$\therefore \qquad f'(c)=\frac{f(2)-f(1)}{2-1}$$

$$\Rightarrow \qquad 4c-5=\frac{(8-10+3)-(2-5+3)}{1}\Rightarrow 4c-5=1$$

$$\therefore \qquad c=\frac{6}{4}=\frac{3}{2}\in(1,2)$$

For $x=\frac{3}{2}$,

$$y=2\left(\frac{3}{2}\right)^2-5\left(\frac{3}{2}\right)+3$$
$$=2\times\frac{9}{4}-\frac{15}{2}+3=\frac{9-15+6}{2}=0$$

Hence, $\left(\frac{3}{2},0\right)$ is the point on the curve $y=2x^2-5x+3$ between the points A (1, 0) and B (2, 1), where tangent is parallel to the chord AB.

Note (*) *Topics related to these question have been removed from latest NCERT.*

Long Answer Type Questions

Q.79 Find the values of p and q, so that $f(x) = \begin{cases} x^2 + 3x + p, & \text{if } x \le 1 \\ qx + 2, & \text{if } x > 1 \end{cases}$ is differentiable at $x = 1$.

Ans. We have, $f(x) = \begin{cases} x^2 + 3x + p, & \text{if } x \le 1 \\ qx + 2, & \text{if } x > 1 \end{cases}$ is differentiable at $x = 1$.

$\therefore \quad Lf'(1) = \lim_{x \to 1^-} \frac{f(x) - f(1)}{x - 1}$ [$Lf'(x)$ = Left hand derivative]

$= \lim_{x \to 1^-} \frac{(x^2 + 3x + p) - (1 + 3 + p)}{x - 1}$

Put $x = 1 - h$, $= \lim_{h \to 0} \frac{[(1-h)^2 + 3(1-h) + p] - [1 + 3 + p]}{(1-h) - 1}$

$= \lim_{h \to 0} \frac{[1 + h^2 - 2h + 3 - 3h + p] - [4 + p]}{-h}$ $[\because (a - b)^2 = a^2 + b^2 - 2ab]$

$= \lim_{h \to 0} \frac{[h^2 - 5h + p + 4 - 4 - p]}{-h} = \lim_{h \to 0} \frac{h[h - 5]}{-h}$

$= \lim_{h \to 0} -[h - 5] = 5$

$Rf'(1) = \lim_{x \to 1^+} \frac{f(x) - f(1)}{x - 1} = \lim_{x \to 1^+} \frac{(qx + 2) - (1 + 3 + p)}{x - 1}$

[$Rf'(x)$ = Right hand derivative]

Put $x = 1 + h$, $= \lim_{h \to 0} \frac{[q(1 + h) + 2] - (4 + p)}{1 + h - 1}$

$= \lim_{h \to 0} \frac{[q + qh + 2 - 4 - p]}{h} = \lim_{h \to 0} \frac{qh + (q - 2 - p)}{h}$

$\Rightarrow \quad q - 2 - p = 0 \Rightarrow p - q = -2$ ...(i)

$\Rightarrow \quad \lim_{h \to 0} \frac{qh + 0}{h} = q$ [for existing the limit]

If $Lf'(1) = Rf'(1)$, then $5 = q \Rightarrow p - 5 = -2 \Rightarrow p = 3$

$\therefore \quad p = 3$ and $q = 5$

Q.80 If $x^m \cdot y^n = (x + y)^{m+n}$, prove that

(i) $\frac{dy}{dx} = \frac{y}{x}$ and (ii) $\frac{d^2y}{dx^2} = 0$

Ans. (i) We have, $x^m \cdot y^n = (x + y)^{m+n}$ ...(i)

Taking log on both sides of Eq. (i), we get

$\log(x^m \cdot y^n) = \log(x + y)^{m+n}$

$\Rightarrow \quad \log x^m + \log y^n = (m + n)\log(x + y)$

$[\because \log m^n = n \log m, \log(m \cdot n) = \log m + \log n]$

$\Rightarrow \quad m \log x + n \log y = (m + n)\log(x + y)$

On differentiating w.r.t. x, we get

$$\frac{m}{x}+\frac{n}{y}\frac{dy}{dx}=(m+n)\frac{1}{x+y}\left[1+\frac{dy}{dx}\right]$$

$$\Rightarrow \quad \frac{n}{y}\frac{dy}{dx}-\frac{(m+n)}{(x+y)}\frac{dy}{dx}=\frac{(m+n)}{(x+y)}-\frac{m}{x}$$

$$\frac{dy}{dx}\left[\frac{n}{y}-\frac{m+n}{x+y}\right]=\frac{m+n}{x+y}-\frac{m}{x}$$

$$\Rightarrow \quad \frac{dy}{dx}\left[\frac{n(x+y)-y(m+n)}{y(x+y)}\right]=\frac{x(m+n)-m(x+y)}{x(x+y)}$$

$$\Rightarrow \quad \frac{dy}{dx}\left[\frac{nx-my}{y(x+y)}\right]=\frac{nx-my}{x(x+y)} \Rightarrow \frac{dy}{dx}=\frac{nx-my}{x(x+y)}\cdot\frac{y(x+y)}{(nx-my)}$$

$$\Rightarrow \quad \frac{dy}{dx}=\frac{y}{x} \qquad \text{...(ii)}$$

Hence proved.

(ii) Further, differentiating Eq. (ii) *i.e.*, $\frac{dy}{dx}=\frac{y}{x}$ on both the sides w.r.t. x, we get

$$\frac{d^2y}{dx^2}=\frac{x\cdot\frac{dy}{dx}-y\cdot 1}{x^2} \qquad \text{[by quotient rule]}$$

$$=\frac{x\cdot\frac{y}{x}-y}{x^2} \qquad \left[\because \frac{dy}{dx}=\frac{y}{x}\right]$$

$$=0$$

Hence proved.

Q.81 If $x=\sin t$ and $y=\sin pt$, then prove that

$$(1-x^2)\frac{d^2y}{dx^2}-x\frac{dy}{dx}+p^2y=0.$$

Ans. We have, $x=\sin t$ and $y=\sin pt$

$$\therefore \quad \frac{dx}{dt}=\cos t \text{ and } \frac{dy}{dt}=\cos pt\cdot p$$

$$\Rightarrow \quad \frac{dy}{dx}=\frac{dy/dt}{dx/dt}=\frac{p\cdot\cos pt}{\cos t} \qquad \text{...(i)}$$

Again, on differentiating both sides w.r.t. x, we get

$$\frac{d^2y}{dx^2}=\frac{\cos t\cdot\frac{d}{dt}(p\cdot\cos pt)\frac{dt}{dx}-p\cos pt\cdot\frac{d}{dt}\cos t\cdot\frac{dt}{dx}}{\cos^2 t} \qquad \text{[using quotient rule]}$$

$$=\frac{[\cos t\cdot p\cdot(-\sin pt)\cdot p-p\cos pt\cdot(-\sin t)]\frac{dt}{dx}}{\cos^2 t}$$

$$=\frac{[-p^2\sin pt\cdot\cos t+p\sin t\cdot\cos pt]\cdot\frac{1}{\cos t}}{\cos^2 t} \qquad \left[\because \frac{dt}{dx}=\frac{1}{\cos t}\right]$$

$$\Rightarrow \quad \frac{d^2y}{dx^2}=\frac{-p^2\sin pt\cdot\cos t+p\cos pt\cdot\sin t}{\cos^3 t} \qquad \text{...(ii)}$$

Now, $(1-x^2)\dfrac{d^2y}{dx^2} - x\dfrac{dy}{dx} + p^2y = 0$

$$\therefore \quad \text{LHS} = (1-\sin^2 t)\frac{[-p^2 \sin pt \cdot \cos t + p\cos pt \cdot \sin t]}{\cos^3 t} - \sin t \cdot \frac{p\cos pt}{\cos t} + p^2 \sin pt$$

$$= \frac{1}{\cos^3 t}\begin{bmatrix} (1-\sin^2 t)(-p^2 \sin pt \cdot \cos t + p\cos pt \cdot \sin t) \\ -p\cos pt \cdot \sin t \cdot \cos^2 t + p^2 \sin pt \cdot \cos^3 t \end{bmatrix}$$

$$= \frac{1}{\cos^3 t}\begin{bmatrix} -p^2 \sin pt \cdot \cos^3 t + p\cos pt \cdot \sin t \cdot \cos^2 t \\ -p\cos pt \cdot \sin t \cdot \cos^2 t + p^2 \sin pt \cdot \cos^3 t \end{bmatrix} \quad [\because 1-\sin^2 t = \cos^2 t]$$

$$= \frac{1}{\cos^3 t} \cdot 0$$

$$= 0$$

Hence Proved.

Q.82 Find the value of $\dfrac{dy}{dx}$, if $y = x^{\tan x} + \sqrt{\dfrac{x^2+1}{2}}$.

Ans. We have, $$y = x^{\tan x} + \sqrt{\frac{x^2+1}{2}} \quad \text{...(i)}$$

Let $u = x^{\tan x}$ and $v = \sqrt{\dfrac{x^2+1}{2}}$,

Taking log on both sides, we get

$$\log u = \tan x \log x \quad \text{...(ii)}$$

and $$v^2 = \frac{x^2+1}{2} \quad \text{...(iii)}$$

On differentiating Eq. (ii) w.r.t. x, we get

$$\frac{1}{u}\cdot\frac{du}{dx} = \tan x \cdot \frac{1}{x} + \log x \cdot \sec^2 x \quad \text{[by product rule]}$$

$$\Rightarrow \quad \frac{du}{dx} = u\left[\frac{\tan x}{x} + \log x \cdot \sec^2 x\right]$$

$$= x^{\tan x}\left[\frac{\tan x}{x} + \log x \cdot \sec^2 x\right] \quad \text{...(iv)}$$

Also, on differentiating Eq. (iii) w.r.t. x, we get

$$2v \cdot \frac{dv}{dx} = \frac{1}{2}(2x) \Rightarrow \frac{dv}{dx} = \frac{1}{4v}\cdot(2x)$$

$$\Rightarrow \quad \frac{dv}{dx} = \frac{1}{4\cdot\sqrt{\dfrac{x^2+1}{2}}}\cdot 2x = \frac{x\cdot\sqrt{2}}{2\sqrt{x^2+1}}$$

$$\Rightarrow \quad \frac{dv}{dx} = \frac{x}{\sqrt{2(x^2+1)}} \quad \text{...(v)}$$

Now, $$y = u + v$$

$$\therefore \quad \frac{dy}{dx} = \frac{du}{dx} + \frac{dv}{dx}$$

$$= x^{\tan x}\left[\frac{\tan x}{x} + \log x \cdot \sec^2 x\right] + \frac{x}{\sqrt{2(x^2+1)}}$$

Objective Type Questions

Choose the correct answer out of the given four options of the exercises from 83 to 96.

Q.83 If $f(x) = 2x$ and $g(x) = \frac{x^2}{2} + 1$, then which of the following can be a discontinuous function?

(a) $f(x) + g(x)$ (b) $f(x) - g(x)$

(c) $f(x) \cdot g(x)$ (d) $\frac{g(x)}{f(x)}$

Ans. ***(d)*** We know that, if f and g be continuous functions, then

(a) $f + g$ is continuous (b) $f - g$ is continuous.

(c) fg is continuous (d) $\frac{f}{g}$ is continuous at these points, where $g(x) \neq 0$.

Here, $$\frac{g(x)}{f(x)} = \frac{\frac{x^2}{2} + 1}{2x} = \frac{x^2 + 2}{4x}$$

which is discontinuous at $x = 0$.

Q.84 The function $f(x) = \frac{4 - x^2}{4x - x^3}$ is

(a) discontinuous at only one point

(b) discontinuous at exactly two points

(c) discontinuous at exactly three points

(d) None of the above

Ans. ***(c)*** We have, $$f(x) = \frac{4 - x^2}{4x - x^3} = \frac{(4 - x^2)}{x(4 - x^2)}$$

$$= \frac{(4 - x^2)}{x(2^2 - x^2)} = \frac{4 - x^2}{x(2 + x)(2 - x)}$$

Clearly, $f(x)$ is discontinuous at exactly three points $x = 0$, $x = -2$ and $x = 2$.

Q.85 The set of points where the function f given by $f(x) = |2x - 1| \sin x$ is differentiable is

(a) R (b) $R - \left(\frac{1}{2}\right)$

(c) $(0, \infty)$ (d) None of these

Ans. ***(b)*** We have, $f(x) = |2x - 1| \sin x$

At $x = \frac{1}{2}$, $f(x)$ is not differentiable.

$\therefore f(x)$ is differentiable in $R - \left(\frac{1}{2}\right)$.

Now, $$Rf'\left(\frac{1}{2}\right) = \lim_{h \to 0} \frac{f\left(\frac{1}{2}+h\right) - f\left(\frac{1}{2}\right)}{h}$$

On puting $x = \frac{1}{2} + h$, we have

$$Rf'\left(\frac{1}{2}\right) = \lim_{h \to 0} \frac{\left|2\left(\frac{1}{2}+h\right) - 1\right| \sin\left(\frac{1}{2}+h\right) - 0}{h}$$

$$= \lim_{h \to 0} \frac{|2h| \sin\left(\frac{1+2h}{2}\right)}{h} = 2 \cdot \sin\frac{1}{2}$$

and $$L\,f'\left(\frac{1}{2}\right) = \lim_{h \to 0} \frac{f\left(\frac{1}{2}-h\right) - f\left(\frac{1}{2}\right)}{-h}$$

Put $x = \frac{1}{2} - h$ $$= \lim_{h \to 0} \frac{\left|2\left(\frac{1}{2}-h\right) - 1\right| - \sin\left(\frac{1}{2}-h\right) - 0}{-h}$$

$$= \lim_{h \to 0} \frac{|-2h| \cdot \sin\left(\frac{1}{2}-h\right)}{-h} = -2\sin\left(\frac{1}{2}\right)$$

$\because$ $$Rf'\left(\frac{1}{2}\right) \neq Lf'\left(\frac{1}{2}\right)$$

So, $f(x)$ is not differentiable at $x = \frac{1}{2}$.

Q.86 The function $f(x) = \cot x$ is discontinuous on the set

(a) $\{x = n\pi : n \in Z\}$
(b) $\{x = 2n\pi : n \in Z\}$
(c) $\left\{x = (2n+1)\frac{\pi}{2}; n \in Z\right\}$
(d) $\left\{x = \frac{n\pi}{2}; n \in Z\right\}$

Ans. ***(a)*** We know that, $f(x) = \cot x$ is continuous in $R - \{x = n\pi \in Z\}$.

Since, $f(x) = \cot x = \frac{\cos x}{\sin x}$ $[\because \sin x = 0$ at $x = n\pi,\ n \in Z]$

Hence, $f(x) = \cot x$ is discontinuous on the set $\{x = n\pi : n \in Z\}$.

Q.87 The function $f(x) = e^{|x|}$ is

(a) continuous everywhere but not differentiable at $x = 0$
(b) continuous and differentiable everywhere
(c) not continuous at $x = 0$
(d) None of the above

Ans. ***(a)*** Let $u(x) = |x|$ and $v(x) = e^x$

$\therefore$ $$f(x) = vou(x) = v[u(x)] = v\,|x| = e^{|x|}$$

Since, $u(x)$ and $v(x)$ are both continuous functions.

So, $f(x)$ is also continuous function but $u(x) = |x|$ is not differentiable at $x = 0$, whereas $v(x) = e^x$ is differentiable at everywhere.

Hence, $f(x)$ is continuous everywhere but not differentiable at $x = 0$.

Q.88 If $f(x) = x^2 \sin\frac{1}{x}$, where $x \neq 0$, then the value of the function f at $x = 0$, so that the function is continuous at $x = 0$, is

(a) 0 (b) -1 (c) 1 (d) None of these

Ans. *(a)* Given, $f(x) = x^2 \sin\left(\frac{1}{x}\right)$, where $x \neq 0$ $\left[\because -1 \le \sin\frac{1}{x} \le 1, \forall\, x \in R\right]$

Both x^2 and $\sin\frac{1}{x}$ is continuous functions, $\forall\, x \in R$.

Hence, value of the function f at $x = 0$, so that it is continuous at $x = 0$ is 0.

Q.89 If $f(x) = \begin{cases} mx + 1, & \text{if } x \le \frac{\pi}{2} \\ \sin x + n, & \text{if } x > \frac{\pi}{2} \end{cases}$ is continuous at $x = \frac{\pi}{2}$, then

(a) $m = 1, n = 0$ (b) $m = \frac{n\pi}{2} + 1$ (c) $n = \frac{m\pi}{2}$ (d) $m = n = \frac{\pi}{2}$

Ans. *(c)* We have, $f(x) = \begin{cases} mx + 1, & \text{if } x \le \frac{\pi}{2} \\ (\sin x + n), & \text{if } x > \frac{\pi}{2} \end{cases}$ is continuous at $x = \frac{\pi}{2}$

$\therefore$ LHL $= \lim_{x \to \frac{\pi^-}{2}} (mx + 1) = \lim_{h \to 0} \left[m\left(\frac{\pi}{2} - h\right) + 1\right] = \frac{m\pi}{2} + 1$ $\left[\because \text{put } x = \frac{\pi}{2} - h\right]$

and RHL $= \lim_{x \to \frac{\pi^+}{2}} (\sin x + n) = \lim_{h \to 0} \left[\sin\left(\frac{\pi}{2} + h\right) + n\right]$ $\left[\because \text{put } x = \frac{\pi}{2} + h\right]$

$= \lim_{h \to 0} \cos h + n = 1 + n$

$\therefore$ LHL = RHL $\left[\text{to be continuous at } x = \frac{\pi}{2}\right]$

$\Rightarrow m \cdot \frac{\pi}{2} + 1 = n + 1$

$\therefore n = m \cdot \frac{\pi}{2} = \frac{m\pi}{2}$

Q.90 If $f(x) = |\sin x|$, then

(a) f is everywhere differentiable
(b) f is everywhere continuous but not differentiable at $x = n\pi, n \in Z$
(c) f is everywhere continuous but not differentiable at $x = (2n + 1)\frac{\pi}{2}, n \in Z$
(d) None of the above

Ans. *(b)* We have, $f(x) = |\sin x|$

Let $f(x) = vou(x) = v[u(x)]$ [where, $u(x) = \sin x$ and $v(x) = |x|$]

$= v(\sin x) = |\sin x|$

where, $u(x)$ and $v(x)$ are both continuous.

Hence, $f(x) = vou(x)$ is also a continuous function but $v(x)$ is not differentiable at $x = 0$.

So, $f(x)$ is not differentiable where $\sin x = 0 \Rightarrow x = n\pi,\ n \in Z$

Hence, $f(x)$ is continuous everywhere but not differentiable at $x = n\pi,\ n \in Z$.

Q.91 If $y = \log\left(\dfrac{1-x^2}{1+x^2}\right)$, then $\dfrac{dy}{dx}$ is equal to

(a) $\dfrac{4x^3}{1-x^4}$ (b) $\dfrac{-4x}{1-x^4}$ (c) $\dfrac{1}{4-x^4}$ (d) $\dfrac{-4x^3}{1-x^4}$

Ans. ***(b)*** We have, $y = \log\left(\dfrac{1-x^2}{1+x^2}\right)$

$$\Rightarrow \quad y = \log(1-x^2) - \log(1+x^2) \qquad \left[\because \log\frac{m}{n} = \log m - \log n\right]$$

On differentiating w.r.t. x, we get

$$\frac{dy}{dx} = \frac{d}{dx}\log(1-x^2) - \frac{d}{dx}\log(1+x^2)$$

$$\Rightarrow \frac{dy}{dx} = \frac{1}{(1-x^2)}\cdot\frac{d}{dx}(1-x^2) - \frac{1}{(1+x^2)}\cdot\frac{d}{dx}(1+x^2) \qquad \left[\because \frac{d}{dx}\log x = \frac{1}{x}\right]$$

$$\Rightarrow \frac{dy}{dx} = \frac{-2x}{1-x^2} - \frac{2x}{1+x^2}$$

$$\Rightarrow \frac{dy}{dx} = -2x\left[\frac{1}{1-x^2} + \frac{1}{1+x^2}\right] \Rightarrow \frac{dy}{dx} = -2x\left[\frac{(1+x^2)+(1-x^2)}{1-x^4}\right]$$

$$\therefore \frac{dy}{dx} = \frac{-2x\cdot 2}{1-x^4} = \frac{-4x}{1-x^4}$$

Q.92 If $y = \sqrt{\sin x + y}$, then $\dfrac{dy}{dx}$ is equal to

(a) $\dfrac{\cos x}{2y-1}$ (b) $\dfrac{\cos x}{1-2y}$ (c) $\dfrac{\sin x}{1-2y}$ (d) $\dfrac{\sin x}{2y-1}$

Ans. ***(a)*** Since, $y = (\sin x + y)^{1/2}$

$$\therefore \frac{dy}{dx} = \frac{1}{2}(\sin x + y)^{-1/2}\cdot\frac{d}{dx}(\sin x + y)$$

$$\Rightarrow \frac{dy}{dx} = \frac{1}{2}\cdot\frac{1}{(\sin x + y)^{1/2}}\cdot\left(\cos x + \frac{dy}{dx}\right)$$

$$\Rightarrow \frac{dy}{dx} = \frac{1}{2y}\left(\cos x + \frac{dy}{dx}\right) \qquad [\because (\sin x + y)^{1/2} = y]$$

$$\Rightarrow \frac{dy}{dx}\left(1 - \frac{1}{2y}\right) = \frac{\cos x}{2y}$$

$$\therefore \frac{dy}{dx} = \frac{\cos x}{2y}\cdot\frac{2y}{2y-1} = \frac{\cos x}{2y-1}$$

Q.93 The derivative of $\cos^{-1}(2x^2 - 1)$ w.r.t. $\cos^{-1} x$ is

(a) 2 (b) $-\frac{-1}{2\sqrt{1-x^2}}$ (b) $\frac{2}{x}$ (d) $1 - x^2$

Ans. **(a)** Let $u = \cos^{-1}(2x^2 - 1)$ and $v = \cos^{-1} x$

Put $x = \cos\theta$

$$u = \cos^{-1}(2\cos^2\theta - 1)$$
$$u = \cos^{-1}(\cos 2\theta) = 2\theta \qquad [\because 2\cos^2\theta - 1 = \cos 2\theta]$$
$$u = 2\cos^{-1} x \qquad [\because \theta = \cos^{-1} x]$$

On differentiating w.r.t. x, we get

$$\frac{du}{dx} = -\frac{2}{\sqrt{1-x^2}} \qquad \left[\because \frac{d}{dx}\cos^{-1} x = \frac{-1}{\sqrt{1-x^2}}\right]$$

and
$$v = \cos^{-1} x$$
$$\frac{dv}{dx} = -\frac{1}{\sqrt{1-x^2}}$$

Now,
$$\frac{du}{dv} = \frac{\frac{du}{dx}}{\frac{dv}{dx}} = \frac{\frac{-2}{\sqrt{1-x^2}}}{-\frac{1}{\sqrt{1-x^2}}}$$
$$\frac{du}{dv} = 2$$

Q.94 If $x = t^2$ and $y = t^3$, then $\frac{d^2y}{dx^2}$ is equal to

(a) $\frac{3}{2}$ (b) $\frac{3}{4t}$ (c) $\frac{3}{2t}$ (d) $\frac{3}{2t}$

Ans. **(b)** We have, $x = t^2$ and $y = t^3$

On differentiating w.r.t. t both Eqs., we get

$$\frac{dx}{dt} = 2t \text{ and } \frac{dy}{dt} = 3t^2$$
$$\therefore \quad \frac{dy}{dx} = \frac{dy/dt}{dx/dt} = \frac{3t^2}{2t} = \frac{3}{2}t$$

On further, differentiating w.r.t. x, we get

$$\frac{d^2y}{dx^2} = \frac{3}{2}\cdot\frac{d}{dt}t\cdot\frac{dt}{dx}$$
$$= \frac{3}{2}\cdot\frac{1}{2t} \qquad \left[\because \frac{dt}{dx} = \frac{1}{2t}\right]$$
$$= \frac{3}{4t}$$

***Q.95** The value of c in Rolle's theorem for the function $f(x) = x^3 - 3x$ in the interval $[0, \sqrt{3}]$ is

(a) 1 (b) – 1 (c) $\frac{3}{2}$ (d) $\frac{1}{3}$

Note (*) *Topics related to these question have been removed from latest NCERT.*

Ans. *(a)* Given, $f(x) = x^3 - 3x$

$\Rightarrow \quad f'(x) = 3x^2 - 3$

$\Rightarrow \quad f'(c) = 3c^2 - 3$

$\because \quad f'(c) = 0$

$\Rightarrow \quad 3c^2 - 3 = 0 \quad \Rightarrow c^2 = \frac{3}{3} = 1 \Rightarrow c = \pm 1$, where $1 \in (0, \sqrt{3})$

$\therefore \quad c = 1$

*** Q.96** For the function $f(x) = x + \frac{1}{x}$, $x \in [1, 3]$, the value of c for mean value theorem is

(a) 1 (b) $\sqrt{3}$ (c) 2 (d) None of these

Ans. *(b)* $\because \quad f'(c) = \frac{f(b) - f(a)}{b - a}$

$\Rightarrow \quad 1 - \frac{1}{c^2} = \frac{\left[3 + \frac{1}{3}\right] - \left[1 + \frac{1}{1}\right]}{3 - 1}$ $\left[\because f'(x) = 1 - \frac{1}{x^2} \text{ and } b = 3, a = 1\right]$

$\Rightarrow \quad \frac{c^2 - 1}{c^2} = \frac{\frac{10}{3} - 2}{2} \Rightarrow \frac{c^2 - 1}{c^2} = \frac{4}{3 \times 2} = \frac{2}{3}$

$\Rightarrow \quad 3(c^2 - 1) = 2c^2 \Rightarrow 3c^2 - 2c^2 = 3 \Rightarrow c^2 = 3 \Rightarrow c = \pm\sqrt{3}$

$\because \quad c = \sqrt{3} \in [1, 3]$

Fill in the blanks in each of the exercises 97 to 101.

Q.97 An example of a function which is continuous everywhere but fails to be differentiable exactly at two points is ....

Ans. $|x| + |x - 1|$ is continuous everywhere but fails to be differentiable exactly at two points $x = 0$ and $x = 1$.
So, there can be more such examples of functions.

Q.98 Derivative of x^2 w.r.t. x^3 is ....

Ans. Let $\quad u = x^2$ and $v = x^3$

On differentiating w.r.t. x, we get

$$\frac{du}{dx} = 2x \text{ and } \frac{dv}{dx} = 3x^2$$

$$\therefore \quad \frac{du}{dv} = \frac{2x}{3x^2} = \frac{2}{3x}$$

Q.99 If $f(x) = |\cos x|$, then $f'\left(\frac{\pi}{4}\right)$ is equal to ....

Ans. We have, $f(x) = |\cos x|$

$\Rightarrow \quad f(x) = \cos x \; 0 < x < \frac{\pi}{2}, \cos x > 0.$

Note (*) *Topics related to these question have been removed from latest NCERT.*

$\therefore \quad f'(x) = (-\sin x)$

$\Rightarrow \quad f'\left(\frac{\pi}{4}\right) = -\sin\frac{\pi}{4} = \frac{-1}{\sqrt{2}} \qquad \left[\because \sin\frac{\pi}{4} = \frac{1}{\sqrt{2}}\right]$

Q.100 If $f(x) = |\cos x - \sin x|$, then $f'\left(\frac{\pi}{3}\right)$ is equal to ....

Ans. We have, $f(x) = |\cos x - \sin x|$

We know that, $\frac{\pi}{4} < x < \frac{\pi}{2}$, $\sin x > \cos x$

$\therefore$ $\cos x - \sin x \le 0$ *i.e.,* $f(x) = -(\cos x - \sin x)$

$$f'(x) = -[-\sin x - \cos x]$$

$$\therefore \quad f'\left(\frac{\pi}{3}\right) = -\left(\frac{-\sqrt{3}}{2} - \frac{1}{2}\right) = \left(\frac{\sqrt{3}+1}{2}\right)$$

Q.101 For the curve $\sqrt{x} + \sqrt{y} = 1$, $\frac{dy}{dx}$ at $\left(\frac{1}{4}, \frac{1}{4}\right)$ is ....

Ans. We have, $\sqrt{x} + \sqrt{y} = 1$

Taking derivative w.r.t. 'x' on both sides, we get

$$\Rightarrow \quad \frac{1}{2\sqrt{x}} + \frac{1}{2\sqrt{y}}\frac{dy}{dx} = 0 \Rightarrow \frac{dy}{dx} = -\frac{\sqrt{y}}{\sqrt{x}}$$

$$\therefore \quad \left(\frac{dy}{dx}\right)_{\left(\frac{1}{4}, \frac{1}{4}\right)} = \frac{-\frac{1}{2}}{\frac{1}{2}} = -1$$

State True or Flase for the statements in each of the exercises 102 to 106.

***Q.102** Rolle's theorem is applicable for the function $f(x) = |x-1|$ in $[0, 2]$.

Ans. *False*

Hence, $f(x) = |x-1|$ in $[0, 2]$ is not differentiable at $x = 1 \in (0, 2)$.

Q.103 If f is continuous on its domain D, then $|f|$ is also continuous on D.

Ans. *True*

Q.104 The composition of two continuous function is a continuous function.

Ans. *True*

Q.105 Trigonometric and inverse trigonometric functions are differentiable in their respective domain.

Ans. *True*

Q.106 If $f \cdot g$ is continuous at $x = a$, then f and g are separately continuous at $x = a$.

Ans. *False*

Let $f(x) = \sin x$ and $g(x) = \cot x$

$$\therefore \quad f(x) \cdot g(x) = \sin x \cdot \frac{\cos x}{\sin x} = \cos x$$

which is continuous at $x = 0$ but $\cot x$ is not continuous at $x = 0$.

Note (*) *Topics related to these question have been removed from latest NCERT.*

Chapter

06 Application of Derivatives

Short Answer Type Questions

Q.1 A spherical ball of salt is dissolving in water in such a manner that the rate of decrease of the volume at any instant is proportional to the surface. Prove that the radius is decreasing at a constant rate.

Thinking Process First, let V be the volume of the ball and S be the surface area of the ball and then by using $\frac{dV}{dt} \propto S$, we can prove the required result.

Ans. We have, rate of decrease of the volume of spherical ball of salt at any instant is $\propto$ surface area.

Let the radius of the spherical ball of the salt be r.

$\therefore$ Volume of the ball $(V) = \frac{4}{3}\pi r^3$

and surface area $(S) = 4\pi r^2$

$$\because \quad \frac{dV}{dt} \propto S \quad \Rightarrow \quad \frac{d}{dt}\left(\frac{4}{3}\pi r^3\right) \propto 4\pi r^2$$

$$\Rightarrow \quad \frac{4}{3}\pi \cdot 3r^2 . \frac{dr}{dt} \propto 4\pi r^2 \quad \Rightarrow \quad \frac{dr}{dt} \propto \frac{4\pi r^2}{4\pi r^2}$$

$$\Rightarrow \quad \frac{dr}{dt} = k \cdot 1 \text{ [where, } k \text{ is the proportionality constant]}$$

$$\Rightarrow \quad \frac{dr}{dt} = k$$

Hence, the radius of ball is decreasing at a constant rate.

Q.2 If the area of a circle increases at a uniform rate, then prove that perimeter varies inversely as the radius.

Ans. Let the radius of circle $= r$ and area of the circle, $A = \pi r^2$

$$\therefore \quad \frac{d}{dt}A = \frac{d}{dt}\pi r^2$$

$$\Rightarrow \quad \frac{dA}{dt} = 2\pi r \cdot \frac{dr}{dt} \qquad \text{...(i)}$$

Since, the area of a circle increases at a uniform rate, then

$$\frac{dA}{dt} = k \qquad \text{...(ii)}$$

where, k is a constant.

From Eqs. (i) and (ii), we get $2\pi r \cdot \frac{dr}{dt} = k$

$$\Rightarrow \quad \frac{dr}{dt} = \frac{k}{2\pi r} = \frac{k}{2\pi} \cdot \left(\frac{1}{r}\right) \quad \text{...(iii)}$$

Let the perimeter, $P = 2\pi r$ [P = circumference of circle]

On differentiating w.r.t. r, we get

$$\therefore \quad \frac{dP}{dt} = \frac{d}{dt} \cdot 2\pi r \Rightarrow \frac{dP}{dt} = 2\pi \cdot \frac{dr}{dt}$$

$$= 2\pi \cdot \frac{k}{2\pi} \cdot \frac{1}{r} = \frac{k}{r} \quad \text{[using Eq. (iii)]}$$

$$\Rightarrow \quad \frac{dP}{dt} \propto \frac{1}{r}$$

Hence proved.

Q.3 A kite is moving horizontally at a height of 151.5 m. If the speed of kite is 10 m/s, how fast is the string being let out, when the kite is 250 m away from the boy who is flying the kite, if the height of boy is 1.5 m?

Ans. We have, height $(h) = 151.5$ m, speed of kite $(v) = 10$ m/s

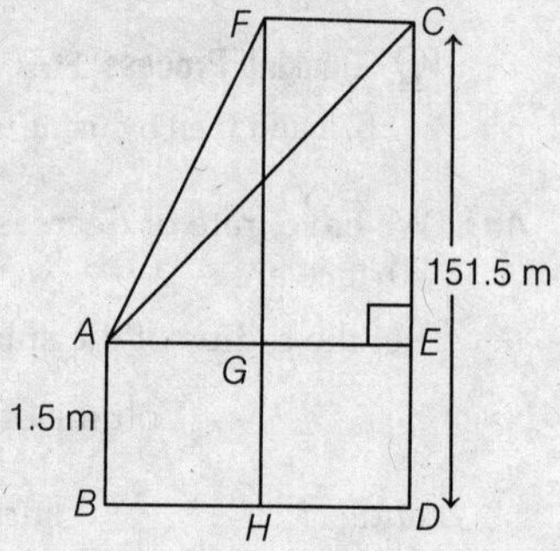

Let CD be the height of kite and AB be the height of boy.

Let $DB = x \text{ m} = EA$ and $AC = 250$ m

$$\therefore \quad \frac{dx}{dt} = 10 \text{ m/s}$$

From the figure, we see that

$$EC = CD - ED \Rightarrow CD - AB \quad [\because ED = AB]$$

$$EC = 151.5 - 1.5 = 150 \text{ m}$$

and $AE = x$ $[\because DB = AE]$

Also, $AC = 250$ m

In right angled ΔCEA,

$$AE^2 + EC^2 = AC^2$$

$$\Rightarrow \quad x^2 + (150)^2 = y^2 \quad \text{... (i)}$$

$$\Rightarrow \quad x^2 + (150)^2 = (250)^2$$

$$\Rightarrow \quad x^2 = (250)^2 - (150)^2$$

$$= (250 + 150)(250 - 150) = 400 \times 100$$

$$\therefore \quad x = 20 \times 10 = 200$$

From Eq. (i), on differentiating w.r.t. t, we get

$$2x \cdot \frac{dx}{dt} + 0 = 2y \frac{dy}{dt}$$

$$\Rightarrow \quad 2y \frac{dy}{dt} = 2x \frac{dx}{dt}$$

$$\therefore \quad \frac{dy}{dt} = \frac{x}{y} \cdot \frac{dx}{dt}$$

$$= \frac{200}{250} \cdot 10 = 8 \text{ m/s} \quad \left[\because \frac{dx}{dt} = 10 \text{ m/s}\right]$$

So, the required rate at which the string is being let out is 8 m/s.

Q.4 Two men A and B start with velocities v at the same time from the junction of two roads inclined at 45° to each other. If they travel by different roads, then find the rate at which they are being separated.

Thinking Process By drawing figure such that men start moving at a point C, A and B are separating points, then draw perpendicular from that point C to AB to get D. Now, get the value of $\angle ACD$ in terms of x and y, then by using $\frac{dy}{dt}$ get desired result.

Ans. Let two men start from the point C with velocity v each at the same time.

Also, $\angle BCA = 45°$

Since, A and B are moving with same velocity v, so they will cover same distance in same time.

Therefore, ΔABC is an isosceles triangle with $AC = BC$.

Now, draw $CD \perp AB$.

Let at any instant t, the distance between them is AB.

Let $AC = BC = x$ and $AB = y$

In ΔACD and ΔDCB,

$$\angle CAD = \angle CBD \quad [\because AC = BC]$$

$$\angle CDA = \angle CDB = 90°$$

$$\therefore \quad \angle ACD = \angle DCB$$

$$\text{or} \quad \angle ACD = \frac{1}{2} \times \angle ACB$$

$$\Rightarrow \quad \angle ACD = \frac{1}{2} \times 45°$$

$$\Rightarrow \quad \angle ACD = \frac{\pi}{8} \quad \left[45° = \frac{\pi}{4}\right]$$

$$\therefore \text{ In } \angle ACD \quad \sin\frac{\pi}{8} = \frac{AD}{AC}$$

$$\Rightarrow \quad \sin\frac{\pi}{8} = \frac{y/2}{x} \quad [\because AD = y/2]$$

$$\Rightarrow \quad \frac{y}{2} = x\sin\frac{\pi}{8}$$

$$\Rightarrow \quad y = 2x\cdot\sin\frac{\pi}{8}$$

Now, on differentiating both sides w.r.t. t, we get

$$\Rightarrow \quad \frac{dy}{dt} = 2\cdot\sin\frac{\pi}{8}\cdot\frac{dx}{dt}$$

$$\Rightarrow \quad \frac{dy}{dt} = 2\cdot\sin\frac{\pi}{8}.v \quad \left[\because v = \frac{dx}{dt}\right] \quad \dots\text{(i)}$$

As we knowthat $\cos\theta = 1 - 2\sin^2\frac{\theta}{2}$

$$\Rightarrow \quad 2\sin^2\frac{\theta}{2} = 1 - \cos\theta$$

$$\Rightarrow \quad \sin^2\frac{\theta}{2} = \frac{1-\cos\theta}{2}$$

$$\Rightarrow \quad \sin\frac{\theta}{2} = \sqrt{\frac{1-\cos\theta}{2}}$$

Put $\theta = \frac{\pi}{4}$

$$\Rightarrow \quad \sin\frac{\pi}{8} = \sqrt{\frac{1-\cos\frac{\pi}{4}}{2}} = \sqrt{\frac{1-\frac{1}{\sqrt{2}}}{2}} \quad \left[\because \cos\frac{\pi}{4} = \frac{1}{\sqrt{2}}\right]$$

$$\Rightarrow \quad \sin\frac{\pi}{8} = \frac{\sqrt{2-\sqrt{2}}}{2} \quad \text{[On multiplying and dividing by } \sqrt{2}\text{]}$$

$$\Rightarrow \quad \frac{dy}{dt} = 2v\cdot\frac{\sqrt{2-\sqrt{2}}}{2} \quad \text{[From Eq. (i)]}\left[\because \sin\frac{\pi}{8} = \frac{\sqrt{2-\sqrt{2}}}{2}\right]$$

$$\therefore \quad = \sqrt{2-\sqrt{2}}\, v\text{unit/s}$$

which is the rate at which A and B are being separated.

Q.5 Find an angle θ, where $0 < \theta < \frac{\pi}{2}$, which increases twice as fast as its sine.

Ans. Let θ increases twice as fast as its sine.

$$\Rightarrow \quad \theta = 2\sin\theta$$

Now, on differentiating both sides w.r.t. t, we get

$$\frac{d\theta}{dt} = 2\cdot\cos\theta\cdot\frac{d\theta}{dt} \Rightarrow 1 = 2\cos\theta$$

$$\Rightarrow \quad \frac{1}{2} = \cos\theta \Rightarrow \cos\theta = \cos\frac{\pi}{3}$$

$$\therefore \quad \theta = \frac{\pi}{3}$$

So, the required angle is $\frac{\pi}{3}$.

***Q.6** Find the approximate value of $(1.999)^5$.

Ans. Let $x = 2$

and $\Delta x = -0.001$ $\quad [\because 2 - 0.001 = 1.999]$

Let $y = x^5$

On differentiating both sides w.r.t. x, we get

$$\frac{dy}{dx} = 5x^4$$

Now,

$$\Delta y = \frac{dy}{dx}\cdot\Delta x = 5x^4 \times \Delta x$$

$$= 5\times 2^4 \times [-0.001] = -80\times 0.001 = -0.080$$

$$\therefore \quad (1.999)^5 = y + \Delta y$$

$$= 2^5 + (-0.080)$$

$$= 32 - 0.080 = 31.920$$

Note (*) *Topics related to these question have been removed from latest NCERT.*

*** Q.7** **Find the approximate volume of metal in a hollow spherical shell whose internal and external radii are 3 cm and 3.0005 cm, respectively.**

Ans. Let internal radius = r and external radius = R

$\therefore$ Volume of hollow spherical shell, $V = \frac{4}{3}\pi(R^3 - r^3)$

$$\Rightarrow \quad V = \frac{4}{3}\pi[(3.0005)^3 - (3)^3] \quad \text{... (i)}$$

Now, we shall use differentiation to get approximate value of $(3.0005)^3$.

Let $\quad (3.0005)^3 = y + \Delta y \quad$ ...(ii)

and $\quad x = 3, \Delta x = 0.0005$

Also, let $\quad y = x^3$

On differentiating both sides w.r.t. x, we get

$$\frac{dy}{dx} = 3x^2$$

$$\therefore \quad \Delta y = \frac{dy}{dx} \times \Delta x = 3x^2 \times 0.0005$$

$$= 3 \times 3^2 \times 0.0005$$

$$= 27 \times 0.0005 = 0.0135$$

Also, $\quad (3.0005)^3 = y + \Delta y \quad$ [using Eq. (ii)]

$$= 3^3 + 0.0135 = 27.0135$$

$$\therefore \quad V = \frac{4}{3}\pi[27.0135 - 27.000] \quad \text{[using Eq. (i)]}$$

$$= \frac{4}{3}\pi[0.0135] = 4\pi \times (0.0045)$$

$$= 0.0180\pi \text{ cm}^3$$

Q.8 **A man, 2 m tall, walks at the rate of $1\frac{2}{3}$ m/s towards a street light which is $5\frac{1}{3}$ m above the ground. At what rate is the tip of his shadow moving and at what rate is the length of the shadow changing when he is $3\frac{1}{3}$ m from the base of the light?**

Ans. Let AB be the street light post and CD be the height of man *i.e.,* $CD = 2$ m.

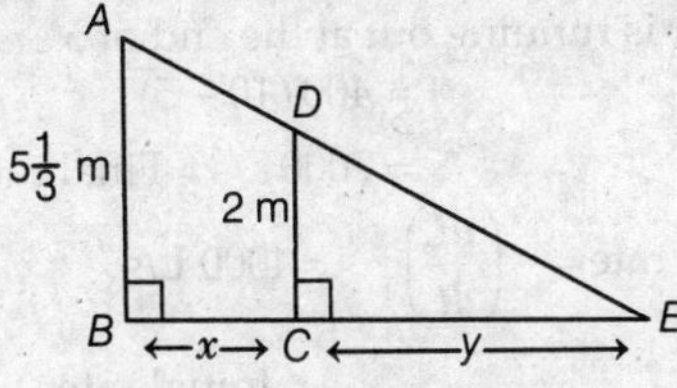

Let $BC = x$ m, $CE = y$ m and $\frac{dx}{dt} = \frac{-5}{3}$ m/s

Note (*) *Topics related to these question have been removed from latest NCERT.*

From ΔABE and ΔDCE, we see that

$$\angle ABE = \angle DCE = 90°$$

$$\angle E \text{ is common}$$

So, $\Delta ABE \sim \Delta DCE$ [by AA similarily]

$$\therefore \quad \frac{AB}{DC} = \frac{BE}{CE} \Rightarrow \frac{\frac{16}{3}}{2} = \frac{x+y}{y} \Rightarrow \frac{16}{6} = \frac{x+y}{y}$$

$$\Rightarrow \quad 16y = 6x + 6y \Rightarrow 10y = 6x \Rightarrow y = \frac{3}{5}x$$

On differentiating both sides w.r.t. t, we get

$$\frac{dy}{dt} = \frac{3}{5} \cdot \frac{dx}{dt} = \frac{3}{5} \cdot \left(-1\frac{2}{3}\right)$$

[since, man is moving towards the light post]

$$= \frac{3}{5} \cdot \left(\frac{-5}{3}\right) = -1 \text{m/s}$$

Let $z = x + y$

Now, on differentiating both sides w.r.t. t, we get

$$\frac{dz}{dt} = \frac{dx}{dt} + \frac{dy}{dt} = -\left(\frac{5}{3} + 1\right) = -\frac{8}{3} = -2\frac{2}{3} \text{m/s}$$

Hence, the tip of shadow is moving at the rate of $2\frac{2}{3}$ m/s towards the light source and length of the shadow is decreasing at the rate of 1 m/s.

Q.9 A swimming pool is to be drained for cleaning. If L represents the number of litres of water in the pool t seconds after the pool has been plugged off to drain and $L = 200(10 - t)^2$. How fast is the water running out at the end of 5 s and what is the average rate at which the water flows out during the first 5 s?

Ans. Let L represents the number of litres of water in the pool t seconds after the pool has been plugged off to drain, then

$$L = 200(10 - t)^2$$

$\therefore$ Rate at which the water is running out $= -\dfrac{dL}{dt}$

[$-ve$ sign shows that water is drain out]

$$\frac{dL}{dt} = -200 \cdot 2(10 - t) \cdot (-1) = 400(10 - t)$$

Rate at which the water is running out at the end of 5 s

$$= 400(10 - 5) \quad [t = 5 \text{ s}]$$

$$= 2000 \text{ L/s} = \text{Final rate}$$

Since, $\quad$ initial rate $= -\left(\dfrac{dL}{dt}\right)_{t=0} = 4000$ L/s

$$\therefore \quad \text{Average rate during 5 s} = \frac{\text{Initial rate} + \text{Final rate}}{2}$$

$$= \frac{4000 + 2000}{2} = 3000 \text{ L/s}$$

Q.10 The volume of a cube increases at a constant rate. Prove that the increase in its surface area varies inversely as the length of the side.

Ans. Let the side of a cube be x unit.

$\therefore$ Volume of cube $(V) = x^3$

On differentiating both side w.r.t. t, we get

$$\frac{dV}{dt} = 3x^2\frac{dx}{dt} = k \text{ [constant]}$$

$$\Rightarrow \quad \frac{dx}{dt} = \frac{k}{3x^2} \quad \text{... (i)}$$

Also, surface area of cube, $S = 6x^2$

On differentiating w.r.t. t, we get

$$\frac{dS}{dt} = 12x\cdot\frac{dx}{dt}$$

$$\Rightarrow \quad \frac{dS}{dt} = 12\,x\cdot\frac{k}{3x^2} \quad \text{[using Eq. (i)]}$$

$$\Rightarrow \quad \frac{dS}{dt} = \frac{12k}{3x} = 4\left(\frac{k}{x}\right)$$

$$\Rightarrow \quad \frac{dS}{dt} \propto \frac{1}{x}$$

Hence, the surface area of the cube varies inversely as the length of the side.

Q.11 If x and y are the sides of two squares such that $y = x - x^2$, then find the rate of change of the area of second square with respect to the area of first square.

Thinking Process First, let A_1 and A_2 be the areas of two squares and get their values in one variable and then by using dA_1/dt and dA_2/dt get the value of dA_2/dA_1.

Ans. Since, x and y are the sides of two squares such that $y = x - x^2$.

$\therefore$ Area of the first square $(A_1) = x^2$

and area of the second square $(A_2) = y^2 = (x - x^2)^2$

$$\therefore \quad \frac{dA_2}{dt} = \frac{d}{dt}(x - x^2)^2 = 2(x - x^2)\left(\frac{dx}{dt} - 2x\cdot\frac{dx}{dt}\right)$$

$$= \frac{dx}{dt}(1-2x)\,2(x - x^2)$$

and

$$\frac{dA_1}{dt} = \frac{d}{dt}x^2 = 2x\cdot\frac{dx}{dt}$$

$$\therefore \quad \frac{dA_2}{dA_1} = \frac{dA_2/dt}{dA_1/dt} = \frac{\frac{dx}{dt}\cdot(1-2x)(2x - 2x^2)}{2x\cdot\frac{dx}{dt}}$$

$$= \frac{(1-2x)\,2x(1-x)}{2x}$$

$$= (1-2x)(1-x)$$

$$= 1 - x - 2x + 2x^2$$

$$= 2x^2 - 3x + 1$$

***Q.12** Find the condition that curves $2x = y^2$ and $2xy = k$ intersect orthogonally.

Thinking Process First, get the intersection point of the curve and then get the slopes of both the curves at that point. Then, by using $m_1 \cdot m_2 = -1$, get the required condition.

Ans. Given, equation of curves are $2x = y^2$ ...(i)

$$2xy = k \quad \text{...(ii)}$$

On solving the Eqs. (i) and (ii) we get the point of intersection of the curves.

On putting value of $2x$ from Eq. (i) in Eq. (ii), we get

$$y^2 \cdot y = k$$

$$\Rightarrow \quad y^3 = k$$

$$y = (k)^{1/3}$$

Put the value of 'y' in Eq. (ii)

$$x = \frac{1}{2}k^{2/3}$$

Hence, the point of intersection of curve will be $\left(\frac{1}{2}k^{2/3}, k^{1/3}\right)$.

From Eqs. (i) and (ii), we will find $\frac{dy}{dx}$ of both curves at intersection points the curves

$$2 = 2y\frac{dy}{dx} \Rightarrow \frac{dy}{dx} = \frac{1}{y} \quad \text{...(iii)}$$

and $$2\left[x \cdot \frac{dy}{dx} + y \cdot 1\right] = 0 \quad \text{[using product rule]}$$

and $$\left(\frac{dy}{dx}\right) = \frac{-2y}{2x} = -\frac{y}{x} \quad \text{...(iv)}$$

$$\Rightarrow \quad \left(\frac{dy}{dx}\right)_{\left(\frac{1}{2}k^{2/3}, k^{1/3}\right)} = \frac{1}{k^{1/3}} \quad \text{From Eq. (iii) [say } m_1]$$

and $$\left(\frac{dy}{dx}\right)_{\left(\frac{1}{2}k^{2/3}, k^{1/3}\right)} = \frac{-k^{1/3}}{\frac{1}{2}k^{2/3}} = -2k^{-1/3} \quad \text{From Eq. (iv) [say } m_2]$$

Since, the curves intersect orthogonally.

i.e., $$m_1 \cdot m_2 = -1$$

$$\Rightarrow \quad \frac{1}{k^{1/3}} \cdot (-2k^{-1/3}) = -1$$

$$\Rightarrow \quad -2k^{-2/3} = -1$$

$$\Rightarrow \quad \frac{2}{k^{2/3}} = 1$$

$$\Rightarrow \quad k^{2/3} = 2$$

$$\therefore \quad k^2 = 8$$

which is the required condition.

Note (*) *Topics related to these question have been removed from latest NCERT.*

***Q.13** Prove that the curves $xy = 4$ and $x^2 + y^2 = 8$ touch each other.

Thinking Process First, find the intersection points of curves and then equate the slopes of both the curves at the obtained point.

Ans. Given, equation of curves are

$$xy = 4 \quad \text{...(i)}$$

and $$x^2 + y^2 = 8 \quad \text{...(ii)}$$

On differentiating Eq. (i) w.r.t. x, we get

$$\Rightarrow \quad x \cdot \frac{dy}{dx} + y = 0 \quad \text{...(iii)}$$

On differentiating Eq. (ii) w.r.t. x, we get

and $$2x + 2y\frac{dy}{dx} = 0 \quad \text{...(iv)}$$

$$\Rightarrow \quad \frac{dy}{dx} = \frac{-y}{x} \quad \text{[from Eq. (iii) say } m_1]$$

and $$\frac{dy}{dx} = \frac{-2x}{2y} = -\frac{x}{y} \quad \text{[from Eq. (iv) say } m_2]$$

Since, both the curves should have same slope.

$$\therefore \quad \frac{-y}{x} = \frac{-x}{y} \Rightarrow -y^2 = -x^2 \quad [\because m_1 = m_2]$$

$$\Rightarrow \quad x^2 = y^2 \quad \text{...(v)}$$

On using the value of x^2 in Eq. (ii), we get

$$y^2 + y^2 = 8$$

$$\Rightarrow \quad y^2 = 4 \Rightarrow y = \pm 2 \quad \text{[from Eq. (i)]}$$

For $y = 2$, $x = \frac{4}{2} = 2$

and for $y = -2$, $x = \frac{4}{-2} = -2$

Thus, the required points of intersection are (2, 2) and (– 2, – 2).

For (2, 2), $$m_1 = \frac{-y}{x} = \frac{-2}{2} = -1$$

and $$m_2 = \frac{-x}{y} = \frac{-2}{2} = -1$$

$$\because \quad m_1 = m_2$$

For (–2,–2), $$m_1 = \frac{-y}{x} = \frac{-(-2)}{-2} = -1$$

and $$m_2 = \frac{-x}{y} = \frac{-(-2)}{-2} = -1$$

Thus, for both the intersection points, we see that slope of both the curves are same.

Hence, the curves touch each other.

Note (✱) *Topics related to these question have been removed from latest NCERT.*

***Q.14** Find the coordinates of the point on the curve $\sqrt{x}+\sqrt{y}=4$ at which tangent is equally inclined to the axes.

Ans. We have,

$$\sqrt{x}+\sqrt{y}=4 \quad \text{...(i)}$$

$$x^{1/2}+y^{1/2}=4$$

On differentiating w.r.t. x, we get

$$\Rightarrow \quad \frac{1}{2}\cdot\frac{1}{x^{1/2}}+\frac{1}{2}\cdot\frac{1}{y^{1/2}}\cdot\frac{dy}{dx}=0$$

$$\therefore \quad \frac{dy}{dx}=-\frac{1}{2}\cdot x^{-1/2}\ 2\cdot y^{1/2}=-\sqrt{\frac{y}{x}}$$

Since, tangent is equally inclined to the axes.

Then, slope will be tan 45° and tan 135°

i.e., $$\frac{dy}{dx}=\pm 1 \Rightarrow -\sqrt{\frac{y}{x}}=\pm 1 \Rightarrow \frac{y}{x}=1 \Rightarrow y=x$$

From Eq. (i), $$\sqrt{y}+\sqrt{y}=4 \quad [\because y=x]$$

$$\Rightarrow \quad 2\sqrt{y}=4 \quad \text{[on squaring both sides of equation]}$$

$$\Rightarrow \quad 4y=16$$

$$\therefore \quad y=4 \text{ and } x=4 \quad [\because y=x]$$

When, $y=4$, then $x=4$

So, the required coordinates are (4, 4).

***Q.15** Find the angle of intersection of the curves $y=4-x^2$ and $y=x^2$.

Ans. We have, $$y=4-x^2 \quad \text{...(i)}$$

and $$y=x^2 \quad \text{... (ii)}$$

On differentiating Eq. (i) w.r.t. x, we get

$$\Rightarrow \quad \frac{dy}{dx}=-2x \quad [\text{say } m_1]$$

On differentiating Eq. (ii) w.r.t. x, we get

and $$\frac{dy}{dx}=2x \quad [\text{say } m_2]$$

$$\Rightarrow \quad m_1=-2x \text{ and } m_2=2x$$

On putting the value of $y=x^2$ in Eq. (i), we get

From Eqs. (i) and (ii), $x^2=4-x^2 \Rightarrow 2x^2=4 \Rightarrow x^2=2 \Rightarrow x=\pm\sqrt{2}$

$$\therefore \quad y=x^2=(\pm\sqrt{2})^2=2$$

So, the points of intersection are $(\sqrt{2}, 2)$ and $(-\sqrt{2}, 2)$.

For point $(\sqrt{2}, 2)$, $m_1=-2x=-2\cdot\sqrt{2}=-2\sqrt{2}$ and $m_2=2x=2\sqrt{2}$

$$\tan\theta=\left|\frac{m_1-m_2}{1+m_1 m_2}\right|=\left|\frac{-2\sqrt{2}-2\sqrt{2}}{1-2\sqrt{2}\cdot 2\sqrt{2}}\right|=\left|\frac{-4\sqrt{2}}{-7}\right|$$

$$\therefore \quad \theta=\tan^{-1}\left(\frac{4\sqrt{2}}{7}\right)$$

Note (*) *Topics related to these question have been removed from latest NCERT.*

***Q.16** Prove that the curves $y^2 = 4x$ and $x^2 + y^2 - 6x + 1 = 0$ touch each other at the point (1, 2).

Ans. We have, $y^2 = 4x$ ...(i)

and $$x^2 + y^2 - 6x + 1 = 0 \quad ...(ii)$$

Since, both the curves touch each other at (1, 2) *i.e.*, curves are passing through (1, 2).

On differentiating Eq. (i) w.r.t. to x, we get

$$2y \cdot \frac{dy}{dx} = 4$$

$$\Rightarrow \quad \frac{dy}{dx} = \frac{4}{2y} \quad ...(iii)$$

On differentiating Eq. (ii) w.r.t. to x, we get

$$2x + 2y\frac{dy}{dx} = 6$$

$$\frac{dy}{dx} = \frac{6 - 2x}{2y} \quad ...(iv)$$

Substitute the points (1, 2) in Eq. (iii) and Eq. (iv) to evaluate the slopes *i.e.*, m_1 and m_2

$$\left(\frac{dy}{dx}\right)_{(1,2)} = \frac{4}{4} = 1 \quad \text{[from Eq. (ii)]}$$

$$\Rightarrow \quad \left(\frac{dy}{dx}\right)_{(1,2)} = \frac{6 - 2 \times 1}{2 \times 2} = \frac{4}{4} = 1 \quad \text{[from Eq. (iv)]}$$

Thus, we see that slope of both the curves are equal to each other *i.e.*, $m_1 = m_2 = 1$ at the point (1, 2).

Hence, both the curves touch each other.

***Q.17** Find the equation of the normal lines to the curve $3x^2 - y^2 = 8$ which are parallel to the line $x + 3y = 4$.

Ans. Given equation of the curve is

$$3x^2 - y^2 = 8 \quad ...(i)$$

On differentiating both sides w.r.t. x, we get

$$6x - 2y\frac{dy}{dx} = 0$$

$$\Rightarrow \quad \frac{dy}{dx} = \frac{6x}{2y} = \frac{3x}{y}$$

$$\Rightarrow \quad m_1 = \frac{3x}{y}$$

and $$\text{slope of normal } (m_2) = \frac{-1}{m_1} = \frac{-y}{3x} \quad ...(ii)$$

Note (*) *Topics related to these question have been removed from latest NCERT.*

Since, slope of normal to the curve should be equal to the slope of line $x + 3y = 4$, which is parallel to curve.

For line, $x + 3y = 4$, $\quad y = \dfrac{4-x}{3} = \dfrac{-x}{3} + \dfrac{4}{3}$

$\Rightarrow$ Slope of the line $(m_2) = \dfrac{-1}{3}$ $\quad [\because y = mx + c]$

$$\therefore \quad m_1 = m_2$$

$$\Rightarrow \quad \frac{-y}{3x} = -\frac{1}{3}$$

$$\Rightarrow \quad -3y = -3x$$

$$\Rightarrow \quad y = x \quad \text{...(iii)}$$

On substituting the value of y in Eq. (i), we get

$$3x^2 - x^2 = 8$$

$$\Rightarrow \quad x^2 = 4$$

$$\Rightarrow \quad x = \pm 2$$

For $x = 2$, $\quad y = 2$ $\quad$ [using Eq. (iii)]

and for $x = -2$, $\quad y = -2$ $\quad$ [using Eq. (iii)]

Thus,the points at which normal to the curve are parallel to the line $x + 3y = 4$ are $(2, 2)$ and $(-2, -2)$.

Required equations of normal are

$$y - 2 = m_2(x-2) \text{ and } y + 2 = m_2(x+2) \quad [\because (y - y_1) = m(x - x_1)]$$

$$\Rightarrow \quad y - 2 = \frac{-1}{3}(x-2) \text{ and } y + 2 = \frac{-1}{3}(x+2)$$

$$\Rightarrow \quad 3y - 6 = -x + 2 \text{ and } 3y + 6 = -x - 2$$

$$\Rightarrow \quad 3y + x = 8 \text{ and } 3y + x = -8$$

So, the required equations are $3y + x = \pm 8$.

***Q.18** At what points on the curve $x^2 + y^2 - 2x - 4y + 1 = 0$, the tangents are parallel to the Y-axis?

Ans. Given, equation of curve which is

$$x^2 + y^2 - 2x - 4y + 1 = 0 \quad \text{... (i)}$$

On differentiating w.r.t. x, we get

$$\Rightarrow \quad 2x + 2y\frac{dy}{dx} - 2 - 4\frac{dy}{dx} = 0$$

$$\Rightarrow \quad \frac{dy}{dx}(2y - 4) = 2 - 2x$$

$$\Rightarrow \quad \frac{dy}{dx} = \frac{2(1-x)}{2(y-2)}$$

Since, the tangents are parallel to the Y-axis *i.e.,* $\tan\theta = \tan 90° = \dfrac{dy}{dx}$.

$$\therefore \quad \frac{1-x}{y-2} = \frac{1}{0} \quad \left[\because \tan 90° = \infty = \frac{1}{0}\right]$$

Note (✱) *Topics related to these question have been removed from latest NCERT.*

$$\Rightarrow \quad y-2=0$$
$$\Rightarrow \quad y=2$$

On substituting $y = 2$ in Eq. (i), we get

$$x^2+2^2-2x-4\times 2+1=0$$
$$\Rightarrow \quad x^2-2x-3=0$$
$$\Rightarrow \quad x^2-3x+x-3=0$$
$$\Rightarrow \quad x(x-3)+1(x-3)=0$$
$$\Rightarrow \quad (x+1)(x-3)=0$$
$$\therefore \quad x=-1,\ x=3$$

So, the required points are $(-1, 2)$ and $(3, 2)$.

***Q.19** Show that the line $\frac{x}{a}+\frac{y}{b}=1$, touches the curve $y=b\cdot e^{-x/a}$ at the point, where the curve intersects the axis of Y.

Ans. We have the equation of line given by $\frac{x}{a}+\frac{y}{b}=1$,which touches the curve $y=b\cdot e^{-x/a}$ at the point, where the curve intersects the Y-axis *i.e.*, $x = 0$.

$$\therefore \quad y=b\cdot e^{-0/a}=b \qquad [\because e^0=1]$$

So, the point of intersection of the curve with Y-axis is $(0, b)$.

Now, slope of the given line at $(0, b)$ is given by

On differentiating equation $\frac{x}{a}+\frac{y}{b}=1$ w.r.t. x, we get

$$\frac{1}{a}\cdot 1+\frac{1}{b}\cdot\frac{dy}{dx}=0$$
$$\Rightarrow \quad \frac{dy}{dx}=\frac{-1}{a}\cdot b$$
$$\Rightarrow \quad \frac{dy}{dx}=-\frac{1}{a}\cdot b=\frac{-b}{a}=m_1 \qquad \text{[say]}$$

Also, the slope of the curve at $(0, b)$ is

On differentiating equation $y=b\cdot e^{-x/a}$ w.r.t. x, we get

$$\Rightarrow \quad \frac{dy}{dx}=b\cdot e^{-x/a}\cdot\frac{-1}{a}$$
$$\Rightarrow \quad \frac{dy}{dx}=\frac{-b}{a}e^{-x/a}$$
$$\Rightarrow \quad \left(\frac{dy}{dx}\right)_{(0,\,b)}=\frac{-b}{a}e^{-0}=\frac{-b}{a}=m_2 \qquad \text{[say]}$$

Since, $$m_1=m_2=\frac{-b}{a}$$

Hence, the line touches the curve at the point, where the curve intersects the of Y-axis.

Note (*) *Topics related to these question have been removed from latest NCERT.*

Q.20 Show that $f(x) = 2x + \cot^{-1} x + \log(\sqrt{1+x^2} - x)$ is increasing in R.

Thinking Process If $f'(x) \geq 0$, then we can say that $f(x)$ is increasing function. Use this condition to show the desired result.

Ans. We have, $f(x) = 2x + \cot^{-1} x + \log(\sqrt{1+x^2} - x)$

$$\therefore \quad f'(x) = 2 + \left(\frac{-1}{1+x^2}\right) + \frac{1}{(\sqrt{1+x^2} - x)}\left(\frac{1}{2\sqrt{1+x^2}} \cdot 2x - 1\right)$$

$$= 2 - \frac{1}{1+x^2} + \frac{1}{(\sqrt{1+x^2} - x)} \cdot \frac{(x - \sqrt{1+x^2})}{\sqrt{1+x^2}}$$

$$= 2 - \frac{1}{1+x^2} - \frac{1}{\sqrt{1+x^2}}$$

$$= \frac{2 + 2x^2 - 1 - \sqrt{1+x^2}}{1+x^2} = \frac{1 + 2x^2 - \sqrt{1+x^2}}{1+x^2}$$

For increasing function, $f'(x) \geq 0 \Rightarrow \dfrac{1 + 2x^2 - \sqrt{1+x^2}}{1+x^2} \geq 0 \Rightarrow 1 + 2x^2 - \sqrt{1+x^2} \geq 0$

i.e., $\qquad 1 + 2x^2 \geq \sqrt{1+x^2}$

On squaring both sides, we get

$$\Rightarrow \quad (1+2x^2)^2 \geq 1 + x^2 \Rightarrow 1 + 4x^4 + 4x^2 \geq 1 + x^2$$

$$\Rightarrow \quad 4x^4 + 3x^2 \geq 0 \Rightarrow x^2(4x^2 + 3) \geq 0$$

which is true for any real value of x.

Hence, $f(x)$ is increasing in R.

Q.21 Show that for $a \geq 1$, $f(x) = \sqrt{3} \sin x - \cos x - 2ax + b$ is decreasing in R.

Thinking Process If $f'(x) \leq 0$, then we can say that $f(x)$ is a decreasing function. So, use this condition to show the result.

Ans. We have, $a \geq 1$, $\qquad f(x) = \sqrt{3} \sin x - \cos x - 2ax + b$

On differentiating w.r.t. x, we get

$$\therefore \quad f'(x) = \sqrt{3} \cos x - (-\sin x) - 2a$$

$$= \sqrt{3} \cos x + \sin x - 2a = 2\left[\frac{\sqrt{3}}{2} \cdot \cos x + \frac{1}{2} \cdot \sin x\right] - 2a$$

$$= 2\left[\cos\frac{\pi}{6} \cdot \cos x + \sin\frac{\pi}{6} \cdot \sin x\right] - 2a \qquad \left[\because \cos\frac{\pi}{6} = \frac{\sqrt{3}}{2}, \sin\frac{\pi}{6} = \frac{1}{2}\right]$$

$$= 2\left(\cos\frac{\pi}{6} - x\right) - 2a \qquad [\because \cos(A - B) = \cos A \cdot \cos B + \sin A \cdot \sin B]$$

$$= 2\left[\left(\cos\frac{\pi}{6} - x\right) - a\right]$$

We know that, $\cos x \in [-1, 1]$ and $a \geq 1$

So, $\qquad 2\left[\cos\left(\frac{\pi}{6} - x\right) - a\right] \leq 0 \therefore f'(x) \leq 0$

Hence, $f(x)$ is a decreasing function in R. **Hence proved.**

Q.22 Show that $f(x) = \tan^{-1}(\sin x + \cos x)$ is an increasing function in $\left(0, \frac{\pi}{4}\right)$.

Ans. We have, $f(x) = \tan^{-1}(\sin x + \cos x)$

On differentiating the function w.r.t. x, we get

$$\therefore \quad f'(x) = \frac{1}{1 + (\sin x + \cos x)^2} \cdot (\cos x - \sin x) \qquad \left[\because \frac{d}{dx} \tan^{-1} x = \frac{1}{1+x^2}\right]$$

$$= \frac{1}{1 + \sin^2 x + \cos^2 x + 2\sin x \cdot \cos x}(\cos x - \sin x)$$

$$[\because (a+b)^2 = a^2 + b^2 + 2ab]$$

$$= \frac{1}{(2 + \sin 2x)}(\cos x - \sin x)$$

$$[\because \sin 2x = 2\sin x \cos x \text{ and } \sin^2 x + \cos^2 x = 1]$$

For $f'(x) \geq 0$,

$$\frac{1}{(2 + \sin 2x)} \cdot (\cos x - \sin x) \geq 0$$

$$\Rightarrow \quad \cos x - \sin x \geq 0 \qquad \left[\because (2 + \sin 2x) \geq 0 \text{ in } \left(0, \frac{\pi}{4}\right)\right]$$

$$\Rightarrow \quad \cos x \geq \sin x$$

which is true, if $x \in \left(0, \frac{\pi}{4}\right)$.

Hence, $f(x)$ is an increasing function in $\left(0, \frac{\pi}{4}\right)$.

***Q.23** At what point, the slope of the curve $y = -x^3 + 3x^2 + 9x - 27$ is maximum? Also, find the maximum slope.

Ans. We have, $y = -x^3 + 3x^2 + 9x - 27$

On differentiating w.r.t. to x, we get

$$\therefore \quad \frac{dy}{dx} = -3x^2 + 6x + 9 = \text{Slope of tangent to the curve}$$

Again differentiating w.r.t. to x, we get

Now, $$\frac{d^2y}{dx^2} = -6x + 6$$

Put $\frac{d^2y}{dx^2} = 0$,

$$-6x + 6 = 0$$

$$\Rightarrow \quad x = \frac{-6}{-6} = 1$$

$$\therefore \quad \frac{d}{dx}\left(\frac{d^2y}{dx^2}\right) = -6 < 0$$

Note (✱) *Topics related to these question have been removed from latest NCERT.*

So, the slope of tangent to the curve is maximum, when $x = 1$.

For $x = 1$, $\left(\frac{dy}{dx}\right)_{(x=1)} = -3 \cdot 1^2 + 6 \cdot 1 + 9 = 12$,

which is maximum slope.

Also, for $x = 1$, $y = -1^3 + 3 \cdot 1^2 + 9 \cdot 1 - 27 = -1 + 3 + 9 - 27 = -16$

So, the required point is (1, – 16) and maximum slope is 12.

Q.24 Prove that $f(x) = \sin x + \sqrt{3}\cos x$ has maximum value at $x = \frac{\pi}{6}$.

Ans. We have, $f(x) = \sin x + \sqrt{3}\cos x$

On differentiating the function w.r.t. to x

$\therefore \quad f'(x) = \cos x + \sqrt{3}(-\sin x) = \cos x - \sqrt{3}\sin x$

For $f'(x) = 0$, $\cos x = \sqrt{3}\sin x$

$\Rightarrow \quad \tan x = \frac{1}{\sqrt{3}} = \tan\frac{\pi}{6}$ $\left[\because \tan\frac{\pi}{6} = \frac{1}{\sqrt{3}}\right]$

$\Rightarrow \quad x = \frac{\pi}{6}$

Again, differentiating $f'(x)$, we get

$$f''(x) = -\sin x - \sqrt{3}\cos x$$

At $x = \frac{\pi}{6}$, $f''(x) = -\sin\frac{\pi}{6} - \sqrt{3}\cos\frac{\pi}{6}$

$= -\frac{1}{2} - \sqrt{3} \cdot \frac{\sqrt{3}}{2}$ $\left[\because \sin\frac{\pi}{6} = \frac{1}{2}, \cos\frac{\pi}{6} = \frac{\sqrt{3}}{2}\right]$

$= -\frac{1}{2} - \frac{3}{2} = -2 < 0$

Hence, at $x = \frac{\pi}{6}$, $f(x)$ has maximum value at $\frac{\pi}{6}$ is the point of local maxima.

Long Answer Type Questions

Q.25 If the sum of lengths of the hypotenuse and a side of a right angled triangle is given, then show that the area of triangle is maximum, when the angle between them is $\frac{\pi}{3}$.

Ans. Let ABC be a triangle with $AC = h$, $AB = x$ and $BC = y$.

Also, $\angle CAB = \theta$

Let $h + x = k$ ...(i)

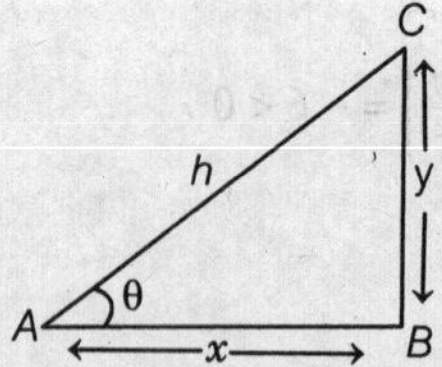

$\therefore \qquad \cos\theta = \dfrac{x}{h} \qquad \left[\because \cos\theta = \dfrac{AB}{AC}\right]$

$\Rightarrow \qquad x = h\cos\theta$

$k - h = h\cos\theta$

$\Rightarrow \qquad h + h\cos\theta = k \qquad$ [using Eq. (i)]

$\Rightarrow \qquad h(1+\cos\theta) = k$

$\Rightarrow \qquad h = \dfrac{k}{(1+\cos\theta)} \qquad$ ...(ii)

Also, $\qquad$ area of $\Delta ABC = \dfrac{1}{2}(AB \cdot BC) \qquad \left[\because \text{area of } \Delta = \dfrac{1}{2} \times \text{height} \times \text{base}\right]$

$$A = \frac{1}{2}\cdot x \cdot y$$

$$= \frac{1}{2} h\cos\theta \cdot h\sin\theta \qquad \left[\because \quad \sin\theta = \frac{y}{h}, \cos\theta \quad \frac{x}{h}\right]$$

$$= \frac{1}{2} h^2 \sin\theta \cdot \cos\theta$$

$$= \frac{2h^2}{4} \sin\theta \cdot \cos\theta$$

$$= \frac{1}{4} h^2 \sin 2\theta \qquad [\because \sin 2\theta = 2\sin\theta\cos\theta] \text{ ...(iii)}$$

Since, $\qquad h = \dfrac{k}{1+\cos\theta}$

Substitute the value of h in Eq. (iii)

$\therefore \qquad A = \dfrac{1}{4}\left(\dfrac{k}{1+\cos\theta}\right)^2 \cdot \sin 2\theta$

$\Rightarrow \qquad A = \dfrac{k^2}{4} \cdot \dfrac{\sin 2\theta}{(1+\cos\theta)^2} \qquad$ ...(iv)

On differentiating the function w.r.t. θ, we get

$$\frac{dA}{d\theta} = \frac{k^2}{4}\left[\frac{(1+\cos\theta)^2 \cdot \cos 2\theta \cdot 2 - \sin 2\theta \cdot 2(1+\cos\theta)\cdot(0-\sin\theta)}{(1+\cos\theta)^4}\right]$$

$$= \frac{k^2}{4}\left\{\frac{2(1+\cos\theta)[(1+\cos\theta)\cdot\cos 2\theta + \sin 2\theta(\sin\theta)}{(1+\cos\theta)^4}\right\}$$

$$= \frac{k^2}{4} \cdot \frac{2}{(1+\cos\theta)^3}[(1+\cos\theta)\cdot\cos 2\theta + 2\sin^2\theta\cdot\cos\theta]$$

$$[\because \sin 2\theta = 2\sin\theta\cos\theta]$$

$$= \frac{k^2}{2(1+\cos\theta)^3}[(1+\cos\theta)(1-2\sin^2\theta) + 2\sin^2\theta\cdot\cos\theta]$$

$$[\because \cos 2\theta = 1 - 2\sin^2\theta]$$

$$= \frac{k^2}{2(1+\cos\theta)^3}[1+\cos\theta - 2\sin^2\theta - 2\sin^2\theta\cdot\cos\theta + 2\sin^2\theta\cdot\cos\theta]$$

$$= \frac{k^2}{2(1+\cos\theta)^3}[(1+\cos\theta) - 2\sin^2\theta]$$

$$= \frac{k^2}{2(1+\cos\theta)^3}[1+\cos\theta - 2 + 2\cos^2\theta] \qquad [\because \sin^2\theta = 1-\cos^2\theta]$$

$$= \frac{k^2}{2(1+\cos\theta)^3}(2\cos^2\theta + \cos\theta - 1) \qquad ...(v)$$

For $\frac{dA}{d\theta} = 0$,

$$\frac{k^2}{2(1+\cos\theta)^3}(2\cos^2\theta + \cos\theta - 1) = 0$$

$$\Rightarrow \quad 2\cos^2\theta + \cos\theta - 1 = 0$$

$$\Rightarrow \quad 2\cos^2\theta + 2\cos\theta - \cos\theta - 1 = 0$$

$$\Rightarrow \quad 2\cos\theta(\cos\theta + 1) - 1(\cos\theta + 1) = 0$$

$$\Rightarrow \quad (2\cos\theta - 1)(\cos\theta + 1) = 0$$

$$\Rightarrow \quad \cos\theta = \frac{1}{2} \text{ or } \cos\theta = -1$$

$$\Rightarrow \quad \theta = \frac{\pi}{3} \qquad \text{[possible]}$$

or $\quad \theta = 2n\pi \pm \pi \qquad$ [not possible]

$$\therefore \quad \theta = \frac{\pi}{3}$$

Again, differentiating w.r.t. θ in Eq. (v), we get

$$\frac{d}{d\theta}\left(\frac{dA}{d\theta}\right) = \frac{d}{d\theta}\left[\frac{k^2}{2(1+\cos\theta)^3}(2\cos^2\theta + \cos\theta - 1)\right]$$

$$\therefore \quad \frac{d^2A}{d\theta^2} = \frac{d}{d\theta}\left[\frac{k^2(2\cos\theta - 1)(1+\cos\theta)}{2(1+\cos\theta)^3}\right] = \frac{d}{d\theta}\left[\frac{k^2}{2}\cdot\frac{(2\cos\theta - 1)}{(1+\cos\theta)^2}\right]$$

$$= \frac{k^2}{2}\left[\frac{(1+\cos\theta)^2\cdot(-2\sin\theta) - 2(1+\cos\theta)\cdot(-\sin\theta)(2\cos\theta - 1)}{(1+\cos\theta)^4}\right]$$

$$= \frac{k^2}{2}\left[\frac{(1+\cos\theta)\cdot[1+\cos\theta](-2\sin\theta) + 2\sin\theta(2\cos\theta - 1)}{(1+\cos\theta)^4}\right]$$

$$= \frac{k^2}{2}\left[\frac{-2\sin\theta - 2\sin\theta\cdot\cos\theta + 4\sin\theta\cdot\cos\theta - 2\sin\theta}{(1+\cos\theta)^3}\right]$$

$$= \frac{k^2}{2}\left[\frac{\sin 2\theta - 4\sin\theta}{(1+\cos\theta)^3}\right] \qquad [\because \sin 2\theta = 2\sin\theta\cos\theta]$$

$$\therefore \left(\frac{d^2A}{d\theta^2}\right)_{\text{at } \theta=\frac{\pi}{3}} = \frac{k^2}{2}\left[\frac{\sin\frac{2\pi}{3} - 4\sin\frac{\pi}{3}}{\left(1+\cos\frac{\pi}{3}\right)^3}\right] = \frac{k^2}{2}\left[\frac{\frac{\sqrt{3}}{2} - \frac{4\sqrt{3}}{2}}{\left(1+\frac{1}{2}\right)^3}\right]$$

$$\left[\because \sin\frac{\pi}{3} = \frac{\sqrt{3}}{2}, \cos\frac{\pi}{3} = \frac{1}{2} \text{ and } \sin\frac{2\pi}{3} = \frac{\sqrt{3}}{2}\right]$$

$$= \frac{k^2}{2}\left[\frac{-3\sqrt{3}\cdot 8}{2\cdot 27}\right]$$

$$= -k^2\left(\frac{2\sqrt{3}}{9}\right) < 0$$

which is less than zero.

Hence, area of the right angled triangle is maximum, when the angle between them is $\frac{\pi}{3}$.

Q.26 Find the points of local maxima, local minima and the points of inflection of the function $f(x) = x^5 - 5x^4 + 5x^3 - 1$. Also, find the corresponding local maximum and local minimum values.

Ans. Given that, $f(x) = x^5 - 5x^4 + 5x^3 - 1$

On differentiating w.r.t. x, we get

$$f'(x) = 5x^4 - 20x^3 + 15x^2$$

For maxima or minima, $f'(x) = 0$

$\Rightarrow \quad 5x^4 - 20x^3 + 15x^2 = 0$

$\Rightarrow \quad 5x^2(x^2 - 4x + 3) = 0$

$\Rightarrow \quad 5x^2(x^2 - 3x - x + 3) = 0$

$\Rightarrow \quad 5x^2[x(x-3) - 1(x-3)] = 0$

$\Rightarrow \quad 5x^2[(x-1)(x-3)] = 0$

$\therefore \quad x = 0, 1, 3$

Sign scheme for $\frac{dy}{dx} = 5x^2(x-1)(x-3)$

$-\infty$ + | 0 | + | 1 | − | 3 | + $+\infty$

So, y has maximum value at $x = 1$ and minimum value at $x = 3$.

At $x = 0$, y has neither maximum nor minimum value.

$\therefore$ Maximum value of $y = 1 - 5 + 5 - 1 = 0$

and minimum value $= (3)^5 - 5(3)^4 + 5(3)^3 - 1$

$= 243 - 81 \times 5 + 27 \times 5 - 1$

$= -28$

Q.27 A telephone company in a town has 500 subscribers on its list and collects fixed charges of ₹ 300 per subscriber per year. The company proposes to increase the annual subscription and it is believed that for every increase of ₹ 1 per one subscriber will discontinue the service. Find what increase will bring maximum profit?

Ans. Consider that company increases the annual subscription by ₹ x.

So, x subscribes will discontinue the service.

$\therefore$ Total revenue of company after the increment is given by

$$R(x) = (500 - x)(300 + x) = 15 \times 10^4 + 500x - 300x - x^2$$
$$= -x^2 + 200x + 150000$$

On differentiating both sides w.r.t. x, we get

$$R'(x) = -2x + 200$$

Now, $R'(x) = 0 \Rightarrow 2x = 200 \Rightarrow x = 100$

$\therefore$ $R''(x) = -2 < 0$

So, $R(x)$ is maximum when $x = 100$.

Hence, the company should increase the subscription fee by ₹ 100, so that it has maximum profit.

***Q.28** If the straight line $x \cos \alpha + y \sin \alpha = p$ touches the curve $\frac{x^2}{a^2} + \frac{y^2}{b^2} = 1$, then prove that $a^2 \cos^2 \alpha + b^2 \sin^2 \alpha = p^2$.

Ans. Given, line is $x \cos \alpha + y \sin \alpha = p$ ... (i)

and curve is $\frac{x^2}{a^2} + \frac{y^2}{b^2} = 1$

$\Rightarrow$ $b^2x^2 + a^2y^2 = a^2b^2$ ...(ii)

Now, differentiating Eq. (ii) w.r.t. x, we get

$$b^2 \cdot 2x + a^2 \cdot 2y \cdot \frac{dy}{dx} = 0$$

$\Rightarrow$ $\frac{dy}{dx} = \frac{-2b^2x}{2a^2y} = \frac{-xb^2}{ya^2}$ ... (iii)

From Eq. (i), $y \sin \alpha = p - x \cos \alpha$

$\Rightarrow$ $y = -x \cot \alpha + \frac{p}{\sin \alpha}$ [$\because y = mx + c$]

Thus, slope of the line is $(-\cot \alpha)$.

So, the given equation of line will be tangent to the Eq. (ii), if $\left(-\frac{x}{y} \cdot \frac{b^2}{a^2}\right) = (-\cot \alpha)$

$\Rightarrow$ $\frac{x}{a^2 \cos \alpha} = \frac{y}{b^2 \sin \alpha} = k$ [say]

$\Rightarrow$ $x = ka^2 \cos \alpha$ and $y = kb^2 \sin \alpha$

So, the line $x \cos \alpha + y \sin \alpha = p$ will touch the curve $\frac{x^2}{a^2} + \frac{y^2}{b^2}$ at point $(ka^2 \cos \alpha, kb^2 \sin \alpha)$.

Note (*) *Topics related to these question have been removed from latest NCERT.*

From Eq. (i), $ka^2 \cos^2 \alpha + kb^2 \sin^2 \alpha = p$

$\Rightarrow \quad a^2 \cos^2 \alpha + b^2 \sin^2 \alpha = \dfrac{p}{k}$

On squaring both sides, we get

$\Rightarrow \quad (a^2 \cos^2 \alpha + b^2 \sin^2 \alpha)^2 = \dfrac{p^2}{k^2}$ ...(iv)

From Eq. (ii), $b^2k^2a^4 \cos^2 \alpha + a^2k^2b^4 \sin^2 \alpha = a^2b^2$ $[\because \cos^2 \theta + \sin^2 \theta = 1]$

$\Rightarrow \quad k^2 (a^2 \cos^2 \alpha + b^2 \sin^2 \alpha) = 1$

$\Rightarrow \quad (a^2 \cos^2 \alpha + b^2 \sin^2 \alpha) = \dfrac{1}{k^2}$ ... (v)

On dividing Eq. (iv) by Eq. (v), we get

$a^2 \cos^2 \alpha + b^2 \sin^2 \alpha = p^2$ **Hence proved.**

Alternate Method

We know that, if a line $y = mx + c$ touches ellipse $\dfrac{x^2}{a^2} + \dfrac{y^2}{b^2} = 1$, then the required condition is $c^2 = a^2m^2 + b^2$ ...(i)

Condition of tangency is obtained by making the equation of curve quadratic in 'x' or 'y' with the help of tangent line.

i.e., $\dfrac{x^2}{a^2} + \dfrac{(mx + c)^2}{b^2} = 1$

$b^2x^2 + a^2(mx + c)^2 = a^2b^2$

Here, given equation of the line is

$x \cos \alpha + y \sin \alpha = p$

$\Rightarrow \quad y = \dfrac{p - x \cos \alpha}{\sin \alpha}$

$= - x \cot \alpha + \dfrac{p}{\sin \alpha}$

$\Rightarrow \quad c = \dfrac{p}{\sin \alpha}$ $(\because y = mx + c)$

and $m = - \cot \alpha$

$\therefore \quad \left(\dfrac{p}{\sin \alpha}\right)^2 = a^2 (- \cot \alpha)^2 + b^2$ [from Eq. (i)]

$\Rightarrow \quad \dfrac{p^2}{\sin^2 \alpha} = a^2 \dfrac{\cos^2 \alpha}{\sin^2 \alpha} + b^2$

$\Rightarrow \quad p^2 = a^2 \cos^2 \alpha + b^2 \sin^2 \alpha$ **Hence proved.**

Q.29 If an open box with square base is to be made of a given quantity of card board of area c^2, then show that the maximum volume of the box is $\dfrac{c^3}{6\sqrt{3}}$ cu units.

Thinking Process First, let the sides of box in x and y then find $\dfrac{dV}{dx}$ in terms c and x. Also, for $\dfrac{dV}{dx} = 0$ get the value of x and if $\dfrac{d^2V}{dx^2} < 0$ at the value of x, then by putting that value of x in the equation of V, get the desired result.

Ans. Let the length of side of the square base of open box be x units and its height be y units.

$\therefore$ Area of the metal used $= x^2 + 4xy$

$$\Rightarrow \quad x^2 + 4xy = c^2 \qquad \text{[given]}$$

$$\Rightarrow \quad y = \frac{c^2 - x^2}{4x} \qquad \text{...(i)}$$

Now, volume of the box $(V) = x^2 y$ $\quad [\because v = l \times b \times h]$

$$\Rightarrow \quad V = x^2 \cdot \left(\frac{c^2 - x^2}{4x}\right) \qquad \text{[From Eq. (i)]}$$

$$= \frac{1}{4} x (c^2 - x^2) = \frac{1}{4}(c^2 x - x^3) \qquad \text{...(ii)}$$

On differentiating both sides w.r.t. x, we get

$$\frac{dV}{dx} = \frac{1}{4}(c^2 - 3x^2) \qquad \text{... (iii)}$$

Now, $\dfrac{dV}{dx} = 0 \Rightarrow c^2 = 3x^2$

$$\Rightarrow \quad x^2 = \frac{c^2}{3}$$

$$\Rightarrow \quad x = \frac{c}{\sqrt{3}} \qquad \text{[on using positive sign]}$$

Again, differentiating Eq. (ii) w.r.t. x, we get

$$\frac{d^2V}{dx^2} = \frac{1}{4}(-6x) = \frac{-3}{2} x < 0$$

$$\therefore \quad \left(\frac{d^2V}{dx^2}\right)_{\text{at } x = \frac{c}{\sqrt{3}}} = -\frac{3}{2} \cdot \left(\frac{c}{\sqrt{3}}\right) < 0$$

Thus, we see that volume (V) is maximum at $x = \dfrac{c}{\sqrt{3}}$.

$$\therefore \text{ Maximum volume of the box, } (V)_{x = \frac{c}{\sqrt{3}}} = \frac{1}{4}\left(c^2 . \frac{c}{\sqrt{3}} - \frac{c^3}{3\sqrt{3}}\right) \qquad \text{[using (ii)]}$$

$$= \frac{1}{4} \cdot \frac{(3c^3 - c^3)}{3\sqrt{3}} = \frac{1}{4} \cdot \frac{2c^3}{3\sqrt{3}} = \frac{c^3}{6\sqrt{3}} \text{ cu units}$$

Q.30 **Find the dimensions of the rectangle of perimeter 36 cm which will sweep out a volume as large as possible, when revolved about one of its sides. Also, find the maximum volume.**

Ans. Let breadth and length of the rectangle be x and y, respectively.

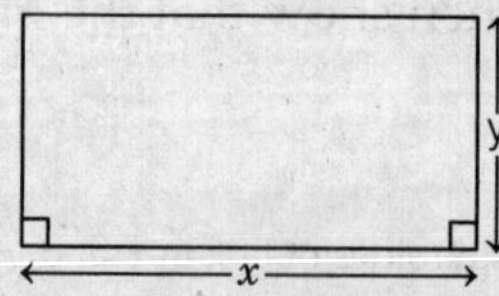

$\because$ Perimeter of the rectangle $= 36$ cm

$$\Rightarrow \quad 2x + 2y = 36$$

$$\Rightarrow \quad x + y = 18$$

$\Rightarrow \quad y = 18 - x \quad$ ... (i)

Let the rectangle is being revolved about its length y.

Then, volume (V) of resultant cylinder $V = \pi x^2 y$

$\Rightarrow \quad = x^2\pi \cdot (18 - x) \quad$ [$\because$ $V = \pi r^2 h$] [using Eq. (i)]

$= \pi(18x^2 - x^3) \quad$ ...(ii)

On differentiating both sides w.r.t. x, we get

$$\frac{dV}{dx} = \pi(36x - 3x^2)$$

Now, $\frac{dV}{dx} = 0$

$\Rightarrow \quad 36x - 3x^2 = 0$

$\Rightarrow \quad 3x(12 - x) = 0$

$\Rightarrow \quad x = 0, x = 12$

$\therefore \quad x = 12 \quad$ [$\because x \neq 0$]

Again, differentiating w.r.t. x, we get

$$\frac{d^2V}{dx^2} = \pi(36 - 6x)$$

$$\Rightarrow \quad \left(\frac{d^2V}{dx^2}\right)_{x=12} = \pi(36 - 6 \times 12) = -36\pi < 0$$

At $x = 12$, volume of the resultant cylinder is the maximum.

So, the dimensions of rectangle are 12 cm and 6 cm, respectively. [using Eq. (i)]

$\therefore$ Maximum volume of resultant cylinder,

$(V)_{x=12} = \pi[18 \cdot (12)^2 - (12)^3] \quad$ [using Eq. (ii)]

$= \pi[12^2(18 - 12)] = \pi \times 144 \times 6 = 864\pi \text{ cm}^3$

Q.31 **If the sum of the surface areas of cube and a sphere is constant, what is the ratio of an edge of the cube to the diameter of the sphere, when the sum of their volumes is minimum?**

Ans. Let length of one edge of cube be x units and radius of sphere be r units.

$\therefore \quad$ Surface area of cube $= 6x^2$

and surface area of sphere $= 4\pi r^2$

Also, $6x^2 + 4\pi r^2 = k \quad$ [constant, given]

$\Rightarrow \quad 6x^2 = k - 4\pi r^2$

$$\Rightarrow \quad x^2 = \frac{k - 4\pi r^2}{6}$$

$$\Rightarrow \quad x = \left[\frac{k - 4\pi r^2}{6}\right]^{1/2} \quad \text{... (i)}$$

Now, volume of cube $= x^3$

and volume of sphere $= \frac{4}{3}\pi r^3$

Let sum of volume of the cube and volume of the sphere be given by

$$S = x^3 + \frac{4}{3}\pi r^3 = \left[\frac{k-4\pi r^2}{6}\right]^{3/2} + \frac{4}{3}\pi r^3 \quad \text{[using Eq. (i)]}$$

On differentiating both sides w.r.t. r, we get

$$\frac{dS}{dr} = \frac{3}{2}\left[\frac{k-4\pi r^2}{6}\right]^{1/2} \cdot \left(\frac{-8\pi r}{6}\right) + \frac{12}{3}\pi r^2$$

$$= -2\pi r\left[\frac{k-4\pi r^2}{6}\right]^{1/2} + 4\pi r^2 \quad \text{... (ii)}$$

$$= -2\pi r\left[\left\{\frac{k-4\pi r^2}{6}\right\}^{1/2} - 2r\right]$$

Now, $\frac{dS}{dr} = 0 \Rightarrow r = 0 \text{ or } 2r = \left(\frac{k-4\pi r^2}{6}\right)^{1/2}$

$\Rightarrow \quad 4r^2 = \frac{k-4\pi r^2}{6} \Rightarrow 24r^2 = k - 4\pi r^2$

$\Rightarrow \quad 24r^2 + 4\pi r^2 = k \Rightarrow r^2[24+4\pi] = k$

$\therefore \quad r = 0 \text{ or } r = \sqrt{\frac{k}{24+4\pi}} = \frac{1}{2}\sqrt{\frac{k}{6+\pi}}$

We know that, $r \neq 0$

$\therefore \quad r = \frac{1}{2}\sqrt{\frac{k}{6+\pi}}$

Again, differentiating w.r.t. r in Eq. (ii), we get

$$\frac{d^2S}{dr^2} = \frac{d}{dr}\left[-2\pi r\left\{\left(\frac{k-4\pi r^2}{6}\right)^{1/2} + 4\pi r^2\right\}\right]$$

$$= -2\pi\left[r \cdot \frac{1}{2}\left(\frac{k-4\pi r^2}{6}\right)^{-1/2} \cdot \left(\frac{-8\pi r}{6}\right) + \left(\frac{k-4\pi r^2}{6}\right)^{1/2} \cdot 1\right] + 4\pi \cdot 2r$$

$$= -2\pi\left[r \cdot \frac{1}{2\sqrt{\frac{k-4\pi r^2}{6}}} \cdot \left(\frac{-8\pi r}{6}\right) + \sqrt{\frac{k-4\pi r^2}{6}}\right] + 8\pi r$$

$$= -2\pi\left[\frac{-8\pi r^2 + 12\left(k - \frac{4\pi r^2}{6}\right)}{12\sqrt{\frac{k-4\pi r^2}{6}}}\right] + 8\pi r$$

$$= -2\pi\left[\frac{-48\pi r^2 + 72k - 48\pi r^2}{72\sqrt{\frac{k-4\pi r^2}{6}}}\right] + 8\pi r = -2\pi\left[\frac{-96\pi r^2 + 72k}{72\sqrt{\frac{k-4\pi r^2}{6}}}\right] + 8\pi r > 0$$

For $r = \frac{1}{2}\sqrt{\frac{k}{6+\pi}}$, then the sum of their volume is minimum.

For $r = \frac{1}{2}\sqrt{\frac{k}{6+\pi}}$, $\qquad x = \left[\dfrac{k - 4\pi \cdot \frac{1}{4}\frac{k}{(6+\pi)}}{6}\right]^{1/2}$ [using Eq. (i)]

$$= \left[\frac{(6+\pi)k - \pi k}{6(6+\pi)}\right]^{1/2} = \left[\frac{k}{6+\pi}\right]^{1/2} = 2r \qquad \left[\because r = \frac{1}{2}\sqrt{\frac{k}{6+\pi}}\right]$$

Since, the sum of their volume is minimum when $x = 2r$.

Hence, the ratio of an edge of cube to the diameter of the sphere is 1:1.

Q.32 If AB is a diameter of a circle and C is any point on the circle, then show that the area of ΔABC is maximum, when it is isosceles.

Ans. Let the radius of circle $= r$ and $\angle BAC = \theta$

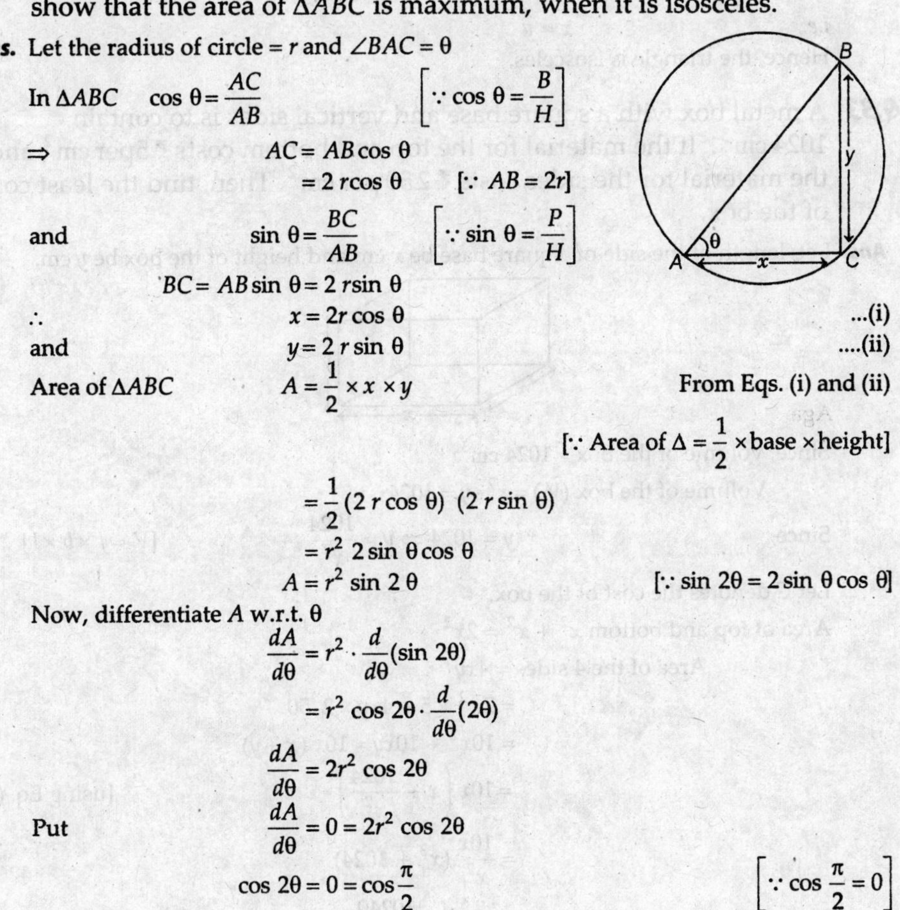

In ΔABC $\qquad \cos\theta = \dfrac{AC}{AB}$ $\qquad \left[\because \cos\theta = \dfrac{B}{H}\right]$

$\Rightarrow \qquad AC = AB\cos\theta$

$\qquad = 2r\cos\theta$ $\qquad [\because AB = 2r]$

and $\qquad \sin\theta = \dfrac{BC}{AB}$ $\qquad \left[\because \sin\theta = \dfrac{P}{H}\right]$

$\qquad BC = AB\sin\theta = 2r\sin\theta$

$\therefore \qquad x = 2r\cos\theta$ ...(i)

and $\qquad y = 2r\sin\theta$ ....(ii)

Area of ΔABC $\qquad A = \frac{1}{2} \times x \times y$ From Eqs. (i) and (ii)

$\left[\because \text{Area of } \Delta = \frac{1}{2} \times \text{base} \times \text{height}\right]$

$$= \frac{1}{2}(2r\cos\theta)(2r\sin\theta)$$

$$= r^2\, 2\sin\theta\cos\theta$$

$$A = r^2 \sin 2\theta \qquad [\because \sin 2\theta = 2\sin\theta\cos\theta]$$

Now, differentiate A w.r.t. θ

$$\frac{dA}{d\theta} = r^2 \cdot \frac{d}{d\theta}(\sin 2\theta)$$

$$= r^2 \cos 2\theta \cdot \frac{d}{d\theta}(2\theta)$$

$$\frac{dA}{d\theta} = 2r^2\cos 2\theta$$

Put $\qquad \dfrac{dA}{d\theta} = 0 = 2r^2\cos 2\theta$

$$\cos 2\theta = 0 = \cos\frac{\pi}{2} \qquad \left[\because \cos\frac{\pi}{2} = 0\right]$$

$$2\theta = \frac{\pi}{2}$$

$$\theta = \frac{\pi}{4}$$

Now, differentiate $\frac{dA}{d\theta}$ w.r.t.θ

$$\frac{d^2A}{d\theta^2} = -4r^2 \sin 2\theta$$

which is $-ve$

i.e., A will be maximum when $\theta = \frac{\pi}{4}$

$$x = 2r\cos\frac{\pi}{4} = 2r \cdot \frac{1}{\sqrt{2}}$$

$$y = 2r\sin\frac{\pi}{4} = 2r \cdot \frac{1}{\sqrt{2}}$$

i.e., $x = y$

Hence, the triangle is isosceles.

Q.33 A metal box with a square base and vertical sides is to contain 1024 cm^3. If the material for the top and bottom costs ₹ 5per cm^2 and the material for the sides costs ₹ 2.50 per cm^2. Then, find the least cost of the box.

Ans. Let length of the side of square base be x cm and height of the box be y cm.

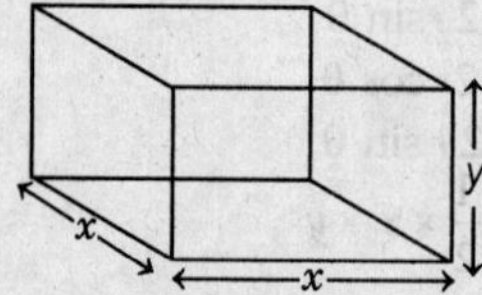

Since, volume of the box = 1024 cm^3

$\therefore$ Volume of the box $(V) = x^2 \cdot y = 1024$

Since, $x^2y = 1024 \Rightarrow y = \frac{1024}{x^2}$ $[V = l \times b \times h]$ ...(i)

Let C denotes the cost of the box.

Area of top and bottom $x^2 + x^2 = 2x^2$

Area of the 4 sides $= 4xy$

$$\therefore \quad C = 2x^2 \times 5 + 4xy \times 2.50$$

$$= 10x^2 + 10xy = 10x(x + y)$$

$$= 10x\left(x + \frac{1024}{x^2}\right) \quad \text{[using Eq. (i)]}$$

$$= \frac{10x}{x^2}(x^3 + 1024)$$

$$\Rightarrow \quad C = 10x^2 + \frac{10240}{x} \quad \text{... (ii)}$$

On differentiating both sides w.r.t. x, we get

$$\frac{dC}{dx} = 20x + 10240\left(-\frac{1}{x^2}\right)$$

$$= 20x - \frac{10240}{x^2} \qquad \text{...(iii)}$$

Now, $\quad \frac{dC}{dx} = 0$

$\Rightarrow \quad 20x = \frac{10240}{x^2}$

$\Rightarrow \quad 20x^3 = 10240$

$\Rightarrow \quad x^3 = 512 = 8^3 \Rightarrow x = 8$

Again, differentiating Eq. (iii) w.r.t. x, we get

$$\frac{d^2C}{dx^2} = 20 - 10240(-2)\cdot\frac{1}{x^3}$$

$$= 20 + \frac{20480}{x^3}$$

$\therefore \quad \left(\frac{d^2C}{dx^2}\right)_{x=8} = 20 + \frac{20480}{512} = 60 > 0$

For $x = 8$, cost is minimum and the corresponding least cost of the box,

$$C(8) = 10\cdot 8^2 + \frac{10240}{8} \qquad \text{[using Eq. (ii)]}$$

$$= 640 + 1280 = 1920$$

$\therefore \quad$ Least cost = ₹ 1920

Q.34 The sum of surface areas of a rectangular parallelopiped with sides x, $2x$ and $\frac{x}{3}$ and a sphere is given to be constant. Prove that the sum of their volumes is minimum, if x is equal to three times the radius of the sphere. Also, find the minimum value of the sum of their volumes.

Ans. We have given that, the sum of the surface areas of a rectangular parallelopiped with sides x, $2x$ and $\frac{x}{3}$ and a sphere is constant.

Let S be the sum of both the surface area.

$\therefore \quad S = 2\left(x\cdot 2x + 2x\cdot\frac{x}{3} + \frac{x}{3}\cdot x\right) + 4\pi r^2 = k \qquad$ (constant)

[SA of sphere = $4\pi r^2$ SA of rectangles = $2(lb + bh + hl)$]

$$k = 2\left[2x^2 + \frac{2x^2}{3} + \frac{x^2}{3}\right] + 4\pi r^2$$

$$= 2[3x^2] + 4\pi r^2 = 6x^2 + 4\pi r^2$$

$\Rightarrow \quad 4\pi r^2 = k - 6x^2 \quad \Rightarrow r^2 = \frac{k - 6x^2}{4\pi}$

$\Rightarrow \quad r = \sqrt{\frac{k - 6x^2}{4\pi}} \qquad$ ... (i)

Let V denotes the volume of both the parallelopiped and the sphere.

Then, $$V = 2x \cdot x \cdot \frac{x}{3} + \frac{4}{3}\pi r^3 = \frac{2}{3}x^3 + \frac{4}{3}\pi r^3$$

[$\because$ V of rectangle $= l \times b \times h$, V of shpere $= \frac{4}{3}\pi r^3$]

$$= \frac{2}{3}x^3 + \frac{4}{3}\pi\left(\frac{k - 6x^2}{4\pi}\right)^{3/2} \quad \text{[using (i)]}$$

$$= \frac{2}{3}x^3 + \frac{4}{3}\pi \cdot \frac{1}{8\pi^{3/2}}(k - 6x^2)^{3/2}$$

$$= \frac{2}{3}x^3 + \frac{1}{6\sqrt{\pi}}(k - 6x^2)^{3/2} \quad \text{...(ii)}$$

On differentiating both sides w.r.t. x, we get

$$\frac{dV}{dx} = \frac{2}{3} \cdot 3x^2 + \frac{1}{6\sqrt{\pi}} \cdot \frac{3}{2}(k - 6x^2)^{1/2} \cdot (-12x)$$

$$= 2x^2 - \frac{12x}{4\sqrt{\pi}}\sqrt{k - 6x^2}$$

$$= 2x^2 - \frac{3x}{\sqrt{\pi}}(k - 6x^2)^{1/2} \quad \text{...(iii)}$$

$\because$ $$\frac{dV}{dx} = 0$$

$\Rightarrow$ $$2x^2 = \frac{3x}{\sqrt{\pi}}(k - 6x^2)^{1/2}$$

$\Rightarrow$ $$4x^4 = \frac{9x^2}{\pi}(k - 6x^2)$$

$\Rightarrow$ $$4\pi x^4 = 9kx^2 - 54x^4$$

$\Rightarrow$ $$4\pi x^4 + 54x^4 = 9kx^2$$

$\Rightarrow$ $$x^4[4\pi + 54] = 9 \cdot k \cdot x^2 \Rightarrow x^2 = \frac{9k}{4\pi + 54}$$

$\Rightarrow$ $$x = 3 \cdot \sqrt{\frac{k}{4\pi + 54}} \quad \text{...(iv)}$$

Again, differentiating Eq. (iii) w.r.t. x, we get

$$\frac{d^2V}{dx^2} = 4x - \frac{3}{\sqrt{\pi}}\left[x \cdot \frac{1}{2}(k - 6x^2)^{-1/2} \cdot (-12x) + (k - 6x^2)^{1/2} \cdot 1\right]$$

$$= 4x - \frac{3}{\sqrt{\pi}}[-6x^2 \cdot (k - 6x^2)^{-1/2} + (k - 6x^2)^{1/2}]$$

$$= 4x - \frac{3}{\sqrt{\pi}}\left[\frac{-6x^2 + k - 6x^2}{\sqrt{k - 6x^2}}\right]$$

$$= 4x - \frac{3}{\sqrt{\pi}}\left[\frac{k - 12x^2}{\sqrt{k - 6x^2}}\right]$$

Now, $\left(\frac{d^2V}{dx^2}\right)_{x=3\cdot\sqrt{\frac{k}{4\pi+54}}} = 4\cdot 3\sqrt{\frac{k}{4\pi+54}} - \frac{3}{\sqrt{\pi}}\left[\frac{k - 12\cdot 9\cdot\frac{k}{4\pi+54}}{\sqrt{k - \frac{6\cdot 9\cdot k}{4\pi+54}}}\right]$

$$= 12\sqrt{\frac{k}{4\pi+54}} - \frac{3}{\sqrt{\pi}}\left[\frac{k - \frac{108k}{4\pi+54}}{\sqrt{k - \frac{54\,k}{4\pi+54}}}\right]$$

$$= 12\sqrt{\frac{k}{4\pi+54}} - \frac{3}{\sqrt{\pi}}\left[\frac{\frac{4k\pi + 54k - 108\,k}{4\pi+54}}{\sqrt{\frac{4k\pi + 54k - 54\,k}{4\pi+54}}}\right]$$

$$= 12\sqrt{\frac{k}{4\pi+54}} - \frac{3}{\sqrt{\pi}}\left[\frac{4k\pi - 54k}{\sqrt{4k\pi}\,\sqrt{4\pi+54}}\right]$$

$$= 12\sqrt{\frac{k}{4\pi+54}} - \frac{3}{\pi}\left[\frac{k\,(2\pi - 27)}{\sqrt{k}\,\sqrt{4\pi+54}}\right]$$

$$\left[\text{since, } (2\pi - 27) < 0 \Rightarrow \frac{d^2V}{dx^2} > 0;\ k > 0\right]$$

For $x = 3\sqrt{\frac{k}{4\pi+54}}$, the sum of volumes is minimum.

Since, $r = \sqrt{\frac{k - 6x^2}{4\pi}}$ [using Eq. (i)]

$$r = \frac{1}{2\sqrt{\pi}}\sqrt{k - 6\cdot\frac{9k}{4\pi+54}} = \frac{1}{2\sqrt{\pi}}\cdot\sqrt{\frac{4k\pi + 54\,k - 54\,k}{4\pi+54}}$$

$$= \frac{1}{2\sqrt{\pi}}\sqrt{\frac{4k\pi}{4\pi+54}} = \frac{\sqrt{k}}{\sqrt{4\pi+54}} = \frac{1}{3}x$$ [using Eq. (iv)]

$\Rightarrow \quad x = 3r$ **Hence proved.**

$\therefore$ Minimum sum of volume,

$$V_{\left(x=3\cdot\sqrt{\frac{k}{4\pi+54}}\right)} = \frac{2}{3}x^3 + \frac{4}{3}\pi r^3$$

$$= \frac{2}{3}(3r)^3 + \frac{4}{3}\pi r^3 = 18r^3 + \frac{4}{3}\pi r^3$$

$$= \left(18 + \frac{4}{3}\pi\right)r^3 = \left(\frac{54 + 4\pi}{3}\right)\left(\frac{k}{4\pi+54}\right)^{3/2}$$

$$= \frac{k^{3/2}}{3(4\pi+54)^{1/2}}$$

Objective Type Questions

Choose the correct answer out of the given four options in each of the exercises from 35 to 39.

Q.35 If the sides of an equilateral triangle are increasing at the rate of 2 cm/s then the rate at which the area increases, when side is 10 cm, is

(a) $10\ \text{cm}^2/\text{s}$ (b) $\sqrt{3}\ \text{cm}^2/\text{s}$ (c) $10\sqrt{3}\ \text{cm}^2/\text{s}$ (d) $\frac{10}{3}\ \text{cm}^2/\text{s}$

Ans. **(c)** Let the side of an equilateral triangle be x cm.

$\therefore$ Area of equilateral triangle, $A = \frac{\sqrt{3}}{4}x^2$ ...(i)

Also, $$\frac{dx}{dt} = 2 \text{ cm/s}$$

On differentiating Eq. (i) w.r.t. t, we get

$$\frac{dA}{dt} = \frac{\sqrt{3}}{4}\cdot 2x \cdot \frac{dx}{dt}$$

$$= \frac{\sqrt{3}}{4}\cdot 2\cdot 10\cdot 2 \quad \left[\because x = 10 \text{ and } \frac{dx}{dt} = 2\right]$$

$$= 10\sqrt{3}\ \text{cm}^2/\text{s}$$

Q.36 A ladder, 5 m long, standing on a horizontal floor, leans against a vertical wall. If the top of the ladder slides downwards at the rate of 10 cm/s, then the rate at which the angle between the floor and the ladder is decreasing when lower end of ladder is 2 m from the wall is

(a) $\frac{1}{10}$ rad/s (b) $\frac{1}{20}$ rad/s

(c) 20 rad/s (d) 10 rad/s

Ans. **(b)** Let the angle between floor and the ladder be θ.

Let $AB = x$ cm and $BC = y$ cm

$$\therefore \sin\theta = \frac{x}{500} \text{ and } \cos\theta = \frac{y}{500} \quad \left[\because \sin\theta = \frac{P}{H} \text{ and } \cos\theta = \frac{B}{H}\right]$$

$$\Rightarrow x = 500\sin\theta \text{ and } y = 500\cos\theta$$

Also, $$\frac{dx}{dt} = 10 \text{ cm/s}$$

$$\Rightarrow 500\cdot\cos\theta\cdot\frac{d\theta}{dt} = 10 \text{ cm/s}$$

$$\Rightarrow \frac{d\theta}{dt} = \frac{10}{500\cos\theta} = \frac{1}{50\cos\theta}$$

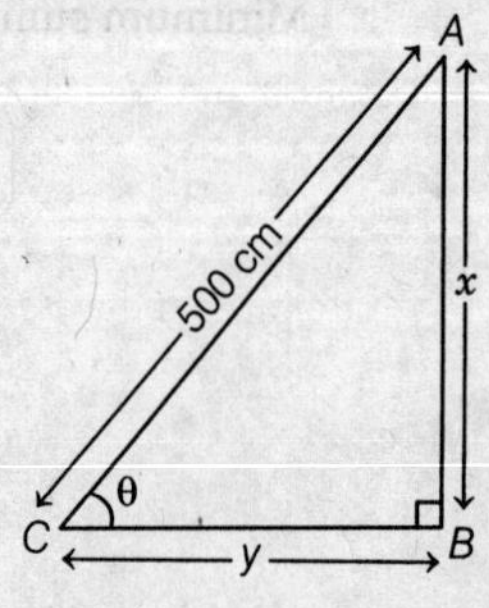

For $y = 2\text{ m} = 200$ cm,

$$\frac{d\theta}{dt} = \frac{1}{50\cdot\frac{y}{500}} = \frac{10}{y}$$

$$= \frac{10}{200} = \frac{1}{20} \text{ rad/s}$$

***Q.37** The curve $y = x^{1/5}$ has at (0, 0)

(a) a vertical tangent (parallel to Y-axis)
(b) a horizontal tangent (parallel to X-axis)
(c) an oblique tangent
(d) no tangent

Ans. ***(a)*** We have, $y = x^{1/5}$

$$\Rightarrow \quad \frac{dy}{dx} = \frac{1}{5}x^{\frac{1}{5}-1} = \frac{1}{5}x^{-4/5}$$

$$\therefore \quad \left(\frac{dy}{dx}\right)_{(0,0)} = \frac{1}{5} \times (0)^{-4/5} = \infty$$

So, the curve $y = x^{1/5}$ has a vertical tangent at (0, 0), which is parallel to Y-axis.

***Q.38** The equation of normal to the curve $3x^2 - y^2 = 8$ which is parallel to the line $x + 3y = 8$ is

(a) $3x - y = 8$ (b) $3x + y + 8 = 0$
(c) $x + 3y \pm 8 = 0$ (d) $x + 3y = 0$

Ans. ***(c)*** We have, the equation of the curve is $3x^2 - y^2 = 8$ ...(i)

Also, the given equation of the line is $x + 3y = 8$.

$$\Rightarrow \quad 3y = 8 - x \quad \Rightarrow \quad y = -\frac{x}{3} + \frac{8}{3} \quad \text{compare with } y = mx + c$$

Thus, slope of the line is $-\frac{1}{3}$ which should be equal to slope of the equation of normal to the curve.

On differentiating Eq. (i) w.r.t. x, we get

$$6x - 2y\frac{dy}{dx} = 0$$

$$\Rightarrow \quad \frac{dy}{dx} = \frac{6x}{2y} = \frac{3x}{y} = \text{Slope of the curve}$$

Now, slope of normal to the curve $= -\dfrac{1}{\left(\dfrac{dy}{dx}\right)} = -\dfrac{1}{\left(\dfrac{3x}{y}\right)} = -\dfrac{y}{3x}$

$$\therefore \quad -\left(\frac{y}{3x}\right) = -\frac{1}{3}$$

$$\Rightarrow \quad -3y = -3x$$

$$\Rightarrow \quad y = x$$

On substituting the value of the given equation of the curve, we get

$$3x^2 - x^2 = 8$$

$$2x^2 = 8$$

$$\Rightarrow \quad x^2 = \frac{8}{2}$$

$$\Rightarrow \quad x = \pm 2$$

Note (✱) *Topics related to these question have been removed from latest NCERT.*

For $x = 2$, $3(2)^2 - y^2 = 8$

$\Rightarrow$ $y^2 = 4$

$\Rightarrow$ $y = \pm 2$

and for $x = -2$, $3(-2)^2 - y^2 = 8$

$\Rightarrow$ $y = \pm 2$

So, the points at which normals are parallel to the given line are $(\pm 2, \pm 2)$.

Hence, the equation of normal at $(\pm 2, \pm 2)$ is

$$y - (\pm 2) = -\frac{1}{3}[x - (\pm 2)] \qquad [\because y - y_1 = m(x - x_1)]$$

$$\Rightarrow \quad 3[y - (\pm 2)] = -[x - (\pm 2)]$$

$$\Rightarrow \quad x + 3y \pm 8 = 0$$

***Q.39** If the curve $ay + x^2 = 7$ and $x^3 = y$, cut orthogonally at (1, 1), then the value of a is

(a) 1 (b) 0 (c) – 6 (d) 6

Ans. *(d)* We have, $ay + x^2 = 7$ and $x^3 = y$

On differentiating w.r.t. x in both equations, we get

$$a \cdot \frac{dy}{dx} + 2x = 0 \quad \text{and} \quad 3x^2 = \frac{dy}{dx}$$

$$\Rightarrow \quad \frac{dy}{dx} = -\frac{2x}{a} \quad \text{and} \quad \frac{dy}{dx} = 3x^2$$

$$\Rightarrow \quad \left(\frac{dy}{dx}\right)_{(1,1)} = \frac{-2}{a} = m_1 \quad \text{and} \quad \left(\frac{dy}{dx}\right)_{(1,1)} = 3 \times 1 = 3 = m_2$$

Since, the curves cut orthogonally at (1, 1).

$$\therefore \quad m_1 \cdot m_2 = -1$$

$$\Rightarrow \quad \left(\frac{-2}{a}\right) \cdot 3 = -1 \qquad \therefore \quad a = 6$$

***Q.40** If $y = x^4 - 10$ and x changes from 2 to 1.99, then what is the change in y?

(a) 0.32 (b) 0.032 (c) 5.68 (d) 5.968

Ans. *(a)* We have, $y = x^4 - 10 \Rightarrow \frac{dy}{dx} = 4x^3$ and $x = 2$

and $\Delta x = 2.00 - 1.99 = 0.01$

$$\therefore \quad \Delta y = \frac{dy}{dx} \times \Delta x = 4x^3 \times \Delta x = 4 \times 2^3 \times 0.01$$

$$= 32 \times 0.01 = 0.32$$

So, the approximate change in y is 0.32.

***Q.41** The equation of tangent to the curve $y(1 + x^2) = 2 - x$, where it crosses X-axis, is

(a) $x + 5y = 2$ (b) $x - 5y = 2$

(c) $5x - y = 2$ (d) $5x + y = 2$

Note (✱) *Topics related to these question have been removed from latest NCERT.*

Ans. *(a)* We have, equation of the curve $y(1+x^2) = 2 - x$ ...(i)

$\therefore \quad y \cdot (0+2x) + (1+x^2) \cdot \dfrac{dy}{dx} = 0 - 1$ [on differentiating w.r.t. x]

$\Rightarrow \quad 2xy + (1+x^2)\dfrac{dy}{dx} = -1$

$\Rightarrow \quad \dfrac{dy}{dx} = \dfrac{-1-2xy}{1+x^2}$ ...(ii)

Since, the given curve crosses through X-axis *i.e.*, $y = 0$.

$\therefore \quad 0(1+x^2) = 2 - x$ [using Eq. (i)]

$\Rightarrow \quad x = 2$

So, the curve passes through the point (2, 0).

$\therefore \quad \left(\dfrac{dy}{dx}\right)_{(2,0)} = \dfrac{-1-2\times 0}{1+2^2} = -\dfrac{1}{5} =$ Slope of the curve

$\therefore$ Slope of tangent to the curve $= -\dfrac{1}{5}$

$\therefore$ Equation of tangent of the curve passing through (2, 0) is

$y - 0 = -\dfrac{1}{5}(x-2)$ $[\because y - y_1 = m(x - x_1)]$

$\Rightarrow \quad y + \dfrac{x}{5} = \dfrac{2}{5} \Rightarrow 5y + x = 2$

***Q.42** The points at which the tangents to the curve $y = x^3 - 12x + 18$ are parallel to X-axis are

(a) (2, − 2), (− 2, − 34)
(b) (2, 34), (− 2, 0)
(c) (0, 34), (− 2, 0)
(d) (2, 2), (− 2, 34)

Ans. *(d)* The given equation of curve is

$y = x^3 - 12x + 18$ ...(i)

$\therefore \quad \dfrac{dy}{dx} = 3x^2 - 12$ [on differentiating w.r.t. x]

So, the slope of line parallel to the X-axis.

$\therefore \quad \left(\dfrac{dy}{dx}\right) = 0$

$\Rightarrow \quad 3x^2 - 12 = 0 \Rightarrow x^2 = \dfrac{12}{3} = 4$

$\therefore \quad x = \pm 2$

For $x = 2$, $y = 2^3 - 12\times 2 + 18 = 2$ [using Eq. (i)]

and for $x = -2$, $y = (-2)^3 - 12(-2) + 18 = 34$ [using Eq. (i)]

So, the points are (2, 2) and (− 2, 34).

***Q.43** The tangent to the curve $y = e^{2x}$ at the point (0, 1) meets X-axis at

(a) (0, 1)
(b) $\left(-\dfrac{1}{2}, 0\right)$
(c) (2, 0)
(d) (0, 2)

Note (*) *Topics related to these question have been removed from latest NCERT.*

Ans. ***(b)*** The equation of curve is $y = e^{2x}$

$\therefore$ $$\frac{dy}{dx} = e^{2x} \cdot 2 = 2 \cdot e^{2x}$$

Since, it passes through the point (0, 1).

$\Rightarrow$ $$\left(\frac{dy}{dx}\right)_{(0,1)} = 2 \cdot e^{2 \cdot 0} = 2 = \text{Slope of tangent to the curve}$$

$\therefore$ Equation of tangent is $y - 1 = 2(x - 0)$ $[\because y - y_1 = m(x - x_1)]$

$\Rightarrow$ $$y = 2x + 1$$

Since, tangent to curve $y = e^{2x}$ at the point (0, 1) meets X-axis *i.e.*, $y = 0$.

$\therefore$ $$0 = 2x + 1 \Rightarrow x = -\frac{1}{2}$$

So, the required point is $\left(\frac{-1}{2}, 0\right)$.

***Q.44** The slope of tangent to the curve $x = t^2 + 3t - 8$ and $y = 2t^2 - 2t - 5$ at the point (2, – 1) is

(a) $\frac{22}{7}$ (b) $\frac{6}{7}$ (c) $-\frac{6}{7}$ (d) -6

Ans. ***(b)*** Equation of curve is given by

$$x = t^2 + 3t - 8 \text{ and } y = 2t^2 - 2t - 5.$$

$\therefore$ $$\frac{dx}{dt} = 2t + 3 \quad \text{and} \quad \frac{dy}{dt} = 4t - 2$$

$\Rightarrow$ $$\frac{dy}{dx} = \frac{\frac{dy}{dt}}{\frac{dx}{dt}} = \frac{4t - 2}{2t + 3} \quad \text{...(i)}$$

Since, the curve passes through the point (2, – 1).

$\therefore$ $$2 = t^2 + 3t - 8$$

and $$-1 = 2t^2 - 2t - 5$$

$\Rightarrow$ $$t^2 + 3t - 10 = 0$$

$\Rightarrow$ $$(t - 2)(t + 5) = 0$$

$\Rightarrow$ $$t = 2, -5$$

and $$2t^2 - 2t - 4 = 0$$

$\Rightarrow$ $$(2t - 4)(t + 1) = 0$$

$\Rightarrow$ $$t = -1, 2 \Rightarrow t = 2$$

$\therefore$ Slope of tangent,

$$\left(\frac{dy}{dx}\right)_{\text{at } t=2} = \frac{4 \times 2 - 2}{2 \times 2 + 3} = \frac{6}{7} \quad \text{[using Eq. (i)]}$$

Note (*) *Topics related to these question have been removed from latest NCERT.*

***Q.45** Two curves $x^3 - 3xy^2 + 2 = 0$ and $3x^2y - y^3 - 2 = 0$ intersect at an angle of

(a) $\frac{\pi}{4}$ (b) $\frac{\pi}{3}$ (c) $\frac{\pi}{2}$ (d) $\frac{\pi}{6}$

Ans. ***(c)*** Equation of two curves are given by

$$x^3 - 3xy^2 + 2 = 0$$

and $$3x^2y - y^3 - 2 = 0$$

On differentiation of both curves with respect to x, we get

$$\Rightarrow \quad 3x^2 - 3\left[x \cdot 2y\frac{dy}{dx} + y^2 \cdot 1\right] + 0 = 0$$

and $$3\left[x^2\frac{dy}{dx} + y \cdot 2x\right] - 3y^2\frac{dy}{dx} - 0 = 0$$

$$\Rightarrow \quad 3x \cdot 2y\frac{dy}{dx} + 3y^2 = 3x^2$$

and $$3y^2\frac{dy}{dx} = 3x^2\frac{dy}{dx} + 6xy$$

$$\Rightarrow \quad \frac{dy}{dx} = \frac{3x^2 - 3y^2}{6xy}$$

and $$\frac{dy}{dx} = \frac{6xy}{3y^2 - 3x^2}$$

$$\Rightarrow \quad \left(\frac{dy}{dx}\right) = \frac{3(x^2 - y^2)}{6xy} \text{ and } \left(\frac{dy}{dx}\right) = \frac{-6xy}{3(x^2 - y^2)}$$

$$\Rightarrow \quad = \frac{x^2 - y^2}{2xy} = m_1 \text{ (say)} = \frac{-2xy}{x^2 - y^2} = m_2 \text{ (say)}$$

$$\therefore \quad m_1m_2 = \frac{x^2 - y^2}{2xy} \cdot \frac{-(2xy)}{x^2 - y^2} = -1$$

Hence, both the curves are intersecting at right angle *i.e.*, making $\frac{\pi}{2}$ with each other.

Q.46 The interval on which the function $f(x) = 2x^3 + 9x^2 + 12x - 1$ is decreasing is

(a) $[-1, \infty)$ (b) $[-2, -1]$ (c) $(-\infty, -2]$ (d) $[-1, 1]$

Ans. ***(b)*** We have, $$f(x) = 2x^3 + 9x^2 + 12x - 1$$

$$\therefore \quad f'(x) = 6x^2 + 18x + 12$$
$$= 6(x^2 + 3x + 2) = 6(x + 2)(x + 1)$$

So, $f'(x) \le 0$, for decreasing.

On drawing number line as below

+ve | − | +ve
(−2) (−1)

We see that $f'(x)$ is decreasing in $[-2, -1]$.

Note (*) *Topics related to these question have been removed from latest NCERT.*

Q.47 If $f: R \rightarrow R$ be defined by $f(x) = 2x + \cos x$, then f

(a) has a minimum at $x = \pi$
(b) has a maximum at $x = 0$
(c) is a decreasing function
(d) is an increasing function

Ans. *(d)* We have, $f(x) = 2x + \cos x$

$\therefore$ $f'(x) = 2 + (-\sin x) = 2 - \sin x$

Since, $f'(x) > 0, \forall x$ $[\because -1 \le \sin x \le 1 \ \forall x \in R]$

Hence, $f(x)$ is an increasing function.

Q.48 If $y = x(x-3)^2$ decreases for the values of x given by

(a) $1 < x < 3$ (b) $x < 0$ (c) $x > 0$ (d) $0 < x < \frac{3}{2}$

Ans. *(a)* We have, $y = x(x-3)^2$

$$\therefore \quad \frac{dy}{dx} = x \cdot 2(x-3) \cdot 1 + (x-3)^2 \cdot 1$$
$$= 2x^2 - 6x + x^2 + 9 - 6x = 3x^2 - 12x + 9$$
$$= 3(x^2 - 3x - x + 3) = 3(x-3)(x-1)$$

+ 1 − 3 +

So, $y = x(x-3)^2$ decreases for $(1, 3)$.

[since, $y' < 0$ for all $x \in (1, 3)$, hence y is decreasing on $(1, 3)$]

Q.49 The function $f(x) = 4\sin^3 x - 6\sin^2 x + 12\sin x + 100$ is strictly

(a) increasing in $\left(\pi, \frac{3\pi}{2}\right)$
(b) decreasing in $\left(\frac{\pi}{2}, \pi\right)$
(c) decreasing in $\left[\frac{-\pi}{2}, \frac{\pi}{2}\right]$
(d) decreasing in $\left[0, \frac{\pi}{2}\right]$

Ans. *(b)* We have, $f(x) = 4\sin^3 x - 6\sin^2 x + 12\sin x + 100$

$$\therefore \quad f'(x) = 12\sin^2 x \cdot \cos x - 12\sin x \cdot \cos x + 12\cos x$$
$$= 12[\sin^2 x \cdot \cos x - \sin x \cdot \cos x + \cos x]$$
$$= 12\cos x[\sin^2 x - \sin x + 1]$$
$$\Rightarrow \quad f'(x) = 12\cos x[\sin^2 x + (1 - \sin x)] \quad \text{...(i)}$$

$\because$ $1 - \sin x \ge 0$ and $\sin^2 x \ge 0$

$\therefore$ $\sin^2 x + 1 - \sin x \ge 0$

Hence, $f'(x) > 0$, when $\cos x > 0$ *i.e.*, $x \in \left(-\frac{\pi}{2}, \frac{\pi}{2}\right)$.

So, $f(x)$ is increasing when $x \in \left(-\frac{\pi}{2}, \frac{\pi}{2}\right)$ and $f'(x) < 0$, when $\cos x < 0$ *i.e.*, $x \in \left(\frac{\pi}{2}, \frac{3\pi}{2}\right)$

Hence, $f(x)$ is decreasing when $x \in \left(\frac{\pi}{2}, \frac{3\pi}{2}\right)$

Since, $\left(\frac{\pi}{2}, \pi\right) \in \left(\frac{\pi}{2}, \frac{3\pi}{2}\right)$

Hence, $f(x)$ is decreasing in $\left(\frac{\pi}{2}, \pi\right)$.

Q.50 Which of the following functions is decreasing on $\left(0, \frac{\pi}{2}\right)$?

(a) $\sin 2x$ (b) $\tan x$ (c) $\cos x$ (d) $\cos 3x$

Ans. *(c)* In the interval $\left(0, \frac{\pi}{2}\right)$, $f(x) = \cos x$

$\Rightarrow$ $f'(x) = -\sin x$

which gives $f'(x) < 0$ in $\left(0, \frac{\pi}{2}\right)$

Hence, $f(x) = \cos x$ is decreasing in $\left(0, \frac{\pi}{2}\right)$.

Q.51 The function $f(x) = \tan x - x$

(a) always increases
(b) always decreases
(c) never increases
(d) sometimes increases and sometimes decreases

Ans. *(a)* We have, $f(x) = \tan x - x$

$\therefore$ $f'(x) = \sec^2 x - 1$ $[\because \sec^2 x \in [1, \infty)]$

$\Rightarrow$ $f'(x) > 0, \forall x \in R$

So, $f(x)$ always increases for all x.

Q.52 If x is real, then the minimum value of $x^2 - 8x + 17$ is

(a) -1 (b) 0 (c) 1 (d) 2

Ans. *(c)* Let $f(x) = x^2 - 8x + 17$

$\therefore$ $f'(x) = 2x - 8$

So, $f'(x) = 0$, gives $x = 4$

Now, $f''(x) = 2 > 0, \forall x$

So, $x = 4$ is the point of local minima.

$\therefore$ Minimum value of $f(x)$ at $x = 4$,

$$f(4) = 4 \times 4 - 8 \times 4 + 17 = 1$$

Q.53 The smallest value of polynomial $x^3 - 18x^2 + 96x$ in [0, 9] is

(a) 126 (b) 0 (c) 135 (d) 160

Ans. *(b)* We have, $f(x) = x^3 - 18x^2 + 96x$

$\therefore$ $f'(x) = 3x^2 - 36x + 96$

So, $f'(x) = 0$

Gives, $3x^2 - 36x + 96 = 0$

$\Rightarrow$ $3(x^2 - 12x + 32) = 0$

$\Rightarrow$ $(x - 8)(x - 4) = 0$

$\Rightarrow$ $x = 8, 4 \in [0, 9]$

We shall now evaluate the value of f at these points and at the end points of the interval [0, 9] *i.e.*, at $x = 4$ and $x = 8$ and at $x = 0$ and at $x = 9$.

$\therefore \quad f(4) = 4^3 - 18 \cdot 4^2 + 96 \cdot 4$

$= 64 - 288 + 384 = 160$

$f(8) = 8^3 - 18 \cdot 8^2 + 96 \cdot 8 = 128$

$f(9) = 9^3 - 18 \cdot 9^2 + 96 \cdot 9$

$= 729 - 1458 + 864 = 135$

and $\quad f(0) = 0^3 - 18 \cdot 0^2 + 96 \cdot 0 = 0$

Thus, we conclude that absolute minimum value of f on [0, 9] is 0 occurring at $x = 0$.

Q.54 The function $f(x) = 2x^3 - 3x^2 - 12x + 4$, has

(a) two points of local maximum
(b) two points of local minimum
(c) one maxima and one minima
(d) no maxima or minima

Ans. *(c)* We have $\quad f(x) = 2x^3 - 3x^2 - 12x + 4$

$\therefore \quad f'(x) = 6x^2 - 6x - 12$

Now, $\quad f'(x) = 0 \Rightarrow 6(x^2 - x - 2) = 0$

$\Rightarrow \quad 6(x+1)(x-2) = 0$

$\Rightarrow \quad x = -1$ and $x = +2$

On number line for $f'(x)$, we get

+ (−1) − (2) +

Hence $x = -1$ is point of local maxima and $x = 2$ is point of local minima.

So, $f(x)$ has one maxima and one minima.

Q.55 The maximum value of $\sin x \cdot \cos x$ is

(a) $\frac{1}{4}$
(b) $\frac{1}{2}$
(c) $\sqrt{2}$
(d) $2\sqrt{2}$

Ans. *(b)* We have, $\quad f(x) = \sin x \cdot \cos x = \frac{1}{2}\sin 2x$

$\therefore \quad f'(x) = \frac{1}{2} \cdot \cos 2x \cdot 2 = \cos 2x$

Now, $\quad f'(x) = 0 \quad \Rightarrow \cos 2x = 0$

$\Rightarrow \quad \cos 2x = \cos\frac{\pi}{2} \Rightarrow x = \frac{\pi}{4}$

Also $\quad f''(x) = \frac{d}{dx}\cos 2x = -\sin 2x \cdot 2 = -2\sin 2x$

$\therefore \quad [f''(x)]_{\text{at } x = \pi/4} = -2 \cdot \sin 2 \cdot \frac{\pi}{4} = -2\sin\frac{\pi}{2} = -2 < 0$

At $\frac{\pi}{4}$, $f(x)$ is maximum and $\frac{\pi}{4}$ is point of maxima.

$\therefore \quad f\left(\frac{\pi}{4}\right) = \frac{1}{2} \cdot \sin 2 \cdot \frac{\pi}{4} = \frac{1}{2}$

Q.56 At $x = \frac{5\pi}{6}$, $f(x) = 2\sin 3x + 3\cos 3x$ is

(a) maximum
(b) minimum
(c) zero
(d) neither maximum nor minimum

Ans. ***(d)*** We have, $f(x) = 2\sin 3x + 3\cos 3x$

$\therefore$ $f'(x) = 2\cdot\cos 3x \cdot 3 + 3(-\sin 3x)\cdot 3$

$\Rightarrow$ $f'(x) = 6\cos 3x - 9\sin 3x$ ...(i)

Now, $f''(x) = -18\sin 3x - 27\cos 3x$

$= -9(2\sin 3x + 3\cos 3x)$

$\therefore$ $f'\left(\frac{5\pi}{6}\right) = 6\cos\left(3\cdot\frac{5\pi}{6}\right) - 9\sin\left(3\cdot\frac{5\pi}{6}\right)$

$= 6\cos\frac{5\pi}{2} - 9\sin\frac{5\pi}{2}$

$[\because \cos(2\pi + \theta) = \cos\theta \sin(2\pi + \theta) = \sin\theta]$

$= 6\cos\left(2\pi + \frac{\pi}{2}\right) - 9\sin\left(2\pi + \frac{\pi}{2}\right) = 6\cos\frac{\pi}{2} - 9\sin\frac{\pi}{2}$

$= 0 - 9 \neq 0$ $\left[\because \cos\frac{\pi}{2} = 0, \sin\frac{\pi}{2} = 1\right]$

So, $x = \frac{5\pi}{6}$ cannot be point of maxima or minima.

Hence, $f(x)$ at $x = \frac{5\pi}{6}$ is neither maximum nor minimum.

Q.57 The maximum slope of curve $y = -x^3 + 3x^2 + 9x - 27$ is

(a) 0 (b) 12 (c) 16 (d) 32

Ans. ***(b)*** We have, $y = -x^3 + 3x^2 + 9x - 27$

$\therefore$ $\frac{dy}{dx} = -3x^2 + 6x + 9 =$ Slope of the curve

and $\frac{d^2y}{dx^2} = -6x + 6 = -6(x-1)$

$\therefore$ $\frac{d^2y}{dx^2} = 0$

$\Rightarrow$ $-6(x-1) = 0 \Rightarrow x = 1 > 0$

Now, $\frac{d^3y}{dx^3} = -6 < 0$

So, the maximum slope of given curve is at $x = 1$.

$\therefore$ $\left(\frac{dy}{dx}\right)_{(x=1)} = -3\cdot 1^2 + 6\cdot 1 + 9 = 12$

Q.58 The function $f(x) = x^x$ has a stationary point at

(a) $x = e$ (b) $x = \frac{1}{e}$ (c) $x = 1$ (d) $x = \sqrt{e}$

Ans. ***(b)*** A stationary point of a function $f(x)$ is a point where the derivative of $f(x)$ is equal to zero

i.e., $f'(x) = 0$ or $\frac{dy}{dx} = 0$

We have, $f(x) = x^x$

Let $y = x^x$

Taking log on both sides

and $$\log y = x \log x$$

$\therefore$ $$\frac{1}{y} \cdot \frac{dy}{dx} = x \cdot \frac{1}{x} + \log x \cdot 1$$

$$\frac{dy}{dx} = (1 + \log x) y$$

$\Rightarrow$ $$\frac{dy}{dx} = (1 + \log x) \cdot x^x \qquad [\because y = x^x]$$

$\therefore$ $$\frac{dy}{dx} = 0$$

$\Rightarrow$ $$(1 + \log x) \cdot x^x = 0$$

$\Rightarrow$ $$\log x = -1$$

$\Rightarrow$ $$\log x = \log e^{-1}$$

$\Rightarrow$ $$x = e^{-1} \Rightarrow x = \frac{1}{e}$$

Hence, $f(x)$ has a stationary point at $x = \frac{1}{e}$.

Q.59 The maximum value of $\left(\frac{1}{x}\right)^x$ is

(a) e (b) e^e (c) $e^{1/e}$ (d) $\left(\frac{1}{e}\right)^{1/e}$

Ans. *(c)* Let $$y = \left(\frac{1}{x}\right)^x$$

Taking log on both sides

$\Rightarrow$ $$\log y = x \cdot \log \frac{1}{x}$$

$\therefore$ $$\frac{1}{y} \cdot \frac{dy}{dx} = x \cdot \frac{1}{\frac{1}{x}} \cdot \left(-\frac{1}{x^2}\right) + \log \frac{1}{x} \cdot 1$$

$$= \left(-1 + \log \frac{1}{x}\right) y$$

$\therefore$ $$\frac{dy}{dx} = \left(\log \frac{1}{x} - 1\right) \cdot \left(\frac{1}{x}\right)^x \qquad \left[\because y = \left(\frac{1}{x}\right)^x\right]$$

Now, $$\frac{dy}{dx} = 0$$

$\Rightarrow$ $$\log \frac{1}{x} = 1 = \log e$$

$\Rightarrow$ $$\frac{1}{x} = e$$

$\therefore$ $$x = \frac{1}{e}$$

Hence, the maximum value of $f\left(\frac{1}{e}\right) = (e)^{1/e}$.

Fill in the blanks in each of the exercises 60 to 64.

***Q.60** The curves $y = 4x^2 + 2x - 8$ and $y = x^3 - x + 13$ touch each other at the point ....

Ans. The curves $y = 4x^2 + 2x - 8$ and $y = x^3 - x + 13$ touch each other at the point (3, 34).

Given, equation of curves are $y = 4x^2 + 2x - 8$ and $y = x^3 - x + 13$

$\therefore \quad \frac{dy}{dx} = 8x + 2 \qquad$ [when $y = 4x^2 + 2x - 8$]

and $\quad \frac{dy}{dx} = 3x^2 - 1 \qquad$ [when $y = x^3 - x + 13$]

So, the slope of both curves should be same

$\therefore \quad 8x + 2 = 3x^2 - 1 \Rightarrow 3x^2 - 8x - 3 = 0$

$\Rightarrow \quad 3x^2 - 9x + x - 3 = 0 \qquad \Rightarrow 3x(x-3) + 1(x-3) = 0$

$\Rightarrow \quad (3x+1)(x-3) = 0$

$\therefore \quad x = -\frac{1}{3}$ and $x = 3$,

For $x = -\frac{1}{3}$, $\quad y = 4\cdot\left(-\frac{1}{3}\right)^2 + 2\cdot\left(\frac{-1}{3}\right) - 8$

$$= \frac{4}{9} - \frac{2}{3} - 8 = \frac{4 - 6 - 72}{9} = -\frac{74}{9}$$

and for $x = 3$, $y = 4\cdot(3)^2 + 2\cdot(3) - 8 = 36 + 6 - 8 = 34$

So, the required points are (3, 34) and $\left(-\frac{1}{3}, \frac{-74}{9}\right)$.

But only point (3, 34) satisfies both equations.

***Q.61** The equation of normal to the curve $y = \tan x$ at (0, 0) is ....

Ans. The equation of normal to the curve $y = \tan x$ at (0, 0) is $x + y = 0$.

$\because \quad y = \tan x \Rightarrow \frac{dy}{dx} = \sec^2 x$

$\Rightarrow \quad \left(\frac{dy}{dx}\right)_{(0,0)} = \sec^2 0 = 1$ and $-\frac{1}{\left(\frac{dy}{dx}\right)} = -\frac{1}{1}$

$\therefore$ Equation of normal to the curve $y = \tan x$ at (0, 0) is

$y - 0 = -1(x - 0) \qquad [\because (y - y_1) = m(x - x_1)]$

$\Rightarrow \quad y + x = 0$

Q.62 The values of a for which the function $f(x) = \sin x - ax + b$ increases on R are ....

Ans. The values of a for which the function $f(x) = \sin x - ax + b$ increases on R are $(-\infty, -1)$.

we have, $\quad f(x) = \sin x - ax + b$

$\because \quad f'(x) = \cos x - a$

and $\quad f'(x) > 0 \Rightarrow \cos x > a$

Since, $\quad \cos x \in [-1, 1]$

$\Rightarrow \quad a < -1 \Rightarrow a \in (-\infty, -1)$

Note (*) *Topics related to these question have been removed from latest NCERT.*

Q.63 The function $f(x)=\dfrac{2x^2-1}{x^4}$, (where, $x>0$) decreases in the interval ....

Ans. The function $f(x)=\dfrac{2x^2-1}{x^4}$, where $x>0$, decreases in the interval $(1,\infty)$.

$\because$ $$f'(x)=\frac{x^4\cdot 4x-(2x^2-1)\cdot 4x^3}{x^8}=\frac{4x^5-8x^5+4x^3}{x^8}$$

$$=\frac{-4x^5+4x^3}{x^8}=\frac{4x^3(-x^2+1)}{x^8}$$

Also, $$f'(x)<0$$

$\Rightarrow$ $$\frac{4x^3(1-x^2)}{x^8}<0\Rightarrow x^2>1$$

$\Rightarrow$ $$x>\pm 1$$

$\therefore$ $$x\in(1,\infty)\cup(-1,0)$$

Q.64 The least value of function $f(x)=ax+\dfrac{b}{x}$ (where, $a>0,\ b>0,\ x>0$) is ....

Ans. The least value of function $f(x)=ax+\dfrac{b}{x}$ (where, $a>0, b>0, x>0$) is $2\sqrt{ab}$.

$\because$ $$f'(x)=a-\frac{b}{x^2}\text{ and } f'(x)=0$$

$\Rightarrow$ $$a=\frac{b}{x^2}$$

$\Rightarrow$ $$x^2=\frac{b}{a}\Rightarrow x=\pm\sqrt{\frac{b}{a}}$$

Now, $$f''(x)=-b\cdot\frac{(-2)}{x^3}=+\frac{2b}{x^3}$$

At $x=\sqrt{\dfrac{b}{a}}$, $$f''(x)=+\frac{2b}{\left(\frac{b}{a}\right)^{3/2}}=\frac{2b\cdot a^{3/2}}{b^{3/2}}$$

$$=2b^{-1/2}\cdot a^{3/2}=2\sqrt{\frac{a^3}{b}}>0 \qquad [\because a,b>0]$$

$\therefore$ Least value of $f(x)$, $$f\left(\sqrt{\frac{b}{a}}\right)=a\cdot\sqrt{\frac{b}{a}}+\frac{b}{\sqrt{\frac{b}{a}}}$$

$$=a\cdot a^{-1/2}\cdot b^{1/2}+b\cdot b^{-1/2}\cdot a^{1/2}$$

$$=\sqrt{ab}+\sqrt{ab}=2\sqrt{ab}$$

Chapter

07

Integrals

Short Answer Type Questions

Verify the following.

Q.1 $\int \frac{2x-1}{2x+3} dx = x - \log|(2x+3)^2| + C$

Ans. Let $I = \int \frac{2x-1}{2x+3} dx = \int \frac{2x+3-3-1}{2x+3} dx = \int 1 dx - 4\int \frac{1}{2x+3} dx$

$= x - 2\log|2x+3| + C$

$= x - \log|(2x+3)^2| + C$ $\quad [\because \log m^n = n \log m]$

Q.2 $\int \frac{2x+3}{x^2+3x} dx = \log|x^2+3x| + C$

Ans. Let $I = \int \frac{2x+3}{x^2+3x} dx$

Put $x^2 + 3x = t$

$\Rightarrow (2x+3)\, dx = dt$

$\therefore I = \int \frac{1}{t} dt = \log|t| + C$

$= \log|(x^2+3x)| + C$

Evaluate the following.

Q.3 $\int \frac{(x^2+2)d}{x+1} x$

Thinking Process First of all divide numerator by denominator, then use the formula $\int \frac{1}{x} dx = \log|x|$ to get the solution.

Ans. Let $I = \int \frac{x^2+2}{x+1} dx$

Divide (x^2+2) by $(x+1)$

Then, $\frac{x^2+2}{x+1} = x - 1 + \frac{3}{x+1}$

Now, $I = \int \left(x - 1 + \frac{3}{x+1} \right) dx = \int (x-1)\, dx + 3\int \frac{1}{x+1} dx$

$= \frac{x^2}{2} - x + 3\log|(x+1)| + C$

Q.4 $\int \frac{e^{6\log x} - e^{5\log x}}{e^{4\log x} - e^{3\log x}} dx$

Ans. Let

$$I = \int \left(\frac{e^{6\log x} - e^{5\log x}}{e^{4\log x} - e^{3\log x}} \right) dx$$

$$= \int \left(\frac{e^{\log x^6} - e^{\log x^5}}{e^{\log x^4} - e^{\log x^3}} \right) dx \quad [\because a \log b = \log b^a]$$

$$= \int \left(\frac{x^6 - x^5}{x^4 - x^3} \right) dx \quad [\because e^{\log x} = x]$$

$$= \int \frac{x^5(x-1)}{x^3(x-1)} dx$$

$$= \int x^2 dx = \frac{x^3}{3} + C$$

Q.5 $\int \frac{(1+\cos x)}{x + \sin x} dx$

Ans. Consider that,

$$I = \int \frac{(1+\cos x)}{(x+\sin x)} dx$$

Let $x + \sin x = t \Rightarrow (1 + \cos x)\, dx = dt$

$\therefore$

$$I = \int \frac{1}{t}\, dt = \log |t| + C$$

$$= \log |(x + \sin x)| + C$$

Q.6 $\int \frac{dx}{1+\cos x}$

Thinking Process Use $\cos x = 2\cos^2 \frac{x}{2} - 1$ and also use formula i.e., $\int \sec^2 x = \tan x + C$ to solve the above problem.

Ans. Let

$$I = \int \frac{dx}{1+\cos x} = \int \frac{dx}{1 + 2\cos^2 \frac{x}{2} - 1}$$

$$= \frac{1}{2} \int \frac{1}{\cos^2 \frac{x}{2}} dx = \frac{1}{2} \int \sec^2 \frac{x}{2} dx \quad \left[\because \frac{1}{\cos x} = \sec x\right]$$

$$= \frac{1}{2} \cdot \tan \frac{x}{2} \cdot 2 + C = \tan \frac{x}{2} + C \quad \left[\because \int \sec^2 x\, dx = \tan x\right]$$

Q.7 $\int \tan^2 x \sec^4 x\, dx$

Thinking Process Use the formula $\sec^2 x = 1 + \tan^2 x$ and put $\tan x = t$ to solve this problem.

Ans. Let

$$I = \int \tan^2 x \sec^4 x\, dx$$

$$= \int \tan^2 x \cdot \sec^2 x \cdot \sec^2 x\, dx$$

$$= \int \tan^2 x (1 + \tan^2 x) \cdot \sec^2 x \cdot dx$$

Put $\tan x = t \Rightarrow \sec^2 x\, dx = dt$

$\therefore$ $$I = \int t^2(1+t^2)\, dt = \int (t^2 + t^4)\, dt$$

$$= \frac{t^3}{3} + \frac{t^5}{5} + C = \frac{\tan^5 x}{5} + \frac{\tan^3 x}{3} + C$$

Q.8 $\int \frac{\sin x + \cos x}{\sqrt{1 + \sin 2x}}\, dx$

Ans. Let $$I = \int \frac{\sin x + \cos x}{\sqrt{1+\sin 2x}}\, dx = \int \frac{(\sin x + \cos x)}{\sqrt{\sin^2 x + \cos^2 x + 2\sin x \cos x}}\, dx$$

$[\because \sin^2 x + \cos^2 x = 1]$

$$= \int \frac{\sin x + \cos x}{\sqrt{(\sin x + \cos x)^2}}\, dx = \int 1\, dx = x + C \quad [\because (a+b)^2 = a^2 + b^2 + 2ab]$$

Q.9 $\int \sqrt{1 + \sin x}\, dx$

Ans. Let $$I = \int \sqrt{1 + \sin x}\, dx$$

$$= \int \sqrt{\left(\sin^2 \frac{x}{2} + \cos^2 \frac{x}{2} + 2\sin\frac{x}{2}\cos\frac{x}{2}\right)}\, dx \qquad \left[\because \sin^2 \frac{x}{2} + \cos^2 \frac{x}{2} = 1\right]$$

$$= \int \sqrt{\left(\sin\frac{x}{2} + \cos\frac{x}{2}\right)^2}\, dx = \int \left(\sin\frac{x}{2} + \cos\frac{x}{2}\right) dx$$

$$= -\cos\frac{x}{2}\cdot 2 + \sin\frac{x}{2}\cdot 2 + C = -2\cos\frac{x}{2} + 2\sin\frac{x}{2} + C$$

Q.10 $\int \frac{x}{\sqrt{x}+1}\, dx$

Ans. Let $$I = \int \frac{x}{\sqrt{x}+1}\, dx$$

Put $\sqrt{x} = t \Rightarrow \frac{1}{2\sqrt{x}}\, dx = dt$

$\Rightarrow$ $dx = 2\sqrt{x}\, dt$

$\therefore$ $$I = 2\int \left(\frac{x\sqrt{x}}{t+1}\right) dt = 2\int \frac{t^2 \cdot t}{t+1}\, dt = 2\int \frac{t^3}{t+1}\, dt$$

$$= 2\int \frac{t^3 + 1 - 1}{t+1}\, dt = 2\int \frac{(t+1)(t^2 - t + 1)}{t+1}\, dt - 2\int \frac{1}{t+1}\, dt$$

$[\because a^3 + b^3 = (a+b)(a^2 - ab + b^2)]$

$$= 2\int (t^2 - t + 1)\, dt - 2\int \frac{1}{t+1}\, dt$$

$$= 2\left[\frac{t^3}{3} - \frac{t^2}{2} + t - \log|(t+1)|\right] + C$$

$$= 2\left[\frac{x\sqrt{x}}{3} - \frac{x}{2} + \sqrt{x} - \log|(\sqrt{x}+1)|\right] + C \qquad [\because t = \sqrt{x}]$$

Q.11 $\int \sqrt{\dfrac{a+x}{a-x}}\, dx$

Thinking Process Here, put $x = a\cos 2\theta$ and also use the formula *i.e.*, $\cos 2\theta = 2\cos^2\theta - 1 = 1 - 2\sin^2\theta$, to get the solution.

Ans. Let $\quad I = \int \sqrt{\dfrac{a+x}{a-x}}\, dx$

Put $\quad x = a\cos 2\theta$

$\Rightarrow \quad dx = -a \cdot \sin 2\theta \cdot 2 \cdot d\theta$

$$\therefore \quad I = -2\int \sqrt{\frac{a + a\cos 2\theta}{a - a\cos 2\theta}} \cdot a\sin 2\theta d\theta$$

$$= -2a\int \sqrt{\frac{1+\cos 2\theta}{1-\cos 2\theta}} \sin 2\theta d\theta = -2a\int \sqrt{\frac{2\cos^2\theta}{2\sin^2\theta}} \sin 2\theta\, d\theta$$

$$= -2a\int \cot\theta \cdot \sin 2\theta d\theta = -2a\int \frac{\cos\theta}{\sin\theta} \cdot 2\sin\theta \cdot \cos\theta\, d\theta$$

$[\because \sin 2\theta = 2\sin\theta\cos\theta]$

$$= -4a\int \cos^2\theta d\theta = -2a\int (1+\cos 2\theta)\, d\theta \quad [\because \cos 2\theta = 2\cos^2\theta - 1]$$

$$= -2a\left[\theta + \frac{\sin 2\theta}{2}\right] + C$$

$\because \quad \cos 2\theta = \dfrac{x}{a} \Rightarrow 2\theta = \cos^{-1}\dfrac{x}{a} \Rightarrow \theta = \dfrac{1}{2}\cos^{-1}\dfrac{x}{a}$

Now, $\quad \sin 2\theta = \dfrac{\sqrt{a^2 - x^2}}{a}$

$$= \sqrt{1 - \frac{x^2}{a^2}}$$

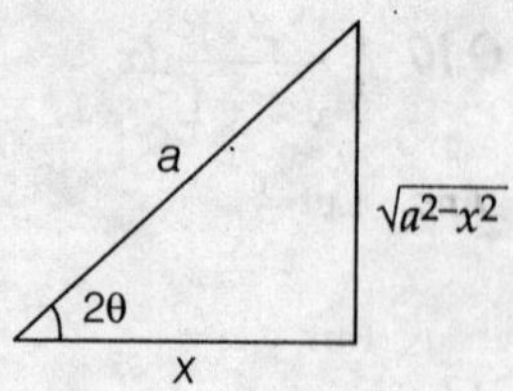

$$\therefore \quad I = -2a\left[\frac{1}{2}\cos^{-1}\frac{x}{a} + \frac{1}{2}\sqrt{1 - \frac{x^2}{a^2}}\right] + C$$

$$= -a\left[\cos^{-1}\left(\frac{x}{a}\right) + \sqrt{1 - \frac{x^2}{a^2}}\right] + C$$

Alternate Method

Let $\quad I = \int \sqrt{\dfrac{a+x}{a-x}}\, dx = \int \sqrt{\dfrac{(a+x)(a+x)}{(a-x)(a+x)}}\, dx$

$$= \int \frac{(a+x)}{\sqrt{a^2 - x^2}} dx$$

$\Rightarrow \quad I = \int \dfrac{a}{\sqrt{a^2 - x^2}} dx + \int \dfrac{x}{\sqrt{a^2 - x^2}} dx$

Let $\quad I = I_1 + I_2 \quad$ ...(i)

Now, $\quad I_1 = \int \dfrac{a}{\sqrt{a^2 - x^2}} dx = a\sin^{-1}\left(\dfrac{x}{a}\right) + C_1$

and $\quad I_2 = \int \frac{x}{\sqrt{a^2 - x^2}} dx$

Put $\quad a^2 - x^2 = t^2 \Rightarrow -2x\,dx = 2t\,dt \Rightarrow x\,dx = -t\,dt$

$\therefore \quad I_2 = -\int \frac{t}{t} dt = -\int 1\,dt$

$= -t + C_2 = -\sqrt{a^2 - x^2} + C_2 \qquad [\because t^2 = a^2 - x^2]$

$\therefore \quad I = a\sin^{-1}\left(\frac{x}{a}\right) + C_1 - \sqrt{a^2 - x^2} + C_2$

$\Rightarrow \quad I = a\sin^{-1}\left(\frac{x}{a}\right) - \sqrt{a^2 - x^2} + C \qquad [\because C = C_1 + C_2]$

Q.12 $\int \frac{x^{1/2}}{1 + x^{3/4}} dx$

Ans. Let $\quad I = \int \frac{x^{1/2}}{1 + x^{3/4}} dx$

Put $\quad x = t^4 \Rightarrow dx = 4t^3 dt$

$\therefore \quad I = 4\int \frac{t^2(t^3)}{1 + t^3} dt = 4\int \left(t^2 - \frac{t^2}{1 + t^3}\right) dt \qquad [\because x^{1/2} = t^2]$

$\Rightarrow \quad I = 4\int t^2 dt - 4\int \frac{t^2}{1 + t^3} dt$

Let $\quad I = I_1 - I_2$

Now, $\quad I_1 = 4\int t^2 dt = 4 \cdot \frac{t^3}{3} + C_1 = \frac{4}{3} x^{3/4} + C_1$

and $\quad I_2 = 4\int \frac{t^2}{1 + t^3} dt$

Again, put $\quad 1 + t^3 = z \Rightarrow 3t^2 dt = dz$

$\Rightarrow \quad t^2 dt = \frac{1}{3} dz$

Now, $\quad I_2 = \frac{4}{3}\int \frac{1}{z} dz$

$= \frac{4}{3}\log|z| + C_2 = \frac{4}{3}\log|(1 + t^3)| + C_2$

$= \frac{4}{3}\log|(1 + x^{3/4})| + C_2$

$\therefore \quad I = \frac{4}{3}x^{3/4} + C_1 - \frac{4}{3}\log|(1 + x^{3/4})| - C_2$

$= \frac{4}{3}[x^{3/4} - \log|(1 + x^{3/4})|] + C \qquad [\because C = C_1 - C_2]$

Q.13 $\int \frac{\sqrt{1 + x^2}}{x^4} dx$

Ans. Let $\quad I = \int \frac{\sqrt{1 + x^2}}{x^4} dx = \int \frac{\sqrt{1 + x^2}}{x} \cdot \frac{1}{x^3} dx$

$= \int \sqrt{\frac{1 + x^2}{x^2}} \cdot \frac{1}{x^3} dx = \int \sqrt{\frac{1}{x^2} + 1} \cdot \frac{1}{x^3} dx$

Put $\quad 1 + \frac{1}{x^2} = t^2 \Rightarrow \frac{-2}{x^3} dx = 2t\,dt$

$$\Rightarrow \quad -\frac{1}{x^3}\cdot dx = t\cdot dt$$

$$\therefore \quad I = -\int t^2 dt = -\frac{t^3}{3} + C = -\frac{1}{3}\left(1+\frac{1}{x^2}\right)^{3/2} + C$$

Q.14 $\int \frac{dx}{\sqrt{16-9x^2}}$

Thinking Process First of all convert the expression in form of $\frac{1}{\sqrt{a^2-x^2}}$, then use the formula, $\int \frac{1}{\sqrt{a^2-x^2}}\,dx = \sin^{-1}\left(\frac{x}{a}\right) + C.$

Ans. Let $I = \int \frac{dx}{\sqrt{16-9x^2}} = \int \frac{dx}{\sqrt{(4)^2-(3x)^2}}dx = \frac{1}{3}\sin^{-1}\left(\frac{3x}{4}\right) + C$

Q.15 $\int \frac{dt}{\sqrt{3t-2t^2}}$

Ans. Let

$$I = \int \frac{dt}{\sqrt{3t-2t^2}} = \frac{1}{\sqrt{2}}\int \frac{dt}{\sqrt{-\left(t^2-\frac{3}{2}t\right)}}$$

$$= \frac{1}{\sqrt{2}}\int \frac{dt}{\sqrt{-\left[\left(t^2 - 2\times t\times\frac{3}{4} + \left(\frac{3}{4}\right)^2 - \left(\frac{3}{4}\right)^2\right)\right]}}$$

$$= \frac{1}{\sqrt{2}}\int \frac{dt}{\sqrt{-\left(t-\frac{3}{4}\right)^2 + \left(\frac{3}{4}\right)^2}}$$

$$= \frac{1}{\sqrt{2}}\int \frac{dt}{\sqrt{\left(\frac{3}{4}\right)^2 - \left(t-\frac{3}{4}\right)^2}} = \frac{1}{\sqrt{2}}\sin^{-1}\left(\frac{t-\frac{3}{4}}{\frac{3}{4}}\right) + C$$

$$\Rightarrow \quad I = \frac{1}{\sqrt{2}}\sin^{-1}\left(\frac{4t-3}{3}\right) + C$$

Q.16 $\int \frac{3x-1}{\sqrt{x^2+9}}dx$

Thinking Process First of all convert to the given integral into two parts, then by using formula *i.e.*, $\int \frac{1}{\sqrt{a^2+x^2}} = \log|x+\sqrt{a^2+x^2}| + C$, get the desired result.

Ans. Let

$$I = \int \frac{3x-1}{\sqrt{x^2+9}}dx$$

$$\Rightarrow \quad I = \int \frac{3x}{\sqrt{x^2+9}}dx - \int \frac{1}{\sqrt{x^2+9}}dx$$

Let $\quad I = I_1 - I_2$

Now, $\quad I_1 = \int \frac{3x}{\sqrt{x^2 + 9}} dx$

Put $\quad x^2 + 9 = t^2 \Rightarrow 2x\,dx = 2t\,dt \Rightarrow xdx = tdt$

$\therefore \quad I_1 = 3\int \frac{t}{t} dt$

$= 3\int dt = 3t + C_1 = 3\sqrt{x^2 + 9} + C_1$

and $\quad I_2 = \int \frac{1}{\sqrt{x^2 + 9}} dx = \int \frac{1}{\sqrt{x^2 + (3)^2}} dx$

$= \log|x + \sqrt{x^2 + 9}| + C_2$

$\therefore \quad I = 3\sqrt{x^2 + 9} + C_1 - \log|x + \sqrt{x^2 + 9}| - C_2$

$= 3\sqrt{x^2 + 9} - \log|x + \sqrt{x^2 + 9}| + C \quad [\because C = C_1 - C_2]$

Q.17 $\int \sqrt{5 - 2x + x^2}\,dx$

Thinking Process First of all convert the given expression into $\sqrt{x^2 + a^2}$ form, then use the formula $\int \sqrt{x^2 + a^2}\,dx = \frac{1}{2}x\sqrt{x^2 + a^2} + \frac{a^2}{2}\log|x + \sqrt{x^2 + a^2}| + C$.

Ans. Let $\quad I = \int \sqrt{5 - 2x + x^2}\,dx = \int \sqrt{x^2 - 2x + 1 + 4}\,dx$

$= \int \sqrt{(x - 1)^2 + (2)^2}\,dx$

$= \frac{x - 1}{2}\sqrt{2^2 + (x - 1)^2} + 2\log|x - 1 + \sqrt{2^2 + (x - 1)^2}| + C$

$= \frac{x - 1}{2}\sqrt{5 - 2x + x^2} + 2\log|x - 1 + \sqrt{5 - 2x + x^2}| + C$

Q.18 $\int \frac{x}{x^4 - 1} dx$

Ans. Let $\quad I = \int \frac{x}{x^4 - 1} dx$

Put $\quad x^2 = t \Rightarrow 2xdx = dt \Rightarrow xdx = \frac{1}{2}dt$

$\therefore \quad I = \frac{1}{2}\int \frac{dt}{t^2 - 1} = \frac{1}{2} \cdot \frac{1}{2}\log\left|\frac{t - 1}{t + 1}\right| + C \quad \left[\because \int \frac{dx}{x^2 - a^2} = \frac{1}{2a}\log\left|\frac{x - a}{x + a}\right| + C\right]$

$= \frac{1}{4}[\log|x^2 - 1| - \log|x^2 + 1|] + C \quad \left[\because \log\frac{m}{n} = \log m - \log n\right]$

Q.19 $\int \frac{x^2}{1 - x^4} dx$

Thinking Process Here, use $\int \frac{1}{1 + x^2} dx = \tan^{-1}x + C$ and $\int \frac{1}{a^2 - x^2} dx = \frac{1}{2a}\log\left|\frac{a + x}{a - x}\right| + C$, to solve this problem.

Ans. Let $\quad I = \int \frac{x^2}{1-x^4} dx$

$$= \int \frac{\left(\frac{1}{2} + \frac{x^2}{2} - \frac{1}{2} + \frac{x^2}{2}\right)}{(1-x^2)(1+x^2)} dx \qquad [\because a^2 - b^2 = (a+b)(a-b)]$$

$$= \int \frac{\frac{1}{2}(1+x^2) - \frac{1}{2}(1-x^2)}{(1-x^2)(1+x^2)} dx = \frac{1}{2}\int \frac{(1+x^2)}{(1-x^2)(1+x^2)} dx - \frac{1}{2}\int \frac{(1-x^2)}{(1-x^2)(1+x^2)} dx$$

$$= \frac{1}{2}\int \frac{1}{1-x^2} dx - \frac{1}{2}\int \frac{1}{1+x^2} dx = \frac{1}{2}\cdot\frac{1}{2}\log\left|\frac{1+x}{1-x}\right| + C_1 - \frac{1}{2}\tan^{-1}x + C_2$$

$$= \frac{1}{4}\log\left|\frac{1+x}{1-x}\right| - \frac{1}{2}\tan^{-1}x + C \qquad [\because C = C_1 + C_2]$$

Q.20 $\int \sqrt{2ax - x^2}\, dx$

Ans. Let, $I = \sqrt{2ax - x^2}\ dx$

$$= \int \sqrt{a^2 - a^2 + 2ax - x^2} \cdot dx$$

$$= \int \sqrt{a^2 - (x^2 - 2ax + a^2)} \cdot dx$$

$$= \int \sqrt{a^2 - (x-a)^2} \cdot dx$$

Put $x - a = t \Rightarrow dx = dt$

$$= \int \sqrt{a^2 - t^2} \cdot dt = \frac{t}{2}\sqrt{a^2 - t^2} + \frac{a^2}{2}\sin^{-1}\frac{t}{a} + C$$

$$\left[\because \int \sqrt{a^2 - x^2}\, dx = \frac{x}{2}\sqrt{a^2 - x^2} + \frac{a^2}{2}\sin\frac{x}{a} + C\right]$$

$$= \frac{x-a}{2}\sqrt{a^2 - (x-a)^2} + \frac{a^2}{2}\sin^{-1}\frac{x-a}{a} + C$$

Q.21 $\int \frac{\sin^{-1} x}{(1-x^2)^{3/2}} dx$

Ans. Let $I = \int \frac{\sin^{-1} x}{(1-x^2)^{3/2}} dx = \int \frac{\sin^{-1} x}{(1-x^2)\sqrt{1-x^2}} dx$

Let $\quad \sin^{-1} x = t \Rightarrow \frac{1}{\sqrt{1-x^2}} dx = dt$

Now $\quad I = \int \frac{t}{(1-x^2)} dt$

$\because \quad \sin^{-1} x = t \Rightarrow x = \sin t \Rightarrow 1 - x^2 = 1 - \sin^2 t = \cos^2 t$

$$= \int \frac{t}{\cos^2 t} dt = \int t \sec^2 t\, dt$$

$$= t\int \sec^2 t\, dt - \int\left[\frac{d}{dt} t \int (\sec^2 t\, dt)\right] dt \qquad \text{[Integration by parts]}$$

$$\left[\because \int uv dx = u\int v dx - \int\left(\frac{du}{dx}\int v dx\right) dx\right]$$

$$= t \cdot \tan t - \int 1 \cdot \tan t \, dt$$

$$= t \tan t + \log |\cos t| + C \quad \left[\because \int \tan x \, dx = -\log|\cos x| + C\right]$$

$$= t\left(\frac{\sin t}{\cos t}\right) + \log|\cos t| + C \quad \left[\because \tan t = \frac{\sin t}{\cos t}\right]$$

$$= \sin^{-1} x \cdot \frac{x}{\sqrt{1-x^2}} + \log|\sqrt{1-x^2}| + C$$

Q.22 $\int \frac{(\cos 5x + \cos 4x)}{1 - 2\cos 3x} dx$

Ans. Let $I = \int \frac{\cos 5x + \cos 4x}{1 - 2\cos 3x} dx = \int \frac{2\cos\frac{9x}{2} \cdot \cos\frac{x}{2}}{1 - 2\left(2\cos^2\frac{3x}{2} - 1\right)} dx$

$$\left[\because \cos C + \cos D = 2\cos\frac{C+D}{2} \cdot \cos\frac{C-D}{2} \text{ and } \cos 2x = 2\cos^2 x - 1\right]$$

$$\therefore \quad I = \int \frac{2\cos\frac{9x}{2} \cdot \cos\frac{x}{2}}{3 - 4\cos^2\frac{3x}{2}} dx = -\int \frac{2\cos\frac{9x}{2} \cdot \cos\frac{x}{2}}{4\cos^2\frac{3x}{2} - 3} dx$$

$$= -\int \frac{2\cos\frac{9x}{2} \cdot \cos\frac{x}{2} \cdot \cos\frac{3x}{2}}{4\cos^3\frac{3x}{2} - 3\cos\frac{3x}{2}} dx \quad \left[\text{multiply and divide by } \cos\frac{3x}{2}\right]$$

$$= -\int \frac{2\cos\frac{9x}{2} \cdot \cos\frac{x}{2} \cdot \cos\frac{3x}{2}}{\cos 3 \cdot \frac{3x}{2}} dx = -\int 2\cos\frac{3x}{2} \cdot \cos\frac{x}{2} dx$$

$$[\because \cos 3x = 4\cos^3 x - 3\cos x]$$

$$= -\int \left\{\cos\left(\frac{3x}{2} + \frac{x}{2}\right) + \cos\left(\frac{3x}{2} - \frac{x}{2}\right)\right\} dx$$

$$[\because 2\cos a \cos b = \cos(a+b) + \cos(a-b)]$$

$$= -\int (\cos 2x + \cos x) \, dx$$

$$= -\left[\frac{\sin 2x}{2} + \sin x\right] + C$$

$$= -\frac{1}{2}\sin 2x - \sin x + C$$

Q.23 $\int \frac{\sin^6 x + \cos^6 x}{\sin^2 x \cos^2 x} dx$

Thinking Process First using identity $a^3 + b^3 = (a+b)^3 - 3ab(a+b)$, where $a = \sin^2 x$ and $b = \cos^2 x$

Ans. Let $I = \int \frac{\sin^6 x + \cos^6 x}{\sin^2 x \cos^2 x} dx = \int \frac{(\sin^2 x)^3 + (\cos^2 x)^3}{\sin^2 x \cos^2 x}$

$$= \int \frac{1 - 3\sin^2 x \cos^2 x}{\sin^2 x \cos^2 x} dx \quad [\because \sin^2 x + \cos^2 x = 1]$$

$$= \int \left[\frac{(\sin^2 x + \cos^2 x) - 3\sin^2 x \cos^2 x}{\sin^2 x \cdot \cos^2 x} \right] dx$$

$$= \int \frac{\sin^2 x}{\sin^2 x \cos^2 x} dx + \int \frac{\cos^2 x}{\sin^2 x \cdot \cos^2 x} dx - \int \frac{3\sin^2 x \cos^2 x}{\sin^2 x \cdot \cos^2 x} dx$$

$$= \int \sec^2 x \, dx + \int \operatorname{cosec}^2 x \, dx - \int 3 \cdot dx$$

$$= \int \sec^2 x \, dx + \int \operatorname{cosec}^2 x \, dx - 3\int dx$$

$$= \tan x - \cot x - 3x + C$$

Q.24 $\int \frac{\sqrt{x}}{\sqrt{a^3 - x^3}} dx$

Ans. Let $I = \int \frac{\sqrt{x}}{\sqrt{a^3 - x^3}} dx = \int \frac{\sqrt{x}}{\sqrt{(a^{3/2})^2 - (x^{3/2})^2}} dx$

Put $x^{3/2} = t \Rightarrow \frac{3}{2} x^{1/2} dx = dt$

$\therefore$ $I = \frac{2}{3} \int \frac{dt}{\sqrt{(a^{3/2})^2 - t^2}} = \frac{2}{3} \sin^{-1} \frac{t}{a^{3/2}} + C$

$$\left[\because \int \frac{1}{\sqrt{a^2 - x^2}} dx = \sin^{-1}\left(\frac{x}{a}\right) + C \right]$$

$$= \frac{2}{3} \sin^{-1} \frac{x^{3/2}}{a^{3/2}} + C = \frac{2}{3} \sin^{-1} \sqrt{\frac{x^3}{a^3}} + C \qquad [\because t = x^{3/2}]$$

Q.25 $\int \frac{\cos x - \cos 2x}{1 - \cos x} dx$

Thinking Process Apply the formula, $\cos C - \cos D = 2\sin \frac{C+D}{2} \cdot \sin \frac{D-C}{2}$ and $\cos x = 1 - 2\sin^2 \frac{x}{2}$ to solve it.

Ans. Let $I = \int \frac{\cos x - \cos 2x}{1 - \cos x} dx = \int \frac{2\sin \frac{3x}{2} \cdot \sin \frac{x}{2}}{1 - 1 + 2\sin^2 \frac{x}{2}} dx$

$$= 2\int \frac{\sin \frac{3x}{2} \cdot \sin \frac{x}{2}}{2\sin^2 \frac{x}{2}} dx = \int \frac{\sin \frac{3x}{2}}{\sin \frac{x}{2}} dx$$

$$= \int \frac{3\sin \frac{x}{2} - 4\sin^3 \frac{x}{2}}{\sin \frac{x}{2}} dx \qquad [\because \sin 3x = 3\sin x - 4\sin^3 x]$$

$$= 3\int dx - 4\int \sin^2 \frac{x}{2} dx = 3\int dx - 4\int \frac{1 - \cos x}{2} dx$$

$$= 3\int dx - 2\int dx + 2\int \cos x \, dx$$

$$= \int dx + 2\int \cos x \, dx = x + 2\sin x + C = 2\sin x + x + C$$

Q.26 $\int \frac{dx}{x\sqrt{x^4 - 1}}$

Ans. Let $I = \int \frac{dx}{x\sqrt{x^4 - 1}}$

Put $x^2 = \sec\theta \Rightarrow \theta = \sec^{-1} x^2$

$\Rightarrow$ $2x\,dx = \sec\theta \cdot \tan\theta\, d\theta$

$$I = \int \frac{\sec\theta \tan\theta}{2\sec\theta\sqrt{\sec^2\theta - 1}}\, d\theta$$

$\therefore$ $I = \frac{1}{2}\int \frac{\sec\theta \cdot \tan\theta}{\sec\theta \tan\theta} d\theta = \frac{1}{2}\int d\theta = \frac{1}{2}\theta + C$ $\quad [\because \sec^2\theta - 1 = \tan^2\theta]$

$$= \frac{1}{2}\sec^{-1}(x^2) + C$$

Evaluate the following.

***Q.27** $\int_0^2 (x^2 + 3)dx$

Ans. We know that $\int_a^b f(x)dx = \lim_{n \to \infty} h \sum_{r=0}^{n-1} f(a + rh)$

For $I = \int_0^2 (x^2 + 3)dx$, we have $a = 0$ and $b = 2$

$$I = \int_0^2 (x^2 + 3(dx)$$

Here, $a = 0$, $b = 2$ and $h = \frac{b - a}{n} = \frac{2 - 0}{n} = \frac{2}{n}$

$\Rightarrow$ $nh = 2$

and $f(x) = (x^2 + 3)$

$\therefore$ $I = \int_0^2 (x^2 + 3)dx = \lim_{h \to 0} h \sum_{r=0}^{n-1} f(a + rh)$

$$= \lim_{h \to 0} h \sum_{r=0}^{n-1} f(rh)$$

$$= \lim_{h \to 0} h \sum_{r=0}^{n-1} (3 + r^2h^2)$$

$$= \lim_{h \to 0} h\left[3n + h^2\left(\frac{(n-1)(n-1+1)(2n-2+1)}{6}\right)\right]$$

$$= \lim_{h \to 0} h\left[3n + h^2\left(\frac{(n^2 - n)(2n - 1)}{6}\right)\right]$$

$$= \lim_{h \to 0} h\left[3n + \frac{h^2}{6}(2n^3 - 3n^2 + n)\right]$$

$$= \lim_{h \to 0}\left[3nh + \frac{2n^3h^3 - 3n^2h^2 \cdot h + nh \cdot h^2}{6}\right]$$

$$= \lim_{h \to 0}\left[3.2 + \frac{2 \cdot 2^3 - 3 \cdot 2^2\, h + 2.h^2}{6}\right] = 6 + \frac{16}{6} = \frac{26}{3}$$

Note (✱) Topics related to these question have been removed from latest NCERT.

***Q.28** $\int_0^2 e^x dx$

Ans. We know that $\int_a^b f(x)dx = \lim_{n \to \infty} h \sum_{r=0}^{n-1} f(a + rh)$

For $I = \int_0^2 e^x \, dx$, we have $a = 0$ and $b = 2$

$$\therefore \quad h = \frac{b-a}{n} = \frac{2-0}{n} = \frac{2}{n} \Rightarrow nh = 2$$

$$\therefore \quad I = \int_0^2 e^x \, dx$$

$$= \lim_{h \to 0} h[1 + e^h + e^{2h} + \dots + e^{(n-1)h}]$$

$$= \lim_{h \to 0} h\left[\frac{1 \cdot (e^h)^n - 1}{e^h - 1}\right]$$

$$= \lim_{h \to 0} h\left(\frac{e^{nh} - 1}{e^h - 1}\right) = \lim_{h \to 0} h\left(\frac{e^2 - 1}{e^h - 1}\right)$$

$$= (e^2 - 1) \cdot \lim \frac{h}{e^h - 1} \qquad \left[\because \lim_{h \to 0} \frac{h}{e^h - 1} = 1\right]$$

$$= e^2 - 1$$

Q.29 $\int_0^1 \frac{dx}{e^x + e^{-x}}$

Ans. Let $$I = \int_0^1 \frac{dx}{e^x + e^{-x}} = \int_0^1 \frac{e^x}{1 + e^{2x}} dx$$

[Divide by e^{-x} in both numerator and denominator]

Put $e^x = t$

$\Rightarrow e^x \, dx = dt$

For limits, when $x = 0$, then $e^\circ = t \Rightarrow t = 1$

and when $x = 1$, then $e^1 = t \Rightarrow t = e$

$$\therefore \quad I = \int_1^e \frac{dt}{1+t^2} = [\tan^{-1} t]_1^e \qquad \left[\because \int \frac{1}{1+t^2} dt = \tan^{-t} t\right]$$

$$= \tan^{-1} e - \tan^{-1} 1$$

$$= \tan^{-1} e - \frac{\pi}{4} \qquad \left[\because \tan \frac{\pi}{4} = 1\right]$$

Q.30 $\int_0^{\pi/2} \frac{\tan x}{1 + m^2 \tan^2 x} dx$

Ans. Let $$I = \int_0^{\pi/2} \frac{\tan x}{1 + m^2 \tan^2 x} dx$$

$$= \int_0^{\pi/2} \frac{\frac{\sin x}{\cos x}}{1 + m^2 \cdot \frac{\sin^2 x}{\cos^2 x}} dx$$

$$= \int_0^{\pi/2} \frac{\frac{\sin x}{\cos x}}{\frac{\cos^2 x + m^2 \sin^2 x}{\cos^2 x}} dx$$

Note (✱) Topics related to these question have been removed from latest NCERT.

$$= \int_0^{\pi/2} \frac{\sin x \cos x}{1 - \sin^2 x + m^2 \sin^2 x}\, dx \qquad [\because \cos^2 x = 1 - \sin^2 x]$$

$$= \int_0^{\pi/2} \frac{\sin x \cos x}{1 - \sin^2 x(1 - m^2)}\, dx$$

Put $\sin^2 x = t$

$\Rightarrow$ $2 \sin x \cos x \, dx = dt$

For limit, when $x = 0$, then $\sin^2 0 = t \Rightarrow t = 0$

when $x = \frac{\pi}{2}$, then $\sin^2 \frac{\pi}{2} = t \Rightarrow t = 1$

$$\therefore \quad I = \frac{1}{2} \int_0^1 \frac{dt}{1 - t(1 - m^2)}$$

$$= \frac{1}{2}\left[-\log |1 - t(1 - m^2)| \cdot \frac{1}{1 - m^2} \right]_0^1$$

$$= \frac{1}{2}\left[-\log |1 - 1 + m^2| \cdot \frac{1}{1 - m^2} + \log |1| \cdot \frac{1}{1 - m^2} \right]$$

$$= \frac{1}{2}\left[-\log |m^2| \cdot \frac{1}{1 - m^2} \right] = \frac{2}{2} \cdot \frac{\log m}{(m^2 - 1)} \qquad [\because \log [1] = 0]$$

$$[\because \log m^n = n \log m]$$

$$= \frac{1}{m^2 - 1} \log m$$

Q.31 $\int_1^2 \frac{dx}{\sqrt{(x-1)(2-x)}}$

Thinking Process First of all convert the given function into $\frac{1}{\sqrt{a^2 - x^2}}$ form, then apply the formula *i.e.*, $\int \frac{1}{\sqrt{a^2 - x^2}} dx = \sin^{-1} \frac{x}{a} + C.$

Ans. Let

$$I = \int_1^2 \frac{dx}{\sqrt{(x-1)(2-x)}} = \int_1^2 \frac{dx}{\sqrt{2x - x^2 - 2 + x}}$$

$$= \int_1^2 \frac{dx}{\sqrt{-(x^2 - 3x + 2)}}$$

$$= \int_1^2 \frac{dx}{\sqrt{-\left[x^2 - 2 \cdot \frac{3}{2} x + \left(\frac{3}{2}\right)^2 + 2 - \left(\frac{3}{2}\right)^2 \right]}}$$

$$= \int_1^2 \frac{dx}{\sqrt{-\left\{ \left(x - \frac{3}{2}\right)^2 - \left(\frac{1}{2}\right)^2 \right\}}}$$

$$= \int_1^2 \frac{dx}{\sqrt{\left(\frac{1}{2}\right)^2 - \left(x - \frac{3}{2}\right)^2}} = \left[\sin^{-1}\left(\frac{x - \frac{3}{2}}{\frac{1}{2}}\right)\right]_1^2$$

$$= [\sin^{-1}(2x - 3)]_1^2 = \sin^{-1} 1 - \sin^{-1}(-1)$$

$$\left[\because \sin^{-1}(1) = \frac{\pi}{2} \text{and } \sin^{-1}(-1) = \sin^{-1}\sin\left(-\frac{\pi}{2}\right)\right]$$

$$= \frac{\pi}{2} + \frac{\pi}{2} = \pi$$

Q.32 $\int_0^1 \frac{x}{\sqrt{1+x^2}} dx$

Ans. Let $I = \int_0^1 \frac{x}{\sqrt{1+x^2}} dx$

Put $1 + x^2 = t^2$

$\Rightarrow$ $2x\,dx = 2t dt$

$\Rightarrow$ $x\,dx = tdt$

$\therefore$ $I = \int_1^{\sqrt{2}} \frac{tdt}{t}$ [when $x = 0$, then $1 + (0)^2 = t^2, t = 1$

when $x = 1$, then $1 + 1^2 = t^2, t = \sqrt{2}$]

$$= [t]_1^{\sqrt{2}} = \sqrt{2} - 1$$

Q.33 $\int_0^{\pi} x \sin x \cos^2 x\, dx$

Thinking Process Here, use the property *i.e.,* $\int_0^a f(x)dx = \int_0^a (a - x)dx$ and $\sin(\pi - x) = \sin x, \cos(\pi - x) = -\cos x$.

Ans. Let $I = \int_0^{\pi} x \sin x \cos^2 x\, dx$ ...(i)

and $I = \int_0^{\pi} (\pi - x)\sin(\pi - x)\cos^2(\pi - x)dx$

$\Rightarrow$ $I = \int_0^{\pi} (\pi - x)\sin x \cos^2 x\, dx$ ...(ii)

On adding Eqs. (i) and (ii), we get

$$2I = \int_0^{\pi} \pi \sin x \cos^2 x\, dx$$

Put $\cos x = t$

$\Rightarrow$ $-\sin x\, dx = dt$

As $x \to 0$, then $t \to 1$

and $x \to \pi$, then $t \to -1$

$\therefore$ $2I = -\pi \int_1^{-1} t^2\, dt \Rightarrow 2I = -\pi\left[\frac{t^3}{3}\right]_1^{-1}$

$\Rightarrow$ $2I = -\frac{\pi}{3}[-1 - 1] \Rightarrow 2I = \frac{2\pi}{3}$

$\therefore$ $I = \frac{\pi}{3}$

Q.34 $\int_0^{1/2} \frac{dx}{(1+x^2)\sqrt{1-x^2}}$

Ans. Let $$I = \int_0^{1/2} \frac{dx}{(1+x^2)\sqrt{1-x^2}}$$

Put $x = \sin\theta \Rightarrow dx = \cos\theta\, d\theta$

As $x \to 0$, then $\theta \to 0$ and $x \to \frac{1}{2}$, then $\theta \to \frac{\pi}{6}$

$$\therefore \quad I = \int_0^{\pi/6} \frac{\cos\theta}{(1+\sin^2\theta)\cos\theta} d\theta = \int_0^{\pi/6} \frac{1}{1+\sin^2\theta} d\theta \quad [\because 1-\sin^2\theta = \cos^2\theta]$$

$$= \int_0^{\pi/6} \frac{1}{\cos^2\theta(\sec^2\theta + \tan^2\theta)} d\theta$$

$$= \int_0^{\pi/6} \frac{\sec^2\theta}{\sec^2\theta + \tan^2\theta} d\theta$$

$$= \int_0^{\pi/6} \frac{\sec^2\theta}{1+\tan^2\theta+\tan^2\theta} d\theta \quad [\because \sec^2\theta = 1+\tan^2\theta]$$

$$= \int_0^{\pi/6} \frac{\sec^2\theta}{1+2\tan^2\theta} d\theta$$

Again, put $\tan\theta = t \Rightarrow \sec^2\theta\, d\theta = dt$

As $\theta \to 0$, then $t \to 0$ and $\theta \to \frac{\pi}{6}$, then $t \to \frac{1}{\sqrt{3}}$

$$\therefore \quad I = \int_0^{1/\sqrt{3}} \frac{dt}{1+2t^2} = \frac{1}{2}\int_0^{1/\sqrt{3}} \frac{dt}{\left(\frac{1}{\sqrt{2}}\right)^2 + t^2}$$

$$= \frac{1}{2}\cdot\frac{1}{1/\sqrt{2}}\left[\tan^{-1}\frac{t}{\frac{1}{\sqrt{2}}}\right]_0^{1/\sqrt{3}} = \frac{1}{\sqrt{2}}[\tan^{-1}(\sqrt{2}t)]_0^{1/\sqrt{3}}$$

$$= \frac{1}{\sqrt{2}}\left[\tan^{-1}\sqrt{\frac{2}{3}} - 0\right] = \frac{1}{\sqrt{2}}\tan^{-1}\left(\sqrt{\frac{2}{3}}\right)$$

Long Answer Type Questions

Q.35 $\int \frac{x^2}{x^4 - x^2 - 12} dx$

Thinking Process Use $\frac{px+q}{(x-a)(x-b)} = \frac{A}{(x-a)} + \frac{B}{x-b}$, where $a \neq b$, then compare the coefficient of x to get the value of A and B.

Ans. Let $$I = \int \frac{x^2}{x^4 - x^2 - 12} dx$$

$$= \int \frac{x^2}{x^4 - 4x^2 + 3x^2 - 12} dx$$

$$= \int \frac{x^2 dx}{x^2(x^2 - 4) + 3(x^2 - 4)}$$

$$= \int \frac{x^2 dx}{(x^2 - 4)(x^2 + 3)}$$

Now, $I = \dfrac{x^2}{(x^2 - 4)(x^2 + 3)}$ [let $x^2 = t$]

$\Rightarrow \dfrac{t}{(t - 4)(t + 3)} = \dfrac{A}{t - 4} + \dfrac{B}{t + 3}$ [using partial fractions]

$\Rightarrow t = A(t + 3) + B(t - 4)$

On comparing the coefficient of t and constant term on both sides, we get

$A + B = 1$ ...(i)

and $3A - 4B = 0$ ...(ii)

Substitute the value of $A = 1 - B$ from Eq. (i) in Eq. (ii)

$\Rightarrow 3(1 - B) - 4B = 0$

$\Rightarrow 3 - 3B - 4B = 0$

$\Rightarrow 7B = 3$

$\Rightarrow B = \dfrac{3}{7}$

If $B = \dfrac{3}{7}$, then $A + \dfrac{3}{7} = 1$

$\Rightarrow A = 1 - \dfrac{3}{7} = \dfrac{4}{7}$

$$\frac{x^2}{(x^2 - 4)(x^2 + 3)} = \frac{4}{7(x^2 - 4)} + \frac{3}{7(x^2 + 3)}$$

$\therefore I = \dfrac{4}{7}\displaystyle\int \frac{1}{x^2 - (2)^2} dx + \frac{3}{7}\int \frac{1}{x^2 + (\sqrt{3})^2} dx$

$$= \frac{4}{7} \cdot \frac{1}{2 \cdot 2} \log\left|\frac{x - 2}{x + 2}\right| + \frac{3}{7} \cdot \frac{1}{\sqrt{3}} \tan^{-1} \frac{x}{\sqrt{3}} + C$$

$$= \frac{1}{7} \log\left|\frac{x - 2}{x + 2}\right| + \frac{\sqrt{3}}{7} \tan^{-1} \frac{x}{\sqrt{3}} + C$$

Q.36 $\displaystyle\int \frac{x^2}{(x^2 + a^2)(x^2 + b^2)} dx$

Ans. Let $I = \displaystyle\int \frac{x^2}{(x^2 + a^2)(x^2 + b^2)} dx$

Now, $\dfrac{x^2}{(x^2 + a^2)(x^2 + b^2)} = \dfrac{t}{(t + a^2)(t + b^2)} = \dfrac{A}{(t + a^2)} + \dfrac{B}{(t + b^2)}$ [let $x^2 = t$]

$t = A(t + b^2) + B(t + a^2)$ [using partial fraction]

On comparing the coefficient of t, and constant term on both side, we get

$A + B = 1$ ...(i)

and $b^2 A + a^2 B = 0$ ...(ii)

Substitute the value of A that is $A = 1 - B$ from Eq. (i) in Eq. (ii)

$\Rightarrow b^2(1 - B) + a^2 B = 0$

$\Rightarrow b^2 - b^2 B + a^2 B = 0$

$\Rightarrow b^2 + (a^2 - b^2) B = 0$

$$\Rightarrow \qquad B = \frac{-b^2}{a^2 - b^2} = \frac{b^2}{b^2 - a^2}$$

From Eq. (i), $A + \dfrac{b^2}{b^2 - a^2} = 1$

$$\Rightarrow \qquad A = \frac{b^2 - a^2 - b^2}{b^2 - a^2} = \frac{-a^2}{b^2 - a^2}$$

$$\therefore \qquad I = \int \frac{-a^2}{(b^2 - a^2)(x^2 + a^2)} dx + \int \frac{b^2}{b^2 - a^2} \cdot \frac{1}{x^2 + b^2} dx$$

$$= \frac{-a^2}{(b^2 - a^2)} \int \frac{1}{x^2 + a^2} dx + \frac{b^2}{b^2 - a^2} \int \frac{1}{x^2 + b^2} dx$$

$$= \frac{-a^2}{b^2 - a^2} \cdot \frac{1}{a} \tan^{-1} \frac{x}{a} + \frac{b^2}{b^2 - a^2} \cdot \frac{1}{b} \tan^{-1} \frac{x}{b}$$

$$= \frac{1}{b^2 - a^2} \left[-a \tan^{-1} \frac{x}{a} + b \tan^{-1} \frac{x}{b} \right]$$

$$= \frac{1}{a^2 - b^2} \left[a \tan^{-1} \frac{x}{a} - b \tan^{-1} \frac{x}{b} \right]$$

Q.37 $\displaystyle\int_0^{\pi} \frac{x}{1 + \sin x}$

Ans. Let $\qquad I = \displaystyle\int_0^{\pi} \frac{x}{1 + \sin x} dx \qquad$ ...(i)

and $\qquad I = \displaystyle\int_0^{\pi} \frac{\pi - x}{1 + \sin(\pi - x)} dx = \int_0^{\pi} \frac{\pi - x}{1 + \sin x} dx \qquad$ ...(ii)

[Applying property $\int_0^a f(x)dx = \int_0^a f(a - x)\, dx$]

On adding Eqs. (i) and (ii), we get

$$2I = \pi \int_0^{\pi} \frac{1}{1 + \sin x} dx$$

$$= \pi \int_0^{\pi} \frac{(1 - \sin x)\, dx}{(1 + \sin x)(1 - \sin x)}$$

$$= \pi \int_0^{\pi} \frac{(1 - \sin x)\, dx}{\cos^2 x}$$

$$= \pi \int_0^{\pi} (\sec^2 x - \tan x \cdot \sec x)\, dx$$

$$= \pi \int_0^{\pi} \sec^2 x\, dx - \pi \int_0^{\pi} \sec x \cdot \tan x\, dx$$

$$= \pi[\tan x]_0^{\pi} - \pi[\sec x]_0^{\pi}$$

$$= \pi[\tan x - \sec x]_0^{\pi}$$

$$= \pi[\tan \pi - \sec \pi - \tan 0 + \sec 0]$$

$$\Rightarrow \qquad 2I = \pi[0 + 1 - 0 + 1]$$

$$2I = 2\pi$$

$$\therefore \qquad I = \pi$$

Q.38 $\int \frac{2x-1}{(x-1)(x+2)(x-3)} dx$

Thinking Process Apply $\frac{px+q}{(x-a)(x-b)(x-c)} = \frac{A}{(x-a)} + \frac{B}{(x-b)} + \frac{C}{(x-c)}$, then get the values of A, B and C and use $\int \frac{1}{x} dx = \log|x| + C$.

Ans. Let $$I = \int \frac{(2x-1)}{(x-1)(x+2)(x-3)} dx$$

Now, $$\frac{2x-1}{(x-1)(x+2)(x-3)} = \frac{A}{(x-1)} + \frac{B}{(x+2)} + \frac{C}{(x-3)}$$

$$\Rightarrow \quad 2x-1 = A(x+2)(x-3) + B(x-1)(x-3) + C(x-1)(x+2)$$

Put $x = 3$, then

$$6-1 = C(3-1)(3+2)$$

$$\Rightarrow \quad 5 = 10C \Rightarrow C = \frac{1}{2}$$

Again, put $x = 1$, then

$$2-1 = A(1+2)(1-3)$$

$$\Rightarrow \quad 1 = -6A \Rightarrow A = -\frac{1}{6}$$

Now, put $x = -2$, then

$$-4-1 = B(-2-1)(-2-3)$$

$$\Rightarrow \quad -5 = 15B \Rightarrow B = -\frac{1}{3}$$

$$\therefore \quad I = -\frac{1}{6}\int \frac{1}{x-1} dx - \frac{1}{3}\int \frac{1}{x+2} dx + \frac{1}{2}\int \frac{1}{x-3} dx$$

$$= -\frac{1}{6}\log|(x-1)| - \frac{1}{3}\log|(x+2)| + \frac{1}{2}\log|(x-3)| + C$$

$$= -\log|(x-1)|^{1/6} - \log|(x+2)|^{1/3} + \log|(x-3)|^{1/2} + C$$

[$\because \log m^n = n\log m$]

$$= \log\left|\frac{\sqrt{x-3}}{(x-1)^{1/6}(x+2)^{1/3}}\right| + C$$

$\left[\because \log mn = \log m + \log n, \quad \log\frac{m}{n} = \log m - \log n\right]$

Q.39 $\int e^{\tan^{-1}x}\left(\frac{1+x+x^2}{1+x^2}\right)dx$

Ans. Let $$I = \int e^{\tan^{-1}x}\left(\frac{1+x+x^2}{1+x^2}\right) dx$$

$$= \int e^{\tan^{-1}x}\left(\frac{1+x^2}{1+x^2} + \frac{x}{1+x^2}\right) dx$$

$$= \int e^{\tan^{-1}x} dx + \int \frac{x\,e^{\tan^{-1}x}}{1+x^2} dx$$

Again let $I = I_1 + I_2$ ...(i)

Now, $I_2 = \int \frac{x\, e^{\tan^{-1} x}}{1+x^2} dx$

Put $\tan^{-1} x = t \Rightarrow x = \tan t$

$\Rightarrow \quad \frac{1}{1+x^2} dx = dt$

$\therefore \quad I_2 = \int \underset{\text{I}}{\tan t} \cdot \underset{\text{II}}{e^t}\, dt$

$= \tan t \cdot e^t - \int \sec^2 t \cdot e^t\, dt + C$ [integration by parts]

$= \tan t \cdot e^t - \int (1+\tan^2 t)\, e^t dt + C$ $[\because \sec^2\theta = 1+\tan^2\theta]$

$\Rightarrow \quad I_2 = \tan t \cdot e^t - \int (1+x^2) \frac{e^{\tan^{-1} x}}{1+x^2} dx + C$

$\Rightarrow \quad I_2 = \tan t \cdot e^t - \int e^{\tan^{-1} x} dx + C$

$\therefore \quad I = \int e^{\tan^{-1} x} dx + \tan t \cdot e^t - \int e^{\tan^{-1} x} dx + C$ [from Eq. (i)]

$= \tan t \cdot e^t + C$

$= x\, e^{\tan^{-1} x} + C$

Q.40 $\int \sin^{-1} \sqrt{\frac{x}{a+x}}\, dx$

Thinking Process First of all put $x = a\tan^2\theta$ and convert the given expression into two parts, then use the formulae for integration by parts i.e.,

$\int \text{I} \cdot \text{II} dx = \text{I} \int \text{II} dx - \int \left(\frac{d}{dx} \text{I} \int \text{II} dx \right) dx.$

Ans. Let $I = \int \sin^{-1} \sqrt{\frac{x}{a+x}}\, dx$

Put $x = a\tan^2\theta$

$\Rightarrow \quad dx = 2a \tan\theta \cdot \sec^2\theta\, d\theta$

$\therefore \quad I = \int \sin^{-1} \sqrt{\frac{a\tan^2\theta}{a + a\tan^2\theta}} (2a\tan\theta \cdot \sec^2\theta)\, d\theta$

$= 2a \int \sin^{-1} \left(\frac{\tan\theta}{\sec\theta} \right) \tan\theta \cdot \sec^2\theta\, d\theta$ $[\because 1+\tan^2\theta = \sec^2\theta]$

$= 2a \int \sin^{-1}(\sin\theta) \tan\theta \cdot \sec^2\theta\, d\theta$

$= 2a \int \underset{\text{I}}{\theta} \cdot \underset{\text{II}}{\tan\theta \sec^2\theta}\, d\theta$

$= 2a \left[\theta \cdot \int \tan\theta \cdot \sec^2\theta\, d\theta - \int \left(\frac{d}{d\theta}\theta \cdot \int \tan\theta \cdot \sec^2\theta\, d\theta \right) d\theta \right]$

$\left[\begin{array}{l} \text{Put } \tan\theta = t \Rightarrow \sec^2\theta\, d\theta = dt \\ \Rightarrow \int \tan\theta \sec^2\theta\, d\theta = \int t\, dt \end{array} \right]$

$= 2a \left[\theta \cdot \frac{\tan^2\theta}{2} - \int \frac{\tan^2\theta}{2} d\theta \right]$ $\left[\because \int (f(x))^n f'(x) dx = \frac{(f(x))^{n+1}}{n+1} + C \right]$

$= a\theta \tan^2\theta - a \int (\sec^2\theta - 1)\, d\theta$

$= a\theta \cdot \tan^2\theta - a\tan\theta + a\theta + C$

$= a \left[\frac{x}{a} \tan^{-1} \sqrt{\frac{x}{a}} - \sqrt{\frac{x}{a}} + \tan^{-1} \sqrt{\frac{x}{a}} \right] + C$

Q.41 $\int_{\pi/3}^{\pi/2} \frac{\sqrt{1+\cos x}}{(1-\cos x)^{5/2}} dx$

Ans. Let $I = \int_{\pi/3}^{\pi/2} \frac{\sqrt{1+\cos x}}{(1-\cos x)^{5/2}} dx = \int_{\pi/3}^{\pi/2} \frac{\left(2\cos^2 \frac{x}{2}\right)^{1/2}}{\left(2\sin^2 \frac{x}{2}\right)^{5/2}} dx$

$\left[\because \cos x = 1 - 2\sin^2 \frac{x}{2} \quad \cos x = 2\cos^2 \frac{x}{2} - 1\right]$

$$= \frac{\sqrt{2}}{4\sqrt{2}} \int_{\pi/3}^{\pi/2} \frac{\cos\left(\frac{x}{2}\right)}{\sin^5\left(\frac{x}{2}\right)} dx \quad = \frac{1}{4}\int_{\pi/3}^{\pi/2} \frac{\cos\left(\frac{x}{2}\right)}{\sin^5\left(\frac{x}{2}\right)} dx$$

Put $\sin\frac{x}{2} = t$

$\Rightarrow \cos\frac{x}{2} \cdot \frac{1}{2} dx = dt$

$\Rightarrow \cos\frac{x}{2} dx = 2dt$

As $x \to \frac{\pi}{3}$, then $t \to \frac{1}{2}$

and $x \to \frac{\pi}{2}$, then $t \to \frac{1}{\sqrt{2}}$

$\therefore$ $$I = \frac{2}{4}\int_{1/2}^{1/\sqrt{2}} \frac{dt}{t^5} = \frac{1}{2}\left[\frac{t^{-5+1}}{-5+1}\right]_{1/2}^{1/\sqrt{2}}$$

$$= \frac{1}{2}\left[\frac{t^{-4}}{-4}\right]_{1/2}^{1/\sqrt{2}} \Rightarrow \frac{-1}{8}\left[\frac{1}{t^4}\right]_{1/2}^{1/\sqrt{2}}$$

$$= -\frac{1}{8}\left[\frac{1}{\left(\frac{1}{\sqrt{2}}\right)^4} - \frac{1}{\left(\frac{1}{2}\right)^4}\right]$$

$$= -\frac{1}{8}(4-16) = \frac{12}{8} = \frac{3}{2}$$

Q.42 $\int e^{-3x} \cos^3 x \, dx$

Ans. Let $I = \int e^{-3x} \cos^3 x \, dx$

$$\overset{\text{II}}{} \quad \overset{\text{I}}{}$$

$$= \cos^3 x \int e^{-3x} dx - \int\left(\frac{d}{dx}\cos^3 x \int e^{-3x} dx\right) dx \qquad \text{[integration by parts]}$$

$$= \cos^3 x \cdot \frac{e^{-3x}}{-3} - \int(-3\cos^2 x \sin x)\frac{e^{-3x}}{-3} dx$$

$$= -\frac{1}{3}\cos^3 x \, e^{-3x} - \int \cos^2 x \sin x \, e^{-3x} dx$$

$$= -\frac{1}{3}\cos^3 x\, e^{-3x} - \int (1-\sin^2 x)\sin x\, e^{-3x} dx$$

$$= -\frac{1}{3}\cos^3 x e^{-3x} - \int \sin x\, e^{-3x} dx + \int \underset{\text{I}}{\sin^3 x} \cdot \underset{\text{II}}{e^{-3x}} dx$$

$$= -\frac{1}{3}\cos^3 x\, e^{-3x} - \int \sin x\, e^{-3x} dx + \sin^3 x \cdot \frac{e^{-3x}}{-3} - \int 3\sin^2 x \cos x \cdot \frac{e^{-3x}}{-3} dx$$

$$= -\frac{1}{3}\cos^3 x\, e^{-3x} - \int \sin x\, e^{-3x} dx - \frac{1}{3}\sin^3 x\, e^{-3x} + \int (1-\cos^2 x)\cos x\, e^{-3x} dx$$

$$\Rightarrow \quad I = -\frac{1}{3}\cos^3 x\, e^{-3x} - \int \underset{\text{I}}{\sin x}\, \underset{\text{II}}{e^{-3x}} dx - \frac{1}{3}\sin^3 x\, e^{-3x} + \int \cos x\, e^{-3x} dx$$

$$\Rightarrow \quad 2I = \frac{e^{-3x}}{13}[\cos^3 x + \sin^3 x] - \left[\sin x \cdot \frac{e^{-3x}}{-3} - \int \cos x \cdot \frac{e^{-3x}}{-3} dx\right] + \int \cos x\, e^{-3x} dx$$

$$- \int \cos^3 x\, e^{-3x} dx \; [\because \cos^3 x\, e^{-3x} dx = I]$$

$$\Rightarrow \quad 2I = \frac{e^{-3x}}{-3}[\cos^3 x + \sin^3 x] + \frac{1}{3}\sin x \cdot e^{-3x} - \frac{1}{3}\int \cos x \cdot e^{-3x} dx + \int \cos x\, e^{-3x} dx$$

$$2I = \frac{e^{-3x}}{-3}[\cos^3 x + \sin^3 x] + \frac{1}{3}\sin x\, e^{-3x} + \frac{2}{3}\int \cos x\, e^{-3x} dx$$

Now, let $\quad I_1 = \int \underset{\text{I}}{\cos x}\, \underset{\text{II}}{e^{-3x}} dx$

$$\Rightarrow \quad I_1 = \cos x \cdot \frac{e^{-3x}}{-3} - \int (-\sin x) \cdot \frac{e^{-3x}}{-3} dx$$

$$\Rightarrow \quad I_1 = \frac{-1}{3}\cos x \cdot e^{-3x} - \frac{1}{3} \int \underset{\text{I}}{\sin x} \cdot \underset{\text{II}}{e^{-3x}} dx$$

$$= -\frac{1}{3}\cos x \cdot e^{-3x} - \frac{1}{3}\left[\sin x \cdot \frac{e^{-3x}}{-3} - \int \cos x \cdot \frac{e^{-3x}}{-3} dx\right]$$

$$= -\frac{1}{3}\cos x \cdot e^{-3x} + \frac{1}{9}\sin x \cdot e^{-3x} - \frac{1}{9}\int \cos x \cdot e^{-3x} dx$$

$$\Rightarrow \quad I_1 + \frac{1}{9}I_1 = -\frac{1}{3}e^{-3x} \cdot \cos x + \frac{1}{9}\sin x \cdot e^{-3x}$$

$$\Rightarrow \quad \left(\frac{10}{9}\right)I_1 = -\frac{1}{3}e^{-3x} \cdot \cos x + \frac{1}{9}\sin x \cdot e^{-3x}$$

$$\Rightarrow \quad I_1 = \frac{-3}{10}e^{-3x} \cdot \cos x + \frac{1}{10}e^{-3x}\sin x$$

Now, $\quad 2I = -\frac{1}{3}e^{-3x}[\sin^3 x + \cos^3 x] + \frac{1}{3}\sin x \cdot e^{-3x} - \frac{1}{5}e^{-3x} \cdot \cos x$

$$+ \frac{1}{15}e^{-3x} \cdot \sin x + C$$

$$\therefore \quad I = -\frac{1}{6}e^{-3x}[\sin^3 x + \cos^3 x] + \frac{1}{5}e^{-3x} \cdot \sin x - \frac{1}{10}e^{-3x} \cdot \cos x + C$$

$$= \frac{e^{-3x}}{24}[\sin 3x - \cos 3x] + \frac{3e^{-3x}}{40}[\sin x - 3\cos x] + C$$

$$\left[\begin{array}{l} \because \quad \sin 3x = 3\sin x - 4\sin^3 x \\ \text{and } \cos 3x = 4\cos^3 x - 3\cos x \end{array}\right]$$

Q.43 $\int \sqrt{\tan x}\, dx$

Ans. Let $I = \int \sqrt{\tan x}\, dx$

Put $\tan x = t^2 \Rightarrow \sec^2 x\, dx = 2t\, dt$

$\therefore$ $I = \int t \cdot \frac{2t}{\sec^2 x} dt = 2\int \frac{t^2}{1+t^4} dt$ $\quad [\because \sec^2 x = 1 + \tan^2 x \Rightarrow 1 + (t^2)^2 \Rightarrow 1 + t^4]$

$$= \int \frac{(t^2+1)+(t^2-1)}{(1+t^4)} dt$$

$$= \int \frac{t^2+1}{1+t^4} dt + \int \frac{t^2-1}{1+t^4} dt$$

$$= \int \frac{1+\frac{1}{t^2}}{t^2+\frac{1}{t^2}} dt + \int \frac{1-\frac{1}{t^2}}{t^2+\frac{1}{t^2}} dt$$

$$= \int \frac{1-\left(-\frac{1}{t^2}\right) dt}{\left(t-\frac{1}{t}\right)^2+2} + \int \frac{1+\left(-\frac{1}{t^2}\right)}{\left(t+\frac{1}{t}\right)^2-2} dt$$

Put $u = t - \frac{1}{t} \Rightarrow du = \left(1+\frac{1}{t^2}\right) dt$

and $v = t + \frac{1}{t} \Rightarrow dv = \left(1-\frac{1}{t^2}\right) dt$

$\therefore$ $I = \int \frac{du}{u^2+(\sqrt{2})^2} + \int \frac{dv}{v^2-(\sqrt{2})^2}$

$$= \frac{1}{\sqrt{2}} \tan^{-1} \frac{u}{\sqrt{2}} + \frac{1}{2\sqrt{2}} \log \left| \frac{v-\sqrt{2}}{v+\sqrt{2}} \right| + C$$

$$= \frac{1}{\sqrt{2}} \tan^{-1}\left(\frac{\tan x - 1}{\sqrt{2\tan x}}\right) + \frac{1}{2\sqrt{2}} \log \left| \frac{\tan x - \sqrt{2\tan x} + 1}{\tan x + \sqrt{2\tan x} + 1} \right| + C$$

$$\left[\because u = t - \frac{1}{t}, v = t + \frac{1}{t} \text{ and } t = \sqrt{\tan x}\right]$$

Q.44 $\int_0^{\pi/2} \frac{dx}{(a^2 \cos^2 x + b^2 \sin^2 x)^2}$

Ans. Let $I = \int_0^{\pi/2} \frac{dx}{(a^2 \cos^2 x + b^2 \sin^2 x)^2}$

Divide numerator and denominator by $\cos^4 x$, we get

$$I = \int_0^{\pi/2} \frac{\sec^4 x\, dx}{(a^2 + b^2 \tan^2 x)^2}$$

$$= \int_0^{\pi/2} \frac{(1+\tan^2 x)\sec^2 x\, dx}{(a^2+b^2\tan^2 x)^2} \quad [\because \sec^2 x = 1 + \tan^2 x]$$

Put $\tan x = t$

$\Rightarrow$ $\sec^2 x\, dx = dt$

As $x \to 0$, then $t \to 0$

and $x \to \frac{\pi}{2}$, then $t \to \infty$ $\quad I = \int_0^\infty \frac{(1+t^2)}{(a^2+b^2t^2)^2} dt$

Now, $\frac{1+t^2}{(a^2+b^2t^2)^2}$ [let $t^2 = u$]

$$\frac{1+u}{(a^2+b^2u)^2} = \frac{A}{(a^2+b^2u)} + \frac{B}{(a^2+b^2u)^2} \quad \text{[using partial fractions]}$$

$\Rightarrow$ $1 + u = A(a^2 + b^2u) + B$

On comparing the coefficient of u and constant term on both sides, we get

$$a^2A + B = 1 \quad \text{...(i)}$$

and $b^2A = 1$ ...(ii)

$\therefore$ $A = \frac{1}{b^2}$

On substituting $A = \frac{1}{b^2}$ in Eq. (i), we get

Now, $\frac{a^2}{b^2} + B = 1$

$\Rightarrow$ $B = 1 - \frac{a^2}{b^2} = \frac{b^2-a^2}{b^2}$

$\therefore$

$$I = \int_0^\infty \frac{(1+t^2)}{(a^2+b^2t^2)^2} dt$$

$$= \frac{1}{b^2}\int_0^\infty \frac{dt}{a^2+b^2t^2} + \frac{b^2-a^2}{b^2}\int_0^\infty \frac{dt}{(a^2+b^2t^2)^2}$$

$$= \frac{1}{b^2}\int_0^\infty \frac{dt}{b^2\left(\frac{a^2}{b^2}+t^2\right)} + \frac{b^2-a^2}{b^2}\int_0^\infty \frac{dt}{(a^2+b^2t^2)^2}$$

$$= \frac{1}{ab^3}\left[\tan^{-1}\left(\frac{tb}{a}\right)\right]_0^\infty + \frac{b^2-a^2}{b^2}\left(\frac{\pi}{4}\cdot\frac{1}{a^3b}\right)$$

$$= \frac{1}{ab^3}[\tan^{-1}\infty - \tan^{-1}0] + \frac{\pi}{4}\cdot\frac{b^2-a^2}{(a^3b^3)}$$

$$= \frac{\pi}{2ab^3} + \frac{\pi}{4}\cdot\frac{b^2-a^2}{(a^3b^3)} = \pi\left(\frac{2a^2+b^2-a^2}{4a^3b^3}\right) = \frac{\pi}{4}\left(\frac{a^2+b^2}{a^3b^3}\right)$$

Q.45 $\int_0^1 x\log(1+2x)\,dx$

Thinking Process Use formula for integration by part *i.e.*, $\int \text{I}\cdot\text{II}\,dx = \text{I}\int\text{II}dx - \int\left(\frac{d}{dx}\text{I}\int\text{II}\,dx\right)dx$ and also use $\int\frac{1}{x} = \log|x| + C$.

Ans. Let $I = \int_0^1 x\log(1+2x)dx$

$$= \left[\log(1+2x)\frac{x^2}{2}\right]_0^1 - \int_0^1 \frac{1}{1+2x}\cdot 2\cdot\frac{x^2}{2}dx \quad \text{[Integration by parts]}$$

$$= \frac{1}{2}[x^2\log(1+2x)]_0^1 - \int_0^1 \frac{x^2}{1+2x}dx$$

$$= \frac{1}{2}[1 \log 3 - 0] - \left[\int_0^1 \left(\frac{x}{2} - \frac{\frac{x}{2}}{1+2x} \right) dx \right]$$

$$= \frac{1}{2}\log 3 - \frac{1}{2}\int_0^1 x\,dx + \frac{1}{2}\int_0^1 \frac{x}{1+2x}\,dx$$

$$= \frac{1}{2}\log 3 - \frac{1}{2}\left[\frac{x^2}{2}\right]_0^1 + \frac{1}{2}\int_0^1 \frac{\frac{1}{2}(2x+1-1)}{(2x+1)}\,dx$$

$$= \frac{1}{2}\log 3 - \frac{1}{2}\left[\frac{1}{2} - 0\right] + \frac{1}{4}\int_0^1 dx - \frac{1}{4}\int_0^1 \frac{1}{1+2x}\,dx$$

$$= \frac{1}{2}\log 3 - \frac{1}{4} + \frac{1}{4}[x]_0^1 - \frac{1}{8}[\log|(1+2x)|]_0^1$$

$$= \frac{1}{2}\log 3 - \frac{1}{4} + \frac{1}{4} - \frac{1}{8}[\log 3 - \log 1]$$

$$= \frac{1}{2}\log 3 - \frac{1}{8}\log 3 = \frac{3}{8}\log 3$$

Q.46 $\int_0^{\pi} x \log \sin x\,dx$

Thinking Process First of all use property of definite integral *i.e.*, $\int_0^a f(x)\,dx = \int_0^a f(a-x)\,dx$, then use $\int_0^{2a} f(x)\,dx = 2\int_0^a f(x)\,dx$..

Ans. Let $I = \int_0^{\pi} x \log \sin x\,dx$ ...(i)

$\Rightarrow$ $I = \int_0^{\pi} (\pi - x) \log \sin(\pi - x)\,dx$

$= \int_0^{\pi} (\pi - x) \log \sin x\,dx$ ...(ii)

On adding Eqs. (i) and (ii), we get

$2I = \pi \int_0^{\pi} \log \sin x\,dx$ ...(iii)

$2I = 2\pi \int_0^{\pi/2} \log \sin x\,dx$ $\left[\because \int_0^{2a} f(x)\,dx = 2\int_0^a f(x)\,dx\right.$

If $f(2a - x) = f(x)$

$\Rightarrow$ $I = \pi \int_0^{\pi/2} \log \sin x\,dx$ ...(iv)

Now, $I = \pi \int_0^{\pi/2} \log \sin(\pi/2 - x)\,dx$ ...(v)

On adding Eqs. (iv) and (v), we get

$2I = \pi \int_0^{\pi/2} (\log \sin x + \log \cos x)\,dx$ $[\because \sin(90 - x) = \cos x]$

$\Rightarrow$ $2I = \pi \int_0^{\pi/2} \log \sin x \cos x\,dx$

$\Rightarrow$ $2I = \pi \int_0^{\pi/2} \log \frac{2 \sin x \cos x}{2}\,dx$

$\Rightarrow$ $2I = \pi \int_0^{\pi/2} (\log \sin 2x - \log 2)\,dx$ $[\because \sin 2x = 2 \sin x \cdot \cos x]$

$\Rightarrow$ $2I = \pi \int_0^{\pi/2} \log \sin 2x\,dx - \pi \int_0^{\pi/2} \log 2\,dx$

Put $2x = t \Rightarrow dx = \frac{1}{2}\,dt$

As $x \to 0$, then $t \to 0$

and $x \to \frac{\pi}{2}$, then $t \to \pi$

$\therefore \quad 2I = \frac{\pi}{2}\int_0^{\pi} \log \sin t\, dt - \frac{\pi^2}{2}\log 2$

$\Rightarrow \quad 2I = \frac{\pi}{2}\int_0^{\pi} \log \sin x\, dx - \frac{\pi^2}{2}\log 2$

$\Rightarrow \quad 2I = I - \frac{\pi^2}{2}\log 2$ [from Eq. (iii)]

$\therefore \quad I = -\frac{\pi^2}{2}\log 2 = \frac{\pi^2}{2}\log\left(\frac{1}{2}\right)$

Q.47 $\int_{-\pi/4}^{\pi/4} \log(\sin x + \cos x)\cdot dx$

Ans. Let $\quad I = \int_{-\pi/4}^{\pi/4} \log(\sin x + \cos x)\, dx$ ...(i)

$\Rightarrow \quad I = \int_{-\pi/4}^{\pi/4} \log\left\{\sin\left(\frac{\pi}{4} - \frac{\pi}{4} - x\right) + \cos\left(\frac{\pi}{4} - \frac{\pi}{4} - x\right)\right\} dx$

$= \int_{-\pi/4}^{\pi/4} \log\{\sin(-x) + \cos(-x)\}\, dx$

$\Rightarrow \quad I = \int_{-\pi/4}^{\pi/4} \log(\cos x - \sin x)\, dx$ ...(ii)

From Eqs. (i) and (ii),

$2I = 2\int_{-\pi/4}^{\pi/4} \log\,(\cos + \sin\, x) + \log\,(\cos\, x - \sin\, x)dx$

$= \int_{-\pi/4}^{\pi/4} \log\,[(\cos\, x + \sin\, x)\,(\cos\, x - \sin\, x)]dx$

[$\because \log mn = \log m + \log n$]

$= \int_{-\pi/4}^{\pi/4} \log \cos^2 x - \sin^2 x)dx$

$\Rightarrow \quad 2I = \int_{-\pi/4}^{\pi/4} \log \cos 2x\, dx$

$\Rightarrow \quad 2I = 2\int_0^{\pi/4} \log \cos 2x\, dx$ ...(iii)

$\left[\because \int_{-a}^{a} f(x)\, dx = 2\int_0^a f(x), \text{ if } f(-x) = f(x)\right]$

Put $\quad 2x = t \Rightarrow dx = \frac{dt}{2}$

As $x \to 0$, then $t \to 0$

and $x \to \frac{\pi}{4}$, then $t \to \frac{\pi}{2}$

$2I = \int_0^{\pi/2} \log \cos t\, dt$ ...(iv)

$\Rightarrow \quad 2I = \int_0^{\pi/2} \log \cos\left(\frac{\pi}{2} - t\right) dt \quad \left[\because \int_0^a f(x)\, dx = \int_0^a f(a-x)\, dx\right]$

$\Rightarrow \quad 2I = \int_0^{\pi/2} \log \sin t\, dt$ ...(v)

On adding Eqs. (iv) and (v), we get

$$4I = \int_0^{\pi/2} \log \sin t \cos t \, dt$$

$$\Rightarrow \quad 4I = \int_0^{\pi/2} \log \frac{\sin 2t}{2} \, dt$$

$$\Rightarrow \quad 4I = \int_0^{\pi/2} \log \sin 2x \, dx - \int_0^{\pi/2} \log 2 \, dx$$

$$\Rightarrow \quad 4I = \int_0^{\pi/2} \log \sin 2x \, dx - \frac{\pi}{2} \log 2$$

$$\Rightarrow \quad 4I = 2\int_0^{\pi/4} \log \sin 2x \, dx - \frac{\pi}{2} \log 2$$

$$\left[\because \int_0^{2a} f(x)\, dx = 2\int_0^{a} f(x)\, dx \text{ if } f(2a-x) = f(x)\right]$$

$$\Rightarrow \quad 4I = 2\int_0^{\pi/4} \log \sin 2\left(\frac{\pi}{4} - x\right) dx - \frac{\pi}{2} \log 2$$

$$= 2\int_0^{\pi/4} \log \sin \left(\frac{\pi}{2} - 2x\right) dx - \frac{\pi}{2} \log 2$$

$$= 2\int_0^{\pi/4} \log \cos 2x \, dx - \frac{\pi}{2} \log 2$$

$$\Rightarrow \quad 4I = 2I - \frac{\pi}{2} \log 2 \qquad \text{[using Eq. (iii)]}$$

$$\Rightarrow \quad 2I = -\frac{\pi}{2} \log 2$$

$$\Rightarrow \quad I = -\frac{\pi}{4} \log 2$$

$$\Rightarrow \quad I = \frac{\pi}{4} \log \left(\frac{1}{2}\right)$$

Objective Type Questions

Choose the correct answer out of the given four options of the exercises from 48 to 60.

Q.48 $\int \frac{\cos 2x - \cos 2\theta}{\cos x - \cos \theta} dx$ is equal to

(a) $2(\sin x + x\cos \theta) + C$ (b) $2(\sin x - x\cos \theta) + C$

(c) $2(\sin x + 2x\cos \theta) + C$ (d) $2(\sin x - 2x\cos \theta) + C$

Thinking Process Use formula $\cos 2\theta = 2\cos^2 \theta - 1$ to get simplest form, then apply $\int \cos x \, dx = \sin x + C$.

Ans. (a) Let $I = \int \frac{\cos 2x - \cos 2\theta}{\cos x - \cos \theta} dx$

$$= \int \frac{(2\cos^2 x - 1) - (2\cos^2 \theta - 1)}{\cos x - \cos \theta} dx = \int \frac{(2\cos^2 x - 1 - 2\cos^2 \theta + 1)}{\cos x - \cos \theta} dx$$

$$= 2\int \frac{\cos^2 x - \cos^2 \theta}{\cos x - \cos \theta} dx$$

$$= 2\int \frac{(\cos x + \cos \theta)(\cos x - \cos \theta)}{(\cos x - \cos \theta)} dx$$

$$= 2\int (\cos x + \cos \theta)\, dx = 2(\sin x + x\cos \theta) + C$$

Q.49 $\int \frac{dx}{\sin(x-a)\sin(x-b)}$ is equal to

(a) $\sin(b-a)\log\left|\frac{\sin(x-b)}{\sin(x-a)}\right| + C$

(b) $\text{cosec}(b-a)\log\left|\frac{\sin(x-a)}{\sin(x-b)}\right| + C$

(c) $\text{cosec}\,(b-a)\log\left|\frac{\sin(x-b)}{\sin(x-a)}\right| + C$

(d) $\sin(b-a)\log\left|\frac{\sin(x-a)}{\sin(x-b)}\right| + C$

Ans. ***(c)*** Let $I = \int \frac{dx}{\sin(x-a)\sin(x-b)}$

Multiplying and dividing by $\sin(b-a)$

$$= \frac{1}{\sin(b-a)}\int \frac{\sin(b-a)}{\sin(x-a)\sin(x-b)}dx$$

$$= \frac{1}{\sin(b-a)}\int \frac{\sin(x-a-x+b)}{\sin(x-a)\sin(x-b)}dx$$

$$= \frac{1}{\sin(b-a)}\int \frac{\sin\{(x-a)-(x-b)\}}{\sin(x-a)\sin(x-b)}dx$$

$$= \frac{1}{\sin(b-a)}\int \frac{\sin(x-a)\cos(x-b)-\cos(x-a)\sin(x-b)}{\sin(x-a)\sin(x-b)}dx$$

$[\because \sin(A-B) = \sin A\cos B - \cos A\sin B]$

$$= \frac{1}{\sin(b-a)}\int [\cot(x-b)-\cot(x-a)]\,dx$$

$$= \frac{1}{\sin(b-a)}[\log|\sin(x-b)| - \log|\sin(x-a)|] + C$$

$$= \frac{1}{\sin(b-a)}\log\left|\frac{\sin(x-b)}{\sin(x-a)}\right| + C \qquad \left[\because \log\frac{m}{n} = \log m - \log n\right]$$

$$= \text{cosec}\,(b-a)\log\left|\frac{\sin(x-b)}{\sin(x-a)}\right| + C$$

Q.50 $\int \tan^{-1}\sqrt{x}\,dx$ is equal to

(a) $(x+1)\tan^{-1}\sqrt{x} - \sqrt{x} + C$

(b) $x\tan^{-1}\sqrt{x} - \sqrt{x} + C$

(c) $\sqrt{x} - x\tan^{-1}\sqrt{x} + C$

(d) $\sqrt{x} - (x+1)\tan^{-1}\sqrt{x} + C$

Thinking Process Use formula for integration by part *i.e.*, $\int \text{I}\cdot\text{II}\,dx = \text{I}\int\text{II}\,dx - \int\left(\frac{d}{dx}\text{I}\int\text{II}\,dx\right)dx.$

Ans. ***(a)*** Let $I = \int 1\cdot\tan^{-1}\sqrt{x}\,dx$

$$= \tan^{-1}\sqrt{x}\cdot x - \frac{1}{2}\int\frac{1}{(1+x)}\cdot\frac{x}{\sqrt{x}}dx \qquad \text{[integration by parts]}$$

$$= x\tan^{-1}\sqrt{x} - \frac{1}{2}\int\frac{\sqrt{x}}{(1+x)}dx$$

Put $x = t^2 \Rightarrow dx = 2t\,dt$

$\therefore$ $I = x\tan^{-1}\sqrt{x} - \frac{1}{2}\int\frac{t\times 2t}{(1+t^2)}dt$

$$= x\tan^{-1}\sqrt{x} - \int \frac{t^2}{1+t^2}\,dt$$

$$= x\tan^{-1}\sqrt{x} - \int \left(1 - \frac{1}{1+t^2}\right) dt$$

$$= x\tan^{-1}\sqrt{x} - \sqrt{x} + \tan^{-1}t + C \qquad [\because x = t^2 \Rightarrow t = \sqrt{x}]$$

$$= x\tan^{-1}\sqrt{x} - \sqrt{x} + \tan^{-1}\sqrt{x} + C$$

$$= (x+1)\tan^{-1}\sqrt{x} - \sqrt{x} + C$$

Q.51 $\int e^x \left(\frac{1-x}{1+x^2}\right)^2 dx$ is equal to

(a) $\frac{e^x}{1+x^2} + C$ (b) $\frac{-e^x}{1+x^2} + C$ (c) $\frac{-e^x}{(1+e^2)^2} + C$ (d) $\frac{-e^x}{(1+x^2)^2} + C$

Ans. **(a)** $\int e^x \left(\frac{1-x}{1+x^2}\right)^2 dx$

$$= \int e^x \frac{1+x^2-2x}{(1+x^2)^2}\,dx \qquad [\because (a-b)^2 = a^2 + b^2 - 2ab]$$

$$= \int e^x \left[\frac{1}{(1+x^2)} - \frac{2x}{(1+x^2)^2}\right] dx$$

$$= \int e^x [f(x) + f'(x)]\,dx$$

$$= e^x f(x) + C$$

$$= \frac{e^x}{1+x^2} + C$$

Q.52 $\int \frac{x^9}{(4x^2+1)^6}\,dx$ is equal to

(a) $\frac{1}{5x}\left(4 + \frac{1}{x^2}\right)^{-5} + C$ (b) $\frac{1}{5}\left(4 + \frac{1}{x^2}\right)^{-5} + C$

(c) $\frac{1}{10x}(1+4)^{-5} + C$ (d) $\frac{1}{10}\left(\frac{1}{x^2} + 4\right)^{-5} + C$

Ans. **(d)** Let $I = \int \frac{x^9}{(4x^2+1)^6}\,dx = \int \frac{x^9}{x^{12}\left(4 + \frac{1}{x^2}\right)^6}\,dx$

$$= \int \frac{dx}{x^3\left(4 + \frac{1}{x^2}\right)^6}$$

Put $4 + \frac{1}{x^2} = t \Rightarrow \frac{-2}{x^3}dx = dt \Rightarrow \frac{1}{x^3}dx = -\frac{1}{2}dt$

$\therefore$ $I = -\frac{1}{2}\int \frac{dt}{t^6} = -\frac{1}{2}\left[\frac{t^{-6+1}}{-6+1}\right] + C$

$$= \frac{1}{10}\left[\frac{1}{t^5}\right] + C = \frac{1}{10}\left(4 + \frac{1}{x^2}\right)^{-5} + C$$

Q.53 If $\int \frac{dx}{(x+2)(x^2+1)} = a \log|1+x^2| + b \tan^{-1} x + \frac{1}{5}\log|x+2| + C$, then

(a) $a = \frac{-1}{10}, b = \frac{-2}{5}$ (b) $a = \frac{1}{10}, b = -\frac{2}{5}$

(c) $a = \frac{-1}{10}, b = \frac{2}{5}$ (d) $a = \frac{1}{10}, b = \frac{2}{5}$

Thinking Process Use method of partial fraction *i.e.*,

$\frac{1}{(x-a)(x^2+bx+c)} = \frac{A}{(x-a)} + \frac{Bx+C}{(x^2+bx+c)}$ to solve the above problem.

Ans. (c) Given that, $\int \frac{dx}{(x+2)(x^2+1)} = a \log|1+x^2| + b \tan^{-1} x + \frac{1}{5}\log|x+2| + C$

Now, $I = \int \frac{dx}{(x+2)(x^2+1)}$

$$\frac{1}{(x+2)(x^2+1)} = \frac{A}{x+2} + \frac{Bx+C}{x^2+1}$$

$\Rightarrow$ $1 = A(x^2+1) + (Bx+C)(x+2)$

$\Rightarrow$ $1 = Ax^2 + A + Bx^2 + 2Bx + Cx + 2C$

$\Rightarrow$ $1 = (A+B)x^2 + (2B+C)x + A + 2C$

$\Rightarrow$ $A + B = 0, A + 2C = 1, 2B + C = 0$

We have, $A = \frac{1}{5}, B = -\frac{1}{5}$ and $C = \frac{2}{5}$

$\therefore$ $\int \frac{dx}{(x+2)(x^2+1)} = \frac{1}{5}\int \frac{1}{x+2}dx + \int \frac{-\frac{1}{5}x + \frac{2}{5}}{x^2+1}dx$

$$= \frac{1}{5}\int \frac{1}{x+2}dx - \frac{1}{5}\int \frac{x}{1+x^2}dx + \frac{1}{5}\int \frac{2}{1+x^2}dx$$

$$= \frac{1}{5}\log|x+2| - \frac{1}{10}\log|1+x^2| + \frac{2}{5}\tan^{-1} x + C$$

$\therefore$ $b = \frac{2}{5}$ and $a = \frac{-1}{10}$ [by comparing]

Q.54 $\int \frac{x^3}{x+1} dx$ is equal to

(a) $x + \frac{x^2}{2} + \frac{x^3}{3} - \log|1-x| + C$ (b) $x + \frac{x^2}{2} - \frac{x^3}{3} - \log|1-x| + C$

(c) $x - \frac{x^2}{2} - \frac{x^3}{3} - \log|1+x| + C$ (d) $x - \frac{x^2}{2} + \frac{x^3}{3} - \log|1+x| + C$

Ans. (d) Let $I = \int \frac{x^3}{x+1} dx$

Now, $\frac{x^3}{x+1} = \frac{x^3+1}{x+1} - \frac{1}{x+1}$

$$= \frac{(x+1)(x^2-x+1)}{(x+1)} - \frac{1}{x+1}$$

$[\because x^3 + y^3 = (x+y)(x^2 - xy + y^2)]$

$$\therefore \quad I = \int \left((x^2 - x + 1) - \frac{1}{(x+1)} \right) dx$$

$$= \frac{x^3}{3} - \frac{x^2}{2} + x - \log|x+1| + C$$

Q.55 $\int \frac{x + \sin x}{1 + \cos x} dx$ is equal to

(a) $\log|1 + \cos x| + C$
(b) $\log|x + \sin x| + C$
(c) $x - \tan\frac{x}{2} + C$
(d) $x \cdot \tan\frac{x}{2} + C$

Ans. ***(d)*** Let $I = \int \frac{x + \sin x}{1 + \cos x} dx$

$$= \int \frac{x}{1+\cos x} dx + \int \frac{\sin x}{1 + \cos x} dx$$

$$= \int \frac{x}{2\cos^2(x/2)} dx + \int \frac{2\sin(x/2)\cos(x/2)}{2\cos^2(x/2)} dx$$

$\left[\because \sin x = 2\sin\frac{x}{2} \cdot \cos\frac{x}{2} \text{ and } \cos x = 2\cos^2\frac{x}{2} - 1\right]$

$$= \frac{1}{2}\int x \sec^2(x/2)\, dx + \int \tan(x/2)\, dx$$

$$= \frac{1}{2}\left[x \cdot \tan(x/2) \cdot 2 - \int \tan\frac{x}{2} \cdot 2\, dx\right] + \int \tan\frac{x}{2} dx$$

$$= x \cdot \tan\frac{x}{2} + C$$

Q.56 If $\frac{x^3 dx}{\sqrt{1+x^2}} = a(1+x^2)^{3/2} + b\sqrt{1+x^2} + C$, then

(a) $a = \frac{1}{3}, b = 1$ (b) $a = \frac{-1}{3}, b = 1$ (c) $a = \frac{-1}{3}, b = -1$ (d) $a = \frac{1}{3}, b = -1$

Ans. ***(d)*** Let $I = \int \frac{x^3}{\sqrt{1+x^2}} dx = a(1+x^2)^{3/2} + b\sqrt{1+x^2} + C$

$$\because \quad I = \int \frac{x^3}{\sqrt{1+x^2}} dx = \int \frac{x^2 \cdot x}{\sqrt{1+x^2}} dx$$

Put $1 + x^2 = t^2 \Rightarrow 2x\, dx = 2t\, dt$

$$\therefore \quad I = \int \frac{t(t^2-1)}{t} dt = \frac{t^3}{3} - t + C$$

$$= \frac{1}{3}(1+x^2)^{3/2} - \sqrt{1+x^2} + C$$

$\therefore \quad a = \frac{1}{3}$ and $b = -1$ [by comparing]

Q.57 $\int_{-\pi/4}^{\pi/4} \frac{dx}{1+\cos 2x}$ is equal to

(a) 1 (b) 2 (c) 3 (d) 4

Ans. *(a)* Let $I = \int_{-\pi/4}^{\pi/4} \frac{dx}{1+\cos 2x}$ z= $\int_{-\pi/4}^{\pi/4} \frac{dx}{2\cos^2 x}$ $[\because \cos 2x = 2\cos^2 x - 1]$

$$= \frac{1}{2}\int_{-\pi/4}^{\pi/4} \sec^2 x \, dx = \int_0^{\pi/4} \sec^2 x \, dx = [\tan x]_0^{\pi/4} = \left[\tan\frac{\pi}{4} - \tan 0\right] = 1$$

$$\left[\int_{-a}^{a} f(x)dx = 2\int_0^a f(x) \text{ if } f(-x) = f(x)\right]$$

Q.58 $\int_0^{\pi/2} \sqrt{1-\sin 2x}\, dx$ is equal to

(a) $2\sqrt{2}$ (b) $2(\sqrt{2}+1)$ (c) 2 (d) $2(\sqrt{2}-1)$

Ans. *(d)* Let $I = \int_0^{\pi/2} \sqrt{1-\sin 2x}\, dx$

$$= \int_0^{\pi/2} \sqrt{\sin^2 x + \cos^2 x - 2\sin x \cos x}\, dx$$

$[\because \sin^2 x + \cos^2 x = 1, \sin 2x = 2\sin x \cdot \cos x]$

$$= \int_0^{\pi/4} \sqrt{(\cos x - \sin x)^2}\, dx + \int_{\pi/4}^{\pi/2} \sqrt{(\sin x - \cos x)^2}\, dx$$

$\left[\because \text{when } 0 < x < \frac{\pi}{4}, \cos x > \sin x \text{ and when } \frac{\pi}{4} < x < \frac{\pi}{2}, \sin x > \cos x\right]$

$$= [\sin x + \cos x]_0^{\pi/4} + [-\cos x - \sin x]_{\pi/4}^{\pi/2}$$

$$= \frac{1}{\sqrt{2}} + \frac{1}{\sqrt{2}} - 0 - 1 + \left(-0 - 1 + \frac{1}{\sqrt{2}} + \frac{1}{\sqrt{2}}\right)$$

$$= 2\sqrt{2} - 2 = 2(\sqrt{2}-1)$$

Fill in the blanks in each of the exercises 59 to 63.

Q.59 $\int_0^{\pi/2} \cos x\, e^{\sin x} dx =$ ....

Ans. Let $I = \int_0^{\pi/2} \cos x\, e^{\sin x} dx$

Put $\sin x = t \Rightarrow \cos x\, dx = dt$

As $x \to 0$, then $t \to 0$

and $x \to \pi/2$, then $t \to 1$

$\therefore$ $I = \int_0^1 e^t dt = [e^t]_0^1$

$= e^1 - e^0 = e - 1$

Q.60 $\int \frac{x+3}{(x+4)^2} e^x dx =$ ....

Ans. Let $I = \int \frac{x+3}{(x+4)^2} e^x\, dx = \int \frac{e^x}{(x+4)} dx - \int \frac{e^x}{(x+4)^2} dx$

$$= \int e^x \left(\frac{1}{(x+4)} - \frac{1}{(x+4)^2}\right) dx$$

$$= e^x \left(\frac{1}{x+4}\right) + C \qquad \left[\because \int e^x \{f(x) + f'(x)\}\, dx = e^x f(x) + C\right]$$

Q.61 If $\int_0^a \frac{1}{1+4x^2}\,dx = \frac{\pi}{8}$, then $a = \ldots\ldots\ldots$.

Ans. Let $I = \int_0^a \frac{1}{1+4x^2}\,dx = \frac{\pi}{8}$

Now, $\int_0^a \frac{1}{4\left(\frac{1}{4}+x^2\right)}\,dx = \frac{1}{4}\int_0^a \frac{1}{\left(x^2+\left(\frac{1}{2}\right)^2\right)}\,dx = \frac{2}{4}[\tan^{-1}2x]_0^a$

$$= \frac{1}{2}\tan^{-1}2a - 0 = \frac{\pi}{8}$$

$$\frac{1}{2}\tan^{-1}2a = \frac{\pi}{8} \Rightarrow \tan^{-1}2a = \frac{\pi}{4}$$

$\Rightarrow$ $2a = \tan\frac{\pi}{4}$ $\left[\because \tan\frac{\pi}{4} = 1\right]$

$\Rightarrow$ $2a = 1 \therefore a = \frac{1}{2}$

Q.62 $\int \frac{\sin x}{3+4\cos^2 x}\,dx = \ldots\ldots\ldots$.

Ans. Let $I = \int \frac{\sin x}{3+4\cos^2 x}\,dx$

Put $\cos x = t \Rightarrow -\sin x\,dx = dt$

$\therefore$ $I = -\int \frac{dt}{3+4t^2} = -\frac{1}{4}\int \frac{dt}{\left(\frac{\sqrt{3}}{2}\right)^2 + t^2}$

$= -\frac{1}{4}\cdot\frac{2}{\sqrt{3}}\tan^{-1}\frac{2t}{\sqrt{3}} + C$ $\left[\because \int \frac{dx}{x^2+a^2} = \frac{1}{a}\tan^{-1}\frac{x}{a} + C\right]$

$$= -\frac{1}{2\sqrt{3}}\tan^{-1}\left(\frac{2\cos x}{\sqrt{3}}\right) + C$$

Q.63 The value of $\int_{-\pi}^{\pi} \sin^3 x \cos^2 x\,dx$ is $\ldots\ldots\ldots$.

Ans. We have, $f(x) = \int_{-\pi}^{\pi} \sin^3 x \cos^2 x\,dx$

Now $f(-x) = \int_{-\pi}^{\pi} \sin^3(-x)\cdot\cos^2(-x)\,dx$

$$= -f(x)$$

Since, $f(x)$ is an odd function.

$\therefore$ $\int_{-\pi}^{\pi} \sin^3 x \cos^2 x\,dx = 0$

Chapter

08 Application of Integrals

Short Answer Type Questions

***Q.1** Find the area of the region bounded by the curves $y^2 = 9x$ and $y = 3x$.

Thinking Process On solving both the equation of curves, get the values of x and then at those values, find the area of the shaded region.

Ans. We have, $y^2 = 9x$ and $y = 3x$

On solving both the equations, we get

$\Rightarrow \quad (3x)^2 = 9x$

$\Rightarrow \quad 9x^2 - 9x = 0$

$\Rightarrow \quad 9x(x - 1) = 0$

$\Rightarrow \quad x = 1, 0$

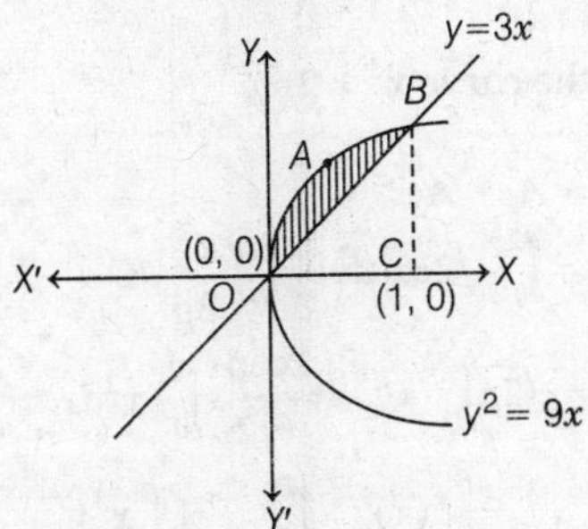

Let A_1 = Area under the curve $y^2 = 9x$

and A_2 = Area under the line $y = 3x$

i.e., A_1 = Area of $OABCO$

and A_2 = Area of $OBCO$

$\therefore$ Required area $= A_1 - A_2$

$$= \int_0^1 \sqrt{9x}\,dx - \int_0^1 3x\,dx$$

$$= 3\left[\frac{x^{3/2}}{3/2}\right]_0^1 - 3\left[\frac{x^2}{2}\right]_0^1 \qquad \left[\because \int x^n\,dx = \frac{x^{n+1}}{n+1}\right]$$

$$= 3\left(\frac{2}{3} - 0\right) - 3\left(\frac{1}{2} - 0\right)$$

$$= 2 - \frac{3}{2} = \frac{1}{2} \text{ sq. unit}$$

Note (*) *Topics related to these question have been removed from latest NCERT.*

***Q.2** Find the area of the region bounded by the parabola $y^2 = 2px, x^2 = 2py$.

Ans. We have, $y^2 = 2px$ ...(i)

and $x^2 = 2py$ ...(ii)

On solving Eqs. (i) and (ii), we get

$\Rightarrow \quad x^2 = 2p\cdot\sqrt{2px}$

On solving both sides, we get

$$x^4 = 4p^2\cdot(2px)$$

$\Rightarrow \quad x^4 = 8p^3x$

$\Rightarrow \quad x^4 - 8p^3x = 0$

$\Rightarrow \quad x(x^3 - 8p^3) = 0$

$\Rightarrow \quad x = 0, 2p$

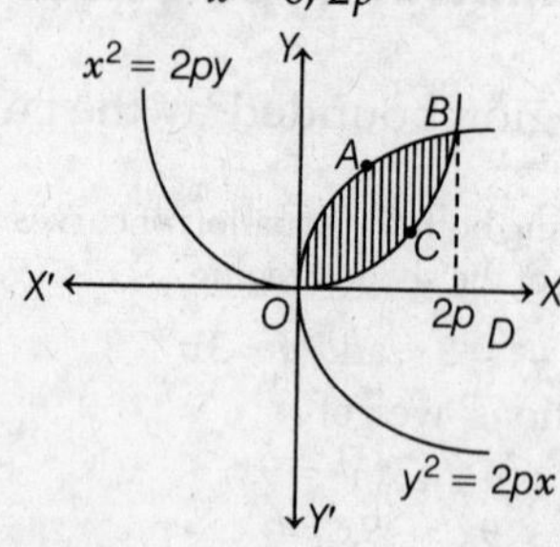

Let A_1 = Area under the curve $y^2 = 2px$
= Area of $OABDO$

A_2 = Area of under the curve $x^2 = 2py$
= Area $OCBDO$

$\therefore$ Required area $= A_1 - A_2$

$$= \int_0^{2p} \sqrt{2px}\, dx - \int_0^{2p} \frac{x^2}{2p}\, dx$$

$$= \sqrt{2p}\int_0^{2p} x^{1/2} dx - \frac{1}{2p}\int_0^{2p} x^2 dx$$

$$= \sqrt{2p}\left[\frac{2(x)^{3/2}}{3}\right]_0^{2p} - \frac{1}{2p}\left[\frac{x^3}{3}\right]_0^{2p} \quad \left[\because \int x^n\, dx = \frac{x^{n+1}}{n+1}\right]$$

$$= \sqrt{2p}\left[\frac{2}{3}\cdot(2p)^{3/2} - 0\right] - \frac{1}{2p}\left[\frac{1}{3}(2p)^3 - 0\right]$$

$$= \sqrt{2p}\left(\frac{2}{3}\cdot 2\sqrt{2}p^{3/2}\right) - \frac{1}{2p}\left(\frac{1}{3}8p^3\right)$$

$$= \sqrt{2p}\left(\frac{4\sqrt{2}}{3}p^{3/2}\right) - \frac{1}{2p}\left(\frac{8}{3}p^3\right)$$

$$= \frac{4\sqrt{2}}{3}\cdot\sqrt{2}p^2 - \frac{8}{6}p^2$$

$$= \frac{(16-8)p^2}{6} = \frac{8p^2}{6}$$

$$= \frac{4p^2}{3} \text{ sq. units}$$

Note (*) *Topics related to these question have been removed from latest NCERT.*

***Q.3** Find the area of the region bounded by the curve $y = x^3$, $y = x + 6$ and $x = 0$.

Ans. We have,
$$y = x^3 \quad \text{...(i)}$$
$$y = x + 6 \quad \text{...(ii)}$$
and
$$x = 0 \quad \text{...(iii)}$$

On solving Eqs. (i) and (ii), we get,
$$x^3 = x + 6$$
$$\Rightarrow \quad x^3 - x - 6 = 0$$
$$\Rightarrow \quad x^2(x - 2) + 2x(x - 2) + 3(x - 2) = 0$$
$$\Rightarrow \quad (x - 2)(x^2 + 2x + 3) = 0$$
$\Rightarrow$ $x = 2$, and two imaginary points

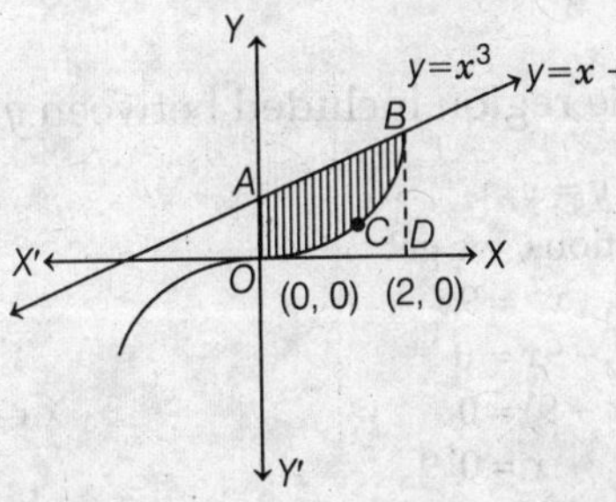

Let A_1 = Area under the line, $y = x + 6$ = Area of $OABDO$
and A_2 = Area under the curve, $y = x^3$ = Area of $OCBDO$

$\therefore$ Area of shaded region $= A_1 - A_2$
$$= \int_0^2 (x + 6)\, dx - \int_0^2 x^3 dx$$
$$= \int_0^2 (x + 6 - x^3)\, dx = \left[\frac{x^2}{2} + 6x - \frac{x^4}{4}\right]_0^2 = \left[\frac{4}{2} + 12 - \frac{16}{4} - 0\right]$$
$$= [2 + 12 - 4] = 10 \text{ sq. units}$$

***Q.4** Find the area of the region bounded by the curve $y^2 = 4x$ and $x^2 = 4y$.

Thinking Process First, by using both the equation get the values of x and then find the area of shaded region at these value of x.

Ans. Given equation of curves are
$$y^2 = 4x \quad \text{...(i)}$$
and
$$x^2 = 4y \quad \text{...(ii)}$$

On solving Eqs. (i) and (ii), we get
$$\left(\frac{x^2}{4}\right)^2 = 4x$$
$$\Rightarrow \quad \frac{x^4}{4 \cdot 4} = 4x$$
$$\Rightarrow \quad x^4 = 64x$$
$$\Rightarrow \quad x^4 - 64x = 0$$
$$\Rightarrow \quad x(x^3 - 4^3) = 0$$
$$\Rightarrow \quad x = 4, 0$$

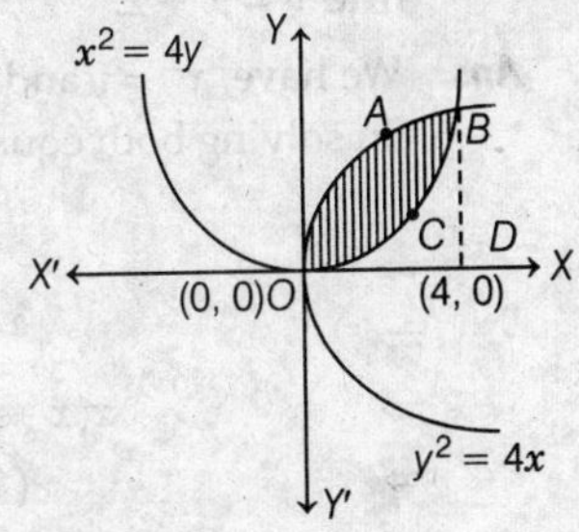

Note (✱) *Topics related to these question have been removed from latest NCERT.*

Let A_1 = Area under the curve $y^2 = 4x$

Area of $OABDO = \int_0^4 \sqrt{4x}\,.\,dx$

and A_2 = Area under the curve $x^2 = 4y$

= Area of $OCBDO = \int_0^4 \frac{x^2}{4} \cdot dx$

$\therefore$ Required area of shaded region = $A_1 - A_2$

$$= \int_0^4 \left(\sqrt{4x} - \frac{x^2}{4}\right) dx = \int_0^4 \left(2\sqrt{x} - \frac{x^2}{4}\right) dx = \left[\frac{2x^{3/2} \cdot 2}{3} - \frac{1}{4} \cdot \frac{x^3}{3}\right]_0^4$$

$$= \frac{2 \cdot 2}{3} \cdot 8 - \frac{1}{4} \cdot \frac{64}{3} - 0$$

$$= \frac{32}{3} - \frac{16}{3} = \frac{16}{3} \text{ sq. units}$$

***Q.5** Find the area of the region included between $y^2 = 9x$ and $y = x$.

Ans. We have, $y^2 = 9x$ and $y = x$

On solving both equations, we get

$\Rightarrow \quad x^2 = 9x$

$\Rightarrow \quad x^2 - 9x = 0$

$\Rightarrow \quad x(x - 9) = 0$

$\Rightarrow \quad x = 0, 9$

$\therefore$ Area of shaded region

= Area under the curve, $(y^2 = 9x)$

− area under the line, $(y = x)$

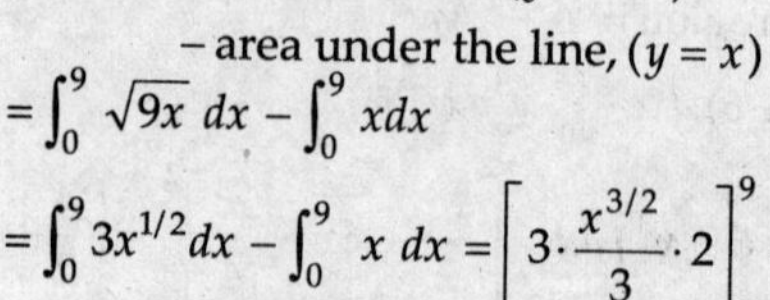

$$= \int_0^9 \sqrt{9x}\, dx - \int_0^9 x dx$$

$$= \int_0^9 3x^{1/2} dx - \int_0^9 x\, dx = \left[3 \cdot \frac{x^{3/2}}{3} \cdot 2\right]_0^9 - \left[\frac{x^2}{2}\right]_0^9$$

$$= \left[\frac{3 \cdot 9^{\frac{3}{2}}}{3} \cdot 2 - 0\right] - \left[\frac{81}{2} - 0\right]$$

$$= 54 - \frac{81}{2} = \frac{108 - 81}{2} = \frac{27}{2} \text{ sq. units}$$

***Q.6** Find the area of the region enclosed by the parabola $x^2 = y$ and the line $y = x + 2$.

Ans. We have, $x^2 = y$ and $y = x + 2$

On solving both equations we get

$x^2 = x + 2$

$\Rightarrow \quad x^2 - x - 2 = 0$

$\Rightarrow \quad x^2 - 2x + x - 2 = 0$

$\Rightarrow \quad x(x - 2) + 1(x - 2) = 0$

$\Rightarrow \quad (x + 1)(x - 2) = 0 \Rightarrow x = -1, 2$

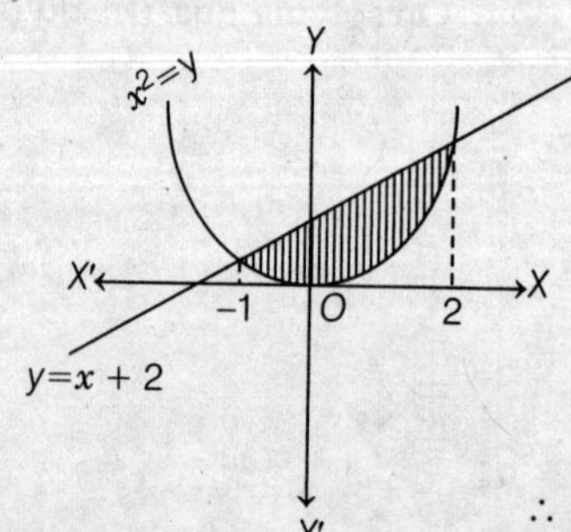

$\therefore$ Re

Note (✱) *Topics related to these question have been removed from latest NCERT.*

quired area of shaded region $= \int_{-1}^{2} (x+2)\,dx - \int_{-1}^{2} x^2\,dx$

$$= \int_{-1}^{2}(x+2-x^2)\,dx = \left[\frac{x^2}{2} + 2x - \frac{x^3}{3}\right]_{-1}^{2}$$

$$= \left[\frac{4}{2} + 4 - \frac{8}{3} - \frac{1}{2} + 2 - \frac{1}{3}\right]$$

$$= 6 + \frac{3}{2} - \frac{9}{3} = \frac{36+9-18}{6}$$

$$= \frac{27}{6} = \frac{9}{2} \text{ sq. units}$$

Q.7 Find the area of the region bounded by line $x = 2$ and parabola $y^2 = 8x$.

Thinking Process First find the area in the first quadrant and then multiply by 2 to get the required result.

Ans. We have, $y^2 = 8x$ and $x = 2$

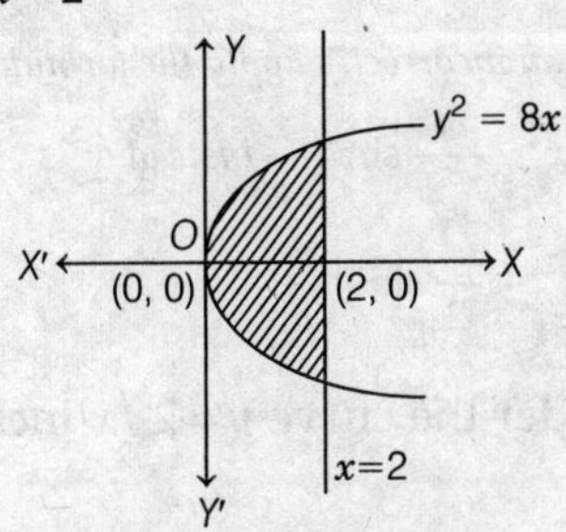

$\therefore$ Area of shaded region $= 2\int_0^2 \sqrt{8x}\,dx = 2\cdot 2\sqrt{2}\int_0^2 x^{1/2}\,dx$

$$= 4\cdot\sqrt{2}\cdot\left[2\cdot\frac{x^{3/2}}{3}\right]_0^2 = 4\sqrt{2}\left[\frac{2}{3}\cdot 2\sqrt{2} - 0\right]$$

$$= \frac{32}{3} \text{ sq. units}$$

Q.8 Sketch the region $\{(x, 0) : y = \sqrt{4 - x^2}\}$ and X-axis. Find the area of the region using integration.

Ans. Given region is $\{(x, 0) : y = \sqrt{4 - x^2}\}$ and X-axis.

We have, $\quad y = \sqrt{4 - x^2} \Rightarrow y^2 = 4 - x^2 \Rightarrow x^2 + y^2 = 4 = (2)^2$

This is the equation of circle with centre (0, 0) and radius 2.

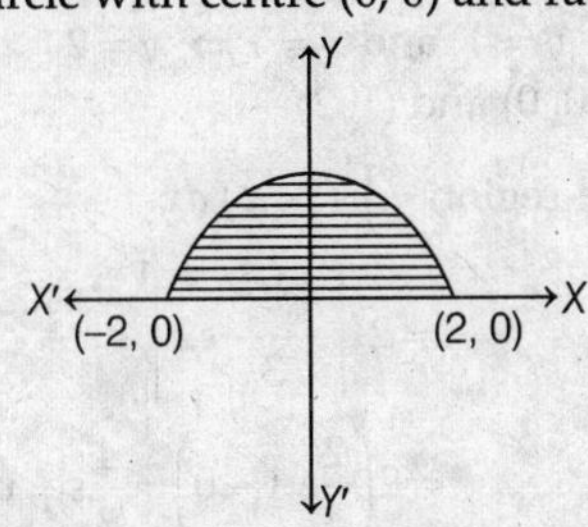

$\therefore$ Area of shaded region $= \int_{-2}^{2} \sqrt{4 - x^2}\, dx = \int_{-2}^{2} \sqrt{2^2 - x^2}\, dx$

$$= \left[\frac{x}{2}\sqrt{2^2 - x^2} + \frac{2^2}{2} \cdot \sin^{-1} \frac{x}{2} \right]_{-2}^{2}$$

$$\left[\because \int \sqrt{a^2 - r^2}\, dx\ \frac{x}{2}\sqrt{a^2 - x^2} + \frac{a^2}{2} \sin^{-1} \frac{x}{a} \right]$$

$$= \frac{2}{2}\sqrt{2^2 - 2^2} + 2\sin^{-1}\frac{2}{2} - \left(\frac{-2}{2}\sqrt{2^2 - (-2)^2} + 2\sin^{-1}\left(\frac{-2}{2}\right) \right)$$

$$= 0 + 2\sin^{-1}(1) - (0 + 2\sin^{-1}(-1))$$

$$= 2\sin^{-1}\left(\sin\frac{\pi}{2}\right) - 2\sin^{-1}\left(-\sin\frac{\pi}{2}\right)$$

$$= 2 \times \frac{\pi}{2} - 2\sin^{-1}\left[\sin\left(-\frac{\pi}{2}\right)\right]$$

$$= \pi - 2\left(-\frac{\pi}{2}\right) = 2\pi \text{ sq. units}$$

Mistake Alert! *Students may incorrectly apply the formula*

$\int \sqrt{a^2 - x^2}\, dx = \frac{x}{2}\sqrt{a^2 - x^2} - \frac{a^2}{2} \sin^{-1} \frac{x}{a}$ *instead of*

$\int \sqrt{a^2 - x^2}\, dx = \frac{x}{2}\sqrt{a^2 - x^2} + \frac{a^2}{2} \sin^{-1} \frac{x}{a}$

Q.9 Calculate the area under the curve $y = 2\sqrt{x}$ included between the lines $x = 0$ and $x = 1$.

Ans. We have, $y = 2\sqrt{x}$ ...(i)

$x = 0$ ...(ii)

and $x = 1$ ...(iii)

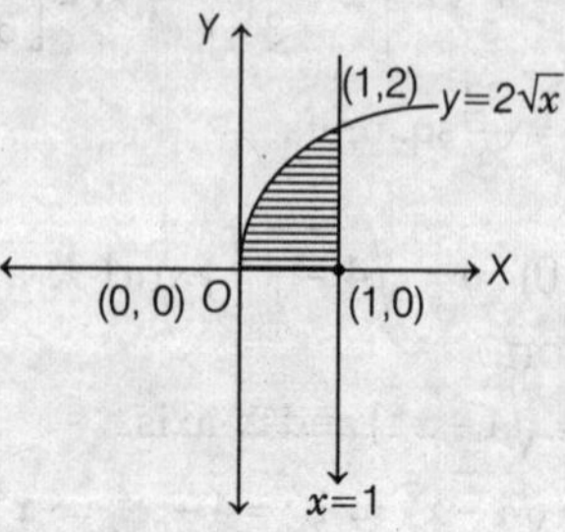

On solving Eqs. (i), (ii) and (iii), we get

$$x = 0 \Rightarrow y = 0 \text{ and } x = 1 \Rightarrow y = 2$$

$\therefore$ Intersecting points are (0, 0) and (1, 2)

$\therefore$ Area of shaded region $= \int_0^1 (2\sqrt{x})\, dx$

$$= 2 \cdot \left[\frac{x^{3/2}}{3} \cdot 2 \right]_0^1$$

$$= 2\left(\frac{2}{3} \cdot 1 - 0 \right) = \frac{4}{3} \text{ sq. units}$$

Q.10 Using integration, find the area of the region bounded by the line $2y = 5x + 7$, X-axis and the lines $x = 2$ and $x = 8$.

Ans. We have, $$2y = 5x + 7$$

$\Rightarrow$ $$y = \frac{5x}{2} + \frac{7}{2}$$

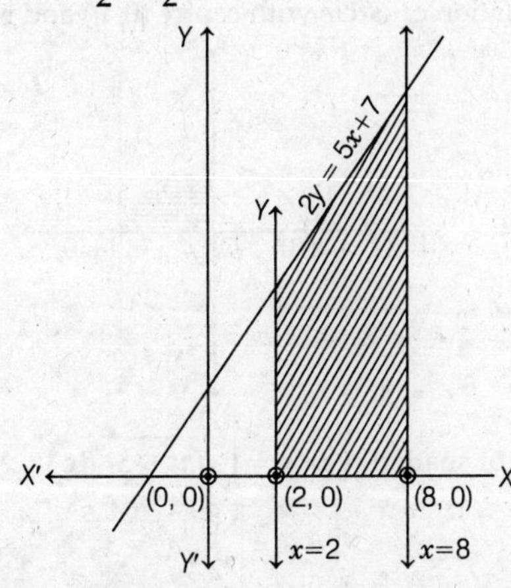

$\therefore$ Area of shaded region $= \frac{1}{2}\int_2^8 (5x + 7)\, dx = \frac{1}{2}\left[5\cdot\frac{x^2}{2} + 7x\right]_2^8$

$$= \frac{1}{2}[5\cdot 32 + 7\cdot 8 - 10 - 14] = \frac{1}{2}[160 + 56 - 24]$$

$$= \frac{192}{2} = 96 \text{ sq. units}$$

Q.11 Draw a rough sketch of the curve $y = \sqrt{x-1}$ in the interval [1, 5]. Find the area under the curve and between the lines $x = 1$ and $x = 5$.

Ans. Given equation of the curve is $y = \sqrt{x - 1}$.

$\Rightarrow$ $$y^2 = x - 1$$

$\therefore$ Area of shaded region $= \int_1^5 (x-1)^{1/2}\, dx = \left[\frac{2\cdot(x-1)^{3/2}}{3}\right]_1^5$

$$= \left[\frac{2}{3}\cdot(5-1)^{3/2} - 0\right] = \frac{16}{3} \text{ sq. units}$$

Q.12 Determine the area under the curve $y = \sqrt{a^2 - x^2}$ included between the lines $x = 0$ and $x = a$.

Ans. Given equation of the curve is $y = \sqrt{a^2 - x^2}$.

$$\Rightarrow \quad y^2 = a^2 - x^2 \quad \Rightarrow \quad y^2 + x^2 = a^2$$

$\therefore$ This is the equation of circle with centre (0, 0) and radius 'a'.

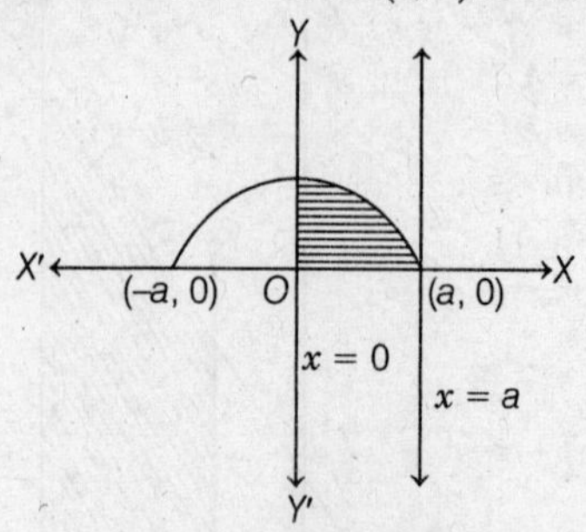

$\therefore$ Required area of shaded region $= \int_0^a \sqrt{a^2 - x^2}\, dx$

$$= \left[\frac{x}{2}\sqrt{a^2 - x^2} + \frac{a^2}{2}\sin^{-1}\frac{x}{a}\right]_0^a$$

$$\left[\because \int \sqrt{a^2 - x^2}\, dx = \frac{x}{2}\sqrt{a^2 - x^2} + \frac{a^2}{2}\sin^{-1}\frac{x}{a}\right]$$

$$= \left[0 + \frac{a^2}{2}\sin^{-1}(1) - 0 - \frac{a^2}{2}\sin^{-1}0\right]$$

$$= \frac{a^2}{2}\cdot\frac{\pi}{2} = \frac{\pi a^2}{4} \text{ sq. units}$$

Mistkae Alert! *Students may incorrectly calculate, area =* $2\int_0^a \sqrt{a^2 - x^2}\, dx$, *instead of area =* $\int_0^a \sqrt{a^2 - x^2}\, dx$.

***Q.13** Find the area of the region bounded by $y = \sqrt{x}$ and $y = x$.

Ans. Given equation of curves are $y = \sqrt{x}$ and $y = x$.

On solving both equations, we get

$$\Rightarrow \quad x = \sqrt{x} \quad \Rightarrow \quad x^2 = x$$

$$\Rightarrow \quad x^2 - x = 0 \quad \Rightarrow \quad x(x-1) = 0$$

$$\Rightarrow \quad x = 0, 1$$

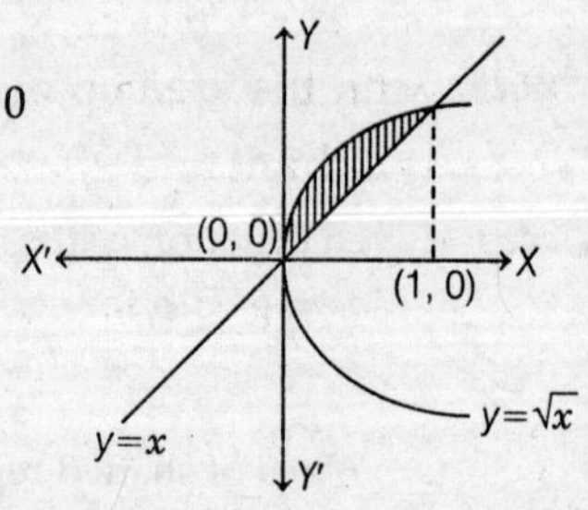

$\therefore$ Required area of shaded region

$$= \int_0^1 (\sqrt{x})dx - \int_0^1 x\, dx$$

$$= \left[2\cdot\frac{x^{3/2}}{3}\right]_0^1 - \left[\frac{x^2}{2}\right]_0^1$$

$$= \frac{2}{3}\cdot 1 - \frac{1}{2} = \frac{4-3}{6} = \frac{1}{6} \text{ sq. unit}$$

Note (✱) *Topics related to these question have been removed from latest NCERT.*

***Q.14** Find the area enclosed by the curve $y = -x^2$ and the straight line $x + y + 2 = 0$.

Ans. We have, $y = -x^2$ and $x + y + 2 = 0$

$\Rightarrow \quad -x - 2 = -x^2$

$\Rightarrow \quad x^2 - x - 2 = 0$

$\Rightarrow \quad x^2 + x - 2x - 2 = 0$

$\Rightarrow \quad x(x+1) - 2(x+1) = 0$

$\Rightarrow \quad (x-2)(x+1) = 0 \Rightarrow x = 2, -1$

On solving both equations, we get

Let A_1 = Area above the line $x + y + 2 = 0$

i.e., A_1 = Area of *CODABC*

and A_2 = Area above the curve $y = -x^2$

i.e., A_2 = Area *ODAOBCO*

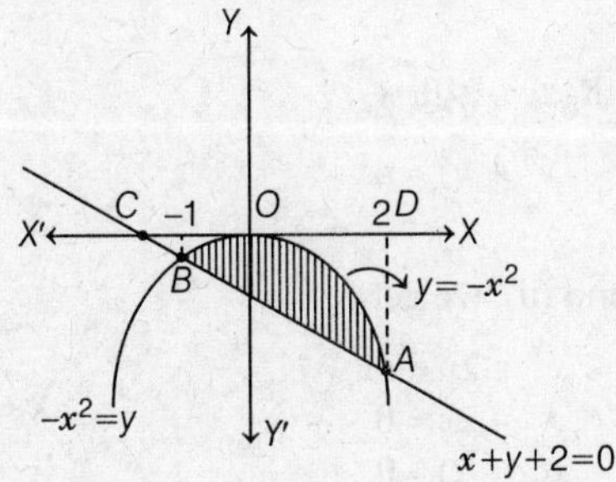

$\therefore$ Area of shaded region $A = A_1 - A_2$

$$= \int_{-1}^{2} (-x-2)dx - \int_{-1}^{2} -x^2 \cdot dx$$

$$= \left| \int_{-1}^{2} (-x - 2 + x^2)\, dx \right| = \left| \int_{-1}^{2} (x^2 - x - 2)\, dx \right|$$

$$= \left| \left[\frac{x^3}{3} - \frac{x^2}{2} - 2x \right]_{-1}^{2} \right| = \left| \left[\frac{8}{3} - \frac{4}{2} - 4 + \frac{1}{3} + \frac{1}{2} - 2 \right] \right|$$

$$= \left| \frac{16 - 12 - 24 + 2 + 3 - 12}{6} \right| = \left| -\frac{27}{6} \right| = \frac{9}{2} \text{ sq. units}$$

***Q.15** Find the area bounded by the curve $y = \sqrt{x}$, $x = 2y + 3$ in the first quadrant and X-axis.

Ans. Given equation of the curves are $y = \sqrt{x}$ and $x = 2y + 3$ in the first quadrant.

On solving both the equations, we get

$y = \sqrt{2y + 3}$

$\Rightarrow \quad y^2 = 2y + 3$

$\Rightarrow \quad y^2 - 2y - 3 = 0$

$\Rightarrow \quad y^2 - 3y + y - 3 = 0$

$\Rightarrow \quad y(y-3) + 1(y-3) = 0$

$\Rightarrow \quad (y+1)(y-3) = 0$

$\Rightarrow \quad y = -1, 3$

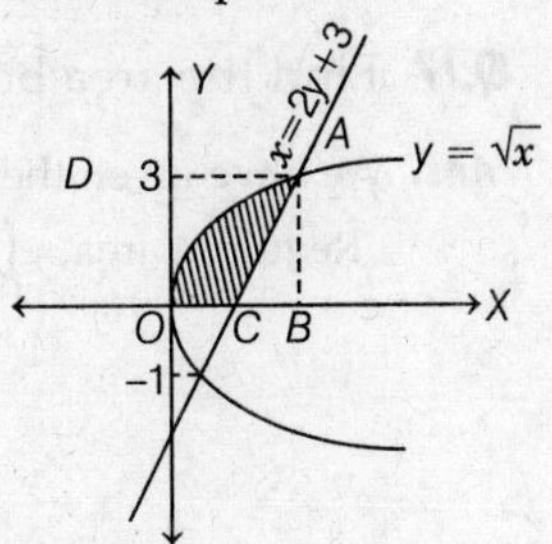

Note (*) *Topics related to these question have been removed from latest NCERT.*

Let A_1 = Area above the straight line $x = 2y + 3$

= Area of $OCADO$

and A_2 = Area above the curve $y = \sqrt{x}$

= Area of $OADO$

$\therefore$ Required area of shaded region = $A_1 - A_2$

$$A = \int_0^3 (2y + 3 - y^2)\,dy = \left[\frac{2y^2}{2} + 3y - \frac{y^3}{3}\right]_0^3$$

$$= [9 + 9 - 9 - 0] = 9 \text{ sq. units}$$

Long Answer Type Questions

***Q.16** Find the area of the region bounded by the curve $y^2 = 2x$ and $x^2 + y^2 = 4x$.

Ans. We have, $\quad y^2 = 2x \quad$ ...(i)

and $\quad x^2 + y^2 = 4x \quad$ ...(ii)

On solving Eqs. (i) and (ii), we get

$$x^2 + 2x = 4x$$

$$\Rightarrow x^2 - 2x = 0$$

$$\Rightarrow x(x - 2) = 0$$

$$\Rightarrow x = 0, 2$$

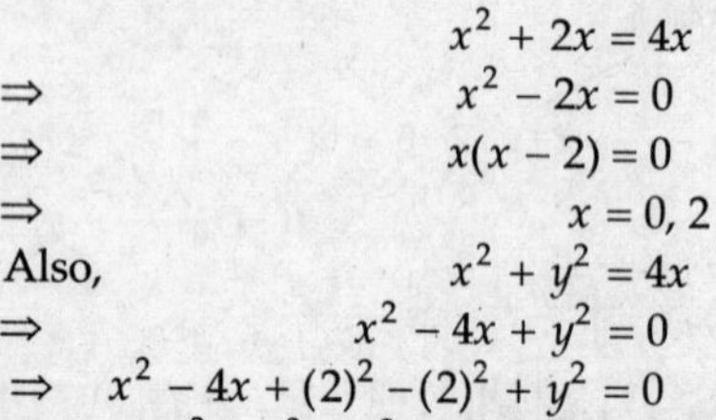

Also, $\quad x^2 + y^2 = 4x$

$$\Rightarrow x^2 - 4x + y^2 = 0$$

$$\Rightarrow x^2 - 4x + (2)^2 - (2)^2 + y^2 = 0$$

$\Rightarrow (x - 2)^2 + y^2 = 2^2$ which is equation of a circle with centre (2, 0) and radius 2 units.

$\therefore$ Required area $= 2 \cdot \int_0^2 [\sqrt{2^2 - (x-2)^2} - \sqrt{2x}]\,dx$

$$= 2\left[\left[\frac{x-2}{2} \cdot \sqrt{2^2 - (x-2)^2} + \frac{2^2}{2}\sin^{-1}\left(\frac{x-2}{2}\right)\right]_0^2 - \left[\sqrt{2} \cdot \frac{x^{3/2}}{3/2}\right]_0^2\right]$$

$$= 2\left[\left(0 + 0 - 1 \cdot 0 + 2 \cdot \frac{\pi}{2}\right) - \frac{2\sqrt{2}}{3}(2^{3/2} - 0)\right]$$

$$= \frac{4\pi}{2} - \frac{8 \cdot 2}{3} = 2\pi - \frac{16}{3} = 2\left(\pi - \frac{8}{3}\right) \text{ sq. units}$$

Q.17 Find the area bounded by the curve $y = \sin x$ between $x = 0$ and $x = 2\pi$.

Ans. We have given the equation of a curve $y = \sin x$ lies between $x = 0$ and $x = 2\pi$.

Required area $= \int_0^{2\pi} \sin x\,dx = \int_0^{\pi} \sin x\,dx + \left|\int_{\pi}^{2\pi} \sin x\,dx\right|$

$$= -[\cos x]_0^{\pi} + \left|[-\cos x]_{\pi}^{2\pi}\right|$$

Note (✱) *Topics related to these question have been removed from latest NCERT.*

$$= -[\cos\pi - \cos 0] + |[-\cos 2\pi + \cos\pi]|$$

$$= -[-1-1] + |-1-1|$$
$$= 2 + 2 = 4 \text{ sq. units}$$

Mistake Alert! *Students may incorrectly take* $\int \sin x\, dx = \cos x$ *instead of* $\int \sin a\, dx = -\cos x$.

***Q.18** Find the area of region bounded by the triangle whose vertices are (−1, 1), (0, 5) and (3, 2), using integration.

- **Concept Insight** If (x_1, y_1) and (x_2, y_2) be the two points, then the equation of line passing through these two points is given by $y - y_1 = \dfrac{y_2 - y_1}{x_2 - x_1}(x - x_1)$.

Ans. Let we have the vertices of a ΔABC as $A(-1, 1)$, $B(0, 5)$ and $C(3, 2)$.

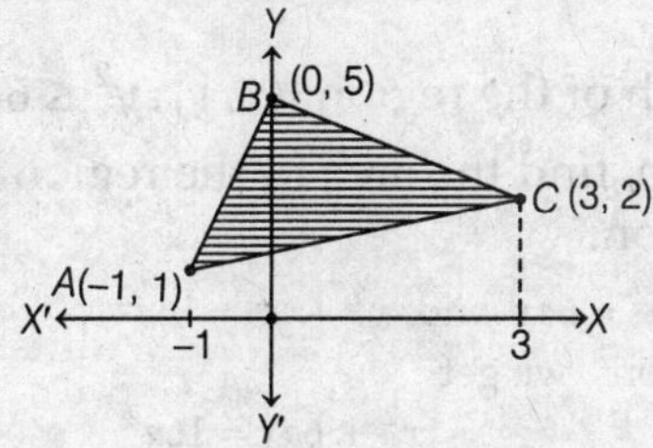

First of all we find the equations of the sides of the triangle by using

$$y - y_1 = \frac{y_2 - y_1}{x_2 - x_1}(x - x_1)$$

$\therefore$ Equation of AB is $y - 1 = \left(\dfrac{5-1}{0+1}\right)(x+1)$

$$\Rightarrow \quad y - 1 = 4x + 4$$

$$\Rightarrow \quad y = 4x + 5 \text{ say } (y_1) \quad \text{...(i)}$$

and equation of BC is $y - 5 = \left(\dfrac{2-5}{3-0}\right)(x-0)$

$$\Rightarrow \quad y - 5 = \frac{-3}{3}(x)$$

$$\Rightarrow \quad y = 5 - x \text{ say } (y_2) \quad \text{...(ii)}$$

Similarly, equation of AC is $y - 1 = \left(\dfrac{2-1}{3+1}\right)(x+1)$

Note (*) *Topics related to these question have been removed from latest NCERT.*

$$\Rightarrow \qquad y - 1 = \frac{1}{4}(x+1)$$

$$\Rightarrow \qquad 4y = x + 5 \qquad \text{...(iii)}$$

$$\Rightarrow \qquad y = \frac{x+5}{4} \text{ (Say } y_3\text{)}$$

$$\therefore \text{Area of shaded region} = \int_{-1}^{0}(y_1 - y_3)\,dx + \int_{0}^{3}(y_2 - y_3)\,dx$$

$$= \int_{-1}^{0}\left[4x + 5 - \left(\frac{x+5}{4}\right)\right]dx + \int_{0}^{3}\left[5 - x - \left(\frac{x+5}{4}\right)\right]dx$$

$$= \left[\frac{4x^2}{2} + 5x - \frac{x^2}{8} - \frac{5x}{4}\right]_{-1}^{0} + \left[5x - \frac{x^2}{2} - \frac{x^2}{8} - \frac{5x}{4}\right]_{0}^{3}$$

$$= \left[0 - \left(4\cdot\frac{1}{2} + 5(-1) - \frac{1}{8} + \frac{5}{4}\right)\right] + \left[\left(15 - \frac{9}{2} - \frac{9}{8} - \frac{15}{4}\right) - 0\right]$$

$$= \left[-2 + 5 + \frac{1}{8} - \frac{5}{4} + 15 - \frac{9}{2} - \frac{9}{8} - \frac{15}{4}\right]$$

$$= 18 + \left(\frac{1 - 10 - 36 - 9 - 30}{8}\right)$$

$$= 18 + \left(-\frac{84}{8}\right) = 18 - \frac{21}{2} = \frac{15}{2} \text{ sq. units}$$

***Q.19** Draw a rough sketch of the region $\{(x, y) : y^2 \le 6ax$ and $x^2 + y^2 \le 16a^2\}$. Also, find the area of the region sketched using method of integration.

Ans. We have, $\quad y^2 = 6ax$ and $x^2 + y^2 = 16a^2$

On solving both equations, we get

$$x^2 + 6ax = 16a^2$$

$$\Rightarrow \qquad x^2 + 6ax - 16a^2 = 0$$

$$\Rightarrow \qquad x^2 + 8ax - 2ax - 16a^2 = 0$$

$$\Rightarrow \qquad x(x + 8a) - 2a(x + 8a) = 0$$

$$\Rightarrow \qquad (x - 2a)(x + 8a) = 0$$

$$\Rightarrow \qquad x = 2a, -8a$$

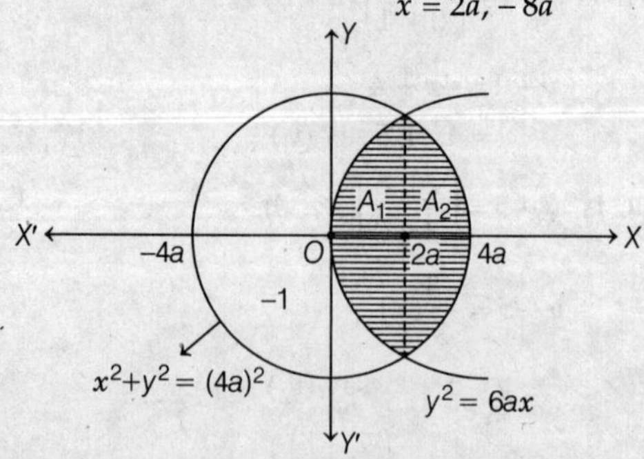

Let A_1 = Area under the curve (parabola) in first quadrant
and A_2 = Area under the curve (circle) in first quadrant

Note (*) *Topics related to these question have been removed from latest NCERT.*

As, A_1 and A_2 symmetrical to X-axis on both sides

Therefore required area = $2(A_1 + A_2)$

where $A_1 = \int_0^{2a} \sqrt{6ax}\, dx$ and $A_2 = \int_{2a}^{4a} \sqrt{(4a)^2 - x^2}\, dx$

$\therefore$ Area of required region $= 2\left[\int_0^{2a} \sqrt{6ax}\, dx + \int_{2a}^{4a} \sqrt{(4a)^2 - x^2}\, dx\right]$

$$= 2\left[\int_0^{2a} \sqrt{6a}\, x^{1/2} dx + \int_{2a}^{4a} \sqrt{(4a)^2 - x^2}\, dx\right]$$

$$= 2\left[\sqrt{6a}\left[\frac{x^{3/2}}{3/2}\right]_0^{2a} + \left(\frac{x}{2}\sqrt{(4a)^2 - x^2} + \frac{(4a)^2}{2}\sin^{-1}\frac{x}{4a}\right)_{2a}^{4a}\right]$$

$$= 2\left[\sqrt{6a}\cdot\frac{2}{3}((2a)^{3/2} - 0) + \frac{4a}{2}\cdot 0 + \frac{16a^2}{2}\cdot\frac{\pi}{2} - \frac{2a}{2}\sqrt{16a^2 - 4a^2} - \frac{16a^2}{2}\cdot\sin^{-1}\frac{2a}{4a}\right]$$

$$= 2\left[\sqrt{6a}\ \frac{2}{3}\cdot 2\sqrt{2}\, a^{3/2} + 0 + 4\pi a^2 - \frac{2a}{2}\cdot 2\sqrt{3}\, a - 8a^2\cdot\frac{\pi}{6}\right]$$

$$= 2\left[\sqrt{12}\cdot\frac{4}{3}a^2 + 4\pi a^2 - 2\sqrt{3}\, a^2 - \frac{4a^2\pi}{3}\right]$$

$$= 2\left[\frac{8\sqrt{3}\, a^2 + 12\pi a^2 - 6\sqrt{3}\, a^2 - 4a^2\pi}{3}\right]$$

$$= \frac{2}{3}a^2[8\sqrt{3} + 12\pi - 6\sqrt{3} - 4\pi]$$

$$= \frac{2}{3}a^2[2\sqrt{3} + 8\pi] = \frac{4}{3}a^2[\sqrt{3} + 4\pi] \text{ sq. units}$$

***Q.20** Compute the area bounded by the lines $x + 2y = 2$, $y - x = 1$ and $2x + y = 7$.

Ans. We have, $x + 2y = 2$ ...(i)

$y - x = 1$ ...(ii)

and $2x + y = 7$ ...(iii)

On solving Eqs. (i) and (ii), we get

$$y - (2 - 2y) = 1 \Rightarrow 3y - 2 = 1 \Rightarrow y = 1$$

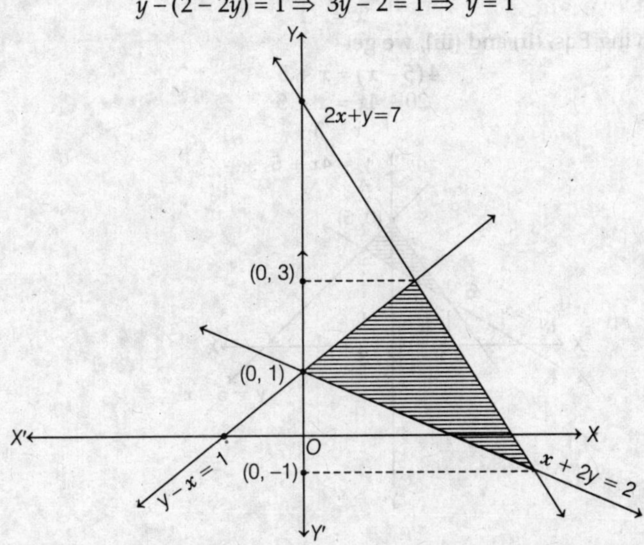

Note (*) *Topics related to these question have been removed from latest NCERT.*

On solving Eqs. (ii) and (iii), we get $2(y-1)+y=7$

$\Rightarrow$ $$2y-2+y=7 \Rightarrow y=3$$

On solving Eqs. (i) and (iii), we get

$$2(2-2y)+y=7$$

$\Rightarrow$ $$4-4y+y=7 \Rightarrow -3y=3 \Rightarrow y=-1$$

$\therefore$ Required area $= -\int_{-1}^{1}(2-2y)\,dy + \int_{-1}^{3}\frac{(7-y)}{2}\,dy - \int_{1}^{3}(y-1)\,dy$

$$= \left[-2y+\frac{2y^2}{2}\right]_{-1}^{1} + \left[\frac{7y}{2}-\frac{y^2}{2\cdot 2}\right]_{-1}^{3} - \left[\frac{y^2}{2}-y\right]_{1}^{3}$$

$$= \left[-2+\frac{2}{2}-2-\frac{2}{2}\right] + \left[\frac{21}{2}-\frac{9}{4}+\frac{7}{2}+\frac{1}{4}\right] - \left[\frac{9}{2}-3-\frac{1}{2}+1\right]$$

$$= [-4] + \left[\frac{42-9+14+1}{4}\right] - \left[\frac{9-6-1+2}{2}\right]$$

$$= -4+12-2 = 6 \text{ sq. units}$$

***Q.21** Find the area bounded by the lines $y=4x+5$, $y=5-x$ and $4y=x+5$.

Ans. Given equations of lines are

$$y=4x+5 \quad \ldots(i)$$

$$y=5-x \quad \ldots(ii)$$

and $$4y=x+5 \quad \ldots(iii)$$

On solving Eqs. (i) and (ii), we get

$$4x+5=5-x$$

$\Rightarrow$ $$x=0$$

On solving Eqs. (i) and (iii), we get

$$4(4x+5)=x+5$$

$\Rightarrow$ $$16x+20=x+5$$

$\Rightarrow$ $$15x=-15$$

$\Rightarrow$ $$x=-1$$

On solving Eqs. (ii) and (iii), we get

$$4(5-x)=x+5$$

$\Rightarrow$ $$20-4x=x+5$$

$\Rightarrow$ $$x=3$$

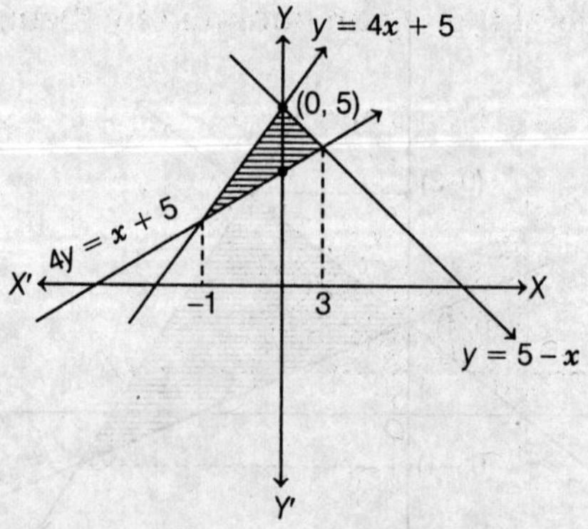

Note (✱) *Topics related to these question have been removed from latest NCERT.*

$\therefore$ Required area $= \int_{-1}^{0}(4x+5)dx + \int_{0}^{3}(5-x)dx - \frac{1}{4}\int_{-1}^{3}(x+5)dx$

$$= \left[\frac{4x^2}{2}+5x\right]_{-1}^{0} + \left[5x - \frac{x^2}{2}\right]_{0}^{3} - \frac{1}{4}\left[\frac{x^2}{2}+5x\right]_{-1}^{3} \quad \left[\because \int x^n dx = \frac{x^{n+1}}{n+1}\right]$$

$$= [0-2+5] + \left[15 - \frac{9}{2} - 0\right] - \frac{1}{4}\left[\frac{9}{2} + 15 - \frac{1}{2} + 5\right]$$

$$= 3 + \frac{21}{2} - \frac{1}{4}\cdot 24$$

$$= -3 + \frac{21}{2} = \frac{15}{2} \text{ sq. units}$$

Q.22 Find the area bounded by the curve $y = 2\cos x$ and the X-axis from $x = 0$ to $x = 2\pi$.

- **Concept Insight** The curve of $\cos x$ has positive region in $\left[0, \frac{\pi}{2}\right]$ and $\left[\frac{3\pi}{2}, 2\pi\right]$ and negative region in $\left[\frac{\pi}{2}, \frac{3\pi}{2}\right]$.

Ans. We have given $y = 2\cos x$

Required area of shaded region $= \int_{0}^{2\pi} 2\cos x dx$

$$= \int_{0}^{\pi/2} 2\cos x\, dx + \left|\int_{\pi/2}^{3\pi/2} 2\cos x\, dx\right| + \int_{3\pi/2}^{2\pi} 2\cos x\, dx$$

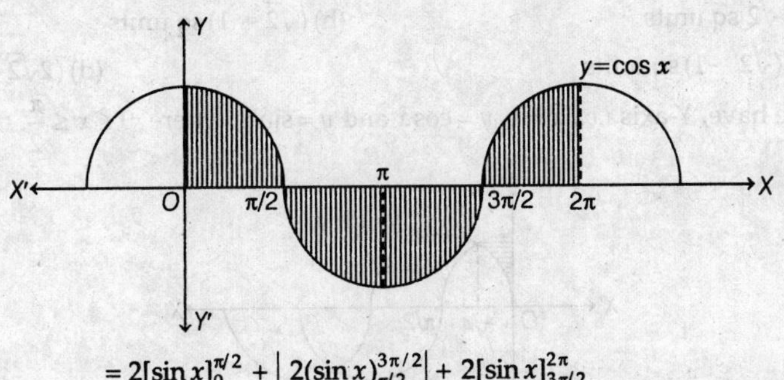

$$= 2[\sin x]_{0}^{\pi/2} + \left|2(\sin x)_{\pi/2}^{3\pi/2}\right| + 2[\sin x]_{3\pi/2}^{2\pi}$$

$$= 2 + 4 + 2 = 8 \text{ sq. units}$$

Q.23 Draw a rough sketch of the given curve $y = 1 + |x+1|$, $x = -3$, $x = 3$, $y = 0$ and find the area of the region bounded by them, using integration.

Ans. We have, $y = 1 + |x+1|$, $x = -3$, $x = 3$ and $y = 0$

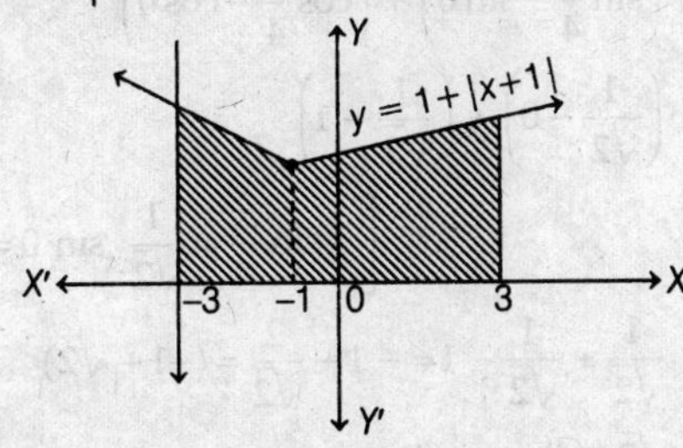

Note (✱) *Topics related to these question have been removed from latest NCERT.*

$\because$ $$y = \begin{cases} -x, & \text{if } x < -1 \\ x+2, & \text{if } x \ge -1 \end{cases}$$

Put $y = -x$ for -3 to -1 and $y = x + 2$ for -1 to 3.

$\therefore$ Area of shaded region $= \int_{-3}^{-1} -x\, dx + \int_{-1}^{3} (x+2)\, dx$

$$= -\left[\frac{x^2}{2}\right]_{-3}^{-1} + \left[\frac{x^2}{2} + 2x\right]_{-1}^{3}$$

$$= -\left[\frac{1}{2} - \frac{9}{2}\right] + \left[\frac{9}{2} + 6 - \frac{1}{2} + 2\right]$$

$$= -[-4] + [8 + 4]$$

$$= 12 + 4 = 16 \text{ sq. units}$$

Objective Type Questions

Choose the correct answer out of the given four options in each of the exercises from 24 to 34.

***Q.24** The area of the region bounded by the Y-axis $y = \cos x$ and $y = \sin x$, where $0 \le x \le \frac{\pi}{2}$, is

(a) $\sqrt{2}$ sq units (b) $(\sqrt{2} + 1)$ sq units

(c) $(\sqrt{2} - 1)$ sq units (d) $(2\sqrt{2} - 1)$ sq units

Ans. ***(c)*** We have, Y-axis *i.e.*, $x = 0$, $y = \cos x$ and $y = \sin x$, where $0 \le x \le \frac{\pi}{2}$.

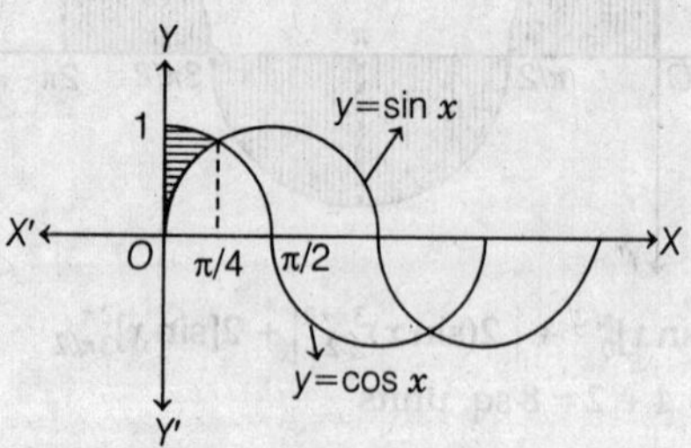

$\therefore$ Required area $= \int_0^{\pi/4} (\cos x - \sin x) dx$

$$= [\sin x]_0^{\pi/4} + [\cos x]_0^{\pi/4}$$

$$= \left(\sin\frac{\pi}{4} - \sin 0\right) + \left(\cos\frac{\pi}{4} - \cos 0\right)$$

$$= \left(\frac{1}{\sqrt{2}} - 0\right) + \left(\frac{1}{\sqrt{2}} - 1\right)$$

$$\left[\because \sin\frac{\pi}{4} = \frac{1}{\sqrt{2}}, \sin 0 = 0, \cos\frac{\pi}{4} = \frac{1}{\sqrt{2}}, \cos 0 = 1\right]$$

$$= \frac{1}{\sqrt{2}} + \frac{1}{\sqrt{2}} - 1 = -1 + \frac{2}{\sqrt{2}} = (-1 + \sqrt{2})$$

$$= (\sqrt{2} - 1) \text{ sq. units}$$

Note (✱) *Topics related to these question have been removed from latest NCERT.*

***Q.25** The area of the region bounded by the curve $x^2 = 4y$ and the straight line $x = 4y - 2$ is

(a) $\frac{3}{8}$ sq unit (b) $\frac{5}{8}$ sq unit (c) $\frac{7}{8}$ sq unit (d) $\frac{9}{8}$ sq units

Thinking Process First, find the points of interesection of line $x = 4y - 2$ and parabola $x^2 = 4y$, then find the proper limits, the required area is given by under the curve parabola w.r.t. X-axis subtracted from area under the line w.r.t. X-axis under the proper limits.

Ans. ***(d)*** Given equation of curve is $x^2 = 4y$ and the straight line $x = 4y - 2$.

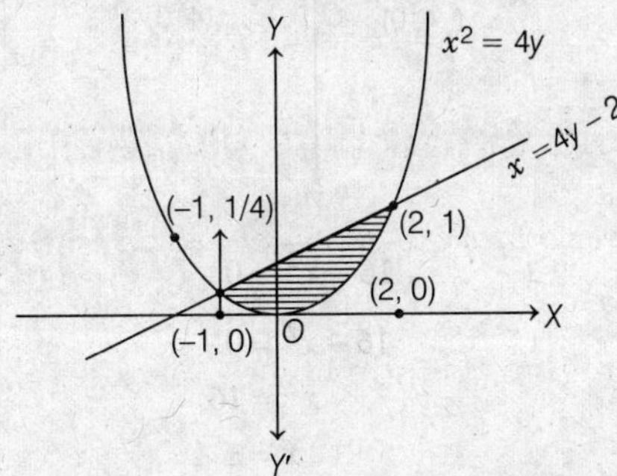

On solving both equations, we get

$$(4y - 2)^2 = 4y$$

$$\Rightarrow \quad 16y^2 + 4 - 16y = 4y$$

$$\Rightarrow \quad 16y^2 - 20y + 4 = 0 \Rightarrow 4y^2 - 5y + 1 = 0$$

$$\Rightarrow \quad 4y^2 - 4y - y + 1 = 0 \Rightarrow 4y(y - 1) - 1(y - 1) = 0$$

$$\Rightarrow \quad (4y - 1)(y - 1) = 0$$

$$\therefore \quad y = 1, \frac{1}{4}$$

For $y = 1$, $x = \sqrt{4 \cdot 1} = 2$ [since, negative value does not satisfy the equation of line]

For $y = \frac{1}{4}$, $x = \sqrt{4 \cdot \frac{1}{4}} = -1$ [positive value does not satisfy the equation of line]

So, the intersection points are (2, 1) and $\left(-1, \frac{1}{4}\right)$.

$$\therefore \text{ Area of shaded region} = \int_{-1}^{2} \left(\frac{x + 2}{4}\right) dx - \int_{-1}^{2} \frac{x^2}{4} dx$$

$$= \frac{1}{4}\left[\frac{x^2}{2} + 2x\right]_{-1}^{2} - \frac{1}{4}\left[\frac{x^3}{3}\right]_{-1}^{2}$$

$$= \frac{1}{4}\left[\frac{4}{2} + 4 - \frac{1}{2} + 2\right] - \frac{1}{4}\left[\frac{8}{3} + \frac{1}{3}\right]$$

$$= \frac{1}{4} \cdot \frac{15}{2} - \frac{1}{4} \cdot \frac{9}{3} = \frac{45 - 18}{24}$$

$$= \frac{27}{24} = \frac{9}{8} \text{ sq. units}$$

Note (✱) *Topics related to these question have been removed from latest NCERT.*

Q.26 The area of the region bounded by the curve $y=\sqrt{16-x^2}$ and X-axis is

(a) 8π sq units
(b) 20π sq units
(c) 16π sq units
(d) 256π sq units

Ans. **(a)** Given equation of curve is $y=\sqrt{16-x^2}$ and the equation of X-axis *i.e.,* $y = 0$.

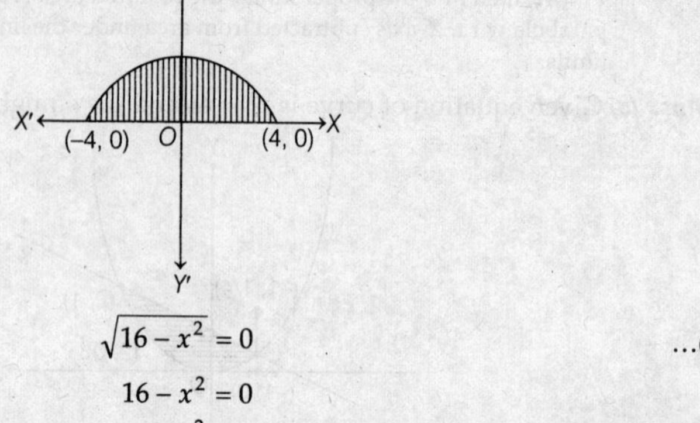

$$\therefore \quad \sqrt{16-x^2}=0 \quad \text{...(i)}$$

$$\Rightarrow \quad 16-x^2=0$$

$$\Rightarrow \quad x^2=16$$

$$\Rightarrow \quad x=\pm 4$$

So, the intersection points are (4, 0) and (–4, 0).

$\therefore$ Required area of curve $=\int_{-4}^{4}(16-x^2)^{1/2}\,dx$

$$=\int_{-4}^{4}\sqrt{(4^2-x^2)}\,dx$$

$$=\left[\frac{x}{2}\sqrt{4^2-x^2}+\frac{4^2}{2}\sin^{-1}\frac{x}{4}\right]_{-4}^{4}$$

$$=\left[\frac{4}{2}\sqrt{4^2-4^2}+8\sin^{-1}\left(\frac{4}{4}\right)\right]-\left[-\frac{4}{2}\sqrt{4^2-(-4)^2}+8\sin^{-1}\left(-\frac{4}{4}\right)\right]$$

$$=[2\times 0+8\sin^{-1}(1)]-[-2\times 0+8\sin^{-1}(-1)]$$

$$=8\sin^{-1}\left(\sin\frac{\pi}{2}\right)-\left[8\sin^{-1}\left(-\sin\frac{\pi}{2}\right)\right]$$

$$=8\times\frac{\pi}{2}-8\sin^{-1}\left[\sin\left(-\frac{\pi}{2}\right)\right]$$

$$=4\pi+8\times\frac{\pi}{2}=8\pi \text{ sq. units.}$$

***Q.27** Area of the region in the first quadrant enclosed by the X-axis, the line $y=x$ and the circle $x^2+y^2=32$ is

(a) 16π sq units (b) 4π sq units (c) 32π sq units (d) 24π sq units

Ans. **(b)** We have, area enclosed by X-axis *i.e.,* $y = 0$, $y = x$ and the circle $x^2 + y^2 = 32$ in first quadrant.

Since, $$x^2+(x)^2=32 \quad [\because y=x]$$

$$\Rightarrow \quad 2x^2=32$$

Note (*) *Topics related to these question have been removed from latest NCERT.*

$$\Rightarrow \qquad x = \pm 4$$

So, the intersection point of circle $x^2 + y^2 = 32$ and line $y = x$ are (4, 4) or (– 4, –4).

and $\qquad x^2 + y^2 = (4\sqrt{2})^2$

Since, $\qquad y = 0$

$\therefore \qquad x^2 + (0)^2 = 32$

$\Rightarrow \qquad x = \pm 4\sqrt{2}$

So, the circle intersects the X-axis at $(\pm 4\sqrt{2}, 0)$.

As we have to find the area of the region in first quadrant we will take only positive value of x and y.

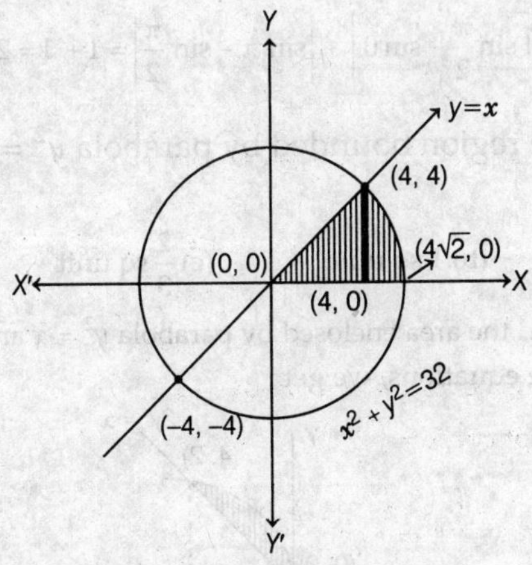

$$\text{Area of shaded region} = \int_0^4 x\,dx + \int_4^{4\sqrt{2}} \sqrt{(4\sqrt{2})^2 - x^2}\,dx$$

$$= \left[\frac{x^2}{2}\right]_0^4 + \left[\frac{x}{2}\sqrt{(4\sqrt{2})^2 - x^2} + \frac{(4\sqrt{2})^2}{2}\sin^{-1}\frac{x}{4\sqrt{2}}\right]_4^{4\sqrt{2}}$$

$$= \frac{16}{2} + \left[\frac{4\sqrt{2}}{2}\cdot 0 + 16\sin^{-1}\frac{(4\sqrt{2})}{(4\sqrt{2})} - \frac{4}{2}\sqrt{(4\sqrt{2})^2 - 16} - 16\sin^{-1}\frac{4}{4\sqrt{2}}\right]$$

$$= 8 + \left[16\cdot\frac{\pi}{2} - 2\cdot\sqrt{16} - 16\cdot\frac{\pi}{4}\right]$$

$$= 8 + [8\pi - 8 - 4\pi] = 4\pi \text{ sq. units}$$

Q.28 Area of the region bounded by the curve $y = \cos x$ between $x = 0$ and $x = \pi$ is

(a) 2 sq units (b) 4 sq units

(c) 3 sq units (d) 1 sq unit

Thinking Process We know that, $\cos x$ curve has positive region in $\left[0, \frac{\pi}{2}\right]$ and negative region in $\left[\frac{\pi}{2}, \pi\right]$.

***Ans.* (a)**

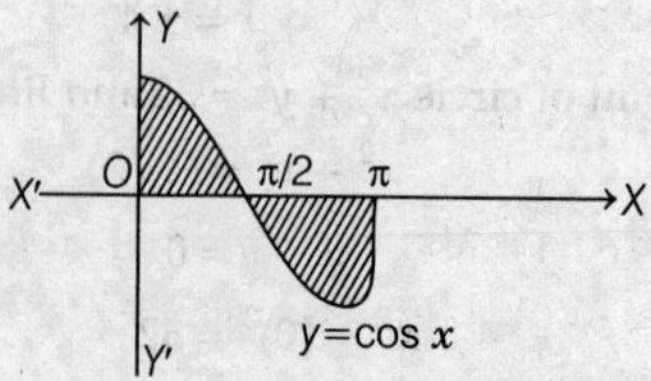

Required area enclosed by the curve $y = \cos x$, $x = 0$ and $x = \pi$ is

$$= \int_0^{\pi/2} \cos x \, dx + \left| \int_{\pi/2}^{\pi} \cos x \, dx \right| \qquad \left[\because \int \cos x \, dx = \sin x\right]$$

$$= [\sin x]_0^{\pi/2} + |[\sin x]_{\pi/2}^{\pi}|$$

$$= \left[\sin \frac{\pi}{2} - \sin 0\right] + \left|\sin \pi - \sin \frac{\pi}{2}\right| = 1 + 1 = 2 \text{ sq. units}$$

***Q.29** The area of the region bounded by parabola $y^2 = x$ and the straight line $2y = x$ is

(a) $\frac{4}{3}$ sq units (b) 1 sq unit (c) $\frac{2}{3}$ sq unit (d) $\frac{1}{3}$ sq unit

***Ans.* (a)** We have to find the area enclosed by parabola $y^2 = x$ and the straight line $2y = x$.
Solving the above equations, we get

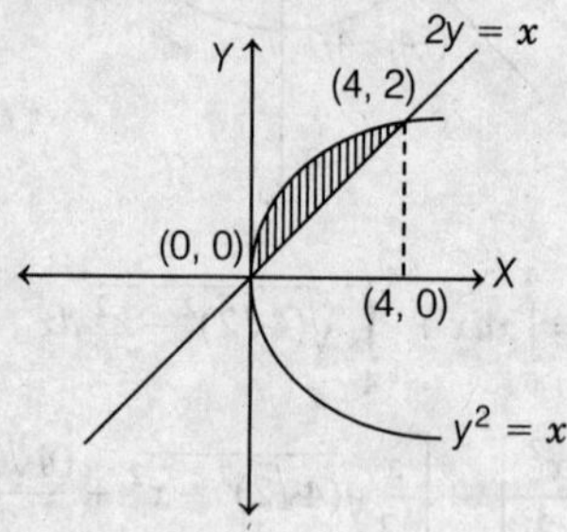

$$\therefore \qquad \left(\frac{x}{2}\right)^2 = x$$

$$\Rightarrow \qquad x^2 = 4x \Rightarrow x(x-4) = 0 \Rightarrow x = 0, 4$$

When, $x = 4 \Rightarrow y = 2$ and when $x = 0 \Rightarrow y = 0$

So, the intersection points are (0, 0) and (4, 2).

Area enclosed by shaded region $= \int_0^4 \left[\sqrt{x} - \frac{x}{2}\right] dx$

$$= \left[\frac{x^{\frac{1}{2}+1}}{\frac{1}{2}+1} - \frac{1}{2}\cdot\frac{x^2}{2}\right]_0^4 = \left[2\cdot\frac{x^{3/2}}{3} - \frac{x^2}{4}\right]_0^4$$

$$= \frac{2}{3}4^{3/2} - \frac{16}{4} - \frac{2}{3}\cdot 0 + \frac{1}{4}\cdot 0$$

$$= \frac{16}{3} - \frac{16}{4} = \frac{64-48}{12} = \frac{16}{12} = \frac{4}{3} \text{ sq. units}$$

Note (*) *Topics related to these question have been removed from latest NCERT.*

Q.30 The area of the region bounded by the curve $y = \sin x$ between the abscissa $x = 0$, $x = \frac{\pi}{2}$ and the X-axis is

(a) 2 sq units (b) 4 sq units (c) 3 sq units (d) 1 sq unit

Ans. ***(d)*** We have to find area of the region bounded by the curve $y = \sin x$ between the abscissa $x = 0$, $x = \frac{\pi}{2}$ and the X-axis *i.e.*, $y = 0$

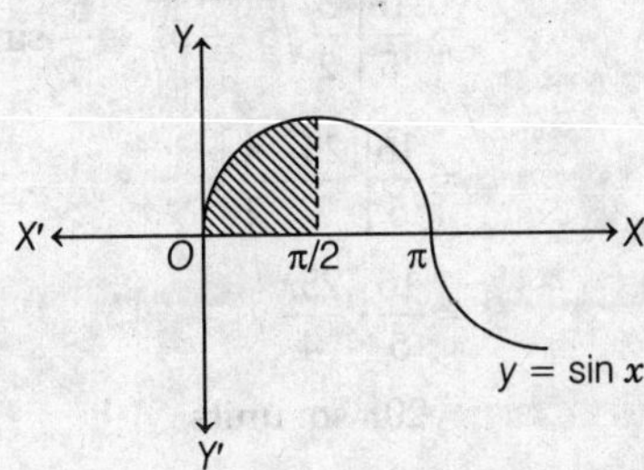

$\therefore$ Required area of shaded region $= \int_0^{\pi/2} \sin x \, dx$ $\quad [\because \int \sin x \, dx = -\cos x]$

$$= -[\cos x]_0^{\pi/2} = -\left[\cos\frac{\pi}{2} - \cos 0\right]$$

$$= -[0 - 1] = 1 \text{ sq. unit}$$

Q.31 The area of the region bounded by the ellipse $\frac{x^2}{25} + \frac{y^2}{16} = 1$ is

(a) 20π sq units (b) $20\pi^2$ sq units (c) $16\pi^2$ sq units (d) 25π sq units

Thinking Process Firstly, find the area in the first quadrant and then multiply by four to get the required result because all four parts are symmetrical.

Ans. ***(a)*** We have, $\frac{x^2}{5^2} + \frac{y^2}{4^2} = 1$

Here, $a = \pm 5$ and $b = \pm 4$

and $\frac{y^2}{4^2} = 1 - \frac{x^2}{5^2}$

$\Rightarrow$ $y^2 = 16\left(1 - \frac{x^2}{25}\right)$

$\Rightarrow$ $y = \sqrt{\frac{16}{25}(25 - x^2)}$

$\Rightarrow$ $y = \frac{4}{5}\sqrt{(5^2 - x^2)}$

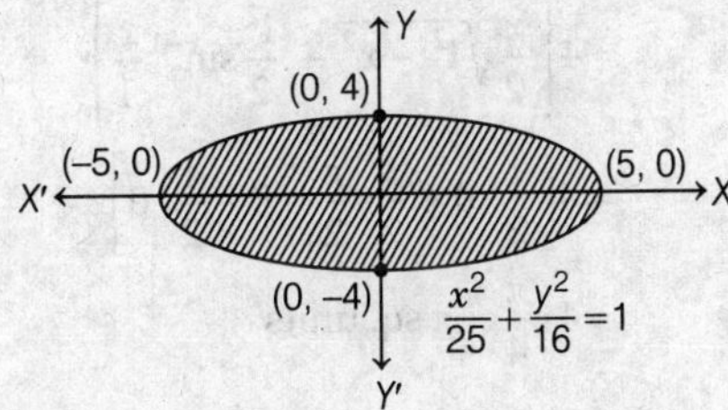

$\therefore$ Area enclosed by ellipse $= 4 \cdot \frac{4}{5}\int_0^5 \sqrt{5^2 - x^2}\,dx$

$$= \frac{16}{5}\left[\frac{x}{2}\sqrt{5^2 - x^2} + \frac{5^2}{2}\sin^{-1}\frac{x}{5}\right]_0^5$$

$$\left[\because \int \sqrt{a^2 - x^2}\,dx = \frac{x}{2}\sqrt{a^2 - x^2} + \frac{a^2}{2}\sin^{-1}\frac{x}{a}\right]$$

$$= \frac{16}{5}\left[\frac{5}{2}\sqrt{5^2 - 5^2} + \frac{5^2}{2}\sin^{-1}\frac{5}{5} - 0 - \frac{25}{2}\cdot 0\right]$$

$$= \frac{16}{5}\left[\frac{25}{2}\cdot\frac{\pi}{2}\right]$$

$$= \frac{16}{5}\cdot\frac{25\pi}{4}$$

$$= 20\pi \text{ sq. units}$$

Q.32 The area of the region bounded by the circle $x^2 + y^2 = 1$ is

(a) 2π sq units (b) π sq units

(c) 3π sq units (d) 4π sq units

Ans. (b) We have, $x^2 + y^2 = 1^2$

$\Rightarrow$ $y^2 = 1 - x^2 \Rightarrow y = \sqrt{1 - x^2}$ $[\because r = \pm 1]$

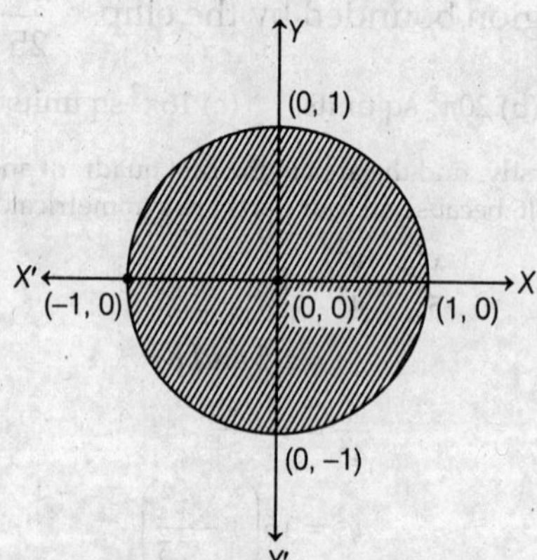

In the given circle, area in each quadrant is same.

Therefore area of circle = 4 × Area in first quadrant

$\therefore$ Area enclosed by circle $= 4\int_0^1 \sqrt{1 - x^2}\,dx = 4\int_0^1 \sqrt{1^2 - x^2}\,dx$

$$= 4\left[\frac{x}{2}\sqrt{1^2 - x^2} + \frac{1^2}{2}\sin^{-1}\frac{x}{1}\right]_0^1$$

$$= 4\left[\frac{1}{2}\cdot 0 + \frac{1}{2}\cdot\frac{\pi}{2} - 0 - \frac{1}{2}\cdot 0\right]$$

$$= 4\cdot\frac{\pi}{4} = \pi \text{ sq. units}$$

Alternate Method

We have, $x^2 + y^2 = 1 \Rightarrow x = \sqrt{1-y^2}$

$\therefore$ Area of circle $= 4\int_0^1 \sqrt{1^2 - y^2}\, dy$

$$= 4\left[\frac{y}{2}\sqrt{1^2 - y^2} + \frac{1^2}{2}\sin^{-1}\frac{y}{1}\right]_0^1$$

$$= 4\left[\frac{1}{2}\sqrt{1^2 - 1^2} + \frac{1}{2}\sin^{-1}\frac{1}{1} - 0\right]$$

$$= 4\left[\frac{1}{2}\sin^{-1}\left(\sin\frac{\pi}{2}\right)\right]$$

$$= 4 \times \frac{\pi}{4} = \pi \text{ sq. units}$$

Q.33 The area of the region bounded by the curve $y = x + 1$ and the lines $x = 2,\ x = 3$, is

(a) $\frac{7}{2}$ sq units (b) $\frac{9}{2}$ sq units (c) $\frac{11}{2}$ sq units (d) $\frac{13}{2}$ sq units

Ans. *(a)* We have the equations $y = x + 1$, $x = 2$ and $x = 3$.

Required area of the shaded region is

$$A = \int_2^3 (x+1)\,dx = \left[\frac{x^2}{2} + x\right]_2^3$$

$$= \left[\frac{9}{2} + 3 - \frac{4}{2} - 2\right] = \left[\frac{5}{2} + 1\right] = \frac{7}{2} \text{ sq. units}$$

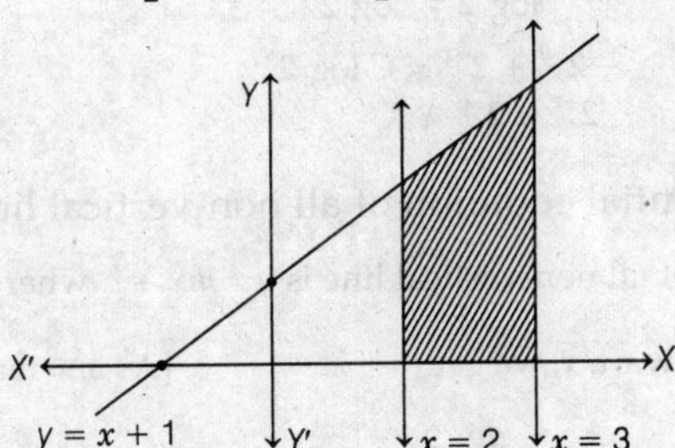

Q.34 The area of the region bounded by the curve $x = 2y + 3$ and the lines $y = 1,\ y = -1$ is

(a) 4 sq units (b) $\frac{3}{2}$ sq units (c) 6 sq units (d) 8 sq units

Ans. *(c)* We have the equations, $x = 2y + 3$, $y = 1$ and $y = -1$.

Required area, $A = \int_{-1}^{1} (2y + 3)\,dy$

$$= \left[\frac{2y^2}{2} + 3y\right]_{-1}^{1}$$

$$= [y^2 + 3y]_{-1}^{1}$$

$$= [1 + 3 - 1 + 3] = 6 \text{ sq. units}$$

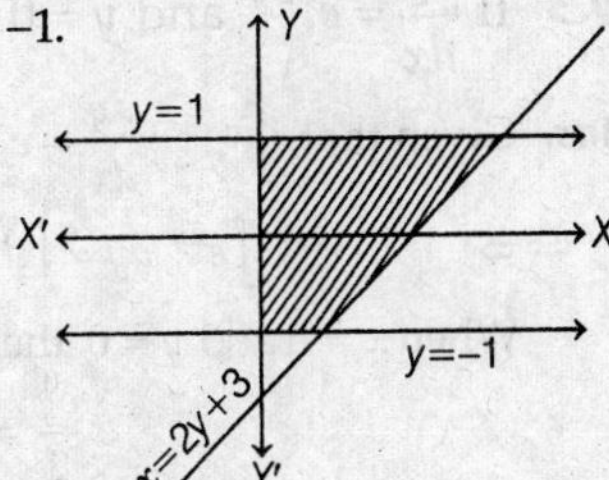

Chapter

09 Differential Equations

Short Answer Type Questions

Q.1 Find the solution of $\frac{dy}{dx} = 2^{y-x}$.

Ans. Given that, $$\frac{dy}{dx} = 2^{y-x}$$

$$\Rightarrow \frac{dy}{dx} = \frac{2^y}{2^x} \qquad \left[\because a^{m-n} = \frac{a^m}{a^n}\right]$$

$$\Rightarrow \frac{dy}{2^y} = \frac{dx}{2^x} \qquad \text{[Applying variable separable method]}$$

On integrationg both sides, we get

$$\int 2^{-y} dy = \int 2^{-x} dx$$

$$\Rightarrow \frac{-2^{-y}}{\log 2} = \frac{-2^{-x}}{\log 2} + C$$

$$\Rightarrow -2^{-y} + 2^{-x} = C \log 2$$

$$\Rightarrow 2^{-x} - 2^{-y} = K \qquad \text{[where, } K = + C \log 2\text{]}$$

***Q.2** Find the differential equation of all non-vertical lines in a plane.

Ans. Since, the family of all non-vertical line is $y = mx + c$, where $m \neq \tan \frac{\pi}{2}$.

On differentiating w.r.t. x, we get $\qquad$ [As $m = \tan \theta$ and $\theta = 90°$ i.e. $\theta = \frac{\pi}{2}$]

$$\frac{dy}{dx} = m$$

Again, differentiating w.r.t. x, we get

$$\frac{d^2y}{dx^2} = 0$$

which is the required differential equation.

Q.3 If $\frac{dy}{dx} = e^{-2y}$ and $y = 0$ when $x = 5$, then find the value of x when $y = 3$.

Ans. Given that, $\frac{dy}{dx} = e^{-2y} \Rightarrow \frac{dy}{e^{-2y}} = dx \qquad$ [applying variable separation]

$$\Rightarrow \int e^{2y}\, dy = \int dx \Rightarrow \frac{e^{2y}}{2} = x + C \qquad \text{... (i)}$$

When $x = 5$ and $y = 0$, then substituting these values in Eq. (i), we get

$$\frac{e^0}{2} = 5 + C$$

Note (*) Topics related to these question have been removed from latest NCERT.

$$\Rightarrow \qquad \frac{1}{2} = 5 + C \Rightarrow C = \frac{1}{2} - 5 = -\frac{9}{2}$$

Eq. (i) becomes $\qquad e^{2y} = 2x - 9$

This is the required equation.

When $y = 3$, then $\qquad e^6 = 2x - 9 \Rightarrow 2x = e^6 + 9$

$$\therefore \qquad x = \frac{(e^6 + 9)}{2} \text{ or } \frac{1}{2}(e^6 + 9)$$

Q.4 Solve $(x^2 - 1)\dfrac{dy}{dx} + 2xy = \dfrac{1}{x^2 - 1}$.

- **Concept Insight** First, write the given differential equation in the form of $\dfrac{dy}{dx} + Py = 0$. Then determine integrating factor by using formula I.F $= e^{\int P dx}$.

Ans. Given differential equation is

$$(x^2 - 1)\frac{dy}{dx} + 2xy = \frac{1}{x^2 - 1}$$

On dividing the above equation by $(x^2 - 1)$, we get

$$\Rightarrow \qquad \frac{dy}{dx} + \left(\frac{2x}{x^2 - 1}\right) y = \frac{1}{(x^2 - 1)^2}$$

which is a linear differential equation.

On comparing it with $\dfrac{dy}{dx} + Py = Q$, we get

$$P = \frac{2x}{x^2 - 1}, \; Q = \frac{1}{(x^2 - 1)^2}$$

$$\text{I.F} = e^{\int P dx} = e^{\int \left(\frac{2x}{x^2 - 1}\right) dx}$$

Put $\qquad x^2 - 1 = t \Rightarrow 2x dx = dt$

$$\therefore \qquad \text{I.F} = e^{\int \frac{dt}{t}} = e^{\log t} = t = (x^2 - 1) \qquad [\because e^{\log x} = x]$$

The complete solution is

$$y \cdot (\text{I.F}) = \int Q \cdot \text{I.F}\, dx + C$$

$$\Rightarrow \qquad y \cdot (x^2 - 1) = \int \frac{1}{(x^2 - 1)^2} \cdot (x^2 - 1)\, dx + C$$

$$\Rightarrow \qquad y \cdot (x^2 - 1) = \int \frac{dx}{(x^2 - 1)} + C$$

$$\Rightarrow \qquad y \cdot (x^2 - 1) = \frac{1}{2} \log \left(\frac{x - 1}{x + 1}\right) + C$$

Q.5 Solve $\dfrac{dy}{dx} + 2xy = y$.

Ans. Given that, $\qquad \dfrac{dy}{dx} + 2xy = y$

$$\Rightarrow \qquad \frac{dy}{dx} + 2xy - y = 0$$

$$\Rightarrow \qquad \frac{dy}{dx} + (2x - 1)\, y = 0$$

which is a linear differential equation.

On comparing it with $\frac{dy}{dx} + Py = Q$, we get

$$P = (2x-1),\ Q = 0$$

$$\text{I.F} = e^{\int P dx} = e^{\int (2x-1)\,dx}$$

$$= e^{\left(\frac{2x^2}{2} - x\right)} = e^{x^2 - x}$$

The general solution is, $y \cdot \text{I.F} = \int Q \cdot \text{I.F} dx + C$

$$y \cdot e^{x^2 - x} = \int Q \cdot e^{x^2 - x}\, dx + C$$

$$\Rightarrow \quad y \cdot e^{x^2 - x} = 0 + C$$

$$\Rightarrow \quad y = C e^{x - x^2}$$

Q.6 Find the general solution of $\frac{dy}{dx} + ay = e^{mx}$.

Ans. Given differential equation is

$$\frac{dy}{dx} + ay = e^{mx}$$

which is a linear differential equation.

On comparing it with $\frac{dy}{dx} + Py = Q$, we get

$$P = a,\ Q = e^{mx}$$

$$\text{I.F} = e^{\int P dx} = e^{\int a dx} = e^{ax}$$

The general solution is $\quad y \cdot e^{ax} = \int e^{mx} \cdot e^{ax}\, dx + C$

$$\Rightarrow \quad y \cdot e^{ax} = \int e^{(m+a)x}\, dx + C$$

$$\Rightarrow \quad y \cdot e^{ax} = \frac{e^{(m+a)x}}{(m+a)} + C$$

$$\Rightarrow \quad (m+a)\,y = \frac{e^{(m+a)x}}{e^{ax}} + \frac{(m+a)\,C}{e^{ax}}$$

$$\Rightarrow \quad (m+a)\,y = e^{mx} + K e^{-ax} \qquad [\because K = (m+a)\,C]$$

Q.7 Solve the differential equation $\frac{dy}{dx} + 1 = e^{x+y}$.

Thinking Process We should substitute $x + y = t$ in this questions.

Ans. Given differential equation is $\frac{dy}{dx} + 1 = e^{x+y}$ ... (i)

Put $x + y = t$; we get

$$1 + \frac{dy}{dx} = \frac{dt}{dx}$$

Eq. (i) becomes $\quad \frac{dt}{dx} = e^t$

$$\Rightarrow \quad \int e^{-t} dt = \int dx \qquad \text{[using variable separable method]}$$

$$\Rightarrow \quad -e^{-t} = x + C$$

$$\Rightarrow \quad \frac{-1}{e^{x+y}} = x + C$$

$$\Rightarrow \quad -1 = (x + C)\, e^{x+y}$$

$$\Rightarrow \quad (x + C)\, e^{x+y} + 1 = 0$$

Q.8 Solve $ydx - xdy = x^2ydx$.

Thinking Process First, divide the given differential equation by x^2ydx to convert it into the form of linear differential equation and then solve it.

Ans. Given that,
$$ydx - xdy = x^2ydx$$
$$\Rightarrow \quad \frac{1}{x^2} - \frac{1}{xy}\cdot\frac{dy}{dx} = 1 \quad \text{[dividing throughout by } x^2ydx]$$
$$\Rightarrow \quad -\frac{1}{xy}\cdot\frac{dy}{dx} + \frac{1}{x^2} - 1 = 0$$
$$\Rightarrow \quad \frac{dy}{dx} - \frac{xy}{x^2} + xy = 0$$
$$\Rightarrow \quad \frac{dy}{dx} - \frac{y}{x} + xy = 0$$
$$\Rightarrow \quad \frac{dy}{dx} + \left(x - \frac{1}{x}\right)y = 0$$
which is a linear differential equation.

On comparing it with $\frac{dy}{dx} + Py = Q$, we get
$$P = \left(x - \frac{1}{x}\right), Q = 0$$
$$\text{I.F} = e^{\int Pdx}$$
$$= e^{\int\left(x - \frac{1}{x}\right)dx}$$
$$= e^{\frac{x^2}{2} - \log x}$$
$$= e^{\frac{x^2}{2}} \cdot e^{-\log x} \quad \left[\because -\log x = \log\left(\frac{1}{x}\right)\right]$$
$$\Rightarrow \quad = e^{\frac{x^2}{2}} \cdot e^{\log\left(\frac{1}{x}\right)}$$
$$= \frac{1}{x} e^{\frac{x^2}{2}} \quad \left[\because e^{\log\left(\frac{1}{x}\right)} = \frac{1}{x} \text{ as } e^{\log x} = x\right]$$
The general solution is y. I.F $= \int Q\cdot \text{I.F}dx \times C$
$$y\cdot\frac{1}{x}e^{x^2/2} = \int 0\cdot\frac{1}{x}e^{x^2/2}\,dx + C$$
$$\Rightarrow \quad y\cdot\frac{1}{x}e^{x^2/2} = C$$
$$\Rightarrow \quad y = C\,x\,e^{-x^2/2}$$

Q.9 Solve the differential equation $\frac{dy}{dx} = 1 + x + y^2 + xy^2$, when $y = 0$ and $x = 0$.

Ans. Given that,
$$\frac{dy}{dx} = 1 + x + y^2 + xy^2$$
$$\Rightarrow \quad \frac{dy}{dx} = (1 + x) + y^2(1 + x)$$
$$\Rightarrow \quad \frac{dy}{dx} = (1 + y^2)(1 + x)$$

$$\Rightarrow \qquad \frac{dy}{1+y^2} = (1+x)\,dx$$

[applying variable separable method]

On integrating both sides, we get

$$\tan^{-1} y = x + \frac{x^2}{2} + C \qquad \text{... (i)}$$

When $y = 0$ and $x = 0$, then substituting these values in Eq. (i), we get

$$\tan^{-1}(0) = 0 + 0 + C$$

$$\Rightarrow \qquad C = 0$$

$$\Rightarrow \qquad \tan^{-1} y = x + \frac{x^2}{2}$$

$$\Rightarrow \qquad y = \tan\left(x + \frac{x^2}{2}\right)$$

Mistake Alert ! *Do not forget to find the value of integral constant in cases of particular solution of differential equations.*

Q.10 Find the general solution of $(x + 2y^3)\dfrac{dy}{dx} = y.$

Ans. Given that, $\qquad (x + 2y^3)\dfrac{dy}{dx} = y$

$$\Rightarrow \qquad y \cdot \frac{dx}{dy} = x + 2y^3$$

$$\Rightarrow \qquad \frac{dx}{dy} = \frac{x}{y} + 2y^2 \qquad \text{[dividing throughout by } y\text{]}$$

$$\Rightarrow \qquad \frac{dx}{dy} - \frac{x}{y} = 2y^2$$

which is a linear differential equation.

On comparing it with $\dfrac{dx}{dy} + Px = Q$, we get

$$P = -\frac{1}{y},\ Q = 2y^2$$

$$\text{I.F} = e^{\int -\frac{1}{y}\,dy} = e^{-\int \frac{1}{y}\,dy}$$

$$\therefore \qquad = e^{-\log y} = e^{\log\left(\frac{1}{y}\right)} = \frac{1}{y} \qquad [\because e^{\log x} = x]$$

The general solution is, $\quad x \cdot \text{I.F} = \int Q \cdot \text{I.F}\,dy + C$

$$x \cdot \frac{1}{y} = \int 2y^2 \cdot \frac{1}{y}\,dy + C$$

$$\Rightarrow \qquad \frac{x}{y} = \frac{2y^2}{2} + C$$

$$\Rightarrow \qquad \frac{x}{y} = y^2 + C$$

$$\Rightarrow \qquad x = y^3 + Cy$$

Q.11 If $y(x)$ is a solution of $\left(\frac{2+\sin x}{1+y}\right)\frac{dy}{dx} = -\cos x$ and $y(0) = 1$, then find the value of $y\left(\frac{\pi}{2}\right)$.

Ans. Given that,
$$\left(\frac{2+\sin x}{1+y}\right)\frac{dy}{dx} = -\cos x$$
$$\Rightarrow \quad \frac{dy}{1+y} = -\frac{\cos x}{2+\sin x}\,dx$$
[applying variable separable method]

On integrating both sides, we get
$$\int \frac{1}{1+y}\,dy = -\int \frac{\cos x}{2+\sin x}\,dx$$
$$\Rightarrow \quad \log(1+y) = -\log(2+\sin x) + \log C$$
[As the equation having log in both sides then constant 'C' can be used as $\log C$]
$$\Rightarrow \quad \log(1+y) + \log(2+\sin x) = \log C$$
$$\Rightarrow \quad \log(1+y)(2+\sin x) = \log C \qquad [\because \log x + \log y = \log(xy)]$$
$$\Rightarrow \quad (1+y)(2+\sin x) = C$$
$$\Rightarrow \quad 1+y = \frac{C}{2+\sin x}$$
$$\Rightarrow \quad y = \frac{C}{2+\sin x} - 1 \qquad \text{... (i)}$$

When $x = 0$ and $y = 1$, then
$$1 = \frac{C}{2} - 1 \qquad [\because \sin 0 = 0]$$
$$\Rightarrow \quad C = 4$$
On putting $C = 4$ in Eq. (i), we get
$$y = \frac{4}{2+\sin x} - 1$$
$$\therefore \quad y\left(\frac{\pi}{2}\right) = \frac{4}{2+\sin\frac{\pi}{2}} - 1 = \frac{4}{2+1} - 1$$
$$= \frac{4}{3} - 1 = \frac{1}{3}$$

Q.12 If $y(t)$ is a solution of $(1+t)\frac{dy}{dt} - ty = 1$ and $y(0) = -1$, then show that $y(1) = -\frac{1}{2}$.

Ans. Given that,
$$(1+t)\frac{dy}{dt} - ty = 1$$
$$\Rightarrow \quad \frac{dy}{dt} - \left(\frac{t}{1+t}\right)y = \frac{1}{1+t}$$
which is a linear differential equation.

On comparing it with $\frac{dy}{dt} + Py = Q$, we get

$$P = -\left(\frac{t}{1+t}\right),\ Q = \frac{1}{1+t}$$

$$\text{I.F} = e^{-\int \frac{t}{1+t}dt} = e^{-\int\left(1-\frac{1}{1+t}\right)dt} = e^{-[t-\log(1+t)]} \qquad [\text{I.F} = e^{\int P\cdot dt}]$$

$$= e^{-t}\cdot e^{\log(1+t)}$$

$$= e^{-t}(1+t) \qquad [\because e^{\log(t+1)} = (1+t)]$$

The general solution is

$$y(t)\cdot\frac{(1+t)}{e^t} = \int\frac{(1+t)\cdot e^{-t}}{(1+t)}dt + C$$

$$\Rightarrow \quad y(t) = \frac{e^{-t}}{(-1)}\cdot\frac{e^t}{1+t} + C', \text{ where } C' = \frac{Ce^t}{1+t}$$

$$\Rightarrow \quad y(t) = -\frac{1}{1+t} + C'$$

When $t = 0$ and $y = -1$, then

$$-1 = -1 + C' \Rightarrow C' = 0$$

$$y(t) = -\frac{1}{1+t} \Rightarrow y(1) = -\frac{1}{2}$$

Mistake Alert ! *Be careful in finding the general solution and particular solution of differential equation. General solution contains an arbitary constant whereas partiular solution contains particular value in place of arbitrary constant.*

Q.13 Form the differential equation having $y = (\sin^{-1}x)^2 + A\cos^{-1}x + B$, where A and B are arbitrary constants, as its general solution.

Thinking Process As we are given two arbitrary constants (A and B) in the differential equation. so, we have to differentiate two times the given differential equation to eliminate the arbitrary constant A and B.

Ans. Given that, $y = (\sin^{-1}x)^2 + A\cos^{-1}x + B$

On differentiating w.r.t. x, we get

$$\frac{dy}{dx} = \frac{2\sin^{-1}x}{\sqrt{1-x^2}} + \frac{(-A)}{\sqrt{1-x^2}}$$

$$\Rightarrow \quad \sqrt{1-x^2}\frac{dy}{dx} = 2\sin^{-1}x - A$$

Again, differentiating w.r.t. x, we get

$$\sqrt{1-x^2}\frac{d^2y}{dx^2} + \frac{dy}{dx}\cdot\frac{-2x}{2\sqrt{1+x^2}} = \frac{2}{\sqrt{1-x^2}}$$

$$\Rightarrow \quad (1-x^2)\frac{d^2y}{dx^2} - \frac{x}{\sqrt{1-x^2}}\cdot\sqrt{1-x^2}\frac{dy}{dx} = 2$$

$$\Rightarrow \quad (1-x^2)\frac{d^2y}{dx^2} - x\frac{dy}{dx} = 2$$

$$\Rightarrow \quad (1-x^2)\frac{d^2y}{dx^2} - x\frac{dy}{dx} - 2 = 0$$

which is the required differential equation.

***Q.14** Form the differential equation of all circles which pass through origin and whose centres lie on Y-axis.

Ans. It is given that, circles pass through origin and their centres lie on Y-axis. Let $(0, k)$ be the centre of the circle and radius is k.

[Equation of circle is $(x-h)^2+(y-k)=r^2$ when (h, k) is centre and r is radius]

So, the equation of circle is

$$(x-0)^2+(y-k)^2=k^2$$

$$\Rightarrow \quad x^2+(y-k)^2=k^2$$

$$\Rightarrow \quad x^2+y^2-2ky=0$$

$$\Rightarrow \quad \frac{x^2+y^2}{2y}=k \quad \ldots \text{(i)}$$

On differentiating Eq. (i) w.r.t. x, we get

$$\frac{2y\left(2x+2y\dfrac{dy}{dx}\right)-(x^2+y^2)\dfrac{2dy}{dx}}{4y^2}=0$$

$$\left[\text{use quotient rule, } \frac{d}{dx}\frac{u}{v}=\frac{v\cdot\dfrac{du}{dx}-u\dfrac{dv}{dx}}{v^2}\right]$$

$$\Rightarrow \quad 4y\left(x+y\frac{dy}{dx}\right)-2(x^2+y^2)\frac{dy}{dx}=0$$

$$\Rightarrow \quad 4xy+4y^2\frac{dy}{dx}-2(x^2+y^2)\frac{dy}{dx}=0$$

$$\Rightarrow \quad [4y^2-2(x^2+y^2)]\frac{dy}{dx}+4xy=0$$

$$\Rightarrow \quad (4y^2-2x^2-2y^2)\frac{dy}{dx}+4xy=0$$

$$\Rightarrow \quad (2y^2-2x^2)\frac{dy}{dx}+4xy=0$$

$$\Rightarrow \quad (y^2-x^2)\frac{dy}{dx}+2xy=0$$

$$\Rightarrow \quad (x^2-y^2)\frac{dy}{dx}-2xy=0$$

Q.15 Find the equation of a curve passing through origin and satisfying the differential equation $(1+x^2)\dfrac{dy}{dx}+2xy=4x^2$.

Ans. Given that, $$(1+x^2)\frac{dy}{dx}+2xy=4x^2$$

$$\Rightarrow \quad \frac{dy}{dx}+\frac{2x}{1+x^2}\cdot y=\frac{4x^2}{1+x^2}$$

which is a linear differential equation.

On comparing it with $\dfrac{dy}{dx}+Py=Q$, we get

$$P=\frac{2x}{1+x^2}, Q=\frac{4x^2}{1+x^2}$$

$$\therefore \quad \text{I.F}=e^{\int Pdx}=e^{\int \frac{2x}{1+x^2}dx}$$

Note (✱) Topics related to these question have been removed from latest NCERT.

Put $1+x^2=t \Rightarrow 2xdx=dt$

$$\text{I.F} = e^{\int \frac{dt}{t}} = e^{\log t} = e^{\log(1+x^2)} = 1+x^2 \qquad [\because e^{\log x}=x]$$

The general solution is

$$y\cdot(1+x^2) = \int \frac{4x^2}{1+x^2}(1+x^2)\,dx + C \quad [y\times\text{I.F} = \int (Q\times\text{I.F})dx + C]$$

$$\Rightarrow \quad y\cdot(1+x^2) = \int 4x^2\,dx + C$$

$$\Rightarrow \quad y\cdot(1+x^2) = 4\frac{x^3}{3} + C \qquad \text{... (i)}$$

Since, the curve passes through origin, then substituting

$x=0$ and $y=0$ in Eq. (i), we get

$C=0$

The required equation of curve is

$$y(1+x^2) = \frac{4x^3}{3}$$

$$\Rightarrow \quad y = \frac{4x^3}{3(1+x^2)}$$

Q.16 Solve $x^2\dfrac{dy}{dx} = x^2+xy+y^2$.

Thinking Process First, consider, $\frac{dy}{dx}$ as equal to $f(x, y)$ then, replace x by λx and y by λy on both sides, we get $F(x, y) = \lambda^0(F(x, y)$.

Now, put $y = vx$ and convert the given equation in terms of v and x, then separate the variables and integrate it.

Further put $v = \frac{y}{x}$ and simplify it to get the result.

Ans. Given that,

$$x^2\frac{dy}{dx} = x^2+xy+y^2$$

$$\Rightarrow \quad \frac{dy}{dx} = 1+\frac{y}{x}+\frac{y^2}{x^2} \qquad \text{... (i)}$$

Let

$$f(x, y) = 1+\frac{y}{x}+\frac{y^2}{x^2}$$

$$f(\lambda x, \lambda y) = 1+\frac{\lambda y}{\lambda x}+\frac{\lambda^2 y^2}{\lambda^2 x^2}$$

$$f(\lambda x, \lambda y) = \lambda^0\left(1+\frac{y}{x}+\frac{y^2}{x^2}\right)$$

$$= \lambda^0 f(x, y)$$

$$\frac{dy}{dx} = \frac{x^2+y+y^2}{x^2} \qquad \text{...(ii)}$$

It is as homogenous equation. So put $y = vx$

$$\frac{dy}{dx} = v + x\frac{dv}{dx}$$

i.e.,

$$v + x\frac{dv}{dx} = \frac{x^2+x\cdot vx+v^2x^2}{x^2}$$

$$v + x\frac{dv}{dx} = 1 + v + v^2$$

which is homogeneous expression of degree 0.

$$\left(v + x\frac{dv}{dx}\right) = 1 + v + v^2$$

$$\Rightarrow \quad x\frac{dv}{dx} = 1 + v + v^2 - v$$

$$\Rightarrow \quad x\frac{dv}{dx} = 1 + v^2 \quad \text{[applying variable separable method]}$$

$$\Rightarrow \quad \frac{dv}{1+v^2} = \frac{dx}{x}$$

On integrating both sides, we get

$$\tan^{-1} v = \log|x| + C \qquad \left[\because y = vx \Rightarrow v = \frac{y}{x}\right]$$

$$\Rightarrow \quad \tan^{-1}\left(\frac{y}{x}\right) = \log|x| + C$$

Q.17 Find the general solution of the differential equation $(1+y^2) + (x - e^{\tan^{-1} y})\dfrac{dy}{dx} = 0.$

Ans. Given, differential equation is

$$(1+y^2) + (x - e^{\tan^{-1} y})\frac{dy}{dx} = 0$$

$$\Rightarrow \quad (1+y^2) = -(x - e^{\tan^{-1} y})\frac{dy}{dx}$$

$$(1+y^2)\frac{dx}{dy} = -x + e^{\tan^{-1} y}$$

$$\Rightarrow \quad (1+y^2)\frac{dx}{dy} + x = e^{\tan^{-1} y}$$

$$\Rightarrow \quad \frac{dx}{dy} + \frac{x}{1+y^2} = \frac{e^{\tan^{-1} y}}{1+y^2} \qquad \text{[dividing throughout by } (1+y^2)\text{]}$$

which is a linear differential equation.

On comparing it with $\dfrac{dx}{dy} + Px = Q$, we get

$$P = \frac{1}{1+y^2},\ Q = \frac{e^{\tan^{-1} y}}{1+y^2}$$

$$\text{I.F} = e^{\int P dy} = e^{\int \frac{1}{1+y^2} dy} = e^{\tan^{-1} y}$$

The general solution is $x \cdot e^{\tan^{-1} y} = \displaystyle\int \frac{e^{\tan^{-1} y}}{1+y^2} \cdot e^{\tan^{-1} y} dy + C$ $\quad [\because x \times \text{I.F} = \int (Q \times \text{I.F}) dy + C]$

$$\Rightarrow \quad x \cdot e^{\tan^{-1} y} = \int \frac{(e^{\tan^{-1} y})^2}{1+y^2} \cdot dy + C$$

Put $\quad \tan^{-1} y = t \Rightarrow \dfrac{1}{1+y^2} dy = dt$

$$\therefore \quad x \cdot e^{\tan^{-1} y} = \int e^{2t}\, dt + C$$

$$\Rightarrow \quad x \cdot e^{\tan\ + y} = \frac{1}{2} e^{2t} + C$$

$$\Rightarrow \quad 2x\, e^{\tan^{-1} y} = e^{2 \tan^{-1} y} + 2\, C$$

$$\Rightarrow \quad 2x\, e^{\tan^{-1} y} = e^{2 \tan^{-1} y} + K \qquad [\because K = 2C]$$

Q.18 Find the general solution of $y^2 dx + (x^2 - x\,y + y^2)\, dy = 0$.

Ans. Given, differential equation is

$$y^2 dx + (x^2 - xy + y^2)\, dy = 0$$

$$\Rightarrow \quad y^2 dx = -(x^2 - xy + y^2)\, dy$$

$$\Rightarrow \quad y^2 \frac{dx}{dy} = -(x^2 - xy + y^2)$$

$$\Rightarrow \quad \frac{dx}{dy} = -\left(\frac{x^2}{y^2} - \frac{x}{y} + 1\right) \qquad \text{... (i)}$$

which is a homogeneous differential equation.

[As the equation is $\frac{dx}{dy}$ from then we put $x = vy$]

Put $\quad \frac{x}{y} = v$ or $x = vy$

$$\Rightarrow \quad \frac{dx}{dy} = v + y\frac{dv}{dy}$$

On substituting these values in Eq. (i), we get

$$v + y\frac{dv}{dy} = -[v^2 - v + 1]$$

$$\Rightarrow \quad y\frac{dv}{dy} = -v^2 + v - 1 - v$$

$$\Rightarrow \quad y\frac{dv}{dy} = -v^2 - 1 \Rightarrow \frac{dv}{v^2 + 1} = -\frac{dy}{y}$$

[applying variable separable method]

On integrating both sides, we get

$$\tan^{-1}(v) = -\log y + C$$

$$\Rightarrow \quad \tan^{-1}\left(\frac{x}{y}\right) + \log y = C \qquad \left[\because v = \frac{x}{y}\right]$$

Q.19 Solve $(x + y)(dx - dy) = dx + dy$.

Ans. Given differential equation is

$$(x + y)(dx - dy) = dx + dy$$

On rearranging the equation, get

$$\Rightarrow \quad (x + y)\left(1 - \frac{dy}{dx}\right) = 1 + \frac{dy}{dx} \qquad \text{...(i)}$$

Put $\quad x + y = z$

On differentiating w.r.t to x, we get

$\Rightarrow$ $$1+\frac{dy}{dx}=\frac{dz}{dx}$$

On substituting these values in Eq. (i), we get

$$z\left(1-\frac{dz}{dx}+1\right)=\frac{dz}{dx} \qquad \left[\because \frac{dy}{dx}=\frac{dz}{dx}-1\right]$$

$\Rightarrow$ $$z\left(2-\frac{dz}{dx}\right)=\frac{dz}{dx}$$

$\Rightarrow$ $$2z-z\frac{dz}{dx}-\frac{dz}{dx}=0$$

$\Rightarrow$ $$2z-(z+1)\frac{dz}{dx}=0$$

$\Rightarrow$ $$\frac{dz}{dx}=\frac{2z}{z+1}$$

$\Rightarrow$ $$\left(\frac{z+1}{z}\right)dz=2\,dx$$

[applying variable separable method]

On integrating both sides, we get

$$\int\left(1+\frac{1}{z}\right)dz=2\int dx$$

$\Rightarrow$ $$z+\log z=2x-\log C$$

$\Rightarrow$ $$(x+y)+\log(x+y)=2x-\log C \qquad [\because z=x+y]$$

$\Rightarrow$ $$2x-x-y=\log C+\log(x+y)$$

$\Rightarrow$ $$x-y=\log|C(x+y)|$$

$\Rightarrow$ $$e^{x-y}=C(x+y)$$

$\Rightarrow$ $$(x+y)=\frac{1}{C}e^{x-y}$$

$\Rightarrow$ $$x+y=Ke^{x-y} \qquad \left[\because K=\frac{1}{C}\right]$$

Q.20 Solve $2(y+3)-xy\frac{dy}{dx}=0$, given that $y(1)=-2$.

Ans. Given that, $$2(y+3)-xy\frac{dy}{dx}=0$$

$\Rightarrow$ $$2(y+3)=xy\frac{dy}{dx}$$

$\Rightarrow$ $$2\frac{dx}{x}=\left(\frac{y}{y+3}\right)dy \Rightarrow 2\cdot\frac{dx}{x}=\left(\frac{y+3-3}{y+3}\right)dy$$

$\Rightarrow$ $$2\cdot\frac{dx}{x}=\left(1-\frac{3}{y+3}\right)dy$$

[applying variable separable method]

On integrating both sides, we get

$$2\log x=y-3\log(y+3)+C \qquad \text{... (i)}$$

When $x=1$ and $y=-2$, then

$$2\log 1=-2-3\log(-2+3)+C$$

$$\Rightarrow \quad 2 \cdot 0 = -2 - 3 \cdot 0 + C \quad [\because \log 1 = 0]$$

$$\Rightarrow \quad C = 2$$

On substituting the value of C in Eq. (i), we get

$$2 \log x = y - 3 \log (y + 3) + 2$$

$$\Rightarrow \quad 2 \log x + 3 \log (y + 3) = y + 2$$

$$\Rightarrow \quad \log x^2 + \log (y + 3)^3 = (y + 2) \quad [\because n \log m = \log m^n]$$

$$\Rightarrow \quad \log x^2 (y + 3)^3 = y + 2 \quad [\because \log x + \log y = \log (xy)]$$

$$\Rightarrow \quad x^2 (y + 3)^3 = e^{y+2}$$

Q.21 Solve the differential equation $dy = \cos x (2 - y \operatorname{cosec} x)\, dx$ given that $y = 2$, when $x = \dfrac{\pi}{2}$.

Ans. Given differential equation,

$$dy = \cos x (2 - y \operatorname{cosec} x) dx$$

$$\Rightarrow \quad \frac{dy}{dx} = \cos x (2 - y \operatorname{cosec} x)$$

$$\Rightarrow \quad \frac{dy}{dx} = 2 \cos x - y \operatorname{cosec} x \cdot \cos x$$

$$\Rightarrow \quad \frac{dy}{dx} = 2 \cos x - y \cot x$$

$$\Rightarrow \quad \frac{dy}{dx} + y \cot x = 2 \cos x$$

which is a linear differential equation.

On comparing it with $\dfrac{dy}{dx} + Py = Q$, we get

$$P = \cot x,\ Q = 2 \cos x$$

$$\text{I.F} = e^{\int P dx} = e^{\int \cot x\, dx} = e^{\log \sin x} = \sin x \quad [\because e^{\log x} = x]$$

The general solution is

$$y \cdot \sin x = \int 2 \cos x \cdot \sin x\, dx + C \quad \left[y \times \text{I.F} = \int (Q \cdot \text{I.F}) dx + C\right]$$

$$\Rightarrow \quad y \cdot \sin x = \int \sin 2x\, dx + C \quad [\because \sin 2x = 2 \sin x \cos x]$$

$$\Rightarrow \quad y \cdot \sin x = -\frac{\cos 2x}{2} + C \quad \text{... (i)}$$

When $x = \dfrac{\pi}{2}$ and $y = 2$, then

On putting the value of x and y in Eq. (i), we get

$$2 . \sin \frac{\pi}{2} = -\frac{\cos \left(2 \times \frac{\pi}{2}\right)}{2} + C$$

$$\Rightarrow \quad 2 \cdot 1 = +\frac{1}{2} + C$$

$$\Rightarrow \quad 2 - \frac{1}{2} = C \Rightarrow \frac{4 - 1}{2} = C$$

$$\Rightarrow \therefore \quad C = \frac{3}{2}$$

On substituting the value of C in Eq. (i), we get

$$y \sin x = -\frac{1}{2} \cos 2x + \frac{3}{2}$$

***Q.22** Form the differential equation by eliminating A and B in $Ax^2 + By^2 = 1$.

Ans. Given differential equation is $Ax^2 + By^2 = 1$

On differentiating both sides w.r.t. x, we get

$$2Ax + 2By\frac{dy}{dx} = 0$$

$$\Rightarrow \quad 2By\frac{dy}{dx} = -2Ax$$

$$\Rightarrow \quad By\frac{dy}{dx} = -Ax \Rightarrow \frac{y}{x}.\frac{dy}{dx} = -\frac{A}{B}$$

Again, differentiating both sides w.r.t. x, we get

$$\frac{y}{x}\cdot\frac{d^2y}{dx^2} + \frac{dy}{dx}\cdot\left(\frac{x\frac{dy}{dx} - y}{x^2}\right) = 0$$

$$\Rightarrow \quad \frac{y}{x}\cdot\frac{d^2y}{dx^2} + \frac{x\left(\frac{dy}{dx}\right)^2 - y\left(\frac{dy}{dx}\right)}{x^2} = 0$$

$$\Rightarrow \quad xy\frac{d^2y}{dx^2} + x\left(\frac{dy}{dx}\right)^2 - y\left(\frac{dy}{dx}\right) = 0$$

$$\Rightarrow \quad xy\,y'' + x\,(y')^2 - y\,y' = 0$$

Hence, the required differential equation is $xy\cdot y'' + x(y')^2 - yy' = 0$

Q.23 Solve the differential equation $(1 + y^2)\tan^{-1} x\,dx + 2y(1 + x^2)\,dy = 0$.

Thinking Process First, separate the variables, then integrate by using integration by parts.

Ans. Given differential equation is

$$(1 + y^2)\tan^{-1} x dx + 2y(1 + x^2)\,dy = 0$$

$$\Rightarrow \quad (1 + y^2)\tan^{-1} x dx = -2y(1 + x^2)\,dy$$

$$\Rightarrow \quad \frac{\tan^{-1} x\,dx}{1 + x^2} = -\frac{2y}{1 + y^2}dy$$

[applying variable separable]

On integrating both sides, we get

$$\int\frac{\tan^{-1} x}{1 + x^2}dx = -\int\frac{2y}{1 + y^2}dy$$

Put $\tan^{-1} x = t$ in LHS, we get

$$\frac{1}{1 + x^2}dx = dt$$

and put $1 + y^2 = u$ in RHS, we get

$$2\,y\,dy = du$$

$$\Rightarrow \quad \int t\,dt = -\int\frac{1}{u}du \Rightarrow \frac{t^2}{2} = -\log u + C$$

$$\Rightarrow \quad \frac{1}{2}(\tan^{-1} x)^2 = -\log(1 + y^2) + C$$

$$\Rightarrow \quad \frac{1}{2}(\tan^{-1} x)^2 + \log(1 + y^2) = C$$

Note (✱) Topics related to these question have been removed from latest NCERT.

***Q.24** Find the differential equation of system of concentric circles with centre (1, 2).

Thinking Process Use the equation of circle $(x-h)^2+(y-k)^2=r^2$ where (h, k) is centre and r is radius.

Ans. The family of concentric circles with centre (1, 2) and radius a is given by

$$(x-1)^2+(y-2)^2=a^2$$

$$\Rightarrow \quad x^2+1-2x+y^2+4-4y=a^2$$

$$\Rightarrow \quad x^2+y^2-2x-4y+5=a^2 \qquad \text{...(i)}$$

On differentiating Eq. (i) w.r.t. x, we get

$$2x+2y\frac{dy}{dx}-2-4\frac{dy}{dx}=0$$

$$\Rightarrow \quad (2y-4)\frac{dy}{dx}+2x-2=0$$

$$\Rightarrow \quad (y-2)\frac{dy}{dx}+(x-1)=0$$

which is the required equation.

Long Answer Type Questions

Q.25 Solve $y+\frac{d}{dx}(xy)=x(\sin x+\log x)$.

Ans. Given differential equation is

$$y+\frac{d}{dx}(xy)=x(\sin x+\log x)$$

$$\Rightarrow \quad y+x\frac{dy}{dx}+y=x(\sin x+\log x) \qquad \text{[applying product rule]}$$

$$\Rightarrow \quad x\frac{dy}{dx}+2y=x(\sin x+\log x)$$

$$\Rightarrow \quad \frac{dy}{dx}+\frac{2}{x}y=\sin x+\log x$$

which is a linear differential equation.

On comparing it with $\frac{dy}{dx}+Py=Q$, we get

$$P=\frac{2}{x},\ Q=\sin x+\log x$$

$$\text{I.F}=e^{\int\frac{2}{x}dx}=e^{2\log x}=x^2$$

The general solution is $y\cdot\text{I.F}=\int Q\cdot\text{IF}dx+C$

$$y\cdot x^2=\int(\sin x+\log x)\,x^2dx+C$$

$$\Rightarrow \quad y\cdot x^2=\int(x^2\sin x+x^2\log x)\,dx+C$$

$$\Rightarrow \quad y\cdot x^2=\int x^2\sin x\,dx+\int x^2\log x\,dx+C$$

$$\Rightarrow \quad y\cdot x^2=I_1+I_2+C \qquad \text{...(i)}$$

Integrate the function in step by step manner

Now, $\quad I_1=\int x^2\sin x\,dx$

Note (✱) Topics related to these question have been removed from latest NCERT.

$$= x^2(-\cos x) + \int 2x \cos x \, dx$$
$$= -x^2 \cos x + [2x(\sin x) - \int 2 \sin x \, dx]$$
$$I_1 = -x^2 \cos x + 2x \sin x + 2\cos x \quad \text{... (ii)}$$

and
$$I_2 = \int x^2 \log x \, dx$$
$$= \log x \cdot \frac{x^3}{3} - \int \frac{1}{x} \cdot \frac{x^3}{3} dx$$
$$= \log x \cdot \frac{x^3}{3} - \frac{1}{3}\int x^2 dx$$
$$= \log x \cdot \frac{x^3}{3} - \frac{1}{3} \cdot \frac{x^3}{3} \quad \text{... (iii)}$$

On substituting the value of I_1 and I_2 in Eq. (i), we get

$$y \cdot x^2 = -x^2 \cos x + 2x \sin x + 2\cos x + \frac{x^3}{3} \log x - \frac{1}{9}x^3 + C$$

$$\therefore \quad y = -\cos x + \frac{2\sin x}{x} + \frac{2\cos x}{x^2} + \frac{x}{3} \log x - \frac{x}{9} + Cx^{-2}$$

[dividing by x^2 on both sides]

Q.26 Find the general solution of $(1 + \tan y)(dx - dy) + 2x\,dy = 0$.

Ans. Given differential equation is $(1 + \tan y)(dx - dy) + 2x\,dy = 0$

On dividing throughout by dy, we get

$$(1 + \tan y)\left(\frac{dx}{dy} - 1\right) + 2x = 0$$

$$\Rightarrow \quad (1 + \tan y)\frac{dx}{dy} - (1 + \tan y) + 2x = 0$$

$$\Rightarrow \quad (1 + \tan y)\frac{dx}{dy} + 2x = (1 + \tan y)$$

$$\Rightarrow \quad \frac{dx}{dy} + \frac{2x}{1 + \tan y} = 1$$

which is a linear differential equation.

On comparing it with $\frac{dx}{dy} + Px = Q$, we get

$$P = \frac{2}{1 + \tan y}, \; Q = 1$$

$$\text{I.F} = e^{\int \frac{2}{1+\tan y} dy} = e^{\int \frac{2\cos y}{\cos y + \sin y} dy} \quad [\because \text{I.F} = e^{\int P dy}]$$

$$= e^{\int \frac{\cos y + \sin y + \cos y - \sin y}{\cos y + \sin y} dy}$$

$$= e^{\int \left(1 + \frac{\cos y - \sin y}{\cos y + \sin y}\right) dy} = e^{y + \log(\cos y + \sin y)}$$

$$= e^y \cdot e^{\log(\cos y + \sin y)}$$

$$= e^y \cdot (\cos y + \sin y) \quad [\because e^{\log x} = x]$$

The general solution is $y \cdot \text{I.F} = \int Q \cdot \text{I.F} dx + C$

$$x \cdot e^y(\cos y + \sin y) = \int 1 \cdot e^y(\cos y + \sin y)\, dy + C$$

$$\Rightarrow \quad x \cdot e^y (\cos y + \sin y) = \int e^y (\sin y + \cos y)\, dy + C$$

$$\Rightarrow \quad x \cdot e^y (\cos y + \sin y) = e^y \sin y + C \quad \left[\because \int e^x \{f(x) + f'(x)\}\, dx = e^x f(x) + C\right]$$

$$\Rightarrow \quad x (\sin y + \cos y) = \sin y + Ce^{-y}$$

Q.27 Solve $\dfrac{dy}{dx} = \cos(x+y) + \sin(x+y)$.

Ans. Given, $\dfrac{dy}{dx} = \cos(x+y) + \sin(x+y)$ ...(i)

Put $x + y = z \Rightarrow 1 + \dfrac{dy}{dx} = \dfrac{dz}{dx}$

On substituting these values in Eq. (i), we get

$$\left(\frac{dz}{dx} - 1\right) = \cos z + \sin z \qquad \left[\frac{dy}{dx} = \frac{dz}{dx} - 1\right]$$

$$\Rightarrow \quad \frac{dz}{dx} = (\cos z + \sin z + 1)$$

$$\Rightarrow \quad \frac{dz}{\cos z + \sin z + 1} = dx$$

On integrating both sides, we get

$$\int \frac{dz}{\cos z + \sin z + 1} = \int 1 dx \qquad \text{[using variable separable]}$$

$$\Rightarrow \quad \int \frac{dz}{\dfrac{1 - \tan^2 z/2}{1 + \tan^2 z/2} + \dfrac{2\tan z/2}{1 + \tan^2 z/2} + 1} = \int dx \qquad \left[\because \cos 2\theta = \frac{1 - \tan^2\theta}{1 + \tan^2\theta}\right]$$

$$\Rightarrow \int \frac{dz}{\dfrac{1 - \tan^2 z/2 + 2\tan z/2 + 1 + \tan^2 z/2}{(1 + \tan^2 z/2)}} = \int dx \qquad \left[\sin 2\theta = \frac{2\tan\theta}{1 + \tan^2\theta}\right]$$

$$\Rightarrow \quad \int \frac{(1 + \tan^2 z/2)\, dz}{2 + 2\tan z/2} = \int dx$$

$$\Rightarrow \quad \int \frac{\sec^2 z/2\, dz}{2(1 + \tan z/2)} = \int dx$$

Put $1 + \tan z/2 = t \Rightarrow \left(\dfrac{1}{2}\sec^2 z/2\right) dz = dt$

$$\Rightarrow \quad \int \frac{dt}{t} = \int dx$$

$$\Rightarrow \quad \log|t| = x + C \qquad \left[\because t = 1 + \tan\frac{z}{2} \text{ and } z = x + y)\right]$$

$$\Rightarrow \quad \log|1 + \tan z/2| = x + C$$

$$\Rightarrow \quad \log\left|1 + \tan\frac{(x+y)}{2}\right| = x + C$$

Q.28 Find the general solution of $\dfrac{dy}{dx} - 3y = \sin 2x$.

Ans. Given, $\dfrac{dy}{dx} - 3y = \sin 2x$

which is a linear differential equation.

On comparing it with $\dfrac{dy}{dx} + Py = Q$, we get

$$P = -3,\ Q = \sin 2x$$
$$\text{I.F} = e^{-3\int dx} = e^{-3x}$$

The general solution is

$$y \cdot e^{-3x} = \int \underset{\text{I}}{\sin 2x}\ \underset{\text{II}}{e^{-3x}}\ dx \qquad [y \cdot \text{I.F} = \int Q \cdot \text{I.F}\ dx + C]$$

Let $y \cdot e^{-3x} = I$ ... (i)

$$\therefore \quad I = \int \underset{\text{II}}{e^{-3x}}\ \underset{\text{I}}{\sin 2x}$$

[applying integration by parts method]

$$\Rightarrow \quad I = \sin 2x \left(\cdot \frac{e^{-3x}}{-3}\right) - \int 2\cos 2x \left(\frac{e^{-3x}}{-3}\right) dx + C_1$$

$$\Rightarrow \quad I = -\frac{1}{3} e^{-3x} \sin 2x + \frac{2}{3} \int \underset{\text{II}}{e^{-3x}}\ \underset{\text{I}}{\cos 2x}\ dx + C_1$$

$$\Rightarrow \quad I = -\frac{1}{3} e^{-3x} \sin 2x + \frac{2}{3}\left(\cos 2x\, \frac{e^{-3x}}{-3} - \int (-2\sin 2x) \frac{e^{-3x}}{-3}\, dx\right) + C_1 + C_2$$

$$\Rightarrow \quad I = -\frac{1}{3} e^{-3x} \sin 2x - \frac{2}{9} \cos 2x\, e^{-3x} - \frac{4}{9} I + C' \qquad [\text{where, } C' = C_1 + C_2]$$

$$\Rightarrow \quad I + \frac{4I}{9} = + e^{-3x}\left(-\frac{1}{3}\sin 2x - \frac{2}{9}\cos 2x\right) + C'$$

$$\Rightarrow \quad \frac{13}{9} I = e^{-3x}\left(-\frac{1}{3}\sin 2x - \frac{2}{9}\cos 2x\right) + C'$$

$$\Rightarrow \quad I = \frac{9}{13} e^{-3x}\left(-\frac{1}{3}\sin 2x - \frac{2}{9}\cos 2x\right) + C \qquad \left[\text{where } C = \frac{9C'}{13}\right]$$

$$\Rightarrow \quad I = \frac{3}{13} e^{-3x}\left(-\sin 2x - \frac{2}{3}\cos 2x\right) + C$$

$$\Rightarrow \quad = \frac{3}{13} e^{-3x} \frac{(-3\sin 2x - 2\cos 2x)}{3} + C$$

$$\Rightarrow \quad = \frac{e^{-3x}}{13}(-3\sin 2x - 2\cos 2x) + C$$

$$\Rightarrow \quad I = -\frac{e^{-3x}}{13}(2\cos 2x + 3\sin 2x) + C$$

On substituting the value of I in Eq. (i), we get

$$y \cdot e^{-3x} = -\frac{e^{-3x}}{13}(2\cos 2x + 3\sin 2x) + C$$

$$\Rightarrow \quad y = -\frac{1}{13}(2\cos 2x + 3\sin 2x) + C e^{3x}$$

Q.29 Find the equation of a curve passing through (2, 1), if the slope of the tangent to the curve at any point (x, y) is $\dfrac{x^2 + y^2}{2xy}$.

Thinking Process The slope of tangent to the curve at a point is given, then rearrange the equation to form a homogeneous differential equation $\dfrac{dy}{dx} = F(x, y)$, then make substitution $y = vx$

Ans. It is given that, the slope of tangent to the curve at point (x, y) is $\frac{x^2+y^2}{2xy}$.

$$\therefore \quad \left(\frac{dy}{dx}\right)_{(x,y)} = \frac{x^2+y^2}{2xy}$$

$$\Rightarrow \quad \frac{dy}{dx} = \frac{1}{2}\left(\frac{x}{y}+\frac{y}{x}\right) \quad \text{... (i)}$$

which is homogeneous differential equation.

Put
$$y = vx$$

$$\Rightarrow \quad \frac{dy}{dx} = v + x\frac{dv}{dx} \quad \text{[differentiating w.r.t. } x\text{]}$$

On substituting these values in Eq. (i), we get

$$v + x\frac{dv}{dx} = \frac{1}{2}\left(\frac{1}{v}+v\right)$$

$$\Rightarrow \quad v + x\frac{dv}{dx} = \frac{1}{2}\left(\frac{1+v^2}{v}\right)$$

$$\Rightarrow \quad x\frac{dv}{dx} = \frac{1+v^2}{2v} - v$$

$$\Rightarrow \quad x\frac{dv}{dx} = \frac{1+v^2-2v^2}{2v}$$

$$\Rightarrow \quad x\frac{dv}{dx} = \frac{1-v^2}{2v}$$

$$\Rightarrow \quad \frac{2v}{1-v^2}\,dv = \frac{dx}{x} \quad \text{[now applying variable separable]}$$

On integrating both sides, we get

$$\int \frac{2v}{1-v^2}\,dv = \int \frac{dx}{x}$$

Put $1 - v^2 = t$ in LHS, we get

$$-2\,vdv = dt$$

$$\Rightarrow \quad -\int \frac{dt}{t} = \int \frac{dx}{x}$$

$$\Rightarrow \quad -\log t = \log x + \log C$$

$$\Rightarrow \quad -\log(1-v^2) = \log x + \log C$$

$$\Rightarrow \quad -\log\left(1-\frac{y^2}{x^2}\right) = \log x + \log C \quad \left[\because v = \frac{y}{x}\right]$$

$$\Rightarrow \quad -\log\left(\frac{x^2-y^2}{x^2}\right) = \log x + \log C$$

$$\Rightarrow \quad \log\left(\frac{x^2}{x^2-y^2}\right) = \log x + \log C \quad \left[\because -\log x = \log\frac{1}{x}\right]$$

$$\Rightarrow \quad \frac{x^2}{x^2-y^2} = Cx \quad \text{...(ii)}$$

Since, the curve passes through the point (2, 1).

$$\therefore \quad \frac{(2)^2}{(2)^2-(1)^2} = C(2) \Rightarrow C = \frac{2}{3} \quad [\because x = 2, y = 1]$$

So, the required solution is $2(x^2 - y^2) = 3x$.

Q.30 Find the equation of the curve through the point (1, 0), if the slope of the tangent to the curve at any point (x, y) is $\frac{y-1}{x^2+x}$.

Ans. It is given that, slope of tangent to the curve at any point (x, y) is $\frac{y-1}{x^2+x}$.

$$\therefore \quad \left(\frac{dy}{dx}\right)_{(x,y)} = \frac{y-1}{x^2+x} \quad \text{[applying variable separable]}$$

$$\Rightarrow \quad \frac{dy}{dx} = \frac{y-1}{x^2+x}$$

$$\Rightarrow \quad \frac{dy}{y-1} = \frac{dx}{x^2+x}$$

On integrating both sides, we get

$$\int \frac{dy}{y-1} = \int \frac{dx}{x^2+x}$$

$$\Rightarrow \quad \int \frac{dy}{y-1} = \int \frac{dx}{x(x+1)}$$

$$\Rightarrow \quad \int \frac{dy}{y-1} = \int \left(\frac{1}{x} - \frac{1}{x+1}\right) dx$$

$$\Rightarrow \quad \log(y-1) = \log x - \log(x+1) + \log C$$

$$\Rightarrow \quad \log(y-1) = \log\left(\frac{xC}{x+1}\right)$$

$$(y-1) = \frac{xC}{x+1}$$

Since, the given curve passes through point (1, 0) i.e., $x = 1, y = 0$

$$\therefore \quad 0 - 1 = \frac{1 \cdot C}{1+1} \Rightarrow C = -2$$

On substituting the value of $C = -2$ in Eq. (i), we get

$$y - 1 = \frac{-2x}{x+1}$$

$$\Rightarrow \quad (y-1)(x+1) = -2x$$

$$\Rightarrow \quad (y-1)(x+1) + 2x = 0$$

Hence, the required equation of the curve is $(y-1)(x+1) + 2x = 0$

Q.31 Find the equation of a curve passing through origin, if the slope of the tangent to the curve at any point (x, y) is equal to the square of the difference of the abscissa and ordinate of the point.

Ans. Slope of tangent to the curve $= \frac{dy}{dx}$

and difference of abscissa and ordinate $= x - y$

According to the question, $\quad \frac{dy}{dx} = (x-y)^2 \quad$ ...(i)

Put $\quad x - y = z$

$$\Rightarrow \quad 1-\frac{dy}{dx}=\frac{dz}{dx}$$

$$\Rightarrow \quad \frac{dy}{dx}=1-\frac{dz}{dx}$$

On substituting these values in Eq. (i), we get

$$1-\frac{dz}{dx}=z^2$$

$$\Rightarrow \quad 1-z^2=\frac{dz}{dx}$$

$$\Rightarrow \quad dx=\frac{dz}{1-z^2}$$

On integrating both sides, we get

$$\int dx=\int\frac{dz}{1-z^2}$$

$$\Rightarrow \quad x=\frac{1}{2}\log\left|\frac{1+z}{1-z}\right|+C$$

$$\Rightarrow \quad x=\frac{1}{2}\log\left|\frac{1+x-y}{1-x+y}\right|+C \qquad [\because x-y=z] \quad \text{... (ii)}$$

Since, the curve passes through the origin.

$$\therefore \quad 0=\frac{1}{2}\log\left|\frac{1+0-0}{1-0+0}\right|+C$$

$$\Rightarrow \quad C=0$$

On substituting the value of C in Eq. (ii), we get

$$x=\frac{1}{2}\log\left|\frac{1+x-y}{1-x+y}\right|$$

$$\Rightarrow \quad 2x=\log\left|\frac{1+x-y}{1-x+y}\right|$$

$$\Rightarrow \quad e^{2x}=\left|\frac{1+x-y}{1-x+y}\right|$$

$$\Rightarrow \quad (1-x+y)\,e^{2x}=1+x-y$$

Hence, the required equation of the curve is $(1-x+y)e^{2x}=1+x-y$.

Q.32 Find the equation of a curve passing through the point (1, 1), if the tangent drawn at any point $P(x, y)$ on the curve meets the coordinate axes at A and B such that P is the mid-point of AB.

Ans. The below figure obtained by the given information

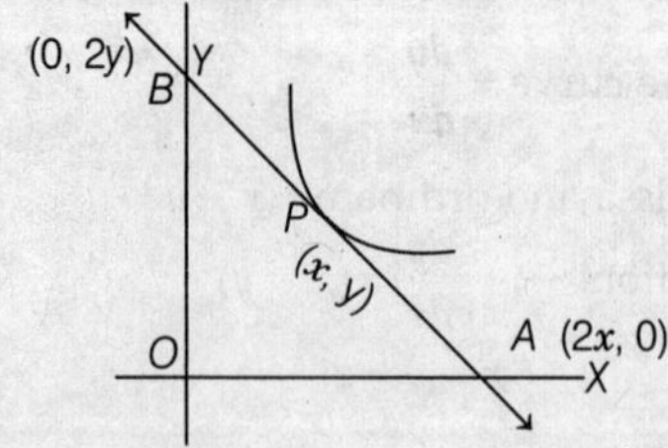

Let the coordinates of the point P are (x, y). It is given that, P is mid-point of AB. So, the coordinates of points A and B are $(2x, 0)$ and $(0, 2y)$, respectively.

[Mid point consider set $\frac{x_1 + x_2}{2}, \frac{y_1 + y_2}{2}$]

$\therefore$ Slope of $AB = \frac{0 - 2y}{2x - 0} = -\frac{y}{x}$ $\left[\text{slope} = \frac{y_2 - y_1}{x_2 - x_1}\right]$

Since, the segment AB is a tangent to the curve at P.

$$\therefore \quad \frac{dy}{dx} = -\frac{y}{x} \quad \text{[applying variable separable method]}$$

$$\Rightarrow \quad \frac{dy}{y} = -\frac{dx}{x}$$

On integrating both sides, we get

$$\log y = -\log x + \log C$$

$$\log y = \log \frac{C}{x} \quad \ldots\text{(i)}$$

Since, the given curve passes through (1, 1).

$$\therefore \quad \log 1 = \log \frac{C}{1}$$

$$\Rightarrow \quad 0 = \log C$$

$$\Rightarrow \quad C = 1$$

On substituting the value of C is Eq. (i), we get

$$\therefore \quad \log y = \log \frac{1}{x}$$

$$\Rightarrow \quad y = \frac{1}{x} \Rightarrow xy = 1$$

Hence, the required equation of the curve is $xy = 1$.

Q.33 Solve $x\frac{dy}{dx} = y(\log y - \log x + 1)$

Ans. Given, $x\frac{dy}{dx} = y(\log y - \log x + 1)$

$$\Rightarrow \quad x\frac{dy}{dx} = y\left(\log \frac{y}{x} + 1\right)$$

$$\Rightarrow \quad \frac{dy}{dx} = \frac{y}{x}\left(\log \frac{y}{x} + 1\right) \quad \ldots\text{(i)}$$

which is a homogeneous equation.

Put $y = vx$

$$\therefore \quad \frac{dy}{dx} = v + x\frac{dv}{dx}$$

On substituting these values in Eq.(i), we get

$$v + x\frac{dv}{dx} = v(\log v + 1) = v \log v + v$$

$$\Rightarrow \quad x\frac{dv}{dx} = v(\log v + 1 - 1)$$

$$\Rightarrow \quad x\frac{dv}{dx} = v(\log v)$$

$$\Rightarrow \quad \frac{dv}{v \log v} = \frac{dx}{x}$$

On integrating both sides, we get

$$\int \frac{dv}{v \log v} = \int \frac{dx}{x}$$

On putting $\log v = u$ in LHS integral, we get

$$\frac{1}{v} \cdot dv = du$$

$$\int \frac{du}{u} = \int \frac{dx}{x}$$

$\Rightarrow \quad \log u = \log x + \log C$

$\Rightarrow \quad \log u = \log Cx$

$\Rightarrow \quad u = Cx$

$\Rightarrow \quad \log v = Cx \quad [\because \log v = u]$

$\Rightarrow \quad \log\left(\frac{y}{x}\right) = Cx \quad \left[\because v = \frac{y}{x}\right]$

Mistake Alert! *Don't forget to put the final values of u and v after integrating the function.*

Objective Type Questions

Choose the correct answer out of the given four options in each of the exercises from 34 to 75.

Q.34 The degree of the differential equation $\left(\frac{d^2y}{dx^2}\right)^2 + \left(\frac{dy}{dx}\right)^2 = x \sin\left(\frac{dy}{dx}\right)$ is

(a) 1 (b) 2 (c) 3 (d) not defined

Ans. ***(d)*** The degree of above differential equation is not defined because when we expand $\sin\left(\frac{dy}{dx}\right)$ we get an infinite series in the increasing powers of $\frac{dy}{dx}$.

Therefore its degree is not defined.

Note If the given differential equation is not polynomial equation in its derivatives, then its degree is not defined.

Q.35 The degree of the differential equation $\left[1 + \left(\frac{dy}{dx}\right)^2\right]^{3/2} = \frac{d^2y}{dx^2}$ is

(a) 4 (b) $\frac{3}{2}$ (c) not defined (d) 2

Ans. ***(d)*** Given that $\left[1 + \left(\frac{dy}{dx}\right)^2\right]^{3/2} = \frac{d^2y}{dx^2}$

On squaring both sides, we get

$$\left[1 + \left(\frac{dy}{dx}\right)^2\right]^3 = \left(\frac{d^2y}{dx^2}\right)^2$$

The highest order derivative is $\frac{d^2y}{dx^2}$ and the highest power raised to $\frac{d^2y}{dx^2}$ is 2.

So, the degree of differential equation is 2.

Q.36 The order and degree of the differential equation $\frac{d^2y}{dx^2}+\left(\frac{dy}{dx}\right)^{1/4}+x^{1/5}=0$ respectively, are

(a) 2 and 4 (b) 2 and 2 (c) 2 and 3 (d) 3 and 3

Ans. *(a)* Given that, $\frac{d^2y}{dx^2}+\left(\frac{dy}{dx}\right)^{1/4}=-x^{1/5}$

$$\Rightarrow \quad \frac{d^2y}{dx^2}+\left(\frac{dy}{dx}\right)^{1/4}=-x^{1/5}$$

$$\Rightarrow \quad \left(\frac{dy}{dx}\right)^{1/4}=-\left(x^{1/5}+\frac{d^2y}{dx^2}\right)$$

On squaring both sides, we get

$$\left(\frac{dy}{dx}\right)^{1/2}=\left(x^{1/5}+\frac{d^2y}{dx^2}\right)^2$$

Again, on squaring both sides, we have

$$\frac{dy}{dx}=\left(x^{1/5}+\frac{d^2y}{dx^2}\right)^4$$

order = 2, degree = 4

Q.37 If $y=e^{-x}(A\cos x+B\sin x)$, then y is a solution of

(a) $\frac{d^2y}{dx^2}+2\frac{dy}{dx}=0$ (b) $\frac{d^2y}{dx^2}-2\frac{dy}{dx}+2y=0$

(c) $\frac{d^2y}{dx^2}+2\frac{dy}{dx}+2y=0$ (d) $\frac{d^2y}{dx^2}+2y=0$

Ans. *(c)* Given that, $y=e^{-x}(A\cos x+B\sin x)$

Solution can be obtained by eliminating A and B.

On differentiating both sides w.r.t., x we get

$$\frac{dy}{dx}=-e^{-x}(A\cos x+B\sin x)+e^{-x}(-A\sin x+B\cos x)$$

$$\frac{dy}{dx}=-y+e^{-x}(-A\sin x+B\cos x)$$

Again, differentiating both sides w.r.t. x, we get

$$\frac{d^2y}{dx^2}=\frac{-dy}{dx}+e^{-x}(-A\cos x-B\sin x)-e^{-x}(-A\sin x+B\cos x)$$

$$\Rightarrow \quad \frac{d^2y}{dx^2}=-\frac{dy}{dx}-y-\left[\frac{dy}{dx}+y\right]$$

$$\Rightarrow \quad \frac{d^2y}{dx^2}=-\frac{dy}{dx}-y-\frac{dy}{dx}-y$$

$$\Rightarrow \quad \frac{d^2y}{dx^2}=-2\frac{dy}{dx}-2y$$

$$\Rightarrow \quad \frac{d^2y}{dx^2}+2\frac{dy}{dx}+2y=0$$

***Q.38** The differential equation for $y = A\cos\alpha\, x + B\sin\alpha\, x$, where A and B are arbitrary constants is

(a) $\dfrac{d^2y}{dx^2} - \alpha^2 y = 0$ (b) $\dfrac{d^2y}{dx^2} + \alpha^2 y = 0$

(c) $\dfrac{d^2y}{dx^2} + \alpha y = 0$ (d) $\dfrac{d^2y}{dx^2} - \alpha y = 0$

Ans. ***(b)*** Given, $y = A\cos \alpha x + B\, x \sin \alpha x$ and differential equation is obtained by eliminating A and B

$$\Rightarrow \quad \frac{dy}{dx} = -\alpha A\sin\alpha x + \alpha B\cos\alpha x$$

Again, differentiating both sides w.r.t. x, we get

$$\frac{d^2y}{dx^2} = -A\alpha^2\cos\alpha x - \alpha^2 B\sin\alpha x$$

$$\Rightarrow \quad \frac{d^2y}{dx^2} = -\alpha^2(A\cos\alpha x + B\sin\alpha x)$$

$$\Rightarrow \quad \frac{d^2y}{dx^2} = -\alpha^2 y \qquad [\because y = A\cos\alpha x + B\sin\alpha x]$$

$$\Rightarrow \quad \frac{d^2y}{dx^2} + \alpha^2 y = 0$$

Q.39 The solution of differential equation $xdy - ydx = 0$ represents

(a) a rectangular hyperbola
(b) parabola whose vertex is at origin
(c) straight line passing through origin
(d) a circle whose centre is at origin

Ans. ***(c)*** Given that, $xdy - ydx = 0$

$$\Rightarrow \quad xdy = ydx$$

$$\Rightarrow \quad \frac{dy}{y} = \frac{dx}{x} \qquad \text{[applying variable separable method]}$$

On integrating both sides, we get

$$\log y = \log x + \log C$$

$$\Rightarrow \quad \log y = \log Cx \Rightarrow y = Cx$$

which is a straight line passing through origin.

Q.40 The integrating factor of differential equation $\cos x\dfrac{dy}{dx} + y\sin x = 1$ is

(a) $\cos x$ (b) $\tan x$ (c) $\sec x$ (d) $\sin x$

Ans. ***(c)*** Given that, $\cos x\dfrac{dy}{dx} + y\sin x = 1$

On dividing both sides by $\cos x$, we get

$$\Rightarrow \quad \frac{dy}{dx} + y\tan x = \sec x \qquad \left[\because \frac{\sin x}{\cos x} = \tan x, \frac{1}{\cos x} = \sec x\right]$$

Note (✱) Topics related to these question have been removed from latest NCERT.

On comparing $\frac{dy}{dx} + Py = 0$

Here, $P = \tan x$ and $Q = \sec x$

$$\text{I.F} = e^{\int P\,dx} = e^{\int \tan x\,dx} = e^{\log \sec x} \qquad [\because e^{\log x} = x]$$

$$\therefore \qquad = \sec x$$

Q.41 The solution of differential equation $\tan y \sec^2 x dx + \tan x \sec^2 y dy = 0$ is

(a) $\tan x + \tan y = k$
(b) $\tan x - \tan y = k$
(c) $\frac{\tan x}{\tan y} = k$
(d) $\tan x \cdot \tan y = k$

Ans. ***(d)*** Given that, $\tan y \sec^2 x\, dx + \tan x \sec^2 y\, dy = 0$

$$\Rightarrow \quad \tan y \sec^2 x\, dx = -\tan x \sec^2 y\, dy$$

$$\Rightarrow \quad \frac{\sec^2 x}{\tan x} dx = \frac{-\sec^2 y}{\tan y} dy \qquad \dots(i)$$

On integrating both sides, we get

$$\int \frac{\sec^2 x}{\tan x} dx = -\int \frac{\sec^2 y}{\tan y} dy$$

Put $\tan x = t$ in LHS integral, we get

$$\sec^2 x\, dx = dt \Rightarrow \sec^2 x\, dx = dt$$

and $\tan y = u$ in RHS integral, we get

$$\sec^2 y dy = du$$

On substituting these values in Eq. (i), we get

$$\int \frac{dt}{t} = -\int \frac{du}{u}$$

$$\log t = -\log u + \log k$$

$$\Rightarrow \quad \log t + \log u = \log k$$

$$\Rightarrow \quad \log(t \cdot u) = \log k \qquad [\because \log x + \log y = \log (xy)]$$

$$\Rightarrow \quad \log(\tan x \tan y) = \log k$$

$$\Rightarrow \quad \tan x \tan y = k$$

Q.42 The family $y = Ax + A^3$ of curves is represented by differential equation of degree

(a) 1 (b) 2 (c) 3 (d) 4

Ans. ***(a)*** Given that, $y = Ax + A^3$

$$\Rightarrow \quad \frac{dy}{dx} = A$$

[we can differentiate above equation only once because it has only one arbitrary constant]

$$\therefore \quad \text{Degree} = 1$$

Q.43 The integrating factor of $\frac{xdy}{dx} - y = x^4 - 3x$ is

(a) x
(b) $\log x$
(c) $\frac{1}{x}$
(d) $-x$

Ans. *(c)* Given that,
$$x\frac{dy}{dx} - y = x^4 - 3x$$
$$\Rightarrow \quad \frac{dy}{dx} - \frac{y}{x} = x^3 - 3$$
which is a linear differential equation

On comparing it with $\frac{dy}{dx} + P\cdot y = Q$

Here, $$P = -\frac{1}{x}, Q = x^3 - 3$$
$$\therefore \quad \text{I.F} = e^{\int P\,dx} = e^{-\int \frac{1}{x}dx} = e^{-\log x} = x^{-1}$$
$$= \frac{1}{x}$$

Q.44 The solution of $\frac{dy}{dx} - y = 1$, $y(0) = 1$ is given by

(a) $xy = -e^x$ (b) $xy = -e^{-x}$
(c) $xy = -1$ (d) $y = 2e^x - 1$

Ans. *(d)* Given that,
$$\frac{dy}{dx} - y = 1$$
$$\Rightarrow \quad \frac{dy}{dx} = 1 + y$$
$$\Rightarrow \quad \frac{dy}{1+y} = dx \quad \text{[using variable separable method]}$$
On integrating both sides, we get
$$\log(1+y) = x + C \quad \text{...(i)}$$
When $x = 0$ and $y = 1$, then
$$\log 2 = 0 + c$$
$$\Rightarrow \quad C = \log 2$$
The required solution is
$$\log(1+y) = x + \log 2 \quad [\because C = \log 2]$$
$$\log(1+y) - \log 2 = x$$
$$\Rightarrow \quad \log\left(\frac{1+y}{2}\right) = x$$
$$\Rightarrow \quad \frac{1+y}{2} = e^x$$
$$\Rightarrow \quad 1 + y = 2e^x$$
$$\Rightarrow \quad y = 2e^x - 1$$

Q.45 The number of solutions of $\frac{dy}{dx} = \frac{y+1}{x-1}$, when $y(1) = 2$ is

(a) none (b) one (c) two (d) infinite

Ans. *(b)* Given that, $\dfrac{dy}{dx} = \dfrac{y+1}{x-1}$

$\Rightarrow \quad \dfrac{dy}{y+1} = \dfrac{dx}{x-1}$ [appliying variable separable method]

On integrating both sides, we get

$$\log(y+1) = \log(x-1) - \log C$$
$$\log(y+1) + \log C = \log(x-1)$$
$$\log(C(y+1)) = \log(x-1)$$
$$C(y+1) = (x-1)$$
$$\Rightarrow \quad C = \frac{x-1}{y+1}$$

When $x = 1$ and $y = 2$, then $C = 0$

So, the required solution is $x - 1 = 0$.

Hence, only one solution exist.

Q.46 Which of the following is a second order differential equation?

(a) $(y')^2 + x = y^2$
(b) $y'y'' + y = \sin x$
(c) $y''' + (y'')^2 + y = 0$
(d) $y' = y^2$

Ans. *(b)* The second order differential equation is $y'y'' + y = \sin x$.

$$\Rightarrow \quad \left(\frac{dy}{dx}\right)\left(\frac{d^2y}{dx^2}\right) + y = \sin x$$

Q.47 The integrating factor of differential equation $(1-x^2)\dfrac{dy}{dx} - xy = 1$ is

(a) $-x$
(b) $\dfrac{x}{1+x^2}$
(c) $\sqrt{1-x^2}$
(d) $\dfrac{1}{2}\log(1-x^2)$

Ans. *(c)* Given that, $(1-x^2)\dfrac{dy}{dx} - xy = 1$

$$\Rightarrow \quad \frac{dy}{dx} - \frac{x}{1-x^2}y = \frac{1}{1-x^2}$$

which is a linear differential equation and on comparing with $\dfrac{dy}{dx} + P \cdot y = Q$, we get

$$P = \frac{-x}{1-x^2}$$
$$Q = \frac{1}{1-x^2}$$

$\therefore \quad \text{I.F} = e^{-\int \frac{x}{1-x^2}dx}$ $\quad [\text{I.F} = e^{\int P \cdot dx}]$

Put $\quad 1 - x^2 = t \Rightarrow -2x\,dx = dt \Rightarrow x\,dx = -\dfrac{dt}{2}$

Now, $\quad \text{I.F} = e^{\frac{1}{2}\int \frac{dt}{t}} = e^{\frac{1}{2}\log t}$

$$= e^{\frac{1}{2}\log(1-x^2)} = \sqrt{1-x^2} \quad [\because 1 - x^2 = t]$$

Q.48 $\tan^{-1} x + \tan^{-1} y = C$ is general solution of the differential equation

(a) $\frac{dy}{dx} = \frac{1+y^2}{1+x^2}$ (b) $\frac{dy}{dx} = \frac{1+x^2}{1+y^2}$

(c) $(1+x^2)dy + (1+y^2)dx = 0$ (d) $(1+x^2)dx + (1+y^2)dy = 0$

Ans. *(c)* Given that, $\tan^{-1} x + \tan^{-1} y = C$

On differentiating w.r.t. x, we get

$$\frac{1}{1+x^2} + \frac{1}{1+y^2} \cdot \frac{dy}{dx} = 0$$

$$\Rightarrow \frac{1}{1+y^2} \cdot \frac{dy}{dx} = -\frac{1}{1+x^2}$$

$$\Rightarrow (1+x^2)\, dy + (1+y^2)\, dx = 0$$

Q.49 The differential equation $y\frac{dy}{dx} + x = C$ represents

(a) family of hyperbolas (b) family of parabolas

(c) family of ellipses (d) family of circles

- **Concept Insight** General equation is $ax^2 + by^2 + 2hxy + 2gx + 2fy + C = 0$ represents the circle when $a = b, h = 0, a \neq 0, b \neq 0$

Ans. *(d)* Given that, $y\frac{dy}{dx} + x = C$

$$\Rightarrow y\frac{dy}{dx} = C - x$$

$$\Rightarrow y \cdot dy = (C - x)\, dx$$

On integrating both sides, we get

$$\frac{y^2}{2} = Cx - \frac{x^2}{2} + K$$

$$\Rightarrow \frac{x^2}{2} + \frac{y^2}{2} = Cx + K$$

$$\Rightarrow \frac{x^2}{2} + \frac{y^2}{2} - Cx = K$$

$$\Rightarrow x^2 + y^2 - 2Cx = 2K$$

which represent family of circles.

Q.50 The general solution of $e^x \cos y dx - e^x \sin y dy = 0$ is

(a) $e^x \cos y = k$ (b) $e^x \sin y = k$

(c) $e^x = k \cos y$ (d) $e^x = k \sin y$

Ans. *(a)* Given that, $e^x \cos y dx - e^x \sin y dy = 0$

$$\Rightarrow e^x \cos y dx = e^x \sin y dy$$

$$\Rightarrow \frac{dx}{dy} = \tan y$$

$$\Rightarrow dx = \tan y dy$$ [applying variable separable method]

On integrating both sides, we get

$$x = \log \sec y + C$$

$\Rightarrow$ $$x - C = \log \sec y$$

$\Rightarrow$ $$\sec y = e^{x - C}$$

$\Rightarrow$ $$\sec y = e^{x} e^{-C}$$

$\Rightarrow$ $$\frac{1}{\cos y} = \frac{e^x}{e^C}$$

$\Rightarrow$ $$e^x \cos y = e^C$$

$\Rightarrow$ $$e^x \cos y = K \quad \text{[where, } K = e^C\text{]}$$

Q.51 The degree of differential equation $\frac{d^2y}{dx^2} + \left(\frac{dy}{dx}\right)^3 + 6y^5 = 0$ is

(a) 1 (b) 2 (c) 3 (d) 5

Ans. ***(a)*** $\frac{d^2y}{dx^2} + \left(\frac{dy}{dx}\right)^3 + 6y^5 = 0$

We know that, the degree of a differential equation is exponent of highest order derivative.

Highest order derivative is $\frac{d^2y}{dx^2}$.

$\therefore$ $$\text{Degree} = 1$$

Q.52 The solution of $\frac{dy}{dx} + y = e^{-x}$, $y(0) = 0$ is

(a) $y = e^x(x - 1)$ (b) $y = xe^{-x}$

(c) $y = xe^{-x} + 1$ (d) $y = (x + 1)e^{-x}$

Ans. ***(b)*** Given that, $$\frac{dy}{dx} + y = e^{-x}$$

On comparing it with $\frac{dy}{dx} + P \cdot y = Q$, we get

$$P = 1, Q = e^{-x}$$

$$\text{I.F} = e^{\int P\,dx} = e^{\int dx} = e^x$$

The general solution is $y \cdot \text{I.F} = \int Q \cdot \text{I.F}\, dx + C$

$$y \cdot e^x = \int e^{-x} e^x dx + C$$

$\Rightarrow$ $$y \cdot e^x = \int dx + C$$

$\Rightarrow$ $$y \cdot e^x = x + C \quad \text{...(i)}$$

When $x = 0$ and $y = 0$, then

$$0 = 0 + C \Rightarrow C = 0$$

Eq. (i) becomes $$y \cdot e^x = x$$

$\Rightarrow$ $$y = x e^{-x}$$

Q.53 The integrating factor of differential equation $\frac{dy}{dx} + y\tan x - \sec x = 0$ is

(a) $\cos x$ (b) $\sec x$ (c) $e^{\cos x}$ (d) $e^{\sec x}$

Ans. ***(b)*** Given that, $\frac{dy}{dx} + y\tan x - \sec x = 0$

$$\Rightarrow \frac{dy}{dx} + y\tan x = \sec x$$

which is a linear differential equation

On comparing it with the equation $\frac{dy}{dx} + Py = Q$, we get

Here, $P = \tan x, Q = \sec x$

$$\text{I.F} = e^{\int P\,dx} = e^{\int \tan x\,dx}$$

$$= e^{(\log \sec x)} \qquad [\because e^{\log x} = x]$$

$$= \sec x$$

Q.54 The solution of differential equation $\frac{dy}{dx} = \frac{1+y^2}{1+x^2}$ is

(a) $y = \tan^{-1} x$ (b) $y - x = k(1 + xy)$
(c) $x = \tan^{-1} y$ (d) $\tan(xy) = k$

Ans. ***(b)*** Given that, $\frac{dy}{dx} = \frac{1+y^2}{1+x^2}$

$$\Rightarrow \frac{dy}{1+y^2} = \frac{dx}{1+x^2} \qquad \text{[apply variable separable method]}$$

On integrating both sides, we get

$$\tan^{-1} y = \tan^{-1} x + C$$

$$\Rightarrow \tan^{-1} y - \tan^{-1} x = C$$

$$\Rightarrow \tan^{-1}\left(\frac{y-x}{1+xy}\right) = C$$

$$\Rightarrow \frac{y-x}{1+xy} = \tan C$$

$$\Rightarrow y - x = \tan C(1 + xy)$$

$$\Rightarrow y - x = k(1 + xy) \qquad [\because \tan C = k]$$

Q.55 The integrating factor of differential equation $\frac{dy}{dx} + y = \frac{1+y}{x}$ is

(a) $\frac{x}{e^x}$ (b) $\frac{e^x}{x}$ (c) xe^x (d) e^x

Ans. ***(b)*** Given that, $\frac{dy}{dx} + y = \frac{1+y}{x}$

$$\Rightarrow \frac{dy}{dx} = \frac{1+y}{x} - y$$

$$\Rightarrow \frac{dy}{dx} = \frac{1+y-xy}{x}$$

$$\Rightarrow \quad \frac{dy}{dx} = \frac{1}{x} + \frac{y(1-x)}{x}$$

$$\Rightarrow \quad \frac{dy}{dx} - \left(\frac{1-x}{x}\right) y = \frac{1}{x}$$

which is a linear differential equation

Now, compare with $\frac{dy}{dx} + P \cdot y = Q$

Here, $$P = \frac{-(1-x)}{x}, Q = \frac{1}{x}$$

$$\text{I.F} = e^{\int P\,dx} = e^{-\int \frac{1-x}{x} dx} = e^{\int \frac{x-1}{x} dx}$$

$$= e^{\int \left(1 - \frac{1}{x}\right) dx} = e^{x - \log x}$$

$$= e^x \cdot e^{\log\left(\frac{1}{x}\right)} = e^x \cdot \frac{1}{x} \qquad [\because e^{\log x} = k]$$

Q.56 $y = ae^{mx} + be^{-mx}$ satisfies which of the following differential equation?

(a) $\frac{dy}{dx} + my = 0$ (b) $\frac{dy}{dx} - my = 0$

(c) $\frac{d^2y}{dx^2} - m^2y = 0$ (d) $\frac{d^2y}{dx^2} + m^2y = 0$

Ans. ***(c)*** Given that, $y = ae^{mx} + be^{-mx}$

To get differential equation we have to eliminate a and b

On differentiating both sides w.r.t. x, we get

$$\frac{dy}{dx} = mae^{mx} - bme^{-mx}$$

Again, differentiating both sides w.r.t. x, we get

$$\frac{d^2y}{dx^2} = m^2ae^{mx} + bm^2e^{-mx}$$

$$\Rightarrow \quad \frac{d^2y}{dx^2} = m^2(ae^{mx} + be^{-mx})$$

$$\Rightarrow \quad \frac{d^2y}{dx^2} = m^2y \qquad [\because y = ae^{mx} + be^{-mx}]$$

$$\Rightarrow \quad \frac{d^2y}{dx^2} - m^2y = 0$$

Q.57 The solution of differential equation $\cos x \sin y\,dx + \sin x \cos y\,dy = 0$ is

(a) $\frac{\sin x}{\sin y} = C$ (b) $\sin x \sin y = C$

(c) $\sin x + \sin y = C$ (d) $\cos x \cos y = C$

Ans. ***(b)*** Given differential equation is

$$\cos x \sin y\,dx + \sin x \cos y\,dy = 0$$

$$\Rightarrow \quad \cos x \sin y\,dx = -\sin x \cos y\,dy$$

$$\Rightarrow \quad \frac{\cos x}{\sin x} dx = -\frac{\cos y}{\sin y} dy \qquad \text{[applying variable separable method]}$$

$$\Rightarrow \quad \cot x\,dx = -\cot y\,dy$$

On integrating both sides, we get

$$\log \sin x = -\log \sin y + \log C$$

$\Rightarrow$ $\log \sin x + \log \sin y = \log C$

$\Rightarrow$ $\log \sin x \sin y = \log C$

$\Rightarrow$ $\sin x \cdot \sin y = C$

Q.58 The solution of $x\dfrac{dy}{dx} + y = e^x$ is

(a) $y = \dfrac{e^x}{x} + \dfrac{k}{x}$ (b) $y = xe^x + Cx$ (c) $y = xe^x + k$ (d) $x = \dfrac{e^y}{y} + \dfrac{k}{y}$

Ans. ***(a)*** Given that, $x\dfrac{dy}{dx} + y = e^x$

$\Rightarrow$ $\dfrac{dy}{dx} + \dfrac{y}{x} = \dfrac{e^x}{x}$

which is a linear differential equation.

On comparing it with $\dfrac{dy}{dx} + P\cdot y = Q$, we get $P = \dfrac{1}{x}$ and $Q = \dfrac{e^x}{x}$

$\therefore$ $\text{I.F} = e^{\int \frac{1}{x} dx} = e^{(\log x)} = x$

The general solution is $y \cdot x = \int \left(\dfrac{e^x}{x} \cdot x\right) dx + k$ $\quad [y \cdot \text{I.F} = \int Q \cdot \text{I.F}\, dx + C]$

$\Rightarrow$ $y \cdot x = \int e^x dx + k$

$\Rightarrow$ $y \cdot x = e^x + k$

$\Rightarrow$ $y = \dfrac{e^x}{x} + \dfrac{k}{x}$

***Q. 59** The differential equation of the family of curves $x^2 + y^2 - 2ay = 0$, where a is arbitrary constant, is

(a) $(x^2 - y^2)\dfrac{dy}{dx} = 2xy$ (b) $2(x^2 + y^2)\dfrac{dy}{dx} = xy$

(c) $2(x^2 - y^2)\dfrac{dy}{dx} = xy$ (d) $(x^2 + y^2)\dfrac{dy}{dx} = 2xy$

Ans. ***(a)*** Given equation of curves is

$$x^2 + y^2 - 2ay = 0$$

$\Rightarrow$ $\dfrac{x^2 + y^2}{y} = 2a$

On differentiating both sides w.r.t. x, we get

$$\frac{y\left(2x + 2y\dfrac{dy}{dx}\right) - (x^2 + y^2)\dfrac{dy}{dx}}{y^2} = 0$$

$\Rightarrow$ $2xy + 2y^2\dfrac{dy}{dx} - (x^2 + y^2)\dfrac{dy}{dx} = 0$

Note (✱) Topics related to these question have been removed from latest NCERT.

$$\Rightarrow \quad (2y^2 - x^2 - y^2)\frac{dy}{dx} = -2xy$$

$$\Rightarrow \quad (y^2 - x^2)\frac{dy}{dx} = -2xy$$

$$\Rightarrow \quad (x^2 - y^2)\frac{dy}{dx} = 2xy$$

***Q.60** The family $Y = Ax + A^3$ of curves will correspond to a differential equation of order

(a) 3 (b) 2
(c) 1 (d) not defined

Ans. ***(c)*** Given family of curves is $y = Ax + A^3$ ...(i)

$$\Rightarrow \quad \frac{dy}{dx} = A$$

Replacing A by $\frac{dy}{dx}$ in Eq. (i), we get

$$y = x\frac{dy}{dx} + \left(\frac{dy}{dx}\right)^3$$

$\therefore$ Order = 1

The order of differential equation is the highest order derivative it contains that is 1.

Q.61 The general solution of $\frac{dy}{dx} = 2x\,e^{x^2 - y}$ is

(a) $e^{x^2 - y} = C$ (b) $e^{-y} + e^{x^2} = C$ (c) $e^y = e^{x^2} + C$ (d) $e^{x^2 + y} = C$

Ans. ***(c)*** Given that, $\frac{dy}{dx} = 2x\,e^{x^2 - y} = 2x\,e^{x^2} \cdot e^{-y}$

$$\Rightarrow \quad e^y\frac{dy}{dx} = 2x\,e^{x^2}$$

$$\Rightarrow \quad e^y\,dy = 2x\,e^{x^2}\,dx \qquad \text{[using variable separable method]}$$

On integrating both sides, we get

$$\int e^y\,dy = 2\int x e^{x^2}\,dx$$

Put $x^2 = t$ in RHS integral, we get

$$2x\,dx = dt$$

$$\int e^y\,dy = \int e^t\,dt$$

$$\Rightarrow \quad e^y = e^t + C$$

$$\Rightarrow \quad e^y = e^{x^2} + C \qquad [\because t = x^2]$$

Q.62 The curve for which the slope of the tangent at any point is equal to the ratio of the abscissa to the ordinate of the point is

(a) an ellipse (b) parabola
(c) circle (d) rectangular hyperbola

Ans. ***(d)*** Slope of tangent to the curve $= \frac{dy}{dx}$

Note (✱) Topics related to these question have been removed from latest NCERT.

and ratio of abscissa to the ordinate $= \dfrac{x}{y}$

According to the question, $\dfrac{dy}{dx} = \dfrac{x}{y}$

$$y\,dy = x\,dx \quad \text{[using variable separable method]}$$

On integrating both sides, we get

$$\frac{y^2}{2} = \frac{x^2}{2} + C$$

$$\Rightarrow \quad \frac{y^2}{2} - \frac{x^2}{2} = C \Rightarrow y^2 - x^2 = 2C$$

$$\Rightarrow \quad y^2 - x^2 = k \quad \text{[where } 2C = k\text{]}$$

which is an equation of rectangular hyperbola.

Q.63 The general solution of differential equation $\dfrac{dy}{dx} = e^{\frac{x^2}{2}} + xy$ is

(a) $y = Ce^{-x^2/2}$ (b) $y = Ce^{x^2/2}$

(c) $y = (x + C)e^{x^2/2}$ (d) $y = (C - x)e^{x^2/2}$

Ans. *(c)* Given that, $\dfrac{dy}{dx} = e^{x^2/2} + xy$

$$\Rightarrow \quad \frac{dy}{dx} - xy = e^{x^2/2}$$

which is a linear differential equation

On comparing it with $\dfrac{dy}{dx} + P \cdot y = Q$, we get

$$P = -x, Q = e^{x^2/2}$$

$$\therefore \quad \text{I.F} = e^{\int -x\,dx} = e^{-x^2/2}$$

The general solution is

$$y \cdot \text{I.F} = \int Q \cdot \text{I.F}\,dx + C$$

$$y \cdot e^{-x^2/2} = \int e^{-x^2/2}\, e^{x^2/2}\,dx + C$$

$$\Rightarrow \quad y\,e^{-x^2/2} = \int 1\,dx + C$$

$$\Rightarrow \quad y \cdot e^{-x^2/2} = x + C$$

$$\Rightarrow \quad y = x\,e^{x^2/2} + C\,e^{x^2/2}$$

$$\Rightarrow \quad y = (x + C)\,e^{x^2/2}$$

Q.64 The solution of equation $(2y - 1)\,dx - (2x + 3)\,dy = 0$ is

(a) $\dfrac{2x - 1}{2y + 3} = k$ (b) $\dfrac{2y + 1}{2x - 3} = k$ (c) $\dfrac{2x + 3}{2y - 1} = k$ (d) $\dfrac{2x - 1}{2y - 1} = k$

Ans. *(c)* Given that, $(2y - 1)\,dx - (2x + 3)\,dy = 0$

$$\Rightarrow \quad (2y - 1)\,dx = (2x + 3)\,dy$$

$$\Rightarrow \quad \frac{dx}{2x+3} = \frac{dy}{2y-1} \quad \text{[using variable separable method]}$$

On integrating both sides, we get

$$\frac{1}{2}\log(2x+3) = \frac{1}{2}\log(2y-1) + \log C$$

$$\Rightarrow \quad \frac{1}{2}[\log(2x+3) - \log(2y-1)] = \log C$$

$$\Rightarrow \quad \frac{1}{2}\log\left(\frac{2x+3}{2y-1}\right) = \log C$$

$$\Rightarrow \quad \left(\frac{2x+3}{2y-1}\right)^{1/2} = C$$

$$\Rightarrow \quad \frac{2x+3}{2y-1} = C^2$$

$$\Rightarrow \quad \frac{2x+3}{2y-1} = k, \text{ where } k = C^2$$

Q.65 The differential equation for which $y = a\cos x + b\sin x$ is a solution, is

(a) $\frac{d^2y}{dx^2} + y = 0$ (b) $\frac{d^2y}{dx^2} - y = 0$

(c) $\frac{d^2y}{dx^2} + (a+b)y = 0$ (d) $\frac{d^2y}{dx^2} + (a-b)y = 0$

Ans. *(a)* Given that, $y = a\cos x + b\sin x$

To get differential equation we have to eliminate a and b.

On differentiating both sides w.r.t. x, we get

$$\frac{dy}{dx} = -a\sin x + b\cos x$$

Again, differentiating w.r.t. x, we get

$$\frac{d^2y}{dx^2} = -a\cos x - b\sin x$$

$$\Rightarrow \quad \frac{d^2y}{dx^2} = -y \quad [\because y = a\cos x + b\sin x]$$

$$\Rightarrow \quad \frac{d^2y}{dx^2} + y = 0$$

Q.66 The solution of $\frac{dy}{dx} + y = e^{-x}$, $y(0) = 0$ is

(a) $y = e^{-x}(x-1)$ (b) $y = xe^{x}$

(c) $y = xe^{-x} + 1$ (d) $y = xe^{-x}$

Ans. *(d)* Given that, $\frac{dy}{dx} + y = e^{-x}$

which is a linear differential equation comparing with $\frac{dy}{dx} + P \cdot y = Q$.

Here, $P = 1$ and $Q = e^{-x}$

$$\text{I.F} = e^{\int dx} = e^{x}$$

The general solution is $y \cdot \text{I.F} = \int Q \cdot \text{I.F}\, dx + C$

$$y \cdot e^x = \int e^{-x} \cdot e^x\, dx + C$$

$$\Rightarrow \quad y\, e^x = \int dx + C$$

$$\Rightarrow \quad y\, e^x = x + C \quad \text{...(i)}$$

When $x = 0$ and $y = 0$ then, $0 = 0 + C \Rightarrow C = 0$

Eq. (i) becomes $y \cdot e^x = x \Rightarrow y = x \cdot e^{-x}$

Q.67 The order and degree of differential equation

$$\left(\frac{d^3y}{dx^3}\right)^2 - 3\frac{d^2y}{dx^2} + 2\left(\frac{dy}{dx}\right)^4 = y^4 \text{ are}$$

(a) 1, 4 (b) 3, 4 (c) 2, 4 (d) 3, 2

Ans. ***(d)*** Given that, $\left(\frac{d^3y}{dx^3}\right)^2 - 3\frac{d^2y}{dx^2} + 2\left(\frac{dy}{dx}\right)^4 = y^4$

$\therefore$ Order = 3 and degree = 2

Highest order derivative is $\frac{d^3y}{dx^2}$ so orders is 3 and power on this is 2 so degree is 2.

Q.68 The order and degree of differential equation $\left[1 + \left(\frac{dy}{dx}\right)^2\right] = \frac{d^2y}{dx^2}$ are

(a) $2, \frac{3}{2}$ (b) 2, 3 (c) 2, 1 (d) 3, 4

Ans. ***(c)*** Given that, $\left[1 + \left(\frac{dy}{dx}\right)^2\right] = \frac{d^2y}{dx^2}$

$\therefore$ Order = 2 and degree = 1

***Q.69** The differential equation of family of curves $y^2 = 4a\,(x + a)$ is

(a) $y^2 = 4\frac{dy}{dx}\left(x + \frac{dy}{dx}\right)$ (b) $2y\frac{dy}{dx} = 4a$

(c) $y\frac{d^2y}{dx^2} + \left(\frac{dy}{dx}\right)^2 = 0$ (d) $2x\frac{dy}{dx} + y\left(\frac{dy}{dx}\right)^2 - y = 0$

Ans. ***(d)*** Given that, $y^2 = 4a\,(x + a)$ .. (i)

On differentiating both sides w.r.t. x, we get

$$2y\frac{dy}{dx} = 4a \Rightarrow 2y\frac{dy}{dx} = 4a$$

$$\Rightarrow \quad y\frac{dy}{dx} = 2a \Rightarrow a = \frac{1}{2}y\frac{dy}{dx} \quad \text{...(ii)}$$

On putting the value of a from Eq. (ii) in Eq. (i), we get

$$y^2 = 2y\frac{dy}{dx}\left(x + \frac{1}{2}y\frac{dy}{dx}\right)$$

Note (✱) Topics related to these question have been removed from latest NCERT.

$$\Rightarrow \qquad y^2 = 2xy\frac{dy}{dx} + y^2\left(\frac{dy}{dx}\right)^2$$

$$\Rightarrow \qquad 2x\frac{dy}{dx} + y\left(\frac{dy}{dx}\right)^2 - y = 0$$

Q.70 Which of the following is the general solution of $\frac{d^2y}{dx^2} - 2\frac{dy}{dx} + y = 0$?

(a) $y = (Ax + B)e^x$ (b) $y = (Ax + B)e^{-x}$
(c) $y = Ae^x + Be^{-x}$ (d) $y = A\cos x + B\sin x$

Ans. *(a)* From option (a) we have,

$$y = (Ax + B)e^x \qquad ...(i)$$

On differentiating both sides w.r.t. x, we get

$$\frac{dy}{dx} = (Ax + B)e^x + e^x(A + 0)$$

$$\Rightarrow \qquad \frac{dy}{dx} = y + Ae^x \qquad ...(ii)$$

Again, differentiating both sides w.r.t. x, we get

$$\frac{d^2y}{dx^2} = \frac{dy}{dx} + Ae^x$$

$$\Rightarrow \qquad \frac{d^2y}{dx^2} = \frac{dy}{dx} + \frac{dy}{dx} - y \qquad \left[\text{from Eq. (ii) } Ae^x = \frac{dy}{dx} - y\right]$$

$$\Rightarrow \qquad \frac{d^2y}{dx^2} - 2\frac{dy}{dx} + y = 0$$

Hence, option (a) is correct.

Q.71 The general solution of $\frac{dy}{dx} + y\tan x = \sec x$ is

(a) $y\sec x = \tan x + C$ (b) $y\tan x = \sec x + C$
(c) $\tan x = y\tan x + C$ (d) $x\sec x = \tan y + C$

Ans. *(a)* Given differential equation is

$$\frac{dy}{dx} + y\tan x = \sec x$$

which is a linear differential equation, comparing it with $\frac{dy}{dx} + Py = Q$, we get

$$P = \tan x,\ Q = \sec x, \qquad [\because e^{\log x} = x]$$

$$\therefore \qquad \text{I.F} = e^{\int \tan x\, dx} = e^{\log|\sec x|} = \sec x$$

The general solution is $y \cdot \text{I.F} = \int (Q \cdot \text{I.F})dx + C$

$$y \cdot \sec x = \int \sec x \cdot \sec x + C$$

$$\Rightarrow \qquad y \cdot \sec x = \int \sec^2 x\, dx + C$$

$$\Rightarrow \qquad y \cdot \sec x = \tan x + C$$

Q.72 The solution of differential equation $\frac{dy}{dx} + \frac{y}{x} = \sin x$ is

(a) $x(y + \cos x) = \sin x + C$ (b) $x(y - \cos x) = \sin x + C$
(c) $xy\cos x = \sin x + C$ (d) $x(y + \cos x) = \cos x + C$

Ans. *(a)* Given differential equation is

$$\frac{dy}{dx} + y\frac{1}{x} = \sin x$$

which is linear differential equation.

On comparing it with $\frac{dy}{dx} + P\cdot y = Q$, we get

$$P = \frac{1}{x} \text{ and } Q = \sin x$$

$$\therefore \quad \text{I.F} = e^{\int \frac{1}{x} dx} = e^{\log x} = x$$

The general solution is $y \cdot \text{I.F} = \int (Q \cdot \text{I.F}) dx + C$

$$y \cdot x = \int x \cdot \sin x \, dx + C \quad \text{...(i)}$$

$$xy = -x \cos x - \int -\cos x \, dx + C$$

[using integration by parts]

$$xy = -x \cos x + \sin x + C$$

$$\Rightarrow \quad x(y + \cos x) = \sin x + C$$

Q.73 The general solution of differential equation $(e^x + 1) ydy = (y + 1) e^x \, dx$ is

(a) $(y + 1) = k(e^x + 1)$

(b) $y + 1 = e^x + 1 + k$

(c) $y = \log \{k(y + 1)(e^x + 1)\}$

(d) $y = \log \left\{\frac{e^x + 1}{y + 1}\right\} + k$

Ans. *(c)* Given differential equation is

$$(e^x + 1) ydy = (y + 1) e^x \, dx$$

$$\Rightarrow \quad \frac{dy}{dx} = \frac{e^x (1 + y)}{(e^x + 1)y} \Rightarrow \frac{dx}{dy} = \frac{(e^x + 1) y}{e^x (1 + y)}$$

$$\Rightarrow \quad \frac{dx}{dy} = \frac{e^x y}{e^x (1 + y)} + \frac{y}{e^x (1 + y)}$$

$$\Rightarrow \quad \frac{dx}{dy} = \frac{y}{1 + y} + \frac{y}{(1 + y) e^x}$$

$$\Rightarrow \quad \frac{dx}{dy} = \frac{y}{1 + y}\left(1 + \frac{1}{e^x}\right)$$

$$\Rightarrow \quad \frac{dx}{dy} = \frac{y}{1 + y}\left(\frac{e^x + 1}{e^x}\right)$$

$$\Rightarrow \quad \left(\frac{y}{1 + y}\right) dy = \left(\frac{e^x}{e^x + 1}\right) dx$$

[using variable separable method]

On integrating both sides, we get

$$\int \frac{y}{1 + y} dy = \int \frac{e^x}{1 + e^x} dx$$

$$\Rightarrow \quad \int \frac{1 + y - 1}{1 + y} dy = \int \frac{e^x}{1 + e^x} dx$$

$\Rightarrow \quad \int 1\,dy - \int \frac{1}{1+y}\,dy = \int \frac{e^x}{1+e^x}\,dx$

$\Rightarrow \quad y - \log|(1+y)| = \log|(1+e^x)| + \log k$

$\Rightarrow \quad y = \log(1+y) + \log(1+e^x) + \log(k)$

$\Rightarrow \quad y = \log\{k(1+y)(1+e^x)\}$ $[\log a + \log b = \log ab]$

Q.74 The solution of differential equation $\frac{dy}{dx} = e^{x-y} + x^2 e^{-y}$ is

(a) $y = e^{x-y} - x^2 e^{-y} + C$

(b) $e^y - e^x = \frac{x^3}{3} + C$

(c) $e^x + e^y = \frac{x^3}{3} + C$

(d) $e^x - e^y = \frac{x^3}{3} + C$

Ans. ***(b)*** Given that, $\frac{dy}{dx} = e^{x-y} + x^2 e^{-y}$

$\Rightarrow \quad \frac{dy}{dx} = e^x e^{-y} + x^2 e^{-y}$

$\Rightarrow \quad \frac{dy}{dx} = \frac{e^x + x^2}{e^y}$

$\Rightarrow \quad e^y\,dy = (e^x + x^2)\,dx$ [using variable separable method]

On integrating both sides, we get

$$\int e^y\,dy = \int (e^x + x^2)\,dx$$

$\Rightarrow \quad e^y = e^x + \frac{x^3}{3} + C$

$\Rightarrow \quad e^y - e^x = \frac{x^3}{3} + C$

Q.75 The solution of differential equation $\frac{dy}{dx} + \frac{2xy}{1+x^2} = \frac{1}{(1+x^2)^2}$ is

(a) $y(1+x^2) = C + \tan^{-1} x$

(b) $\frac{y}{1+x^2} = C + \tan^{-1} x$

(c) $y \log(1+x^2) = C + \tan^{-1} x$

(d) $y(1+x^2) = C + \sin^{-1} x$

Ans. ***(a)*** Given that, $\frac{dy}{dx} + \frac{2xy}{1+x^2} = \frac{1}{(1+x^2)^2} \Rightarrow \frac{dy}{dx} + P\cdot y = Q$

Here, $P = \frac{2x}{1+x^2}$ and $Q = \frac{1}{(1+x^2)^2}$

which is a linear differential equation.

$\therefore \quad \text{I.F} = e^{\int \frac{2x}{1+x^2}dx}$ $[\because \text{I.F} = e^{\int P\,dx}]$

Put $\quad 1 + x^2 = t \Rightarrow 2x\,dx = dt$

$\therefore \quad \text{I.F} = e^{\int \frac{dt}{t}} = e^{\log t} = e^{\log(1+x^2)} = 1 + x^2$

The general solution is $y \cdot \text{I.F} = \int Q \cdot \text{I.F}\,dx \times C$

$$y\cdot(1+x^2) = \int (1+x^2)\frac{1}{(1+x^2)^2}\,dx + C$$

$$\Rightarrow \quad y(1+x^2) = \int \frac{1}{1+x^2}\, dx + C$$

$$\Rightarrow \quad y(1+x^2) = \tan^{-1} x + C$$

Q.76 Fill in the blanks. (i to xi)

(i) The degree of the differential equation $\frac{d^2y}{dx^2} + e^{dy/dx} = 0$ is ....

(ii) The degree of the differential equation $\sqrt{1+\left(\frac{dy}{dx}\right)^2} = x$ is ....

(iii) The number of arbitrary constants in the general solution of a differential equation of order three is ....

(iv) $\frac{dy}{dx} + \frac{y}{x \log x} = \frac{1}{x}$ is an equation of the type ....

(v) General solution of the differential equation of the type $\frac{dx}{dy} + P_1 x = Q_1$ is given by ....

(vi) The solution of the differential equation $\frac{xdy}{dx} + 2y = x^2$ is ....

(vii) The solution of $(1+x^2)\frac{dy}{dx} + 2xy - 4x^2 = 0$ is ....

(viii) The solution of the differential equation $ydx + (x + xy)\, dy = 0$ is ....

(ix) General solution of $\frac{dy}{dx} + y = \sin x$ is ....

(x) The solution of differential equation $\cot y\, dx = xdy$ is ....

(xi) The integrating factor of $\frac{dy}{dx} + y = \frac{1+y}{x}$ is ....

Ans. (i) Given differential equation is

$$\frac{d^2y}{dx^2} + e^{\frac{dy}{dx}} = 0$$

Since the given differential equation is not a polynomial Degree of this equation is not defined.

(ii) Given differential equation is $\sqrt{1+\left(\frac{dy}{dx}\right)^2} = x \Rightarrow 1+\left(\frac{dy}{dx}\right)^2 = x^2$

So, degree of this equation is two.

(iii) There are three arbitrary constants.

(iv) Given differential equation is $\frac{dy}{dx} + \frac{y}{x \log x} = \frac{1}{x}$

The equation is of the type $\frac{dy}{dx} + Py = Q$ i.e. linear differential equation.

(v) Given differential equation is

$$\frac{dx}{dy} + P_1 x = Q_1$$

On comparing with $\frac{dx}{dy} + Px = Q$

$$P = P_1 \quad Q = Q_1$$

The general solution is

$$x \cdot \text{I.F} = \int Q(\text{I.F})\,dy + C \text{ i.e., } x\,e^{\int P\,dy} = \int Q\{e^{\int P dy}\}\,dy + C$$

(vi) Given differential equation is

$$x\frac{dy}{dx} + 2y = x^2 \Rightarrow \frac{dy}{dx} + \frac{2y}{x} = x$$

This equation of the form $\frac{dy}{dx} + Py = Q$.

$$\text{I.F} = e^{\int P dx} \text{ when } P = \frac{2}{x} \text{ and } Q = x$$

$$\therefore \quad \text{I.F} = e^{\int \frac{2}{x}dx} = e^{2\log x} = x^2 \qquad (m\log n = \log n^m)$$

The general solution is $y \times \text{I.F} = \int Q \times \text{I.F})dx + C$

$$yx^2 = \int x \cdot x^2\,dx + C$$

$$\Rightarrow \quad yx^2 = \frac{x^4}{4} + C$$

$$\Rightarrow \quad y = \frac{x^2}{4} + Cx^{-2}$$

(vii) Given differential equation is

$$(1+x^2)\frac{dy}{dx} + 2xy - 4x^2 = 0$$

$$\Rightarrow \quad \frac{dy}{dx} + \frac{2xy}{1+x^2} - \frac{4x^2}{1+x^2} = 0$$

$$\Rightarrow \quad \frac{dy}{dx} + \frac{2x}{1+x^2}y = \frac{4x^2}{1+x^2}$$

Which is linear differential equation of the form $\frac{dy}{dx} + Py = Q$

$$\therefore \quad P = \frac{2x}{1+x^2},\ Q = \frac{4x^2}{1+x^2}$$

Put $\quad 1+x^2 = t \Rightarrow 2xdx = dt$

$$\therefore \quad \text{I.F} = e^{\int \frac{dt}{t}} = e^{\log t} = e^{\log(1+x^2)} = 1+x^2 \qquad [\because \text{I.F} = e^{\int P d x}]$$

The general solution is, $y \cdot \text{I.F} = \int Q.\text{I.F}dx + C$

$$y\cdot(1+x^2)=\int(1+x^2)\frac{4x^2}{(1+x^2)}dx+C$$

$\Rightarrow$ $$(1+x^2)\,y=\int 4x^2\,dx+C$$

$\Rightarrow$ $$(1+x^2)y=4\frac{x^3}{3}+C$$

$\Rightarrow$ $$3y(1+x^2)=4x^3+C$$

(viii) Given differential equation is

$\Rightarrow$ $$ydx+(x+xy)\,dy=0$$

$\Rightarrow$ $$ydx+x\,(1+y)\,dy=0$$

$\Rightarrow$ $$\frac{dx}{-x}=\left(\frac{1+y}{y}\right)dy$$ (using variable separable method)

$\Rightarrow$ $$\int\frac{1}{x}\,dx=-\int\left(\frac{1}{y}+1\right)dy$$ [on integrating]

$\Rightarrow$ $$\log(x)=-\log(y)-y+\log A$$

$$\log(x)+\log(y)+y=\log A$$

$$\log(xy)+y=\log A$$

$\Rightarrow$ $$\log xy+\log e^y=\log A$$

$\Rightarrow$ $$xy\,e^y=A$$

$\Rightarrow$ $$xy=Ae^{-y}$$

(ix) Given differential equation is

$$\frac{dy}{dx}+y=\sin x$$

Which is a linear differential equation of the form $\frac{dy}{dx}+Py=Q$

$\therefore\ P=1,\ Q=\sin x$

$$\text{I.F}=e^{\int I\cdot dx}=e^x$$

The general solution is $y\cdot\text{I.F}=\int Q\cdot\text{I.F}dx+C$

$$y\cdot e^x=\int e^x\sin x\,dx+C \quad \text{...(i)}$$

Let $$I=\int e^x\sin x\,dx$$

$$I=\sin x\,e^x-\int\cos x\,e^x\,dx$$

$$=\sin x\,|e^x-\cos x|\,e^x+\int(-\sin x)\,e^x\,dx$$

$$=\sin\,|xe^x-\cos|\,xe^x-I$$

$$2I=e^x(\sin x-\cos x)$$

$$I=\frac{1}{2}e^x(\sin x-\cos x)$$

From Eq. (i), we get

$$y\cdot e^x=\frac{e^x}{2}(\sin x-\cos x)+C$$

$\Rightarrow$ $$y=\frac{1}{2}(\sin x-\cos x)+C\cdot e^{-x}$$

(x) Given differential equation is

$$\cot y\, dx = x dy$$

$\Rightarrow \quad \frac{1}{x} dx = \tan y\, dy$ [applying variable separable method]

On integrating both sides, we get

$$\Rightarrow \quad \int \frac{1}{x}\, dx = \int \tan y\, d y$$

$$\Rightarrow \quad \log (x) = \log (\sec y) + \log C$$

$$\log x - \log (\sec y) = \log C$$

$$\Rightarrow \quad \log \left(\frac{x}{\sec y} \right) = \log C$$

$$\Rightarrow \quad \frac{x}{\sec y} = C$$

$$\Rightarrow \quad x = C \sec y$$

(xi) Given differential equation is

$$\frac{dy}{dx} + y = \frac{1+y}{x}$$

$$\frac{dy}{dx} + y = \frac{1}{x} + \frac{y}{x}$$

$$\Rightarrow \quad \frac{dy}{dx} + y\left(1 - \frac{1}{x}\right) = \frac{1}{x}$$

which is a linear differential equation

On comparing it with $\frac{dy}{dx} + Py = Q$, we get

$p = 1 - \frac{1}{x}, Q = \frac{1}{x},$ $\quad I \cdot F = e^{\int P d x}$

$$= e^{x - \log x}$$

$= e^x \cdot e^{-\log x} = \frac{e^x}{x}$ $\quad [\because e^{\log x} = x]$

Hence, $\text{I.F} = \frac{e^x}{x}$

Q.77 State True or False for the following

(i) Integrating factor of the differential of the form $\frac{dx}{dy} + P_1\, x = Q_1$ is given by $e^{\int P_1 dy}$.

(ii) Solution of the differential equation of the type $\frac{dx}{dy} + P_1 x = Q_1$ is given by $x \cdot \text{IF} = \int (\text{IF}) \times Q_1\, dy$.

(iii) Correct substitution for the solution of the differential equation of the type $\frac{dy}{dx} = f(x, y)$, where $f(x, y)$ is a homogeneous function of zero degree is $y = vx$.

(iv) Correct substitution for the solution of the differential equation of the type $\frac{dx}{dy} = g(x, y)$, where $g(x, y)$ is a homogeneous function of the degree zero is $x = vy$.

(v) Number of arbitrary constants in the particular solution of a differential equation of order two is two.

(vi) The differential equation representing the family of circles $x^2 + (y - a)^2 = a^2$ will be of order two.

(vii) The solution of $\frac{dy}{dx} = \left(\frac{y}{x}\right)^{1/3}$ is $y^{2/3} - x^{2/3} = c$

* (viii) Differential equation representing the family of curves $y = e^x(A\cos x + B\sin x)$ is $\frac{d^2y}{dx^2} - 2\frac{dy}{dx} + 2y = 0$.

(ix) The solution of the differential equation $\frac{dy}{dx} = \frac{x + 2y}{x}$ is $x + y = kx^2$.

(x) Solution of $\frac{xdy}{dx} = y + x\tan\frac{y}{x}$ is $\sin\left(\frac{y}{x}\right) = cx$

(xi) The differential equation of all non horizontal lines in a plane is $\frac{d^2x}{dy^2} = 0$.

Ans. (i) True

Given differential equation,

$$\frac{dx}{dy} + P_1 x = Q_1$$

$$\therefore \qquad \text{I.F} = e^{\int p_1\, dy}$$

(ii) True

(iii) True

(iv) True

(v) False

There is no arbitrary constant in the particular solution of a differential equation.

(vi) False

We know that, order of the differential equation = number of arbitrary constant

Here, number of arbitrary constant = 1.

So order is one.

Note (*) Topics related to these question have been removed from latest NCERT.

(vii) True

Given differential equation, $\frac{dy}{dx} = \left(\frac{y}{x}\right)^{1/3}$

$$\Rightarrow \quad \frac{dy}{dx} = \frac{y^{1/3}}{x^{1/3}}$$

$$\Rightarrow \quad y^{-1/3}\, dy = x^{-1/3}\, dx \qquad \text{[using variable separable method]}$$

On integrating both sides, we get

$$\int y^{-1/3}\, dy = \int x^{-1/3}\, dx$$

$$\Rightarrow \quad \frac{y^{-1/3+1}}{\frac{-1}{3}+1} = \frac{x^{-1/3+1}}{\frac{-1}{3}+1} + \alpha$$

$$\Rightarrow \quad \frac{3}{2} y^{2/3} = \frac{3}{2} x^{2/3} + \alpha$$

$$\Rightarrow \quad y^{2/3} - x^{2/3} = C' \qquad \left[\text{where, } \frac{2}{3}\alpha = C'\right]$$

(viii) True

Given that, $y = e^x (A\cos x + B\sin x)$

On differentiating w.r.t. x, we get

$$\frac{dy}{dx} = e^x(-A\sin x + B\cos x) + e^x(A\cos x + B\sin x)$$

$$= e^x(-A\sin x + B\cos x) + y$$

$$\Rightarrow \quad \frac{dy}{dx} - y = e^x(-A\sin x + B\cos x)$$

Again differentiating w.r.t. x, we get

$$\frac{d^2y}{dx^2} - \frac{dy}{dx} = e^x(-A\cos x - B\sin x) + e^x(-A\sin x + B\cos x)$$

$$\frac{d^2y}{dx^2} - \frac{dy}{dx} = -e^x[A\cos x + B\sin x] + \frac{dy}{dx} - y$$

$\Rightarrow$

$$\Rightarrow \quad \frac{d^2y}{dx^2} - 2\frac{dy}{dx} + 2y = 0$$

(ix) True

Given that, $\frac{dy}{dx} = \frac{x+2y}{x} \Rightarrow \frac{dy}{dx} = 1 + \frac{2}{x}.y$

$$\Rightarrow \quad \frac{dy}{dx} - \frac{2}{x}y = 1 \qquad \text{...(i)}$$

On comparing Eq. (i) with $\frac{dy}{dx} + P\cdot y = Q$, we get

$$P = -\frac{2}{x},\; Q = 1$$

$$\text{I.F} = e^{\int \frac{-2}{x}dx} = e^{-2\log x} = x^{-2} \qquad [\because \text{I.F} = e^{\int P\,dx}]$$

The general solution, $y \times \text{I.F} = \int (Q \times \text{I.F.})dx + C$

$$y \cdot x^{-2} = \int x^{-2}\cdot 1\,dx + k$$

$$\Rightarrow \quad \frac{y}{x^2} = \frac{x^{-2+1}}{-2+1} + k$$

$$\Rightarrow \quad \frac{y}{x^2} = \frac{-1}{x} + k$$

$$\Rightarrow \quad y = -x + kx^2$$

$$\Rightarrow \quad x + y = kx^2$$

(x) True

Given differential equation,

$$\frac{xdy}{dx} = y + x \tan\left(\frac{y}{x}\right)$$

$$\Rightarrow \quad \frac{dy}{dx} = \frac{y}{x} + \tan\left(\frac{y}{x}\right). \quad \text{...(i)}$$

Put $\frac{y}{x} = v$ *i.e.*, $y = vx$ [It is a homogeneous equation]

$$\Rightarrow \quad \frac{dy}{dx} = v + \frac{xdv}{dx}$$

On substituting these values in Eq. (i), we get

$$\frac{xdv}{dx} + v = v + \tan v$$

$$\Rightarrow \quad \frac{dx}{x} = \frac{dv}{\tan v} \quad \text{[applying variable separable method]}$$

On integrating both sides, we get

$$\int \frac{1}{x}\, dx = \int \frac{1}{\tan v}\, dx$$

$$\Rightarrow \quad \log(x) = \log(\sin v) + \log C'$$

$$\Rightarrow \quad \log\left(\frac{x}{\sin v}\right) = \log C'$$

$$\Rightarrow \quad \frac{x}{\sin v} = C'$$

$$\Rightarrow \quad \frac{x}{\sin v} = \frac{1}{C'}$$

$$\Rightarrow \quad \sin v = Cx \quad \left[\text{where, } C = \frac{1}{C'}\right]$$

$$\Rightarrow \quad \sin \frac{y}{x} = Cx$$

(xi) True

Let any non-horizontal line in a plane is given by

$$y = mx + c \Rightarrow \frac{y}{m} = x + \frac{c}{m}$$

On differentiating w.r.t. y, we get

$$\frac{1}{m} = \frac{dx}{dy}$$

Again, differentiating w.r.t. y, we get $\frac{d^2x}{dy^2} = 0$

Chapter

10 Vector Algebra

Short Answer Type Questions

Q.1 Find the unit vector in the direction of sum of vectors $\mathbf{a} = 2\hat{\mathbf{i}} - \hat{\mathbf{j}} + \hat{\mathbf{k}}$ and $\mathbf{b} = 2\hat{\mathbf{j}} + \hat{\mathbf{k}}$.

Thinking Process We know that, unit vector in the direction of a vector **a** is $\frac{\mathbf{a}}{|\mathbf{a}|}$ and magnitudes of $\mathbf{a} = |\mathbf{a}| = \sqrt{(a_1^2) + (a_2^2) + (a_3^2)}$

So, first we will find the sum of vectors and then we will use this concept.

Ans. Let **c** denote the sum of **a** and **b**.

We have,
$$\mathbf{c} = \mathbf{a} + \mathbf{b}$$
$$= 2\hat{\mathbf{i}} - \hat{\mathbf{j}} + \hat{\mathbf{k}} + 2\hat{\mathbf{j}} + \hat{\mathbf{k}} = 2\hat{\mathbf{i}} + \hat{\mathbf{j}} + 2\hat{\mathbf{k}}$$

$\therefore$ Unit vector in the direction of
$$\mathbf{c} = \frac{\mathbf{c}}{|\mathbf{c}|} = \frac{2\hat{\mathbf{i}} + \hat{\mathbf{j}} + 2\hat{\mathbf{k}}}{\sqrt{2^2 + 1^2 + 2^2}} = \frac{2\hat{\mathbf{i}} + \hat{\mathbf{J}} + 2\hat{\mathbf{k}}}{\sqrt{9}}$$
$$\mathbf{c} = \frac{2\hat{\mathbf{i}} + \hat{\mathbf{j}} + 2\hat{\mathbf{k}}}{3}$$

Q.2 If $\mathbf{a} = \hat{\mathbf{i}} + \hat{\mathbf{j}} + 2\hat{\mathbf{k}}$ and $\mathbf{b} = 2\hat{\mathbf{i}} + \hat{\mathbf{j}} + 2\hat{\mathbf{k}}$, then find the unit vector in the direction of

(i) $6\mathbf{b}$

(ii) $2\mathbf{a} - \mathbf{b}$

Ans. Here, $\mathbf{a} = \hat{\mathbf{i}} + \hat{\mathbf{j}} + 2\hat{\mathbf{k}}$ and $\mathbf{b} = 2\hat{\mathbf{i}} + \hat{\mathbf{j}} - 2\hat{\mathbf{k}}$

(i) Since,
$$6\mathbf{b} = 12\hat{\mathbf{i}} + 6\hat{\mathbf{j}} - 12\hat{\mathbf{k}}$$

$\therefore$ Unit vector in the direction of $6\mathbf{b} = \frac{6\mathbf{b}}{|6\mathbf{b}|}$
$$= \frac{12\hat{\mathbf{i}} + 6\hat{\mathbf{j}} - 12\hat{\mathbf{i}}}{\sqrt{12^2 + 6^2 + (-12)^2}} = \frac{6(2\hat{\mathbf{i}} + \hat{\mathbf{j}} - 2\hat{\mathbf{k}})}{\sqrt{324}}$$
$$= \frac{6(2\hat{\mathbf{i}} + \hat{\mathbf{j}} - 2\hat{\mathbf{k}})}{18} = \frac{2\hat{\mathbf{i}} + \mathbf{j} - 2\hat{\mathbf{k}}}{3}$$

(ii) Since, $2\mathbf{a} - \mathbf{b} = 2(\hat{\mathbf{i}} + \hat{\mathbf{j}} + 2\hat{\mathbf{k}}) - (2\mathbf{i} + \hat{\mathbf{j}} - 2\hat{\mathbf{k}})$
$$= 2\hat{\mathbf{i}} + 2\hat{\mathbf{j}} + 4\hat{\mathbf{k}} - 2\hat{\mathbf{i}} - \hat{\mathbf{j}} + 2\hat{\mathbf{k}} = \hat{\mathbf{j}} + 6\hat{\mathbf{k}}$$

$\therefore$ Unit vector in the direction of $2\mathbf{a} - \mathbf{b} = \frac{2\mathbf{a} - \mathbf{b}}{|2\mathbf{a} - \mathbf{b}|} = \frac{\hat{\mathbf{j}} + 6\hat{\mathbf{k}}}{\sqrt{1 + 36}} = \frac{1}{\sqrt{37}}(\hat{\mathbf{j}} + 6\hat{\mathbf{k}})$

Q.3 Find a unit vector in the direction of $\mathbf{PQ}$, where P and Q have coordinates (5, 0, 8) and (3, 3, 2), respectively.

Thinking Process First find the vector $\mathbf{PQ}$ by using the formula $(x_2 - x_1)\hat{\mathbf{i}} + (y_2 - y_1)\hat{\mathbf{j}} + (z_2 - z_1)\hat{\mathbf{k}}$, then required unit vector is given by $\frac{\mathbf{PQ}}{|\mathbf{PQ}|}$.

Ans. Since, the coordinates of P and Q are (5, 0, 8) and (3, 3, 2), respectively.

$\therefore \quad \mathbf{PQ} = \mathbf{OQ} - \mathbf{OP}$

$$= (3\hat{\mathbf{i}} + 3\hat{\mathbf{j}} + 2\hat{\mathbf{k}}) - (5\hat{\mathbf{i}} + 0\hat{\mathbf{j}} + 8\hat{\mathbf{k}})$$

$$= -2\mathbf{i} + 3\mathbf{j} - 6\mathbf{k}$$

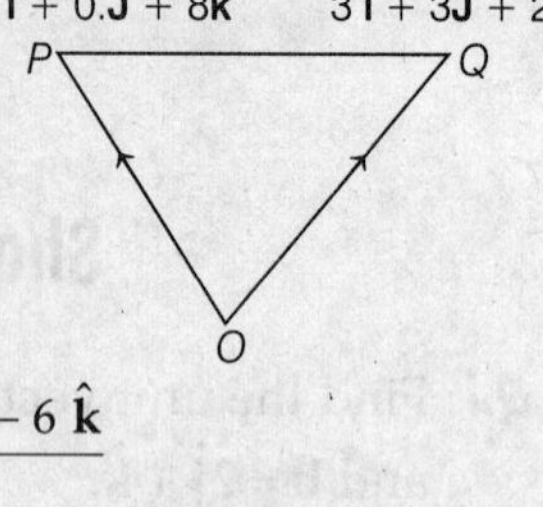

$\therefore$ Unit vector in the direction of $\mathbf{PQ} = \frac{\mathbf{PQ}}{|\mathbf{PQ}|}$

$$= \frac{-2\hat{\mathbf{i}} + 3\hat{\mathbf{j}} - 6\hat{\mathbf{k}}}{\sqrt{(-2)^2 + (3)^2 + (-6)^2}}$$

$$= \frac{-2\hat{\mathbf{i}} + 3\hat{\mathbf{j}} - 6\hat{\mathbf{k}}}{\sqrt{49}} = \frac{-2\hat{\mathbf{i}} + 3\hat{\mathbf{j}} - 6\hat{\mathbf{k}}}{7}$$

Q.4 If $\mathbf{a}$ and $\mathbf{b}$ are the position vectors of $\mathbf{A}$ and $\mathbf{B}$ respectively, then find the position vector of a point $\mathbf{C}$ in $\mathbf{BA}$ produced such that $\mathbf{BC} = 1.5\mathbf{BA}$.

Ans. Since, $\mathbf{OA} = \mathbf{a}$ and $\mathbf{OB} = \mathbf{b}$

$\therefore \quad \mathbf{BA} = \mathbf{OA} - \mathbf{OB} = \mathbf{a} - \mathbf{b}$

[multiply by 1.5 on both sides of equation]

and $\quad 1.5\,\mathbf{BA} = 1.5(\mathbf{a} - \mathbf{b})$

Since, $\quad \mathbf{BC} = 1.5\mathbf{BA} = 1.5(\mathbf{a} - \mathbf{b})$

$$\mathbf{OC} - \mathbf{OB} = 1.5\mathbf{a} - 1.5\mathbf{b}$$

$$\mathbf{OC} = 1.5\mathbf{a} - 1.5\mathbf{b} + \mathbf{b} \quad [\because \mathbf{OB} = \mathbf{b}]$$

$$= 1.5\mathbf{a} - 0.5\mathbf{b}$$

$$= \frac{3\mathbf{a} - \mathbf{b}}{2}$$

Graphically, explanation of the above solution is given below

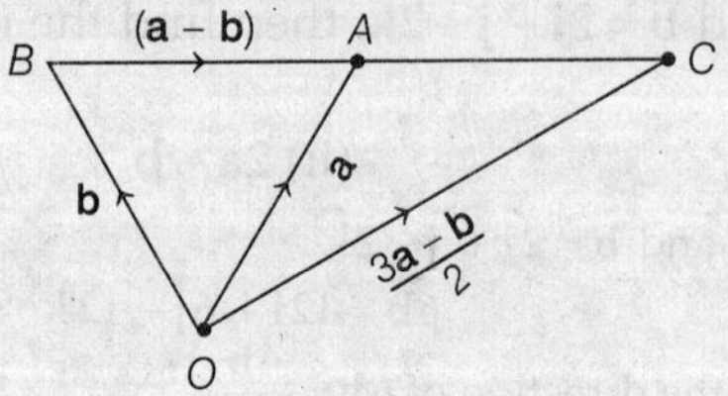

Note If $P_1(x_1, y_1, z_1)$ and $P_2(x_2, y_2, z_2)$ are any two points, then the vector joining P_1 and P_2 is the vector $P_1 P_2$.

$$\mathbf{OP}_2 - \mathbf{OP}_1 = (x_2\hat{\mathbf{i}} + y_2\hat{\mathbf{j}} + z_2\hat{\mathbf{k}}) - (x_1\hat{\mathbf{i}} + y_1\hat{\mathbf{j}} + z_1\hat{\mathbf{k}})$$

Q.5 Using vectors, find the value of k, such that the points $(k, -10, 3)$, $(1, -1, 3)$ and $(3, 5, 3)$ are collinear.

Thinking Process Here, use the following stepwise approach first, get the values of $|\mathbf{AB}|, |\mathbf{BC}|$ and $|\mathbf{AC}|$ and then use the concept that three points are collinear, if $|\mathbf{AB}| + |\mathbf{BC}| = |\mathbf{AC}|$ such that.

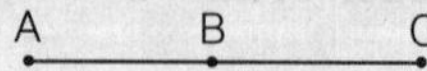

Ans. Let the points are $A(k, -10, 3)$, $B(1, -1, 3)$ and $C(3, 5, 3)$.

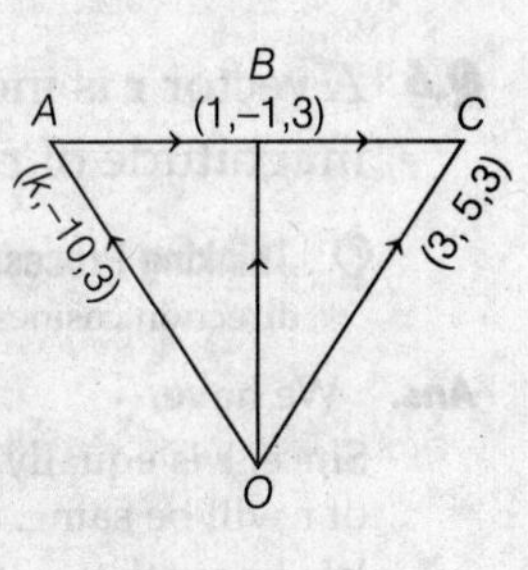

$$\mathbf{OA} = k\hat{\mathbf{i}} - 10\hat{\mathbf{j}} + 3\hat{\mathbf{k}}$$
$$\mathbf{OB} = \hat{\mathbf{i}} - \hat{\mathbf{j}} + 3\hat{\mathbf{k}}$$
$$\mathbf{OC} = 3\hat{\mathbf{i}} + 5\hat{\mathbf{j}} + 3\hat{\mathbf{k}}$$

So, $\mathbf{AB} = \mathbf{OB} - \mathbf{OA}$

$$= (\hat{\mathbf{i}} - \hat{\mathbf{j}} + 3\,\hat{\mathbf{k}}) - (k\hat{\mathbf{i}} - 10\hat{\mathbf{j}} + 3\hat{\mathbf{k}})$$
$$= (1-k)\hat{\mathbf{i}} + (-1+10)\hat{\mathbf{j}} + (3-3)\,\hat{\mathbf{k}}$$
$$= (1-k)\hat{\mathbf{i}} + 9\hat{\mathbf{j}} + 0\,\hat{\mathbf{k}}$$

$\therefore$ $|\mathbf{AB}| = \sqrt{(1-k)^2 + (9)^2 + 0} = \sqrt{(1-k)^2 + 81}$

Similarly, $\mathbf{BC} = \mathbf{OC} - \mathbf{OB}$

$$= (3\hat{\mathbf{i}} + 5\hat{\mathbf{j}} + 3\,\hat{\mathbf{k}}) - (\hat{\mathbf{i}} - \hat{\mathbf{j}} + 3\,\hat{\mathbf{k}})$$
$$= 2\hat{\mathbf{i}} + 6\hat{\mathbf{j}} + 0\,\hat{\mathbf{k}}$$

$\therefore$ $|\mathbf{BC}| = \sqrt{2^2 + 6^2 + 0} = 2\sqrt{10}$

and $\mathbf{AC} = \mathbf{OC} - \mathbf{OA}$

$$= (3\hat{\mathbf{i}} + 5\hat{\mathbf{j}} + 3\hat{\mathbf{k}}) - (k\hat{\mathbf{i}} - 10\hat{\mathbf{j}} + 3\,\hat{\mathbf{k}})$$
$$= (3-k)\hat{\mathbf{i}} + 15\hat{\mathbf{j}} + 0\,\hat{\mathbf{k}}$$

$\therefore$ $|\mathbf{AC}| = \sqrt{(3-k)^2 + 225}$

If A, B and C are collinear, then sum of modulus of any two vectors will be equal to the modulus of third vectors

For $|\mathbf{AB}| + |\mathbf{BC}| = |\mathbf{AC}|$,

$$\sqrt{(1-k)^2 + 81} + 2\sqrt{10} = \sqrt{(3-k)^2 + 225}$$

$\Rightarrow$ $\sqrt{(3-k)^2 + 225} - \sqrt{(1-k)^2 + 81} = 2\sqrt{10}$

$\Rightarrow$ $\sqrt{9 + k^2 - 6k + 225} - \sqrt{1 + k^2 - 2k + 81} = 2\sqrt{10}$ $\quad [\because (a-b)^2 = a^2 + b^2 - 2ab]$

$\Rightarrow$ $\sqrt{k^2 - 6k + 234} - 2\sqrt{10} = \sqrt{k^2 - 2k + 82}$ $\quad$ [on squaring on both sides]

$\Rightarrow$ $k^2 - 6k + 234 + 40 - 2\sqrt{k^2 - 6k + 234} \cdot 2\sqrt{10} = k^2 - 2k + 82$

$[\because (a-b)^2 = a^2 + b^2 - 2ab]$

$\Rightarrow$ $k^2 - 6k + 234 + 40 - k^2 + 2k - 82 = 4\sqrt{10}\sqrt{k^2 + 234 - 6k}$

$\Rightarrow$ $-4k + 192 = 4\sqrt{10}\sqrt{k^2 + 234 - 6k}$

$\Rightarrow$ $-k + 48 = \sqrt{10}\,\sqrt{k^2 + 234 - 6k}$

On squaring both sides, we get

$$48 \times 48 + k^2 - 96k = 10(k^2 + 234 - 6k)$$

$\Rightarrow$ $k^2 - 96k - 10k^2 + 60k = -48 \times 48 + 2340$

$\Rightarrow$ $-9k^2 - 36k = -48 \times 48 + 2340$

$\Rightarrow$ $(k^2 + 4k) = +16 \times 16 - 260$ $\quad$ [on dividing by 9 in both sides]

$\Rightarrow$ $k^2 + 4k = -4$

$$k^2 + 4k + 4 = 0$$

$\Rightarrow$ $(k+2)^2 = 0$

$\therefore$ $k = -2$

Q.6 A vector **r** is inclined at equal angles to the three axes. If the magnitude of **r** is $2\sqrt{3}$ units, then find the value of **r**.

Thinking Process If a vector **r** is inclined at equal angles to the three axes, then direction cosines of vector **r** will be same and then use, $\mathbf{r} = \mathbf{r}|\mathbf{r}|$.

Ans. We have, $|\mathbf{r}| = 2\sqrt{3}$

Since, **r** is equally inclined to the three axes, so direction cosines of the unit vector of r will be same. *i.e.*, $l = m = n$.

We know that,

$$l^2 + m^2 + n^2 = 1 \qquad [\because l = m = n]$$

$$\Rightarrow \quad l^2 + l^2 + l^2 = 1$$

$$\Rightarrow \quad l^2 = \frac{1}{3}$$

$$\Rightarrow \quad l = \pm\left(\frac{1}{\sqrt{3}}\right)$$

So,
$$\mathbf{r} = \pm\frac{1}{\sqrt{3}}\hat{\mathbf{i}} \pm \frac{1}{\sqrt{3}}\hat{\mathbf{j}} \pm \frac{1}{\sqrt{3}}\hat{\mathbf{k}}$$

$$\therefore \quad \mathbf{r} = \hat{\mathbf{r}}|\mathbf{r}| \qquad \left[\because \hat{\mathbf{r}} = \frac{\mathbf{r}}{|\mathbf{r}|}\right]$$

$$= \left[\pm\frac{1}{\sqrt{3}}\hat{\mathbf{i}} \pm \frac{1}{\sqrt{3}}\hat{\mathbf{j}} \pm \frac{1}{\sqrt{3}}\hat{\mathbf{k}}\right] 2\sqrt{3} \qquad [\because |r| = 2\sqrt{3}]$$

$$= \pm 2\hat{\mathbf{i}} \pm 2\hat{\mathbf{j}} \pm 2\hat{\mathbf{k}} = \pm 2(\hat{\mathbf{i}} + \hat{\mathbf{j}} + \hat{\mathbf{k}})$$

Q.7 If a vector **r** has magnitude 14 and direction ratios 2, 3 and – 6. Then, find the direction cosines and components of **r**, given that **r** makes an acute angle with *X*-axis.

Ans. Here, $|\mathbf{r}| = 14$, $\mathbf{a} = 2k$, $\mathbf{b} = 3k$ and $\mathbf{c} = -6k$

∴Direction cosines *l*, *m* and *n* are

$$l = \frac{\mathbf{a}}{|\mathbf{r}|} = \frac{2k}{14} = \frac{k}{7}$$

$$m = \frac{\mathbf{b}}{|\mathbf{r}|} = \frac{3k}{14}$$

and
$$n = \frac{\mathbf{c}}{|\mathbf{r}|} = \frac{-6k}{14} = \frac{-3k}{7}$$

Also, we know that

$$l^2 + m^2 + n^2 = 1$$

$$\Rightarrow \quad \frac{k^2}{49} + \frac{9k^2}{196} + \frac{9k^2}{49} = 1$$

$$\Rightarrow \quad \frac{4k^2 + 9k^2 + 36k^2}{196} = 1$$

$$\Rightarrow \quad k^2 = \frac{196}{49} = 4$$

$$\Rightarrow \quad k = \pm 2$$

So, the direction cosines (l, m, n) are $\frac{2}{7}, \frac{3}{7}$ and $\frac{-6}{7}$.

[since, **r** makes an acute angle with *X*-axis and take $k = 2$]

$\because \quad \mathbf{r} = \hat{\mathbf{r}} \cdot |\mathbf{r}|$

$\therefore \quad \mathbf{r} = (l\hat{\mathbf{i}} + m\hat{\mathbf{j}} + n\hat{\mathbf{k}})|\mathbf{r}|$

$$= \left(\frac{2}{7}\hat{\mathbf{i}} + \frac{3}{7}\hat{\mathbf{j}} - \frac{6}{7}\hat{\mathbf{k}}\right) \cdot 14$$

$$= +4\hat{\mathbf{i}} + 6\hat{\mathbf{j}} - 12\hat{\mathbf{k}}$$

Q.8 Find a vector of magnitude 6, which is perpendicular to both the vectors $2\hat{\mathbf{i}} - \hat{\mathbf{j}} + 2\hat{\mathbf{k}}$ and $4\hat{\mathbf{i}} - \hat{\mathbf{j}} + 3\hat{\mathbf{k}}$.

Thinking Process First, we will use this concept any vector is perpendicular to both the vectors **a** and **b** is given by $\mathbf{a} \times \mathbf{b} = \begin{vmatrix} \hat{\mathbf{i}} & \hat{\mathbf{j}} & \hat{\mathbf{k}} \\ a_1 & a_2 & a_3 \\ b_1 & b_2 & b_3 \end{vmatrix}$ and then we will find the vector with magnitude 6.

Ans. Let $\mathbf{a} = 2\hat{\mathbf{i}} - \hat{\mathbf{j}} + 2\hat{\mathbf{k}}$ and $\mathbf{b} = 4\hat{\mathbf{i}} - \hat{\mathbf{j}} + 3\hat{\mathbf{k}}$

So, any vector is perpendicular to both the vectors **a** and **b** is given by

$$\mathbf{a} \times \mathbf{b} = \begin{vmatrix} \hat{\mathbf{i}} & \hat{\mathbf{j}} & \hat{\mathbf{k}} \\ 2 & -1 & 2 \\ 4 & -1 & 3 \end{vmatrix}$$

$$= \hat{\mathbf{i}}(-3 + 2) - \hat{\mathbf{j}}(6 - 8) + \hat{\mathbf{k}}(-2 + 4)$$

$$= -\hat{\mathbf{i}} + 2\hat{\mathbf{j}} + 2\hat{\mathbf{k}} = \mathbf{r} \quad \text{[say]}$$

A vector of magnitude 6 in the direction of **r**

$$= \frac{\mathbf{r}}{|\mathbf{r}|} \cdot 6 = \frac{-\hat{\mathbf{i}} + 2\hat{\mathbf{j}} + 2\hat{\mathbf{k}}}{\sqrt{(-1)^2 + 2^2 + 2^2}} \cdot 6$$

$$= \frac{-6}{3}\hat{\mathbf{i}} + \frac{12}{3}\hat{\mathbf{j}} + \frac{12}{3}\hat{\mathbf{k}}$$

$$= -2\hat{\mathbf{i}} + 4\hat{\mathbf{j}} + 4\hat{\mathbf{k}}$$

Mistake Alert! *Students forget to multiply the unit vector to given magnitude that is 6.*

Q.9 Find the angle between the vectors $2\hat{\mathbf{i}} - \hat{\mathbf{j}} + \hat{\mathbf{k}}$ and $3\hat{\mathbf{i}} + 4\hat{\mathbf{j}} - \hat{\mathbf{k}}$.

Thinking Process If **a** and **b** are two vectors, making angle θ with each other, then $\cos\theta = \frac{\mathbf{a} \cdot \mathbf{b}}{|\mathbf{a}||\mathbf{b}|}$, using this concept we will find θ.

Ans. Let $\mathbf{a} = 2\hat{\mathbf{i}} - \hat{\mathbf{j}} + \hat{\mathbf{k}}$ and $\mathbf{b} = 3\hat{\mathbf{i}} + 4\hat{\mathbf{j}} - \hat{\mathbf{k}}$

We know that, angle between two vectors **a** and **b** is given by

$$\cos\theta = \frac{\mathbf{a} \cdot \mathbf{b}}{|\mathbf{a}||\mathbf{b}|}$$

$$[\because (a_1\hat{\mathbf{i}} + b_1\hat{\mathbf{j}} + c_1\hat{\mathbf{k}}) \cdot (a_2\hat{\mathbf{i}} + b_2\hat{\mathbf{j}} + c_2\hat{\mathbf{k}}) = a_1a_2 + b_1b_2 + c_1c_2]$$

$$= \frac{(2\hat{\mathbf{i}} - \hat{\mathbf{j}} + \hat{\mathbf{k}})(3\hat{\mathbf{i}} + 4\hat{\mathbf{j}} - \hat{\mathbf{k}})}{\sqrt{4 + 1 + 1}\sqrt{9 + 16 + 1}}$$

$$= \frac{6 - 4 - 1}{\sqrt{6}\sqrt{26}} = \frac{1}{2\sqrt{39}}$$

$$\therefore \quad \theta = \cos^{-1}\left(\frac{1}{2\sqrt{39}}\right)$$

Q.10 If $\mathbf{a}+\mathbf{b}+\mathbf{c}=0$, then show that $\mathbf{a}\times\mathbf{b}=\mathbf{b}\times\mathbf{c}=\mathbf{c}\times\mathbf{a}$. Interpret the result geometrically.

Ans. Since, $\mathbf{a}+\mathbf{b}+\mathbf{c}=0$

$\Rightarrow \quad \mathbf{b}=-\mathbf{c}-\mathbf{a}$

Now, $\mathbf{a}\times\mathbf{b}=\mathbf{a}\times(-\mathbf{c}-\mathbf{a})$

$=\mathbf{a}\times(-\mathbf{c})+\mathbf{a}\times(-\mathbf{a})=-\mathbf{a}\times\mathbf{c}$ $\quad [\because a\times a=0]$

$\Rightarrow \quad \mathbf{a}\times\mathbf{b}=\mathbf{c}\times\mathbf{a}$ $\quad$ ...(i)

Also, $\mathbf{b}\times\mathbf{c}=(-\mathbf{c}-\mathbf{a})\times\mathbf{c}$

$=(-\mathbf{c}\times\mathbf{c})+(-\mathbf{a}\times\mathbf{c})=-\mathbf{a}\times\mathbf{c}$ $\quad [\because c\times c=0]$

$\Rightarrow \quad \mathbf{b}\times\mathbf{c}=\mathbf{c}\times\mathbf{a}$ $\quad$ ...(ii)

From Eqs. (i) and (ii), we get

$$\mathbf{a}\times\mathbf{b}=\mathbf{b}\times\mathbf{c}=\mathbf{c}\times\mathbf{a}$$

Geometrical interpretation of the result

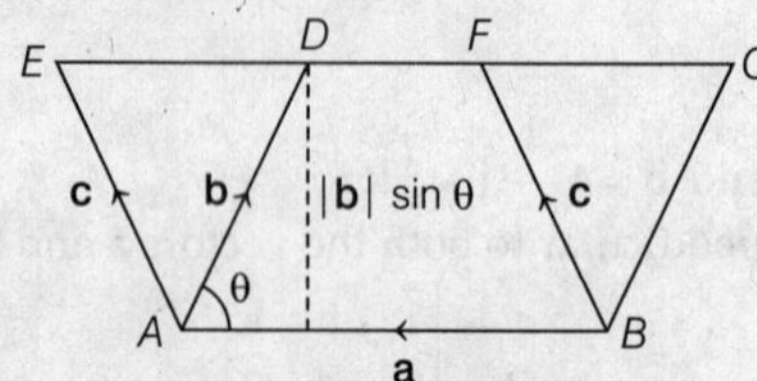

If *ABCD* is a parallelogram such that $\mathbf{AB}=\mathbf{a}$ and $\mathbf{AD}=\mathbf{b}$ and these adjacent sides are making angle θ between each other, then we say that

Area of parallelogram $ABCD=|\mathbf{a}||\mathbf{b}||\sin\theta|=|\mathbf{a}\times\mathbf{b}|$

Since, parallelogram on the same base and between the same parallels are equal in area.

We can say that, $\quad |\mathbf{a}\times\mathbf{b}|=|\mathbf{a}\times\mathbf{c}|=|\mathbf{b}\times\mathbf{c}|$

This also implies that, $\quad \mathbf{a}\times\mathbf{b}=\mathbf{a}\times\mathbf{c}=\mathbf{b}\times\mathbf{c}$

So, area of the parallelograms formed by taking any two sides represented by **a**, **b** and **c** as adjacent are equal.

Note We know that diagonal of parallelogram is $\mathbf{AC}=\mathbf{AB}+\mathbf{AD}$, $\mathbf{BD}=\mathbf{BC}+\mathbf{BA}$.

Q.11 Find the sine of the angle between the vectors $\mathbf{a}=3\hat{\mathbf{i}}+\hat{\mathbf{j}}+2\hat{\mathbf{k}}$ and $\mathbf{b}=2\hat{\mathbf{i}}-2\hat{\mathbf{j}}+4\hat{\mathbf{k}}$.

Thinking Process We know that, if θ is the angle between **a** and **b** then $\sin\theta=\dfrac{|\mathbf{a}\times\mathbf{b}|}{|\mathbf{a}||\mathbf{b}|}$

Ans. Given vector are

$\mathbf{a}=3\hat{\mathbf{i}}+\hat{\mathbf{j}}+2\hat{\mathbf{k}}$ and $\mathbf{b}=2\hat{\mathbf{i}}-2\hat{\mathbf{j}}+4\hat{\mathbf{k}}$

Let θ be the angle between the vector **a** and **b** then $\sin\theta=\dfrac{|\mathbf{a}\times\mathbf{b}|}{|\mathbf{a}||\mathbf{b}|}$

$$\therefore \quad \mathbf{a}\times\mathbf{b}=\begin{vmatrix}\hat{\mathbf{i}} & \hat{\mathbf{j}} & \hat{\mathbf{k}}\\ 3 & 1 & 2\\ 2 & -2 & 4\end{vmatrix}=\hat{\mathbf{i}}(4+4)-\hat{\mathbf{j}}(12-4)+\hat{\mathbf{k}}(-6-2)$$

$$=8\hat{\mathbf{i}}-8\hat{\mathbf{j}}-8\hat{\mathbf{k}}$$

$$|\mathbf{a}\times\mathbf{b}|=\sqrt{(8)^2+(8)^2+(-8)^2}=\sqrt{64+64+64}=\sqrt{192}$$

$$\therefore \quad \sin\theta=\frac{\sqrt{192}}{\sqrt{24}\times\sqrt{14}}=\frac{8\sqrt{3}}{\sqrt{8}\times\sqrt{3}\times\sqrt{2}\times\sqrt{7}}=\frac{\sqrt{8}}{\sqrt{2}\times\sqrt{7}}=\frac{\sqrt{4}}{\sqrt{7}}=\frac{2}{\sqrt{7}}$$

$$\left[\because |\mathbf{a}|=\sqrt{3^2+1^2+2^2}=\sqrt{14}, |\mathbf{b}|=\sqrt{2^2+(-2)^2+(4)^2}=\sqrt{24}\right]$$

Alternate Method

$$\cos\theta = \frac{\mathbf{a}\cdot\mathbf{b}}{|\mathbf{a}||\mathbf{b}|}$$

$$\cos\theta = \frac{(3\hat{\mathbf{i}}+\hat{\mathbf{j}}+2\hat{\mathbf{k}})\cdot(2\hat{\mathbf{i}}-2\hat{\mathbf{j}}+4\hat{\mathbf{k}})}{\sqrt{(3)^2+(1)^2+(2)^2}\cdot\sqrt{(2)^2+(-2)^2+(4)^2}}$$

$$= \frac{6-2+8}{\sqrt{14}\sqrt{24}} = \frac{12}{2\sqrt{14}\sqrt{6}} = \frac{6}{\sqrt{84}} = \frac{6}{2\sqrt{21}} = \frac{3}{\sqrt{21}}$$

$\therefore$ $$\sin\theta = \sqrt{1-\cos^2\theta}$$

$$= \sqrt{1-\frac{9}{21}} = \sqrt{\frac{12}{21}} = \frac{2\sqrt{3}}{\sqrt{3}\sqrt{7}} = \frac{2}{\sqrt{7}}$$

Q.12 If A, B, C and D are the points with position vectors $\hat{\mathbf{i}}+\hat{\mathbf{j}}-\hat{\mathbf{k}}$, $2\hat{\mathbf{i}}-\hat{\mathbf{j}}+3\hat{\mathbf{k}}$, $2\hat{\mathbf{i}}-3\hat{\mathbf{k}}$ and $3\hat{\mathbf{i}}-2\hat{\mathbf{j}}+\hat{\mathbf{k}}$ respectively, then find the projection of **AB** along **CD**.

Thinking Process We will use the concept that projection of **a** along **b** is $\frac{\mathbf{a}\cdot\mathbf{b}}{|\mathbf{b}|}$.

Ans. Here, $\mathbf{OA} = \hat{\mathbf{i}}+\hat{\mathbf{j}}-\hat{\mathbf{k}}$, $\mathbf{OB} = 2\hat{\mathbf{i}}-\hat{\mathbf{j}}+3\hat{\mathbf{k}}$, $\mathbf{OC} = 2\hat{\mathbf{i}}-3\hat{\mathbf{k}}$ and $\mathbf{OD} = 3\hat{\mathbf{i}}-2\hat{\mathbf{j}}+\hat{\mathbf{k}}$

$\therefore$ $$\mathbf{AB} = \mathbf{OB}-\mathbf{OA} = (2-1)\hat{\mathbf{i}}+(-1-1)\hat{\mathbf{j}}+(3+1)\hat{\mathbf{k}}$$
$$= \hat{\mathbf{i}}-2\hat{\mathbf{j}}+4\hat{\mathbf{k}}$$

and $$\mathbf{CD} = \mathbf{OD}-\mathbf{OC} = (3-2)\hat{\mathbf{i}}+(-2-0)\hat{\mathbf{j}}+(1+3)\hat{\mathbf{k}}$$
$$= \hat{\mathbf{i}}-2\hat{\mathbf{j}}+4\hat{\mathbf{k}}$$

So, the projection of **AB** along $\mathbf{CD} = \mathbf{AB}\cdot\frac{\mathbf{CD}}{|\mathbf{CD}|}$

$$= \frac{(\hat{\mathbf{i}}-2\hat{\mathbf{j}}+4\hat{\mathbf{k}})\cdot(\hat{\mathbf{i}}-2\hat{\mathbf{j}}+4\hat{\mathbf{k}})}{\sqrt{1^2+2^2+4^2}}$$

$$= \frac{1+4+16}{\sqrt{21}} = \frac{21}{\sqrt{21}}$$

$$= \sqrt{21} \text{ units}$$

Mistake Alert! *Here, student commonly confuse in projection of **a** on **b** is $\frac{\mathbf{a}\cdot\mathbf{b}}{|\mathbf{b}|}$ and projection of **b** on **a** is $\frac{\mathbf{a}\cdot\mathbf{b}}{|\mathbf{a}|}$. So, student be careful in use of projection.*

Q.13 Using vectors, find the area of the ΔABC with vertices $A(1, 2, 3)$, $B(2, -1, 4)$ and $C(4, 5, -1)$.

Thinking Process We know that,

Area of $\Delta ABC = \frac{1}{2}|\mathbf{AB}\times\mathbf{AC}|$. So, here we will use this concept.

Ans. Here, $$\mathbf{AB} = (2-1)\hat{\mathbf{i}}+(-1-2)\hat{\mathbf{j}}+(4-3)\hat{\mathbf{k}}$$

$[\because \mathbf{AB} = (x_2-x_1)\hat{\mathbf{i}}+(y_2-y_1)\hat{\mathbf{j}}+(z_2-z_1)\hat{\mathbf{k}}]$

$$= \hat{\mathbf{i}}-3\hat{\mathbf{j}}+\hat{\mathbf{k}}$$

and $$\mathbf{AC} = (4-1)\hat{\mathbf{i}}+(5-2)\hat{\mathbf{j}}+(-1-3)\hat{\mathbf{k}}$$
$$= 3\hat{\mathbf{i}}+3\hat{\mathbf{j}}-4\hat{\mathbf{k}}$$

C (4, 5, –1)

A (1, 2, 3) B (2, –1, 4)

$$\therefore \quad \mathbf{AB} \times \mathbf{AC} = \begin{vmatrix} \hat{\mathbf{i}} & \hat{\mathbf{j}} & \hat{\mathbf{k}} \\ 1 & -3 & 1 \\ 3 & 3 & -4 \end{vmatrix}$$

$$= \hat{\mathbf{i}}(12 - 3) - \hat{\mathbf{j}}(-4 - 3) + \hat{\mathbf{k}}(3 + 9)$$

$$= 9\hat{\mathbf{i}} + 7\hat{\mathbf{j}} + 12\hat{\mathbf{k}}$$

and

$$|\mathbf{AB} \times \mathbf{AC}| = \sqrt{9^2 + 7^2 + 12^2}$$

$$= \sqrt{81 + 49 + 144}$$

$$= \sqrt{274}$$

$$\therefore \quad \text{Area of } \Delta ABC = \frac{1}{2}|\mathbf{AB} \times \mathbf{AC}|$$

$$= \frac{1}{2}\sqrt{274} \text{ sq units}$$

Q.14 Using vectors, prove that the parallelogram on the same base and between the same parallels are equal in area.

Ans. Let $ABCD$ and $ABFE$ are parallelograms on the same base AB and between the same parallel lines AB and DF.

Here, $AB \parallel CD$ and $AE \parallel BF$

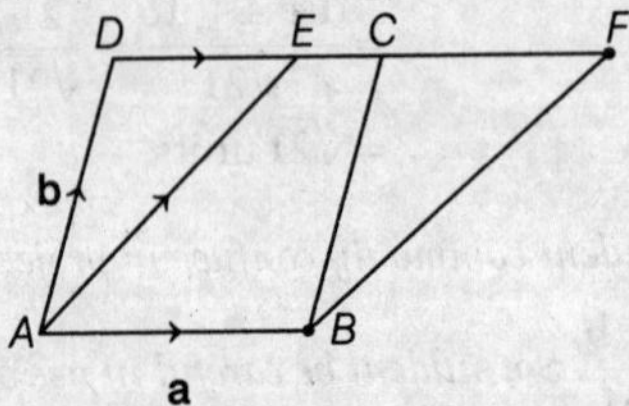

Let $\mathbf{AB} = \mathbf{a}$ and $\mathbf{AD} = \mathbf{b}$

$\therefore$ Area of parallelogram $ABCD = \mathbf{a} \times \mathbf{b}$

Now, area of parallelogram $ABFE = \mathbf{AB} \times \mathbf{AE}$

$= \mathbf{AB} \times (\mathbf{AD} + \mathbf{DE})$ [In ΔADE, $\mathbf{AE} = \mathbf{AD} + \mathbf{DE}$]

$= \mathbf{AB} \times (\mathbf{b} + k\mathbf{a})$ [let $\mathbf{DE} = k\mathbf{a}$, where k is a scalar]

$= \mathbf{a} \times (\mathbf{b} + k\mathbf{a})$ [$\because AB = a$]

$= (\mathbf{a} \times \mathbf{b}) + (\mathbf{a} \times k\mathbf{a})$

$= (\mathbf{a} \times \mathbf{b}) + k(\mathbf{a} \times \mathbf{a})$

$= (\mathbf{a} \times \mathbf{b})$ [$\because \mathbf{a} \times \mathbf{a} = 0$]

= Area of parallelogram $ABCD$

Hence proved.

Long Answer Type Questions

Q.15 Prove that in any ΔABC, $\cos A = \frac{b^2 + c^2 - a^2}{2bc}$, where a, b and c are the magnitudes of the sides opposite to the vertices A, B and C, respectively.

Ans. Here, components of C are $c\cos A$ and $c\sin A$ is drawn.

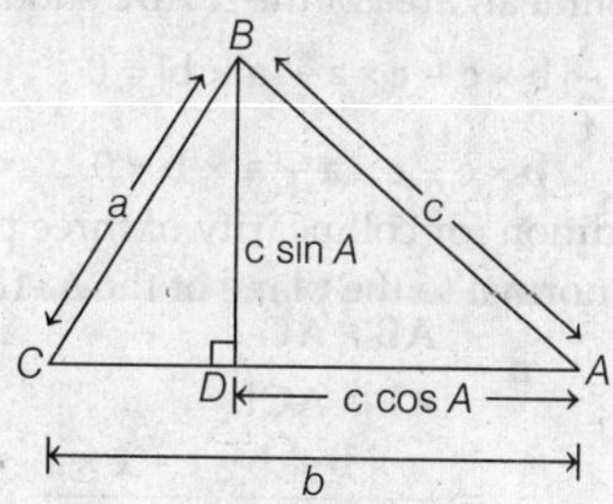

Since, $\quad \mathbf{CD} = b - c\cos A \quad [\because CD = CA - DA]$

In ΔBDC, $\quad \mathbf{BC}^2 = \mathbf{CD}^2 + \mathbf{BD}^2$

$$a^2 = (b - c\cos A)^2 + (c\sin A)^2$$

$$\Rightarrow \quad a^2 = b^2 + c^2\cos^2 A - 2bc\cos A + c^2\sin^2 A$$

$$\Rightarrow \quad 2bc\cos A = b^2 - a^2 + c^2(\cos^2 A + \sin^2 A)$$

$$\therefore \quad \cos A = \frac{b^2 + c^2 - a^2}{2bc}$$

Q.16 If **a**, **b** and **c** determine the vertices of a triangle, show that $\frac{1}{2}[\mathbf{b}\times\mathbf{c} + \mathbf{c}\times\mathbf{a} + \mathbf{a}\times\mathbf{b}]$ gives the vector area of the triangle. Hence, deduce the condition that the three points **a**, **b** and **c** are collinear. Also, find the unit vector normal to the plane of the triangle.

Thinking Process Here, we will use the following two concepts.

(i) If **a**, **b** and **c** are collinear, then the area of the triangle formed by the vectors will be zero.

(ii) We know that, $\mathbf{a} \times \mathbf{b} = |\mathbf{a}||\mathbf{b}|\sin\theta\,\hat{\mathbf{n}}$.

Ans. Since, **a**, **b** and **c** are the vertices of a ΔABC as shown.

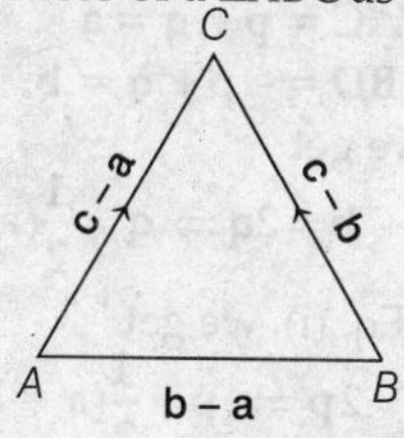

$$\therefore \quad \text{Area of } \Delta ABC = \frac{1}{2}|\mathbf{AB} \times \mathbf{AC}|$$

Now, $\quad \mathbf{AB} = \mathbf{b} - \mathbf{a} \quad$ and $\quad \mathbf{AC} = \mathbf{c} - \mathbf{a}$

$$\therefore \quad \text{Area of } \Delta ABC = \frac{1}{2} |(\mathbf{b} - \mathbf{a}) \times (\mathbf{c} - \mathbf{a})|$$

$$= \frac{1}{2} |\mathbf{b} \times \mathbf{c} - \mathbf{b} \times \mathbf{a} - \mathbf{a} \times \mathbf{c} + \mathbf{a} \times \mathbf{a}|$$

$$= \frac{1}{2} |\mathbf{b} \times \mathbf{c} + \mathbf{a} \times \mathbf{b} + \mathbf{c} \times \mathbf{a} + 0| \qquad [\because a \times a = 0]$$

$$= \frac{1}{2} |\mathbf{b} \times \mathbf{c} + \mathbf{a} \times \mathbf{b} + \mathbf{c} \times \mathbf{a}| \qquad \text{...(i)}$$

For three points to be collinear, area of the Δ *ABC* should be equal to zero.

$$\Rightarrow \quad \frac{1}{2}[\mathbf{b} \times \mathbf{c} + \mathbf{c} \times \mathbf{a} + \mathbf{a} \times \mathbf{b}] = 0$$

$$\Rightarrow \quad \mathbf{b} \times \mathbf{c} + \mathbf{c} \times \mathbf{a} + \mathbf{a} \times \mathbf{b} = 0 \qquad \text{...(ii)}$$

This is the required condition for collinearity of three points **a**, **b** and **c**.

Let $\hat{\mathbf{n}}$ be the unit vector normal to the plane of the Δ *ABC*.

$$\therefore \quad \hat{\mathbf{n}} = \frac{\mathbf{AB} \times \mathbf{AC}}{|\mathbf{AB} \times \mathbf{AC}|}$$

$$= \frac{\mathbf{a} \times \mathbf{b} + \mathbf{b} \times \mathbf{c} + \mathbf{c} \times \mathbf{a}}{|\mathbf{a} \times \mathbf{b} + \mathbf{b} \times \mathbf{c} + \mathbf{c} \times \mathbf{a}|}$$

Note Unit vector $\hat{\mathbf{n}}$ which is perpendicular to both the vecotrs **a** and **b** is

$$\hat{\mathbf{n}} = \pm \frac{\mathbf{a} \times \mathbf{b}}{|\mathbf{a} \times \mathbf{b}|}$$

Q.17 Show that area of the parallelogram whose diagonals are given by **a** and **b** is $\frac{|\mathbf{a} \times \mathbf{b}|}{2}$. Also, find the area of the parallelogram, whose diagonals are $2\hat{\mathbf{i}} - \hat{\mathbf{j}} + \hat{\mathbf{k}}$ and $\hat{\mathbf{i}} + 3\hat{\mathbf{j}} - \hat{\mathbf{k}}$.

Thinking Process If **p** and **q** are adjacent sides of a parallelogram, then the area formed by parallelogram $= |\mathbf{p} \times \mathbf{q}|$ and then we will obtained the desired result.

Ans. Let *ABCD* be a parallelogram such that

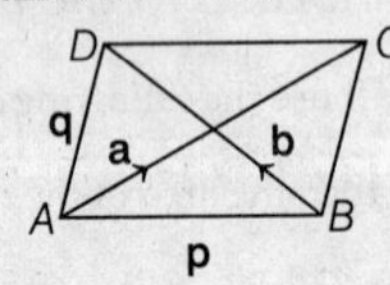

$$\mathbf{AB} = \mathbf{p}, \mathbf{AD} = \mathbf{q} \Rightarrow \mathbf{BC} = \mathbf{q} \quad \text{[opposite side of parallelogram]}$$

By triangle law of addition, we get

$$\mathbf{AC} = \mathbf{p} + \mathbf{q} = \mathbf{a} \qquad \text{[say] ...(i)}$$

Similarly,

$$\mathbf{BD} = -\mathbf{p} + \mathbf{q} = \mathbf{b} \qquad \text{[say] ...(ii)}$$

On adding Eqs. (i) and (ii), we get

$$\mathbf{a} + \mathbf{b} = 2\mathbf{q} \Rightarrow \mathbf{q} = \frac{1}{2}(\mathbf{a} + \mathbf{b})$$

On subtracting Eq. (ii) from Eq. (i), we get

$$\mathbf{a} - \mathbf{b} = 2\mathbf{p} \Rightarrow \mathbf{p} = \frac{1}{2}(\mathbf{a} - \mathbf{b})$$

Since, area of a parallelogram $ABCD = |\mathbf{p} \times \mathbf{q}|$

Now, $\quad \mathbf{p} \times \mathbf{q} = \frac{1}{4}(\mathbf{a} - \mathbf{b}) \times (\mathbf{a} + \mathbf{b}) \quad \left[p = \frac{1}{2}(a - b) \text{ and } q = \frac{1}{2}(a + b)\right]$

$$= \frac{1}{4}(\mathbf{a} \times \mathbf{a} + \mathbf{a} \times \mathbf{b} - \mathbf{b} \times \mathbf{a} - \mathbf{b} \times \mathbf{b})$$

$$= \frac{1}{4}[\mathbf{a} \times \mathbf{b} + \mathbf{a} \times \mathbf{b}] \quad [\because \mathbf{a} \times \mathbf{a} = 0 \text{ and } \mathbf{b} \times \mathbf{b} = 0]$$

$$= \frac{1}{2}(\mathbf{a} \times \mathbf{b})$$

So, area of a parallelogram $ABCD = |\mathbf{p} \times \mathbf{q}| = \frac{1}{2}|\mathbf{a} \times \mathbf{b}|$

Now, area of a parallelogram, whose diagonals are $2\hat{\mathbf{i}} - \hat{\mathbf{j}} + \hat{\mathbf{k}}$ and $\hat{\mathbf{i}} + 3\hat{\mathbf{j}} - \hat{\mathbf{k}}$.

$$= \frac{1}{2}|(2\hat{\mathbf{i}} - \hat{\mathbf{j}} + \hat{\mathbf{k}}) \times (\hat{\mathbf{i}} + 3\hat{\mathbf{j}} - \hat{\mathbf{k}})|$$

$$= \frac{1}{2}\begin{vmatrix} \hat{\mathbf{i}} & \hat{\mathbf{j}} & \hat{\mathbf{k}} \\ 2 & -1 & 1 \\ 1 & 3 & -1 \end{vmatrix}$$

$$= \frac{1}{2}|[\hat{\mathbf{i}}(1 - 3) - \hat{\mathbf{j}}(-2 - 1) + \hat{\mathbf{k}}(6 + 1)]|$$

$$= \frac{1}{2}|-2\hat{\mathbf{i}} + 3\hat{\mathbf{j}} + 7\hat{\mathbf{k}}| = \frac{1}{2}\sqrt{4 + 9 + 49} = \frac{1}{2}\sqrt{62} \text{ sq. units}$$

Q.18 If $\mathbf{a} = \hat{\mathbf{i}} + \hat{\mathbf{j}} + \hat{\mathbf{k}}$ and $\mathbf{b} = \hat{\mathbf{j}} - \hat{\mathbf{k}}$, then find a vector $\mathbf{c}$ such that $\mathbf{a} \times \mathbf{c} = \mathbf{b}$ and $\mathbf{a} \cdot \mathbf{c} = 3$.

Thinking Process We know that, for any two vectors

$$\mathbf{a} \times \mathbf{b} = \begin{vmatrix} \hat{\mathbf{i}} & \hat{\mathbf{j}} & \hat{\mathbf{k}} \\ a_1 & a_2 & a_3 \\ b_1 & b_2 & b_3 \end{vmatrix}$$

and $\mathbf{a} \cdot \mathbf{b} = a_1 b_1 + a_2 b_2 + a_3 b_3$, where $\mathbf{a} = a_1 \hat{\mathbf{i}} + a_2 \hat{\mathbf{j}} + a_3 \hat{\mathbf{k}}$ and $\mathbf{b} = b_1 \hat{\mathbf{i}} + b_2 \hat{\mathbf{j}} + b_3 \hat{\mathbf{k}}$. So, we will use this concept.

Ans. Let $\quad \mathbf{c} = x\hat{\mathbf{i}} + y\hat{\mathbf{j}} + z\hat{\mathbf{k}}$

Also, $\quad \mathbf{a} = \hat{\mathbf{i}} + \hat{\mathbf{j}} + \hat{\mathbf{k}}$ and $\mathbf{b} = \hat{\mathbf{j}} - \hat{\mathbf{k}}$

For $\mathbf{a} \times \mathbf{c} = \mathbf{b}$,

$$\begin{vmatrix} \hat{\mathbf{i}} & \hat{\mathbf{j}} & \hat{\mathbf{k}} \\ 1 & 1 & 1 \\ x & y & z \end{vmatrix} = \hat{\mathbf{j}} - \hat{\mathbf{k}}$$

$\Rightarrow \quad \hat{\mathbf{i}}(z - y) - \hat{\mathbf{j}}(z - x) + \hat{\mathbf{k}}(y - x) = \hat{\mathbf{j}} - \hat{\mathbf{k}}$

On comparing both sides, we get

$\therefore \quad z - y = 0 \quad$ ...(i)

$x - z = 1 \quad$ ...(ii)

$x - y = 1 \quad$ ...(iii)

Also, $\quad \mathbf{a} \cdot \mathbf{c} = 3 \quad$ [given]

$(\hat{\mathbf{i}} + \hat{\mathbf{j}} + \hat{\mathbf{k}}) \cdot (x\hat{\mathbf{i}} + y\hat{\mathbf{j}} + z\hat{\mathbf{k}}) = 3$

$\Rightarrow \quad x + y + z = 3 \quad$ ...(iv)

On adding Eqs. (ii) and (iii), we get

$$2x - y - z = 2 \quad \ldots(v)$$

On adding Eqs. (iv) and (v), we get

$$3x = 5$$

$$x = \frac{5}{3}$$

On putting the value of x in Eq. (ii) to evaluate the value of y, z, we get

$$\therefore \quad y = \frac{5}{3} - 1 = \frac{2}{3} \text{ and } z = \frac{2}{3}$$

Now,

$$\vec{c} = \frac{5}{3}\hat{i} + \frac{2}{3}\hat{j} + \frac{2}{3}\hat{k}$$

$$= \frac{1}{3}(5\hat{i} + 2\hat{j} + 2\hat{k})$$

Objective Type Questions

Choose the correct answer out of the given four options in each of the exercises from 19 to 33.

Q.19 The vector in the direction of the vector $\hat{i} - 2\hat{j} + 2\hat{k}$ that has magnitude 9 is

(a) $\hat{i} - 2\hat{j} + 2\hat{k}$ (b) $\frac{\hat{i} - 2\hat{j} + 2\hat{k}}{3}$ (c) $3(\hat{i} - 2\hat{j} + 2\hat{k})$ (d) $9(\hat{i} - 2\hat{j} + 2\hat{k})$

Ans. *(c)* Let $\mathbf{a} = \hat{i} - 2\hat{j} + 2\hat{k}$

Any vector in the direction of a vector $\mathbf{a}$ is given by $\frac{\mathbf{a}}{|\mathbf{a}|}$.

$$= \frac{\hat{i} - 2\hat{j} + 2\hat{k}}{\sqrt{1^2 + 2^2 + 2^2}} = \frac{\hat{i} - 2\hat{j} + 2\hat{k}}{3}$$

$\therefore$ Vector in the direction of $\mathbf{a}$ with magnitude $9 = 9 \cdot \frac{\hat{i} - 2\hat{j} + 2\hat{k}}{3}$

$$= 3(\hat{i} - 2\hat{j} + 2\hat{k})$$

Q.20 The position vector of the point which divides the join of points $2\mathbf{a} - 3\mathbf{b}$ and $\mathbf{a} + \mathbf{b}$ in the ratio 3 : 1, is

(a) $\frac{3\mathbf{a} - 2\mathbf{b}}{2}$ (b) $\frac{7\mathbf{a} - 8\mathbf{b}}{4}$ (c) $\frac{3\mathbf{a}}{4}$ (d) $\frac{5\mathbf{a}}{4}$

Ans. *(d)* Since, the position vector of a point R dividing the line segment joining the points P and Q, whose position vectors are $\mathbf{p}$ and $\mathbf{q}$ in the ratio $m : n$ internally, is given by $\frac{m\mathbf{q} + n\mathbf{p}}{m + n}$.

Let $\mathbf{OP} = 2\mathbf{a} - 3\mathbf{b}$, $\mathbf{OQ} = \mathbf{a} + \mathbf{b}$

Let the position vector of the point R divides the join of points $2\mathbf{a} - 3\mathbf{b}$ and $\mathbf{a} + \mathbf{b}$.

$$\therefore \quad \text{Position vector } R = \frac{3(\mathbf{a} + \mathbf{b}) + 1(2\mathbf{a} - 3\mathbf{b})}{3 + 1}$$

$$R = \frac{5\mathbf{a}}{4}$$

Q.21 The vector having initial and terminal points as (2, 5, 0) and (–3, 7, 4), respectively is

(a) $-\hat{i}+12\hat{j}+4\hat{k}$ (b) $5\hat{i}+2\hat{j}-4\hat{k}$ (c) $-5\hat{i}+2\hat{j}+4\hat{k}$ (d) $\hat{i}+\hat{j}+\hat{k}$

Ans. *(c)* Required vector $=(-3-2)\hat{i}+(7-5)\hat{j}+(4-0)\hat{k}$

[Position vector = *PV* of terminal point – *PV* of initial point]

$=-5\hat{i}+2\hat{j}+4\hat{k}$

Q.22 The angle between two vectors **a** and **b** with magnitudes $\sqrt{3}$ and 4, respectively and $\mathbf{a}\cdot\mathbf{b}=2\sqrt{3}$ is

(a) $\frac{\pi}{6}$ (b) $\frac{\pi}{3}$ (c) $\frac{\pi}{2}$ (d) $\frac{5\pi}{2}$

Ans. *(b)* Here, $|\mathbf{a}|=\sqrt{3}$, $|\mathbf{b}|=4$ and $\mathbf{a}\cdot\mathbf{b}=2\sqrt{3}$ [given]

We know that, $\mathbf{a}\cdot\mathbf{b}=|\mathbf{a}||\mathbf{b}|\cos\theta$

$\Rightarrow\quad 2\sqrt{3}=\sqrt{3}\cdot 4\cdot\cos\theta$

$\Rightarrow\quad \cos\theta=\frac{2\sqrt{3}}{4\sqrt{3}}=\frac{1}{2}=\cos\frac{\pi}{3}$ $\left[\because \cos\frac{\pi}{3}=\frac{1}{2}\right]$

$\therefore\quad \theta=\frac{\pi}{3}$

Q.23 Find the value of λ such that the vectors $\mathbf{a}=2\hat{i}+\lambda\hat{j}+\hat{k}$ and $\mathbf{b}=\hat{i}+2\hat{j}+3\hat{k}$ are orthogonal.

(a) 0 (b) 1 (c) $\frac{3}{2}$ (d) $\frac{-5}{2}$

Thinking Process Two non-zero vectors are orthogonal, if their dot product is zero. So, by using this concept, we will get the value of λ.

Ans. *(d)* Since, two non-zero vectors **a** and **b** are orthogonal *i.e.*, $\mathbf{a}\cdot\mathbf{b}=0$.

$\therefore\quad (2\hat{i}+\lambda\hat{j}+\hat{k})\cdot(\hat{i}+2\hat{j}+3\hat{k})=0$

$\Rightarrow\quad 2+2\lambda+3=0$

$\therefore\quad \lambda=\frac{-5}{2}$

Q.24 The value of λ for which the vectors $3\hat{i}-6\hat{j}+\hat{k}$ and $2\hat{i}-4\hat{j}+\lambda\hat{k}$ are parallel, is

(a) $\frac{2}{3}$ (b) $\frac{3}{2}$ (c) $\frac{5}{2}$ (d) $\frac{2}{5}$

Ans. *(a)* Since, two vectors are parallel *i.e.*, angle between them is zero.

$\therefore\quad (3\hat{i}-6\hat{j}+\hat{k})\cdot(2\hat{i}-4\hat{j}+\lambda\hat{k})=|3\hat{i}-6\hat{j}+\hat{k}|\cdot|2\hat{i}-4\hat{j}+\lambda\hat{k}|$

$[\because \mathbf{a}\cdot\mathbf{b}=|\mathbf{a}||\mathbf{b}|\cos 0^\circ \Rightarrow \mathbf{a}\cdot\mathbf{b}=|\mathbf{a}||\mathbf{b}|]\ [\therefore \cos 0^\circ=1]$

$\Rightarrow\quad 6+24+\lambda=\sqrt{9+36+1}\sqrt{4+16+\lambda^2}$

$\Rightarrow\quad 30+\lambda=\sqrt{46}\sqrt{20+\lambda^2}$

$\Rightarrow\quad 900+\lambda^2+60\lambda=46(20+\lambda^2)$ [on squaring both sides]

$\Rightarrow\quad \lambda^2+60\lambda-46\lambda^2=920-900$

$\Rightarrow\quad -45\lambda^2+60\lambda-20=0$

$$\Rightarrow \quad -45\lambda^2 + 30\lambda + 30\lambda - 20 = 0$$
$$\Rightarrow \quad -15\lambda(3\lambda - 2) + 10(3\lambda - 2) = 0$$
$$\Rightarrow \quad (10 - 15\lambda)(3\lambda - 2) = 0$$
$$\therefore \quad \lambda = \frac{2}{3}, \frac{2}{3}$$

Alternate Method

Let $\mathbf{a} = 3\hat{\mathbf{i}} - 6\hat{\mathbf{j}} + \hat{\mathbf{k}}$ and $\hat{\mathbf{b}} = 2\hat{\mathbf{i}} - 4\hat{\mathbf{j}} + \lambda\hat{\mathbf{k}}$

Since, $\mathbf{a} \parallel \mathbf{b}$

$$\Rightarrow \quad \frac{3}{2} = \frac{-6}{-4} = \frac{1}{\lambda} \Rightarrow \lambda = \frac{2}{3}$$

Mistake Alert! *In these questions students commonly use* $\mathbf{a} \cdot \mathbf{b} = 0$*, but if they are parallel, then use* $\frac{a_1}{a_2} = \frac{b_1}{b_2} = \frac{c_1}{c_2}$.

Q.25 The vectors from origin to the points A and B are $\mathbf{a} = 2\hat{\mathbf{i}} - 3\hat{\mathbf{j}} + 2\hat{\mathbf{k}}$ and $\mathbf{b} = 2\hat{\mathbf{i}} + 3\hat{\mathbf{j}} + \hat{\mathbf{k}}$ respectively, then the area of ΔOAB is equal to

(a) 340 (b) $\sqrt{25}$ (c) $\sqrt{229}$ (d) $\frac{1}{2}\sqrt{229}$

Ans. (d) $\therefore$ Area of $\Delta OAB = \frac{1}{2}|\mathbf{OA} \times \mathbf{OB}|$

$$= \frac{1}{2}|(2\hat{\mathbf{i}} - 3\hat{\mathbf{j}} + 2\hat{\mathbf{k}}) \times (2\hat{\mathbf{i}} + 3\hat{\mathbf{j}} + \hat{\mathbf{k}})|$$
$$= \frac{1}{2}\begin{vmatrix} \hat{\mathbf{i}} & \hat{\mathbf{j}} & \hat{\mathbf{k}} \\ 2 & -3 & 2 \\ 2 & 3 & 1 \end{vmatrix}$$
$$= \frac{1}{2}|[\hat{\mathbf{i}}(-3-6) - \hat{\mathbf{j}}(2-4) + \hat{\mathbf{k}}(6+6)]|$$
$$= \frac{1}{2}|-9\hat{\mathbf{i}} + 2\hat{\mathbf{j}} + 12\hat{\mathbf{k}}|$$
$$\therefore \quad \text{Area of } \Delta OAB = \frac{1}{2}\sqrt{(81 + 4 + 144)} = \frac{1}{2}\sqrt{229}$$

Q.26 For any vector $\mathbf{a}$, the value of $(\mathbf{a} \times \hat{\mathbf{i}})^2 + (\mathbf{a} \times \hat{\mathbf{j}})^2 + (\mathbf{a} \times \hat{\mathbf{k}})^2$ is

(a) $\mathbf{a}^2$ (b) $3\mathbf{a}^2$ (c) $4\mathbf{a}^2$ (d) $2\mathbf{a}^2$

Ans. (d) Let $\mathbf{a} = x\hat{\mathbf{i}} + y\hat{\mathbf{j}} + z\hat{\mathbf{k}}$

$$\therefore \quad \mathbf{a}^2 = x^2 + y^2 + z^2 \quad \text{...(i)}$$
$$\therefore \quad \mathbf{a} \times \hat{\mathbf{i}} = \begin{vmatrix} \hat{\mathbf{i}} & \hat{\mathbf{j}} & \hat{\mathbf{k}} \\ x & y & z \\ 1 & 0 & 0 \end{vmatrix}$$
$$= \hat{\mathbf{i}}[0] - \hat{\mathbf{j}}[-z] + \hat{\mathbf{k}}[-y]$$
$$= z\hat{\mathbf{j}} - y\hat{\mathbf{k}}$$
$$\therefore \quad (\mathbf{a} \times \hat{\mathbf{i}})^2 = (z\hat{\mathbf{j}} - y\hat{\mathbf{k}}) \cdot (z\hat{\mathbf{i}} - y\hat{\mathbf{k}})$$
$$= y^2 + z^2$$

Similarly, $(\mathbf{a} \times \hat{\mathbf{j}})^2 = x^2 + z^2$

and $(\mathbf{a} \times \hat{\mathbf{k}})^2 = x^2 + y^2$

$\therefore \; (\mathbf{a} \times \hat{\mathbf{i}})^2 + (\mathbf{a} \times \hat{\mathbf{j}})^2 + (\mathbf{a} \times \hat{\mathbf{k}})^2 = y^2 + z^2 + x^2 + z^2 + x^2 + y^2$

$= 2(x^2 + y^2 + z^2) = 2\mathbf{a}^2$ [using ...(i)]

Q.27 If $|\mathbf{a}| = 10, |\mathbf{b}| = 2$ and $\mathbf{a} \cdot \mathbf{b} = 12$, then the value of $|\vec{\mathbf{a}} \times \vec{\mathbf{b}}|$ is

(a) 5 (b) 10 (c) 14 (d) 16

Thinking Process We know that, $|\mathbf{a} \times \mathbf{b}| = |\mathbf{a}||\mathbf{b}||\sin\theta|\,\hat{\mathbf{n}}$ and $\mathbf{a} \cdot \mathbf{b} = |\mathbf{a}||\mathbf{b}|\cos\theta$. So, we will use these formulae to get the value of $|\mathbf{a} \times \mathbf{b}|$.

Ans. ***(d)*** Here, $|\mathbf{a}| = 10, |\mathbf{b}| = 2$ and $\mathbf{a} \cdot \mathbf{b} = 12$ [given]

$\therefore \; \mathbf{a} \cdot \mathbf{b} = |\mathbf{a}||\mathbf{b}|\cos\theta$

$12 = 10 \times 2\cos\theta$

$\Rightarrow \; \cos\theta = \frac{12}{20} = \frac{3}{5}$

$\therefore \; \sin\theta = \sqrt{1 - \cos^2\theta}$

$= \sqrt{1 - \frac{9}{25}}$

$\sin\theta = \pm\frac{4}{5}$

Now, $|\mathbf{a} \times \mathbf{b}| = |\mathbf{a}||\mathbf{b}||\sin\theta|$

$= 10 \times 2 \times \frac{4}{5} = 16$

***Q.28** The vectors $\lambda\hat{\mathbf{i}} + \hat{\mathbf{j}} + 2\hat{\mathbf{k}}, \hat{\mathbf{i}} + \lambda\hat{\mathbf{j}} - \hat{\mathbf{k}}$ and $2\hat{\mathbf{i}} - \hat{\mathbf{j}} + \lambda\hat{\mathbf{k}}$ are coplanar, if

(a) $\lambda = -2$ (b) $\lambda = 0$
(c) $\lambda = 1$ (d) $\lambda = -1$

Thinking Process To be a coplanar vectors the cross product is 0.

Ans. ***(a)*** Let $\mathbf{a} = \lambda\hat{\mathbf{i}} + \hat{\mathbf{j}} + 2\hat{\mathbf{k}}$, $\mathbf{b} = \hat{\mathbf{i}} + \lambda\hat{\mathbf{j}} - \hat{\mathbf{k}}$ and $\mathbf{c} = 2\hat{\mathbf{i}} - \hat{\mathbf{j}} + \lambda\hat{\mathbf{k}}$

For **a**, **b** and **c** to be coplanar,

$$\begin{vmatrix} \lambda & 1 & 2 \\ 1 & \lambda & -1 \\ 2 & -1 & \lambda \end{vmatrix} = 0$$

$\Rightarrow \; \lambda(\lambda^2 - 1) - 1(\lambda + 2) + 2(-1 - 2\lambda) = 0$

$\Rightarrow \; \lambda^3 - \lambda - \lambda - 2 - 2 - 4\lambda = 0$

$\Rightarrow \; \lambda^3 - 6\lambda - 4 = 0$

$\Rightarrow \; (\lambda + 2)(\lambda^2 - 2\lambda - 2) = 0$

$\Rightarrow \; \lambda = -2 \text{ or } \lambda = \frac{2 \pm \sqrt{12}}{2}$

$\Rightarrow \; \lambda = -2 \text{ or } \lambda = \frac{2 \pm 2\sqrt{3}}{2} = 1 \pm \sqrt{3}$

Note (*) *Topics related to these question have been removed from latest NCERT.*

Q.29 If **a**, **b** and **c** are unit vectors such that $\mathbf{a}+\mathbf{b}+\mathbf{c}=0$, then the value of $\mathbf{a}\cdot\mathbf{b}+\mathbf{b}\cdot\mathbf{c}+\mathbf{c}\cdot\mathbf{a}$ is

(a) 1 (b) 3

(c) $-\frac{3}{2}$ (d) None of these

Ans. *(c)* We have, $\mathbf{a}+\mathbf{b}+\mathbf{c}=0$

and $\mathbf{a}^2=1, \mathbf{b}^2=1, \mathbf{c}^2=1$

On multiplying by $(a+b+c)$ both sides in Eq. (i), we get

$\because \quad (\mathbf{a}+\mathbf{b}+\mathbf{c})\cdot(\mathbf{a}+\mathbf{b}+\mathbf{c})=0$

$\Rightarrow \quad \mathbf{a}\cdot\mathbf{a}+\mathbf{a}\cdot\mathbf{b}+\mathbf{a}\cdot\mathbf{c}+\mathbf{b}\cdot\mathbf{a}+\mathbf{b}\cdot\mathbf{b}+\mathbf{b}\cdot\mathbf{c}+\mathbf{c}\cdot\mathbf{a}+\mathbf{c}\cdot\mathbf{b}+\mathbf{c}\cdot\mathbf{c}=0$

$\Rightarrow \quad \mathbf{a}^2+\mathbf{b}^2+\mathbf{c}^2+2(\mathbf{a}\cdot\mathbf{b}+\mathbf{b}\cdot\mathbf{c}+\mathbf{c}\cdot\mathbf{a})=0$

$[\because \mathbf{a}\cdot\mathbf{b}=\mathbf{b}\cdot\mathbf{a}, \mathbf{b}\cdot\mathbf{c}=\mathbf{c}\cdot\mathbf{b} \text{ and } \mathbf{c}\cdot\mathbf{a}=\mathbf{a}\cdot\mathbf{c}]$

$\Rightarrow \quad 1+1+1+2(\mathbf{a}\cdot\mathbf{b}+\mathbf{b}\cdot\mathbf{c}+\mathbf{c}\cdot\mathbf{a})=0 \quad [\because a^2=b^2=c^2=1]$

$\Rightarrow \quad \mathbf{a}\cdot\mathbf{b}+\mathbf{b}\cdot\mathbf{c}+\mathbf{c}\cdot\mathbf{a}=-\frac{3}{2}$

Q.30 The projection vector of **a** on **b** is

(a) $\left(\frac{\mathbf{a}\cdot\mathbf{b}}{|\mathbf{b}|^2}\right)\mathbf{b}$ (b) $\frac{\mathbf{a}\cdot\mathbf{b}}{|\mathbf{a}|}$ (c) $\frac{\mathbf{a}\cdot\mathbf{b}}{|\mathbf{a}|}$ (d) $\left(\frac{\mathbf{a}\cdot\mathbf{b}}{|\mathbf{a}|^2}\right)\mathbf{b}$

Ans. *(a)* Projection vector of **a** on **b** is given by $=\frac{\mathbf{a}\cdot\mathbf{b}}{|\mathbf{b}|^2}\cdot\mathbf{b}$

Q.31 If **a**, **b** and **c** are three vectors such that $\mathbf{a}+\mathbf{b}+\mathbf{c}=0$ and $|\mathbf{a}|=2, |\mathbf{b}|=3$ and $|\mathbf{c}|=5$, then the value of $\mathbf{a}\cdot\mathbf{b}+\mathbf{b}\cdot\mathbf{c}+\mathbf{c}\cdot\mathbf{a}$ is

(a) 0 (b) 1 (c) –19 (d) 38

Ans. *(c)* Here, $\mathbf{a}+\mathbf{b}+\mathbf{c}=0$ and $\mathbf{a}^2=4, \mathbf{b}^2=9, \mathbf{c}^2=25$

or $\mathbf{a}\cdot\mathbf{a}=4$ $\mathbf{b}\cdot\mathbf{b}=9$ and $\mathbf{c}\cdot\mathbf{c}=25$

$\therefore \quad (\mathbf{a}+\mathbf{b}+\mathbf{c})\cdot(\mathbf{a}+\mathbf{b}+\mathbf{c})=0$

$\Rightarrow \mathbf{a}^2+\mathbf{a}\cdot\mathbf{b}+\mathbf{a}\cdot\mathbf{c}+\mathbf{b}\cdot\mathbf{a}+\mathbf{b}^2+\mathbf{b}\cdot\mathbf{c}+\mathbf{c}\cdot\mathbf{a}+\mathbf{c}\cdot\mathbf{b}+\mathbf{c}^2=0$

$\Rightarrow \quad \mathbf{a}\cdot\mathbf{a}+\mathbf{b}\cdot\mathbf{b}+\mathbf{c}\cdot\mathbf{c}+2(\mathbf{a}\cdot\mathbf{b}+\mathbf{b}\cdot\mathbf{c}+\mathbf{c}\cdot\mathbf{a})=0 \quad [\because \mathbf{a}\cdot\mathbf{b}=\mathbf{b}\cdot\mathbf{a}]$

$\Rightarrow \quad 4+9+25+2(\mathbf{a}\cdot\mathbf{b}+\mathbf{b}\cdot\mathbf{c}+\mathbf{c}\cdot\mathbf{a})=0$

$\Rightarrow \quad \mathbf{a}\cdot\mathbf{b}+\mathbf{b}\cdot\mathbf{c}+\mathbf{c}\cdot\mathbf{a}=\frac{-38}{2}=-19$

Q.32 If $|\mathbf{a}|=4$ and $-3\le\lambda\le 2$, then the range of $|\lambda\,\mathbf{a}|$ is

(a) [0, 8] (b) [–12, 8]

(c) [0, 12] (d) [8, 12]

Ans. *(c)* We have, $|\mathbf{a}|=4$ and $-3\le\lambda\le 2$

$\therefore \quad |\lambda\mathbf{a}|=|\lambda||\mathbf{a}|=\lambda|4|$

$\Rightarrow \quad |\lambda\mathbf{a}|=|-3|4=12, \text{ at } \lambda=-3$

$|\lambda\mathbf{a}|=|0|4=0, \text{ at } \lambda=0$

and $\quad |\lambda\mathbf{a}|=|2|4=8, \text{ at } \lambda=2$

So, the range of $|\lambda\mathbf{a}|$ is [0, 12].

Alternate Method

Since,
$$-3 \le \lambda \le 2$$
$$0 \le |\lambda| \le 3$$
$$\Rightarrow \quad 0 \le 4|\lambda| \le 12$$
$$|\lambda\, \mathbf{a}| \in [0, 12] \qquad [\because |\lambda \mathbf{a}| = |\lambda|\,|\mathbf{a}| = 4|\lambda|]$$

Q.33 The number of vectors of unit length perpendicular to the vectors $\mathbf{a} = 2\hat{\mathbf{i}} + \hat{\mathbf{j}} + 2\hat{\mathbf{k}}$ and $\mathbf{b} = \hat{\mathbf{j}} + \hat{\mathbf{k}}$ is

(a) one (b) two

(c) three (d) infinite

Ans. *(b)* The number of vectors of unit length perpendicular to the vectors **a** and **b** is **c** (say)

i.e., $$\mathbf{c} = \pm(\mathbf{a} \times \mathbf{b}).$$

So, there will be two vectors of unit length perpendicular to the vectors **a** and **b**.

Fill in the blanks in each of the exercises 34 to 40.

Q.34 The vector **a** + **b** bisects the angle between the non-collinear vectors **a** and **b**, if...... .

Ans. If vector **a** + **b** bisects the angle between the non-collinear vectors, then $\theta_1 = \theta_2$

then
$$\mathbf{a}\cdot(\mathbf{a} + \mathbf{b}) = |\mathbf{a}| \cdot |\mathbf{a} + \mathbf{b}| \cos\theta$$
$$\mathbf{a}\cdot(\mathbf{a} + \mathbf{b}) = a\sqrt{a^2 + b^2}\cos\theta$$
$$\Rightarrow \quad \cos\theta = \frac{\mathbf{a}\cdot(\mathbf{a} + \mathbf{b})}{a\sqrt{a^2 + b^2}} \qquad \text{...(i)}$$

and
$$\mathbf{b}\cdot(\mathbf{a} + \mathbf{b}) = |\mathbf{b}| \cdot |\mathbf{a} + \mathbf{b}| \cos\theta$$
$$\mathbf{b}\cdot(\mathbf{a} + \mathbf{b}) = b\sqrt{a^2 + b^2}\cos\theta \qquad \text{[since, } \theta \text{ should be same]}$$
$$\Rightarrow \quad \cos\theta = \frac{\mathbf{b}\cdot(\mathbf{a} + \mathbf{b})}{b\sqrt{a^2 + b^2}} \qquad \text{...(ii)}$$

From Eqs. (i) and (ii), we get
$$\frac{\mathbf{a}\cdot(\mathbf{a} + \mathbf{b})}{a\sqrt{a^2 + b^2}} = \frac{\mathbf{b}\cdot(\mathbf{a} + \mathbf{b})}{b\sqrt{a^2 + b^2}}$$
$$\Rightarrow \quad \frac{\mathbf{a}}{|\mathbf{a}|} = \frac{\mathbf{b}}{|\mathbf{b}|}$$

$\therefore$ $\mathbf{a} = \mathbf{b} \Rightarrow$ **a** and **b** are equal vectors.

Q.35 If $\mathbf{r}\cdot\mathbf{a} = 0$, $\mathbf{r}\cdot\mathbf{b} = 0$ and $\mathbf{r}\cdot\mathbf{c} = 0$ for some non-zero vector **r**, then the value of $\mathbf{a}\cdot(\mathbf{b} \times \mathbf{c})$ is...... .

Ans. Since, **r** is a non-zero vector. Then $\mathbf{r}\cdot\mathbf{a} = \mathbf{r}\cdot\mathbf{b} = \mathbf{r}\cdot\mathbf{c} = 0$

So, we can say that **a**, **b** and **c** are in a same plane.

$\therefore$
$$\mathbf{a}\cdot(\mathbf{b} \times \mathbf{c}) = 0$$
[since, angle between **a**, **b** and **c** are zero *i.e.*, $\theta = 0$]

Q.36 The vectors $\mathbf{a} = 3\hat{\mathbf{i}} - 2\hat{\mathbf{j}} + 2\hat{\mathbf{k}}$ and $\mathbf{b} = -\hat{\mathbf{i}} - 2\hat{\mathbf{k}}$ are the adjacent sides of a parallelogram. The angle between its diagonals is...... .

Ans. We have, $\mathbf{a} = 3\hat{\mathbf{i}} - 2\hat{\mathbf{j}} + 2\hat{\mathbf{k}}$ and $\mathbf{b} = -\hat{\mathbf{i}} - 2\hat{\mathbf{k}}$

$\therefore$ $\mathbf{a} + \mathbf{b} = 2\hat{\mathbf{i}} - 2\hat{\mathbf{j}}$ and $\mathbf{a} - \mathbf{b} = 4\hat{\mathbf{i}} - 2\hat{\mathbf{j}} + 4\hat{\mathbf{k}}$

Now, let θ is the acute angle between the diagonals $\mathbf{a} + \mathbf{b}$ and $\mathbf{a} - \mathbf{b}$.

$$\therefore \quad \cos\theta = \frac{(\mathbf{a} + \mathbf{b})\cdot(\mathbf{a} - \mathbf{b})}{|\mathbf{a} + \mathbf{b}||\mathbf{a} - \mathbf{b}|}$$

$$= \frac{(2\hat{\mathbf{i}} - 2\hat{\mathbf{j}})\cdot(4\hat{\mathbf{i}} - 2\hat{\mathbf{j}} + 4\hat{\mathbf{k}})}{\sqrt{8}\sqrt{16 + 4 + 16}} = \frac{8 + 4}{2\sqrt{2}\cdot 6} = \frac{1}{\sqrt{2}} = \cos\frac{\pi}{4} \quad \left[\because \cos\frac{\pi}{4} = \frac{1}{\sqrt{2}}\right]$$

$$\therefore \quad \theta = \frac{\pi}{4}$$

Q.37 The values of k, for which $|k\,\mathbf{a}| < |\mathbf{a}|$ and $k\,\mathbf{a} + \frac{1}{2}\mathbf{a}$ is parallel to $\mathbf{a}$ holds true are ....

Ans. We have, $|k\,\mathbf{a}| < |\mathbf{a}|$ and $k\mathbf{a} + \frac{1}{2}\mathbf{a}$ is parallel to $\mathbf{a}$.

$$\therefore \quad |k\,\mathbf{a}| < |\mathbf{a}| \Rightarrow |k||\mathbf{a}| < |\mathbf{a}|$$

$$\Rightarrow \quad |k| < 1 \Rightarrow -1 < k < 1$$

Also, since $k\mathbf{a} + \frac{1}{2}\mathbf{a}$ is parallel to $\mathbf{a}$, then we see that at $k = \frac{-1}{2}$, $k\mathbf{a} + \frac{1}{2}\mathbf{a}$ becomes a null vector and then it will not be parallel to $\mathbf{a}$.

So, $k\,\mathbf{a} + \frac{1}{2}\mathbf{a}$ is parallel to $\vec{a}$ holds true when $k \in]-1, 1[$, $k \neq \frac{-1}{2}$.

Q.38 The value of the expression $|\mathbf{a} \times \mathbf{b}|^2 + (\mathbf{a} \cdot \mathbf{b})^2$ is ....

Ans. We know that $|\mathbf{a} \times \mathbf{b}| = |\mathbf{a}||\mathbf{b}| = \sin\theta\,\hat{\mathbf{n}}$

$$|\mathbf{a} \times \mathbf{b}| = |\mathbf{a}||\mathbf{b}| = \cos\theta$$

On putting the values in expression, we get

$$|\mathbf{a} \times \mathbf{b}|^2 + (\mathbf{a} \cdot \mathbf{b})^2 = |\mathbf{a}|^2|\mathbf{b}|^2\sin^2\theta + |\mathbf{a}|^2\cdot|\mathbf{b}|^2\cos^2\theta \quad [\because |\hat{\mathbf{n}}| = 1]$$

$$= |\mathbf{a}||^2|\mathbf{b}|^2[\sin^2\theta + \cos^2\theta]$$

$$= |\mathbf{a}||^2|\mathbf{b}|^2 \quad [\because \sin^2\theta + \cos^2\theta = 1]$$

Alternate Method

$$|\mathbf{a} \times \mathbf{b}|^2 + (\mathbf{a} \cdot \mathbf{b})^2 = |\mathbf{a}|^2|\mathbf{b}|^2\sin^2\theta + (\mathbf{a} \cdot \mathbf{b})^2$$

$$= |\mathbf{a}|^2|\mathbf{b}|^2(1 - \cos^2\theta) + (\mathbf{a} \cdot \mathbf{b})^2$$

$$= |\mathbf{a}|^2|\mathbf{b}|^2 - |\mathbf{a}|^2|\mathbf{b}|^2\cos^2\theta + (\mathbf{a} \cdot \mathbf{b})^2$$

$$= |\mathbf{a}|^2|\mathbf{b}|^2 - (\mathbf{a} \cdot \mathbf{b})^2 + (\mathbf{a} \cdot \mathbf{b})^2$$

$$|\mathbf{a} \times \mathbf{b}|^2 + (\mathbf{a} \cdot \mathbf{b})^2 = |\mathbf{a}|^2|\mathbf{b}|^2$$

Q.39 If $|\mathbf{a} \times \mathbf{b}|^2 + |\mathbf{a} \cdot \mathbf{b}|^2 = 144$ and $|\mathbf{a}| = 4$, then $|\mathbf{b}|$ is equal to ......

Ans. Given, $|\mathbf{a} \times \mathbf{b}|^2 + |\mathbf{a} \cdot \mathbf{b}|^2 = 144$ ...(i)

and $|\mathbf{a}| = 4$

On putting the values of $|\mathbf{a} \times \mathbf{b}|$ and $|\mathbf{a} \cdot \mathbf{b}|$, in Eq. (i), we get

$$= |\mathbf{a}|^2|\mathbf{b}|^2\sin^2\theta + |\mathbf{a}|^2|\mathbf{b}|^2\cos^2\theta = 144$$

$$|\mathbf{a}|^2\,|\mathbf{b}|^2 = 144 \qquad [\because \sin^2\theta + \cos^2\theta = 1]$$
$$(4)^2 \cdot \mathbf{b}^2 = 144 \Rightarrow b^2 = 9$$
$$|\mathbf{b}| = 3$$

Q.40 If **a** is any non-zero vector, then $(\mathbf{a}\cdot\hat{\mathbf{i}})\cdot\hat{\mathbf{i}} + (\mathbf{a}\cdot\hat{\mathbf{j}})\cdot\hat{\mathbf{j}} + (\mathbf{a}\cdot\hat{\mathbf{k}}).\,\hat{\mathbf{k}}$ is equal to ......

Ans. Let $\mathbf{a} = a_1\hat{\mathbf{i}} + a_2\hat{\mathbf{j}} + a_3\hat{\mathbf{k}}$

$\therefore$ $\mathbf{a}\cdot\hat{\mathbf{i}} = a_1$, $\mathbf{a}\cdot\hat{\mathbf{j}} = a_2$ and $\mathbf{a}\cdot\hat{\mathbf{k}} = a_3$

Now, $(\mathbf{a}\cdot\hat{\mathbf{i}})\hat{\mathbf{i}} + (\mathbf{a}\cdot\hat{\mathbf{j}})\hat{\mathbf{j}} + (\mathbf{a}\cdot\hat{\mathbf{k}})\hat{\mathbf{k}} = a_1\hat{\mathbf{i}} + a_2\hat{\mathbf{j}} + a_3\hat{\mathbf{k}} = \mathbf{a}$

State True or False for the statements in each of the exercises 41 to 45.

Q.41 If $|\mathbf{a}| = |\mathbf{b}|$, then necessarily it implies $\mathbf{a} = \pm\,\mathbf{b}$.

Ans. *True*

If $|\mathbf{a}| = |\mathbf{b}| \Rightarrow \mathbf{a} = \pm\,\mathbf{b}$

So, it is a true statement.

Q.42 Position vector of a point **P** is a vector whose initial point is origin.

Ans. *True*

Since, **P** = **OP** = displacement of vector **P** from origin.

Q.43 If $|\mathbf{a}+\mathbf{b}| = |\mathbf{a}-\mathbf{b}|$, then the vectors **a** and **b** are orthogonal

Ans. *True*

Since, $|\mathbf{a}+\mathbf{b}| = |\mathbf{a}-\mathbf{b}|$

$\Rightarrow |\mathbf{a}+\mathbf{b}|^2 = |\mathbf{a}-\mathbf{b}|^2$

$\Rightarrow |\mathbf{a}^2| + |\mathbf{b}^2| + 2\mathbf{a}\cdot\mathbf{b} = |\mathbf{a}|^2| + |\mathbf{b}|^2 - 2\mathbf{a}\cdot\mathbf{b}$

$\Rightarrow 4|\mathbf{a}||\mathbf{b}| = 0$

$\Rightarrow |\mathbf{a}||\mathbf{b}| = 0$

Hence, **a** and **b** are orthogonal. $[\because \mathbf{a}\cdot\mathbf{b} = |\mathbf{a}|\cdot|\mathbf{b}|\cos 90° = 0]$

Q.44 The formula $(\mathbf{a}+\mathbf{b})^2 = \mathbf{a}^2 + \mathbf{b}^2 + 2\mathbf{a}\times\mathbf{b}$ is valid for non-zero vectors **a** and **b**.

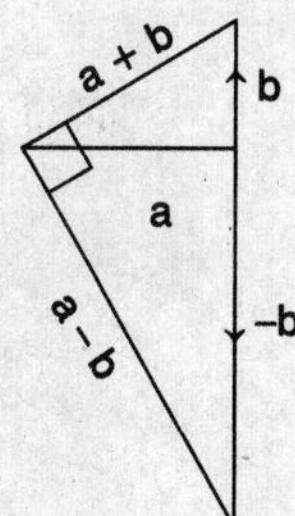

Ans. *False*

$$(\mathbf{a}+\mathbf{b})^2 = (\mathbf{a}+\mathbf{b})\cdot(\mathbf{a}+\mathbf{b})$$
$$= \mathbf{a}^2 + \mathbf{b}^2 + 2\mathbf{a}\cdot\mathbf{b}$$

Q.45 If **a** and **b** are adjacent sides of a rhombus, then $\mathbf{a}\cdot\mathbf{b} = 0$.

Ans. *False*

If $\mathbf{a}\cdot\mathbf{b} = 0$, then $\mathbf{a}\cdot\mathbf{b} = |\mathbf{a}||\mathbf{b}|\cos 90°$ [$\because$ **a** and **b** are non-zero vectors]

Since, angle between **a** and **b** is 90°, which is not possible in a rhombus.

Hence, angle between adjacent sides in a rhombus is not equal to 90°.

Chapter 11

Three Dimensional Geometry

Short Answer Type Questions

Q.1 Find the position vector of a point A in space such that **OA** is inclined at 60° to OX and at 45° to OY and $|\mathbf{OA}| = 10$ units.

Ans. Given, **OA** is inclined at 60° to OX and at 45° to OY. Let **OA** makes angle α with OZ.

Then $\cos^2 60° + \cos^2 45° + \cos^2 \alpha = 1$ $\quad [\because l^2 + m^2 + n^2 = 1]$

$$\Rightarrow \left(\frac{1}{2}\right)^2 + \left(\frac{1}{\sqrt{2}}\right)^2 + \cos^2 \alpha = 1$$

$$\Rightarrow \frac{1}{4} + \frac{1}{2} + \cos^2 \alpha = 1$$

$$\Rightarrow \cos^2 \alpha = 1 - \left(\frac{1}{2} + \frac{1}{4}\right)$$

$$\Rightarrow \cos^2 \alpha = 1 - \left(\frac{3}{4}\right)$$

$$\Rightarrow \cos^2 \alpha = \frac{1}{4}$$

$$\Rightarrow \cos \alpha = \frac{1}{2} = \cos 60°$$

$$\therefore \alpha = 60°$$

Now,
$$\mathbf{OA} = |\mathbf{OA}| \left(\frac{1}{2}\hat{\mathbf{i}} + \frac{1}{\sqrt{2}}\hat{\mathbf{j}} + \frac{1}{2}\hat{\mathbf{k}}\right)$$

$$= 10\left(\frac{1}{2}\hat{\mathbf{i}} + \frac{1}{\sqrt{2}}\hat{\mathbf{j}} + \frac{1}{2}\hat{\mathbf{k}}\right) \quad [\because |\mathbf{OA}| = 10]$$

$$= 5\hat{\mathbf{k}} + 5\sqrt{2}\hat{\mathbf{j}} + 5\hat{\mathbf{k}}$$

Q.2 Find the vector equation of the line which is parallel to the vector $3\hat{\mathbf{i}} - 2\hat{\mathbf{j}} + 6\hat{\mathbf{k}}$ and which passes through the point (1, – 2, 3).

Thinking Process Here, we use the formula $\mathbf{r} = \mathbf{b} + \lambda\,\mathbf{a}$, where $\mathbf{r}$ is the equation of the line which passes through **b** and parallel to **a**.

Ans. Let $\mathbf{a} = 3\hat{\mathbf{i}} - 2\hat{\mathbf{j}} + 6\hat{\mathbf{k}}$ and $\mathbf{b} = \hat{\mathbf{i}} - 2\hat{\mathbf{j}} + 3\hat{\mathbf{k}}$

So, vector equation of the line, which is parallel to the vector $\mathbf{a} = 3\hat{\mathbf{i}} - 2\hat{\mathbf{j}} + 6\hat{\mathbf{k}}$ and passes through the vector $\mathbf{b} = \hat{\mathbf{i}} - 2\hat{\mathbf{j}} + 3\hat{\mathbf{k}}$ is $\mathbf{r} = \mathbf{b} + \lambda\mathbf{a}$.

$\therefore \quad \mathbf{r} = (\hat{\mathbf{i}} - 2\hat{\mathbf{j}} + 3\hat{\mathbf{k}}) + \lambda(3\hat{\mathbf{i}} - 2\hat{\mathbf{j}} + 6\hat{\mathbf{k}})$

$\Rightarrow \quad (x\,\hat{\mathbf{i}} + y\,\hat{\mathbf{j}} + z\,\hat{\mathbf{k}}) - (\hat{\mathbf{i}} - 2\hat{\mathbf{j}} + 3\,\hat{\mathbf{k}}) = \lambda(3\hat{\mathbf{i}} - 2\hat{\mathbf{j}} + 6\hat{\mathbf{k}})$

$\Rightarrow \quad (x-1)\hat{\mathbf{i}} + (y+2)\hat{\mathbf{j}} + (z-3)\hat{\mathbf{k}} = \lambda(3\hat{\mathbf{i}} - 2\hat{\mathbf{j}} + 6\hat{\mathbf{k}})$

Q.3 Show that the lines $\frac{x-1}{2} = \frac{y-2}{3} = \frac{z-3}{4}$ and $\frac{x-4}{5} = \frac{y-1}{2} = z$ intersect. Also, find their point of intersection.

- **Concept Insight** If L_1 and L_2 are two skew lines with equations $L_1 : \frac{x-x_1}{a_1} = \frac{y-y_1}{b_1} = \frac{z-z_1}{c_1}$ and $L_2 : \frac{x-x_2}{a_2} = \frac{y-y_2}{b_2} = \frac{z-z_2}{c_2}$

  Then, the shortest distance between these lines is

$$\left| \frac{\begin{vmatrix} x_2-x_1 & y_2-y_1 & z_2-z_1 \\ a_1 & b_1 & c_1 \\ a_1 & b_2 & c_2 \end{vmatrix}}{\sqrt{(b_1c_2-b_2c_1)^2 + (c_1a_2-c_2a_1)^2 + (a_1b_2-a_2b_1)^2}} \right|$$

Ans. Given lines are $\frac{x-1}{2} = \frac{y-2}{3} = \frac{z-3}{4}$ and $\frac{x-4}{5} = \frac{y-1}{2} = z$

On comparing the given equations of lines with

$$\frac{x-x_1}{a_1} = \frac{y-y_1}{b_1} = \frac{z-z_1}{c_1} \text{ and } \frac{x-x_2}{a_1} = \frac{y-y_2}{b_2} = \frac{z-z_2}{c_2}$$

We get, $\quad x_1 = 1, y_1 = 2, z_1 = 3;\ a_1 = 2, b_1 = 3, c_1 = 4$

$x_2 = 4, y_2 = 1, z_2 = 0;\ a_2 = 5, b_2 = 2, c_2 = 1$

If two lines intersect, then shortest distance between them should be zero.

$\therefore$ Shortest distance between two given lines

$$= \frac{\begin{vmatrix} x_2-x_1 & y_2-y_1 & z_2-z_1 \\ a_1 & b_1 & c_1 \\ a_2 & b_2 & c_2 \end{vmatrix}}{\sqrt{(b_1c_2-b_2c_1)^2 + (c_1a_2-c_2a_1)^2 + (a_1b_2-a_2b_1)^2}}$$

$$= \frac{\begin{vmatrix} 4-1 & 1-2 & 0-3 \\ 2 & 3 & 4 \\ 5 & 2 & 1 \end{vmatrix}}{\sqrt{(3\cdot1-2\cdot4)^2 + (4\cdot5-1\cdot2)^2 + (2\cdot2-5\cdot3)^2}} = \frac{\begin{vmatrix} 3 & -1 & -3 \\ 2 & 3 & 4 \\ 5 & 2 & 1 \end{vmatrix}}{\sqrt{25+324+121}}$$

$$= \frac{3(3-8) + 1(2-20) - 3(4-15)}{\sqrt{470}}$$

$$= \frac{-15-18+33}{\sqrt{470}} = \frac{0}{\sqrt{470}} = 0$$

Hence, the given two lines are intersecting.

For finding their point of intersection for first line,

$$\frac{x-1}{2}=\frac{y-2}{3}=\frac{z-3}{4}=\lambda$$

$\Rightarrow$ $x=2\lambda+1, y=3\lambda+2$ and $z=4\lambda+3$

Since, the lines are intersecting. So, let us put these values in the equation of another line.

Thus, $$\frac{2\lambda+1-4}{5}=\frac{3\lambda+2-1}{2}=\frac{4\lambda+3}{1}$$

$\Rightarrow$ $$\frac{2\lambda-3}{5}=\frac{3\lambda+1}{2}=\frac{4\lambda+3}{1}$$

$\Rightarrow$ $$\frac{2\lambda-3}{5}=\frac{4\lambda+3}{1}$$

$\Rightarrow$ $2\lambda-3=20\lambda+15$

$\Rightarrow$ $18\lambda=-18$

$\lambda=-1$

So, the required point of intersection is

$$x=2(-1)+1=-1$$
$$y=3(-1)+2=-1$$
$$z=4(-1)+3=-1$$

Thus, the lines intersect at $(-1,-1,-1)$.

Q.4 Find the angle between the lines

$\mathbf{r}=3\hat{i}-2\hat{j}+6\hat{k}+\lambda(2\hat{i}+\hat{j}+2\hat{k})$ and $\mathbf{r}=(2\hat{j}-5\hat{k})+\mu(6\hat{i}+3\hat{j}+2\hat{k})$.

Thinking Process We know that, $\cos\theta=\frac{|\mathbf{b}_1\cdot\mathbf{b}_2|}{|\mathbf{b}_1|\cdot|\mathbf{b}_2|}$, *where,* θ *is the angle between the lines* $\mathbf{a}_1+\lambda\,\mathbf{b}_1$ and $\mathbf{a}_2+\mu\mathbf{b}_2$.

Ans. Given lines are $\mathbf{r}=3\hat{i}-2\hat{j}+6\hat{k}+\lambda(2\hat{i}+\hat{j}+2\hat{k})$

and $\mathbf{r}=(2\hat{i}-5\hat{k})+\mu(6\hat{i}+3\hat{j}+2\hat{k})$

On comparing the given equation with $\mathbf{r}=\mathbf{a}_1+\lambda\mathbf{b}_1$ and $\mathbf{r}=\mathbf{a}_2+\lambda\mathbf{b}_2$, we get

$$\mathbf{a}_1=3\hat{i}-2\hat{j}+6\hat{k},\ \mathbf{b}_1=2\hat{i}+\hat{j}+2\hat{k}$$

and $$\mathbf{a}_2=2\hat{j}-5\hat{k},\ \mathbf{b}_2=6\hat{i}+3\hat{j}+2\hat{k}$$

If θ is angle between the given lines, then

$$\cos\theta=\frac{|\mathbf{b}_1\cdot\mathbf{b}_2|}{|\mathbf{b}_1|\cdot|\mathbf{b}_2|}$$

$$=\frac{|(2\hat{i}+\hat{j}+2\hat{k})\cdot(6\hat{i}+3\hat{j}+2\hat{k})|}{|2\hat{i}+\hat{j}+2\hat{k}||6\hat{i}+3\hat{j}+2\hat{k}|}$$

$$=\frac{|12+3+4|}{\sqrt{9}\sqrt{49}}=\frac{19}{21}$$

$\therefore$ $$\theta=\cos^{-1}\frac{19}{21}$$

Q.5 Prove that the line through $A(0,-1,-1)$ and $B(4,5,1)$ intersects the line through $C(3,9,4)$ and $D(-4,4,4)$.

Ans. We know that, the cartesian equation of a line that passes through two points (x_1, y_1, z_1) and (x_2, y_2, z_2) is

$$\frac{x - x_1}{x_2 - x_1} = \frac{y - y_1}{y_2 - y_1} = \frac{z - z_1}{z_2 - z_1}$$

Therefore, the cartesian equation of line passes through $A(0, -1, -1)$ and $B(4, 5, 1)$ is

$$\frac{x - 0}{4 - 0} = \frac{y + 1}{5 + 1} = \frac{z + 1}{1 + 1}$$

$$\Rightarrow \quad \frac{x}{4} = \frac{y + 1}{6} = \frac{z + 1}{2} \quad \text{...(i)}$$

and cartesian equation of the line passes through $C(3, 9, 4)$ and $D(-4, 4, 4)$ is

$$\frac{x - 3}{-4 - 3} = \frac{y - 9}{4 - 9} = \frac{z - 4}{4 - 4}$$

$$\Rightarrow \quad \frac{x - 3}{-7} = \frac{y - 9}{-5} = \frac{z - 4}{0} \quad \text{...(ii)}$$

If the lines intersect, then shortest distance between both of them should be zero.

$\therefore$ Shortest distance between the lines

$$= \frac{\begin{vmatrix} x_2 - x_1 & y_2 - y_1 & z_2 - z_1 \\ a_1 & b_1 & c_1 \\ a_2 & b_2 & c_2 \end{vmatrix}}{\sqrt{(b_1c_2 - b_2c_1)^2 + (c_1a_2 - c_2a_1)^2 + (a_1b_2 - a_2b_1)^2}}$$

$$= \frac{\begin{vmatrix} 3 - 0 & 9 + 1 & 4 + 1 \\ 4 & 6 & 2 \\ -7 & -5 & 0 \end{vmatrix}}{\sqrt{(6 \cdot 0 + 10)^2 + (-14 - 0)^2 + (-20 + 42)^2}}$$

$$= \frac{\begin{vmatrix} 3 & 10 & 5 \\ 4 & 6 & 2 \\ -7 & -5 & 0 \end{vmatrix}}{\sqrt{100 + 196 + 484}}$$

$$= \frac{3(0 + 10) - 10(14) + 5(-20 + 42)}{\sqrt{780}}$$

$$= \frac{30 - 140 + 110}{\sqrt{780}} = 0$$

Hence, the given lines intersect.

Q. 6 Prove that the lines $x = py + q, z = ry + s$ and $x = p'y + q', z = r'y + s'$ are perpendicular, if $pp' + rr' + 1 = 0$.

- **Concept insight** We know that if two lines $L_1: \frac{x - x_1}{a_1} = \frac{y - y_1}{b_1} = \frac{z - z_1}{c_1}$ and $L_2: \frac{x - x_2}{a_2} = \frac{y - y_2}{b_2} = \frac{z - z_2}{c_2}$ are perpendicular to each other then $a_1a_2 + b_1b_2 + c_1c_2 = 0$

Ans. We have, $\quad x = py + q \Rightarrow y = \frac{x - q}{p} \quad$ ..(i)

and $\quad z = ry + s \Rightarrow y = \frac{z - s}{r} \quad$ ...(ii)

$\Rightarrow \quad \dfrac{x-q}{p}=\dfrac{y}{1}=\dfrac{z-s}{r}$ [using Eqs. (i) and (ii)] ...(iii)

Similarly, $\dfrac{x-q'}{p'}=\dfrac{y}{1}=\dfrac{z-s'}{r'}$ ...(iv)

From Eqs. (iii) and (iv), we get

$$a_1 = p, b_1 = 1, c_1 = r$$

and $a_2 = p', b_2 = 1, c_2 = r'$

If these given lines are perpendicular to each other, then

$$a_1 a_2 + b_1 b_2 + c_1 c_2 = 0$$

$\Rightarrow \quad pp' + 1 + rr' = 0$

which is the required condition.

***Q.7 Find the equation of a plane which bisects perpendicularly the line joining the points $A(2, 3, 4)$ and $B(4, 5, 8)$ at right angles.**

Ans. Since, the equation of a plane is bisecting perpendicularly the line joining the points $A(2, 3, 4)$ and $B(4, 5, 8)$ at right angles.

So, mid-point of AB is $\left(\dfrac{2+4}{2}, \dfrac{3+5}{2}, \dfrac{4+8}{2}\right)$ *i.e.*, $(3, 4, 6)$.

Also, $\mathbf{N} = (4-2)\hat{\mathbf{i}} + (5-3)\hat{\mathbf{j}} + (8-4)\hat{\mathbf{k}} = 2\hat{\mathbf{i}} + 2\hat{\mathbf{j}} + 4\hat{\mathbf{k}}$

So, the required equation of the plane is $(\mathbf{r}-\mathbf{a})\cdot\mathbf{N} = 0$. $[\because \mathbf{r} = x\hat{\mathbf{i}} + y\hat{\mathbf{j}} + z\hat{\mathbf{k}}]$

$\Rightarrow \quad [(x-3)\hat{\mathbf{i}} + (y-4)\hat{\mathbf{j}} + (z-6)\hat{\mathbf{k}}]\cdot(2\hat{\mathbf{i}} + 2\hat{\mathbf{j}} + 4\hat{\mathbf{k}}) = 0$ $[\because \mathbf{a} = 3\hat{\mathbf{i}} + 4\hat{\mathbf{j}} + 6\hat{\mathbf{k}}]$

$\Rightarrow \quad 2x - 6 + 2y - 8 + 4z - 24 = 0$

$\Rightarrow \quad 2x + 2y + 4z = 38$

$\Rightarrow \quad x + y + 2z = 19$

Which is the required equation of a plane.

***Q.8 Find the equation of a plane which is at a distance $3\sqrt{3}$ units from origin and the normal to which is equally inclined to coordinate axis.**

Ans. Since, normal to the plane is equally inclined to the coordinate axis.

$\therefore \quad \cos^2\alpha + \cos^2\alpha + \cos^2\alpha = 1$ $[\because l^2 + m^2 + n^2 = 1]$

$\Rightarrow \quad 3\cos^2\alpha = 1 \Rightarrow \cos^2\alpha = \dfrac{1}{3}$

$\Rightarrow \quad \cos\alpha = \dfrac{1}{\sqrt{3}} \Rightarrow \cos\alpha = \cos\beta = \cos\gamma = \dfrac{1}{\sqrt{3}}$

So, the normal is $\mathbf{N} = \dfrac{1}{\sqrt{3}}\hat{\mathbf{i}} + \dfrac{1}{\sqrt{3}}\hat{\mathbf{j}} + \dfrac{1}{\sqrt{3}}\hat{\mathbf{k}}$ and plane is at a distance of $3\sqrt{3}$ units from origin.

The equation of plane is $\mathbf{r}\cdot\mathbf{N} = 3\sqrt{3}$ $\left[\because \mathbf{N} = \dfrac{\mathbf{N}}{|\mathbf{N}|}\right]$

[since, vector equation of the plane at a distance p from the origin is $\mathbf{r}\cdot\mathbf{N} = p$]

$\Rightarrow \quad (x\hat{\mathbf{i}} + y\mathbf{j} + z\hat{\mathbf{k}})\cdot\dfrac{\left(\dfrac{1}{\sqrt{3}}\hat{\mathbf{i}} + \dfrac{1}{\sqrt{3}}\hat{\mathbf{j}} + \dfrac{1}{\sqrt{3}}\hat{\mathbf{k}}\right)}{1} = 3\sqrt{3}$

Note (✱) Topics related to these question have been removed from latest NCERT.

$$\Rightarrow \quad \frac{x}{\sqrt{3}} + \frac{y}{\sqrt{3}} + \frac{z}{\sqrt{3}} = 3\sqrt{3}$$

$$\therefore \quad x + y + z = 3\sqrt{3} \cdot \sqrt{3} = 9$$

So, the required equation of plane is $x + y + z = 9$.

*** Q.9** If the line drawn from the point (–2, – 1, – 3) meets a plane at right angle at the point (1, – 3, 3), then find the equation of the plane.

Ans. Since, the line drawn from the point (–2, – 1, – 3) meets a plane at right angle at the point (1, – 3, 3). So, the plane passes through the point (1, – 3, 3) and normal to plane is $(-2\hat{i} - \hat{j} - 3\hat{k}) - (\hat{i} - 3\hat{j} + 3\hat{k})$ or $(-3\hat{i} + 2\hat{j} - 6\hat{k})$.

$$\Rightarrow \quad \mathbf{a} = \hat{i} - 3\hat{j} + 3\hat{k} \text{ and } \mathbf{N} = -3\hat{i} + 2\hat{j} - 6\hat{k}$$

So, the equation of required plane is $(\mathbf{r} - \mathbf{a}) \cdot \mathbf{N} = 0$

$$\Rightarrow \quad [(x\hat{i} + y\hat{j} + z\hat{k}) - (\hat{i} - 3\hat{j} + 3\hat{k})] \cdot (-3\hat{i} + 2\hat{j} - 6\hat{k}) = 0$$

$$\Rightarrow \quad [(x - 1)\hat{i} + (y + 3)\hat{j} + (z - 3)\hat{k}] \cdot (-3\hat{i} + 2\hat{j} - 6\hat{k}) = 0$$

$$\Rightarrow \quad -3x + 3 + 2y + 6 - 6z + 18 = 0$$

$$\Rightarrow \quad -3x + 2y - 6z = -27$$

$$\Rightarrow \quad 3x - 2y + 6z - 27 = 0$$

which is the required equation of the plane.

***Q.10** Find the equation of the plane through the points (2, 1, 0), (3, – 2, – 2) and (3, 1, 7).

Thinking Process Here, apply the equation of the plane passing through the points $(x_1, y_1, z_1), (x_2, y_2, z_2)$ and (x_3, y_3, z_3) is given by $\begin{vmatrix} x - x_1 & y - y_1 & z - z_1 \\ x_2 - x_1 & y_2 - y_1 & z_2 - z_1 \\ x_3 - x_1 & y_3 - y_1 & z_3 - z_1 \end{vmatrix} = 0.$

Ans. We know that, the equation of a plane passing through three non-collinear points $(x_1, y_1, z_1), (x_2, y_2, z_2)$ and (x_3, y_3, z_3) is

$$\begin{vmatrix} x - x_1 & y - y_1 & z - z_1 \\ x_2 - x_1 & y_2 - y_1 & z_2 - z_1 \\ x_3 - x_1 & y_3 - y_1 & z_3 - z_1 \end{vmatrix} = 0$$

$$\Rightarrow \quad \begin{vmatrix} x - 2 & y - 1 & z - 0 \\ 3 - 2 & -2 - 1 & -2 - 0 \\ 3 - 2 & 1 - 1 & 7 - 0 \end{vmatrix} = 0$$

$$\Rightarrow \quad \begin{vmatrix} x - 2 & y - 1 & z \\ 1 & -3 & -2 \\ 1 & 0 & 7 \end{vmatrix} = 0$$

$$\Rightarrow \quad (x - 2)(-21 + 0) - (y - 1)(7 + 2) + z(3) = 0$$

$$\Rightarrow \quad -21x + 42 - 9y + 9 + 3z = 0$$

$$\Rightarrow \quad -21x - 9y + 3z = -51$$

$$\Rightarrow \quad 7x + 3y - z = 17$$

Hence, the required equation of plane is $7x + 3y - z = 17$.

Note (✱) Topics related to these question have been removed from latest NCERT.

Q.11 Find the equations of the two lines through the origin which intersect the line $\frac{x-3}{2}=\frac{y-3}{1}=\frac{z}{1}$ at angles of $\frac{\pi}{3}$ each.

Ans. Given equation of the line is $\frac{x-3}{2}=\frac{y-3}{1}=\frac{z}{1}=\lambda$

$$\Rightarrow \quad x=(2\lambda+3),\ y=(\lambda+3) \text{ and } z=\lambda \quad \ldots\text{(i)}$$

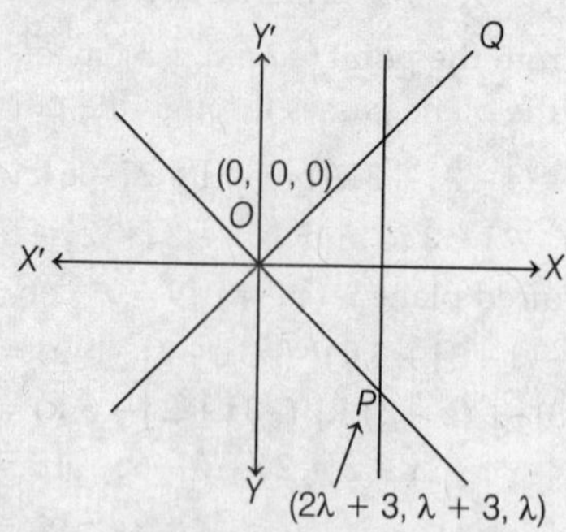

So, DR's of the line are 2, 1, 1 and DC's of the given line are $\frac{2}{\sqrt{6}}, \frac{1}{\sqrt{6}}, \frac{1}{\sqrt{16}}$.

Since, the required lines make angle $\frac{\pi}{3}$ with the given line.

$$\therefore \quad \cos\theta=\frac{a_1a_2+b_1b_2+c_1c_2}{\sqrt{a_1^2+b_1^2+c_1^2}\sqrt{a_2^2+b_2^2+c_2^2}}$$

$$\Rightarrow \quad \cos\frac{\pi}{3}=\frac{(4\lambda+6)+(\lambda+3)+(\lambda)}{\sqrt{6}\sqrt{(2\lambda+3)^2+(\lambda+3)^2+\lambda^2}}$$

$$\Rightarrow \quad \frac{1}{2}=\frac{6\lambda+9}{\sqrt{6}\sqrt{(4\lambda^2+9+12\lambda+\lambda^2+9+6\lambda+\lambda^2)}}$$

$$\Rightarrow \quad \frac{\sqrt{6}}{2}=\frac{6\lambda+9}{\sqrt{6\lambda^2+18\lambda+18}}$$

$$\Rightarrow \quad 6\sqrt{(\lambda^2+3\lambda+3)}=6(2\lambda+3)$$

$$\Rightarrow \quad (\lambda^2+3\lambda+3)=(4\lambda^2+9+12\lambda) \quad \text{[on squarring both sides]}$$

$$\Rightarrow \quad \lambda^2+3\lambda+3=4\lambda^2+9+12\lambda$$

$$\Rightarrow \quad 3\lambda^2+9\lambda+6=0$$

$$\Rightarrow \quad \lambda^2+3\lambda+2=0$$

$$\Rightarrow \quad \lambda(\lambda+2)+1(\lambda+2)=0$$

$$\Rightarrow \quad (\lambda+1)(\lambda+2)=0$$

$$\therefore \quad \lambda=-1,-2$$

So, the DC's are 1, 2, −1 and −1, 1, −2 [from Eq. (i)]

Also, both the required lines passes through origin.

So, the equations of required lines are $\frac{x}{1}=\frac{y}{2}=\frac{z}{-1}$ and $\frac{x}{-1}=\frac{y}{1}=\frac{z}{-2}$.

Q.12 Find the angle between the lines whose direction cosines are given by the equation $l+m+n=0$ and $l^2+m^2-n^2=0$.

Ans. Given, $$l+m+n=0 \quad \text{...(i)}$$
and $$l^2+m^2-n^2=0 \quad \text{...(ii)}$$

On eliminating n from both the equations, we get
$$l^2+m^2-(l-m)^2=0$$
$\Rightarrow$ $l^2+m^2-l^2-m^2+2ml=0 \Rightarrow 2lm=0$ $\quad [\because (a-b)^2=a^2+b^2-2ab]$

$\Rightarrow$ $lm=0 \Rightarrow (-m-n)m=0$ $\quad [\because l=-m-n]$

$\Rightarrow$ $(m+n)m=0$

$\Rightarrow$ $m=-n$ and $m=0$

$\Rightarrow$ $l=0, l=-n$ $\quad$ [from Eq. (i)]

Thus, Dr's two lines are proportional to $0, -n, n$ and $-n, 0, n$ *i.e.*, $0, -1, 1$ and $-1, 0, 1$.

So, the vector parallel to these given lines are $\mathbf{a}=-\hat{\mathbf{j}}+\hat{\mathbf{k}}$ and $\mathbf{b}=-\hat{\mathbf{i}}+\hat{\mathbf{k}}$

Now, $$\cos\theta=\frac{\mathbf{a}\cdot\mathbf{b}}{|\mathbf{a}||\mathbf{b}|}=\frac{1}{\sqrt{2}\sqrt{2}}$$

$\Rightarrow$ $$\cos\theta=\frac{1}{2}=\cos\frac{\pi}{3} \quad \left[\because \cos\frac{\pi}{3}=\frac{1}{2}\right]$$

$\therefore$ $$\theta=\frac{\pi}{3}$$

Q.13 If a variable line in two adjacent positions has direction cosines l, m, n and $l+\delta l, m+\delta m, n+\delta n$, then show that the small angle $\delta\theta$ between the two positions is given by $\delta\theta^2=\delta l^2+\delta m^2+\delta n^2$.

Ans. We have l, m, n and $l+\delta l, m+\delta m, n+\delta n$ as direction cosines of a variable line in two different positions.

$\therefore$ $$l^2+m^2+n^2=1 \quad \text{... (i)}$$
and $$(l+\delta l)^2+(m+\delta m)^2+(n+\delta n)^2=1 \quad \text{...(ii)}$$
$\Rightarrow l^2+m^2+n^2+\delta l^2+\delta m^2+\delta n^2+2(l\,\delta l+m\,\delta m+n\,\delta n)=1$
$$[\because (a+b)^2=a^2+b^2+2ab]$$
$\Rightarrow$ $\delta l^2+\delta m^2+\delta n^2=-2(l\,\delta l+m\,\delta m+n\,\delta n)$ $\quad [\because l^2+m^2+n^2=1]$

$\Rightarrow$ $$l\,\delta l+m\,\delta m+n\,\delta n=\frac{-1}{2}(\delta l^2+\delta m^2+\delta n^2) \quad \text{...(iii)}$$

Now, **a** and **b** are unit vectors along a line with direction cosines l, m, n and $(l+\delta l), (m+\delta m), (n+\delta n)$, respectively.

$\therefore$ $$\mathbf{a}=l\hat{\mathbf{i}}+m\hat{\mathbf{j}}+n\hat{\mathbf{k}}$$
and $\mathbf{b}=(l+\delta l)\hat{\mathbf{i}}+(m+\delta m)\hat{\mathbf{j}}+(n+\delta n)\hat{\mathbf{k}}$

$\Rightarrow$ $$\cos\delta\theta=\frac{\mathbf{a}\cdot\mathbf{b}}{|\mathbf{a}||\mathbf{b}|}=\mathbf{a}\cdot\mathbf{b} \quad [\because |\mathbf{a}|=|\mathbf{b}|=1]$$

$\Rightarrow$ $$\cos\delta\theta=l(l+\delta l)+m(m+\delta m)+n(n+\delta n)$$
$$=(l^2+m^2+n^2)+(l\,\delta l+m\,\delta m+n\,\delta n)$$
$$=1-\frac{1}{2}(\delta l^2+\delta m^2+\delta n^2) \quad \text{[using Eq. (iii) and (i)]}$$

$\Rightarrow$ $$2(1-\cos\delta\theta)=(\delta l^2+\delta m^2+\delta n^2)$$

$$\Rightarrow \quad 2\cdot 2\sin^2\frac{\delta\theta}{2} = \delta l^2 + \delta m^2 + \delta n^2 \quad \left[\because 1-\cos\theta = 2\sin^2\frac{\theta}{2}\right]$$

$$\Rightarrow \quad 4\left(\frac{\delta\theta}{2}\right)^2 = \delta l^2 + \delta m^2 + \delta n^2 \quad \left[\text{since, } \frac{\delta\theta}{2} \text{ is small, then } \sin\frac{\delta\theta}{2} = \frac{\delta\theta}{2}\right]$$

$$\therefore \quad \delta\theta^2 = \delta l^2 + \delta m^2 + \delta n^2$$

***Q.14** O is the origin and A is (a, b, c). Find the direction cosines of the line OA and the equation of plane through A at right angle to OA.

Ans. Since, O is origin and A is (a, b, c)

$\therefore$ DC's of line OA are $\frac{a}{\sqrt{a^2+b^2+c^2}}, \frac{b}{\sqrt{a^2+b^2+c^2}}$ and $\frac{c}{\sqrt{a^2+b^2+c^2}}$.

Also, $\mathbf{n} = \mathbf{OA} = \mathbf{a} = a\hat{\mathbf{i}} + b\hat{\mathbf{j}} + c\hat{\mathbf{k}}$

The equation of plane passes through (a, b, c) and perpendicular to OA is given by

$$[\mathbf{r} - \mathbf{a}]\cdot\mathbf{n} = 0$$

$$\Rightarrow \quad \mathbf{r}\cdot\mathbf{n} = \mathbf{a}\cdot\mathbf{n}$$

$$\Rightarrow \quad [(x\hat{\mathbf{i}} + y\hat{\mathbf{j}} + z\hat{\mathbf{k}})\cdot(a\hat{\mathbf{i}} + b\hat{\mathbf{j}} + c\hat{\mathbf{k}})] = (a\hat{\mathbf{i}} + b\hat{\mathbf{j}} + c\hat{\mathbf{k}})\cdot(a\hat{\mathbf{i}} + b\hat{\mathbf{j}} + c\hat{\mathbf{k}})$$

$$\Rightarrow \quad ax + by + cz = a^2 + b^2 + c^2$$

***Q.15** Two systems of rectangular axis have the same origin. If a plane cuts them at distances a, b, c and a', b', c' respectively from the origin, prove that $\frac{1}{a^2}+\frac{1}{b^2}+\frac{1}{c^2} = \frac{1}{a'^2}+\frac{1}{b'^2}+\frac{1}{c'^2}$.

Ans. Consider OX, OY, OZ and ox, oy, oz are two system of rectangular axes.

Let their corresponding equation of plane be

$$\frac{x}{a}+\frac{y}{b}+\frac{z}{c} = 1 \quad \text{...(i)}$$

and

$$\frac{x}{a'}+\frac{y}{b'}+\frac{z}{c'} = 1 \quad \text{...(ii)}$$

Also, the length of perpendicular from origin to Eqs. (i) and (ii) must be same.

$$\therefore \quad \frac{\frac{0}{a}+\frac{0}{b}+\frac{0}{c}-1}{\sqrt{\frac{1}{a^2}+\frac{1}{b^2}+\frac{1}{c^2}}} = \frac{\frac{0}{a'}+\frac{0}{b'}+\frac{0}{c'}-1}{\sqrt{\frac{1}{a'^2}+\frac{1}{b'^2}+\frac{1}{c'^2}}}$$

$$\Rightarrow \quad \sqrt{\frac{1}{a'^2}+\frac{1}{b'^2}+\frac{1}{c'^2}} = \sqrt{\frac{1}{a^2}+\frac{1}{b^2}+\frac{1}{c^2}}$$

$$\Rightarrow \quad \frac{1}{a^2}+\frac{1}{b^2}+\frac{1}{c^2} = \frac{1}{a'^2}+\frac{1}{b'^2}+\frac{1}{c'^2}$$

Hence proved.

Note (✱) Topics related to these question have been removed from latest NCERT.

Long Answer Type Questions

Q.16 Find the foot of perpendicular from the point (2, 3, –8) to the line $\frac{4-x}{2} = \frac{y}{6} = \frac{1-z}{3}$. Also, find the perpendicular distance from the given point to the line.

Ans. We have, equation of line as $\frac{4-x}{2} = \frac{y}{6} = \frac{1-z}{3}$

$$\Rightarrow \quad \frac{x-4}{-2} = \frac{y}{6} = \frac{z-1}{-3} = \lambda$$

$$\Rightarrow \quad x = -2\lambda + 4,\ y = 6\lambda \text{ and } z = -3\lambda + 1$$

Let the coordinates of L be $(4 - 2\lambda, 6\lambda, 1 - 3\lambda)$ and direction ratios of PL are proportional to $(4 - 2\lambda - 2, 6\lambda - 3, 1 - 3\lambda + 8)$ *i.e.*, $(2 - 2\lambda, 6\lambda - 3, 9 - 3\lambda)$.

Also, direction ratios are proportional to –2, 6, – 3. Since, PL is perpendicular to given line.

$$\therefore \quad -2(2 - 2\lambda) + 6(6\lambda - 3) - 3(9 - 3\lambda) = 0$$

$$\Rightarrow \quad -4 + 4\lambda + 36\lambda - 18 - 27 + 9\lambda = 0$$

$$\Rightarrow \quad 49\lambda = 49 \Rightarrow \lambda = 1$$

So, the coordinates of L are $(4 - 2\lambda, 6\lambda, 1 - 3\lambda)$ *i.e.*, (2, 6, – 2).

P (2, 3, – 8)

L

$\frac{4-x}{2} = \frac{y}{6} = \frac{1-z}{3}$

Now, length of $PL = \sqrt{(2-2)^2 + (6-3)^2 + (-2+8)^2}$

$$= \sqrt{0 + 9 + 36} = 3\sqrt{5} \text{ units}$$

Mistake Alert! *Remember that if the equation of line are not in standard form, then firstly convert this in standard form of equation of lines i.e.* $\frac{x-x_1}{a_1} = \frac{y-y_1}{b_1} = \frac{z-z_1}{c_1}$

Q.17 Find the distance of a point (2, 4, – 1) from the line

$$\frac{x+5}{1} = \frac{y+3}{4} = \frac{z-6}{-9}.$$

Ans. We have, equation of the line as $\frac{x+5}{1} = \frac{y+3}{4} = \frac{z-6}{-9} = \lambda$

$$\Rightarrow \quad x = \lambda - 5,\ y = 4\lambda - 3,\ z = 6 - 9\lambda$$

Let the coordinates of L be $(\lambda - 5, 4\lambda - 3, 6 - 9\lambda)$, then Dr's of PL are $(\lambda - 7, 4\lambda - 7, 7 - 9\lambda)$.

Also, the direction ratios of given line are proportional to 1, 4, –9.

Since, PL is perpendicular to the given line.

$$\therefore \quad (\lambda - 7)\cdot 1 + (4\lambda - 7)\cdot 4 + (7 - 9\lambda)\cdot(-9) = 0$$

$\Rightarrow \quad \lambda - 7 + 16\lambda - 28 + 81\lambda - 63 = 0$

$\Rightarrow \quad 98\lambda = 98 \Rightarrow \lambda = 1$

So, the coordinates of L are $(-4, 1, -3)$.

$\therefore$ Required distance, $PL = \sqrt{(-4-2)^2 + (1-4)^2 + (-3+1)^2}$

$= \sqrt{36 + 9 + 4} = 7$ units

***Q.18** Find the length and the foot of perpendicular from the point $\left(1, \frac{3}{2}, 2\right)$ to the plane $2x - 2y + 4z + 5 = 0$.

Ans. Equation of the given plane is $2x - 2y + 4z + 5 = 0$ ... (i)

$\Rightarrow \quad \mathbf{n} = 2\hat{\mathbf{i}} - 2\hat{\mathbf{j}} + 4\hat{\mathbf{k}}$

So, the equation of line through $\left(1, \frac{3}{2}, 2\right)$ and parallel to $\mathbf{n}$ is given by

$$\frac{x-1}{2} = \frac{y - \frac{3}{2}}{-2} = \frac{z-2}{4} = \lambda$$

$\Rightarrow \quad x = 2\lambda + 1, y = -2\lambda + \frac{3}{2}$ and $z = 4\lambda + 2$

If this point lies on the given plane, then

$2(2\lambda + 1) - 2\left(-2\lambda + \frac{3}{2}\right) + 4(4\lambda + 2) + 5 = 0$ [using Eq. (i)]

$\Rightarrow \quad 4\lambda + 2 + 4\lambda - 3 + 16\lambda + 8 + 5 = 0$

$\Rightarrow \quad 24\lambda = -12 \Rightarrow \lambda = \frac{-1}{2}$

$\therefore$ Required foot of perpendicular

$$= \left[2 \times \left(\frac{-1}{2}\right) + 1, -2 \times \left(\frac{-1}{2}\right) + \frac{3}{2}, 4 \times \left(\frac{-1}{2}\right) + 2\right] \text{ i.e., } \left(0, \frac{5}{2}, 0\right)$$

$\therefore$ Required length of perpendicular $= \sqrt{(1-0)^2 + \left(\frac{3}{2} - \frac{5}{2}\right)^2 + (2-0)^2}$

$= \sqrt{1 + 1 + 4} = \sqrt{6}$ units

***Q.19** Find the equation of the line passing through the point (3, 0, 1) and parallel to the planes $x + 2y = 0$ and $3y - z = 0$.

Ans. Equation of the given two planes are $x + 2y = 0$ and $3y - z = 0$.

Let $\vec{n}_1$ and $\vec{n}_2$ are the normals to the two planes, respectively.

$\therefore \quad \mathbf{n}_1 = \hat{\mathbf{i}} + 2\hat{\mathbf{j}}$ and $\mathbf{n}_2 = 3\hat{\mathbf{j}} - \hat{\mathbf{k}}$

Since, required line is parallel to the given two planes.

Therefore, $\mathbf{b} = \mathbf{n}_1 \times \mathbf{n}_2 = \begin{vmatrix} \hat{\mathbf{i}} & \hat{\mathbf{j}} & \hat{\mathbf{k}} \\ 1 & 2 & 0 \\ 0 & 3 & -1 \end{vmatrix}$

$= \hat{\mathbf{i}}(-2) - \hat{\mathbf{j}}(-1) + \hat{\mathbf{k}}(3) = -2\hat{\mathbf{i}} + \hat{\mathbf{j}} + 3\hat{\mathbf{k}}$

Note (*) Topics related to these question have been removed from latest NCERT.

So, the equation of the lines through the point (3, 0, 1) and parallel to the given two planes are

$$(x-3)\,\hat{\mathbf{i}}+(y-0)\hat{\mathbf{j}}+(z-1)\,\hat{\mathbf{k}}+\lambda(-2\hat{\mathbf{i}}+\hat{\mathbf{j}}+3\,\hat{\mathbf{k}})$$
$$(x-3)\,\hat{\mathbf{i}}+y\,\hat{\mathbf{j}}+(z-1)\,\hat{\mathbf{k}}+\lambda(-2\,\hat{\mathbf{i}}+\hat{\mathbf{j}}+3\,\hat{\mathbf{k}})$$

***Q.20** Find the equation of the plane through the points (2, 1, – 1), (–1, 3, 4) and perpendicular to the plane $x-2y+4z=10$.

Ans. The equation of the plane passing through (2, 1, – 1) is

$$a(x-2)+b(y-1)+c(z+1)=0 \quad \text{...(i)}$$

Since, this passes through (–1, 3, 4).

$$\therefore \quad a(-1-2)+b(3-1)+c(4+1)=0$$
$$\Rightarrow \quad -3a+2b+5c=0 \quad \text{... (ii)}$$

Since, the plane (i) is perpendicular to the plane $x-2y+4z=10$.

$$\therefore \quad 1\cdot a-2\cdot b+4\cdot c=0$$
$$\Rightarrow \quad a-2b+4c=0 \quad \text{...(iii)}$$

On solving Eqs. (ii) and (iii), we get

$$\frac{a}{8+10}=\frac{-b}{-17}=\frac{c}{4}=\lambda$$
$$\Rightarrow \quad a=18\lambda,\ b=17\lambda,\ c=4\lambda$$

From Eq. (i),

$$18\lambda(x-2)+17\lambda(y-1)+4\lambda(z+1)=0$$
$$\Rightarrow \quad 18x-36+17y-17+4z+4=0$$
$$\Rightarrow \quad 18x+17y+4z-49=0$$
$$\Rightarrow \quad 18x+17y+4z=49$$

which is the required equation of the plane.

Q.21 Find the shortest distance between the lines gives by $\mathbf{r}=(8+3\lambda)\hat{\mathbf{i}}-(9+16\lambda)\hat{\mathbf{j}}+(10+7\lambda)\hat{\mathbf{k}}$ and $\mathbf{r}=15\hat{\mathbf{i}}+29\hat{\mathbf{j}}+5\hat{\mathbf{k}}+\mu(3\hat{\mathbf{i}}+8\hat{\mathbf{j}}-5\hat{\mathbf{k}})$.

- **Concept Insight** The shortest distance between two lines $\mathbf{r}=\mathbf{a}_1+\lambda\mathbf{b}_1$ and $\mathbf{r}=\mathbf{a}_2+\mu\mathbf{b}_2$ is given by $\left|\frac{(\mathbf{b}_1\times\mathbf{b}_1)\cdot(\mathbf{a}_2-\mathbf{a}_1)}{|\mathbf{b}_1\times\mathbf{b}_2|}\right|$

Ans. We have,

$$\mathbf{r}=(8+3\lambda)\,\hat{\mathbf{i}}-(9+16\lambda)\hat{\mathbf{j}}+(10+7\lambda)\,\hat{\mathbf{k}})$$
$$=8\hat{\mathbf{i}}-9\hat{\mathbf{j}}+10\,\hat{\mathbf{k}}+3\lambda\hat{\mathbf{i}}-16\lambda\hat{\mathbf{j}}+7\lambda\,\hat{\mathbf{k}}$$
$$=8\hat{\mathbf{i}}-9\hat{\mathbf{j}}+10\,\hat{\mathbf{k}}+\lambda(3\hat{\mathbf{i}}-16\hat{\mathbf{j}}+7\hat{\mathbf{k}})$$

On comparing with $\mathbf{r}=\mathbf{a}_1+\lambda\mathbf{b}_1$, we get

$$\mathbf{a}_1=8\hat{\mathbf{i}}-9\hat{\mathbf{j}}+10\,\hat{\mathbf{k}} \text{ and } \mathbf{b}=3\hat{\mathbf{i}}-16\,\hat{\mathbf{j}}+7\hat{\mathbf{k}} \quad \text{...(i)}$$

Also $\mathbf{r}=15\hat{\mathbf{i}}+29\,\hat{\mathbf{j}}+5\,\hat{\mathbf{k}}+\mu(3\hat{\mathbf{i}}+8\hat{\mathbf{j}}-5\,\hat{\mathbf{k}})$

On comparing with $\mathbf{r}=\mathbf{a}_2+\mu\mathbf{b}_2$, we get

$$\mathbf{a}_2=15\,\hat{\mathbf{i}}+29\,\hat{\mathbf{j}}+5\,\hat{\mathbf{k}} \text{ and } \mathbf{b}_2=3\,\hat{\mathbf{i}}+8\hat{\mathbf{j}}-5\,\hat{\mathbf{k}} \quad \text{... (ii)}$$

Note (✱) Topics related to these question have been removed from latest NCERT.

Now, shortest distance betwen two lines is given by $\left|\dfrac{(\mathbf{b}_1 \times \mathbf{b}_2)\cdot(\mathbf{a}_2 - \mathbf{a}_1)}{|\mathbf{b}_1 \times \mathbf{b}_2|}\right|$

$$\therefore \quad \mathbf{b}_1 \times \mathbf{b}_2 = \begin{vmatrix} \hat{\mathbf{i}} & \mathbf{j} & \hat{\mathbf{k}} \\ 3 & -16 & 7 \\ 3 & 8 & -5 \end{vmatrix}$$

$$= \hat{\mathbf{i}}(80 - 56) - \mathbf{j}(-15 - 21) + \hat{\mathbf{k}}(24 + 48)$$

$$= 24\hat{\mathbf{i}} + 36\hat{\mathbf{j}} + 72\hat{\mathbf{k}}$$

Now, $$|\mathbf{b}_1 \times \mathbf{b}_2| = \sqrt{(24)^2 + (36)^2 + (72)^2}$$

$$= 12\sqrt{2^2 + 3^2 + 6^2} = 84$$

and $$(\mathbf{a}_2 - \mathbf{a}_1) = (15 - 8)\hat{\mathbf{i}} + (29 + 9)\hat{\mathbf{j}} + (5 - 10)\hat{\mathbf{k}}$$

$$= 7\hat{\mathbf{i}} + 38\hat{\mathbf{j}} - 5\hat{\mathbf{k}}$$

$$\therefore \quad \text{Shortest distance} = \left|\frac{(24\hat{\mathbf{i}} + 36\hat{\mathbf{j}} + 72\hat{\mathbf{k}})\cdot(7\hat{\mathbf{i}} + 38\hat{\mathbf{j}} - 5\hat{\mathbf{k}})}{84}\right|$$

$$= \left|\frac{168 + 1368 - 360}{84}\right| = \left|\frac{1176}{84}\right| = 14 \text{ units}$$

***Q.22** **Find the equation of the plane which is perpendicular to the plane $5x + 3y + 6z + 8 = 0$ and which contains the line of intersection of the planes $x + 2y + 3z - 4 = 0$ and $2x + y - z + 5 = 0$.**

Ans. The equation of a plane through the line of intersection of the planes $x + 2y + 3z - 4 = 0$ and $2x + y - z + 5 = 0$ is

$$(x + 2y + 3z - 4) + \lambda(2x + y - z + 5) = 0$$

$$\Rightarrow \quad x(1 + 2\lambda) + y(2 + \lambda) + z(-\lambda + 3) - 4 + 5\lambda = 0 \quad \text{...(i)}$$

Also, this is perpendicular to the plane $5x + 3y + 6z + 8 = 0$.

$$\therefore \quad 5(1 + 2\lambda) + 3(2 + \lambda) + 6(3 - \lambda) = 0 \quad [\because a_1a_2 + b_1b_2 + c_1c_2 = 0]$$

$$\Rightarrow \quad 5 + 10\lambda + 6 + 3\lambda + 18 - 6\lambda = 0$$

$$\therefore \quad \lambda = -\frac{29}{7}$$

From Eq. (i),

$$x\left[1 + 2\left(\frac{-29}{7}\right)\right] + y\left(2 - \frac{29}{7}\right) + z\left(\frac{29}{7} + 3\right) - 4 + 5\left(\frac{-29}{7}\right) = 0$$

$$\Rightarrow \quad x(7 - 58) + y(14 - 29) + z(29 + 21) - 28 - 145 = 0$$

$$\Rightarrow \quad -51x - 15y + 50z - 173 = 0$$

Hence, the required equation of plane is $51x + 15y - 50z + 173 = 0$.

***Q.23** **The plane $ax + by = 0$ is rotated about its line of intersection with the plane $z = 0$ through an angle α, then prove that the equation of the plane in its new position is $ax + by \pm (\sqrt{a^2 + b^2}\tan\alpha)z = 0$.**

Note (✱) Topics related to these question have been removed from latest NCERT.

Ans. Given equation of the plane is $ax + by = 0$ ... (i)

$\therefore$ Equation of the plane in its new position is

$$\frac{ax \cos\alpha}{\sqrt{a^2+b^2}} + \frac{by \cos\alpha}{\sqrt{b^2+a^2}} \pm z \sin\alpha = 0$$

$$\Rightarrow \quad \frac{ax}{\sqrt{a^2+b^2}} + \frac{by}{\sqrt{b^2+a^2}} \pm z \tan\alpha = 0 \qquad \text{[on dividing by } \cos\alpha]$$

$$\Rightarrow \quad ax + by \pm z \tan\alpha \sqrt{a^2+b^2} = 0 \qquad \text{[on multiplying with } \sqrt{a^2+b^2}]$$

Alternate Method

Given,planes are $ax + by = 0$ ...(i)

and $z = 0$ ...(ii)

Therefore, the equation of any plane passing through the line of intersection of planes (i) and (ii) may be taken as $ax + by + k = 0$. ...(iii)

Then, direction cosines of a normal to the plane (iii) are $\frac{a}{\sqrt{a^2+b^2+k^2}}, \frac{b}{\sqrt{a^2+b^2+k^2}}, \frac{c}{\sqrt{a^2+b^2+k^2}}$ and direction cosines of the normal to the plane (i) are $\frac{a}{\sqrt{a^2+b^2}}, \frac{b}{\sqrt{a^2+b^2}}, 0.$

Since, the angle between the planes (i) and (ii) is α,

$$\therefore \quad \cos\alpha = \frac{a\cdot a + b\cdot b + k\cdot 0}{\sqrt{a^2+b^2+k^2}\sqrt{a^2+b^2}} = \sqrt{\frac{a^2+b^2}{a^2+b^2+k^2}}$$

$$\Rightarrow \quad k^2\cos^2\alpha = a^2(1-\cos^2\alpha) + b^2(1-\cos^2\alpha)$$

$$\Rightarrow \quad k^2 = \frac{(a^2+b^2)\sin^2\alpha}{\cos^2\alpha} \Rightarrow k = \pm\sqrt{a^2+b^2}\tan\alpha$$

On putting this value in plane (iii), we get the equation of the plane as

$$ax + by + z\sqrt{a^2+b^2}\tan\alpha = 0$$

***Q.24** Find the equation of the plane through the intersection of the planes $\mathbf{r}\cdot(\hat{\mathbf{i}} + 3\hat{\mathbf{j}}) - 6 = 0$ and $\mathbf{r}\cdot(3\hat{\mathbf{i}} - \hat{\mathbf{j}} - 4\hat{\mathbf{k}}) = 0$, whose perpendicular distance from origin is unity.

Ans. We have, $\mathbf{n}_1 = (\hat{\mathbf{i}} + 3\hat{\mathbf{j}}), d_1 = 6$ and $\mathbf{n}_2 = (3\hat{\mathbf{i}} - \hat{\mathbf{j}} - 4\hat{\mathbf{k}}), d_2 = 0$

On using the relation, $\mathbf{r}\cdot(\mathbf{n}_1 + \lambda\mathbf{n}_2) = d_1 + d_2\lambda$

$$\Rightarrow \quad \mathbf{r}\cdot[(\hat{\mathbf{i}} + 3\hat{\mathbf{j}}) + \lambda(3\hat{\mathbf{i}} - \hat{\mathbf{j}} - 4\hat{\mathbf{k}})] = 6 + 0\cdot\lambda$$

$$\Rightarrow \quad \mathbf{r}\cdot[(1+3\lambda)\hat{\mathbf{i}} + (3-\lambda)\hat{\mathbf{j}} + \hat{\mathbf{k}}(-4\lambda)] = 6 \qquad \text{...(i)}$$

On dividing both sides by $\sqrt{(1+3\lambda)^2 + (3-\lambda)^2 + (-4\lambda)^2}$, we get

$$\frac{\mathbf{r}\cdot[(1+3\lambda)\hat{\mathbf{i}} + (3-\lambda)\hat{\mathbf{j}} + \hat{\mathbf{k}}(-4\lambda)]}{\sqrt{(1+3\lambda)^2 + (3-\lambda)^2 + (-4\lambda)^2}} = \frac{6}{\sqrt{(1+3\lambda)^2 + (3-\lambda)^2 + (-4\lambda)^2}}$$

[From Eq. (i)]

Since, the perpendicular distance from origin is unity.

$$\therefore \quad \frac{6}{\sqrt{(1+3\lambda)^2 + (3-\lambda)^2 + (-4\lambda)^2}} = 1$$

Note (✱) Topics related to these question have been removed from latest NCERT.

$$\Rightarrow \quad (1+3\lambda)^2+(3-\lambda)^2+(-4\lambda)^2=36$$
$$\Rightarrow \quad 1+9\lambda^2+6\lambda+9+\lambda^2-6\lambda+16\lambda^2=36$$
$$\Rightarrow \quad 26\lambda^2+10=36$$
$$\Rightarrow \quad \lambda^2=1$$
$$\therefore \quad \lambda=\pm 1$$

On using Eq. (i), the required equation of plane is

$$\hat{\mathbf{r}}\cdot[(1\pm 3)\hat{\mathbf{i}}+(3\mp 1)\hat{\mathbf{j}}+(\mp 4)\hat{\mathbf{k}}]=6$$
$$\Rightarrow \quad \mathbf{r}\cdot[(1+3)\hat{\mathbf{i}}+(3-1)\hat{\mathbf{j}}+(-4)\hat{\mathbf{k}}]=6$$
and
$$\mathbf{r}\cdot[(1-3)\hat{\mathbf{i}}+(3+1)\hat{\mathbf{j}}+4\hat{\mathbf{k}}]=6$$
$$\Rightarrow \quad \mathbf{r}\cdot(4\hat{\mathbf{i}}+2\hat{\mathbf{j}}-4\hat{\mathbf{k}})=6$$
and
$$\mathbf{r}\cdot(-2\hat{\mathbf{i}}+4\hat{\mathbf{j}}+4\hat{\mathbf{k}})=6$$
$$\Rightarrow \quad 4x+2y-4z-6=0$$
and
$$-2x+4y+4z-6=0$$

***Q.25** Show that the points $(\hat{\mathbf{i}}-\hat{\mathbf{j}}+3\hat{\mathbf{k}})$ and $3(\hat{\mathbf{i}}+\hat{\mathbf{j}}+\hat{\mathbf{k}})$ are equidistant from the plane $\mathbf{r}\cdot(5\hat{\mathbf{i}}+2\hat{\mathbf{j}}-7\hat{\mathbf{k}})+9=0$ and lies on opposite side of it.

Ans. To show that these given points $(\hat{\mathbf{i}}-\hat{\mathbf{j}}+3\hat{\mathbf{k}})$ and $3(\hat{\mathbf{i}}+\hat{\mathbf{j}}+\hat{\mathbf{k}})$ are equidistant from the plane $\mathbf{r}\cdot(5\hat{\mathbf{i}}+2\hat{\mathbf{j}}-7\hat{\mathbf{k}})+9=0$, we first find out the mid- point of the points which is $2\hat{\mathbf{i}}+\hat{\mathbf{j}}+3\hat{\mathbf{k}}$.

On substituting **r** by the mid-point in plane, we get

$$\text{LHS}=(2\hat{\mathbf{i}}+\hat{\mathbf{j}}+3\hat{\mathbf{k}})\cdot(5\hat{\mathbf{i}}+2\hat{\mathbf{j}}-7\hat{\mathbf{k}})+9$$
$$=10+2-21+9=0=\text{RHS}$$

Hence, the two points lie on opposite sides of the plane are equidistant from the plane.

Q.26 $\mathbf{AB}=3\hat{\mathbf{i}}-\hat{\mathbf{j}}+\hat{\mathbf{k}}$ and $\mathbf{CD}=-3\hat{\mathbf{i}}+2\hat{\mathbf{j}}+4\hat{\mathbf{k}}$ are two vectors. The position vectors of the points A and C are $6\hat{\mathbf{i}}+7\hat{\mathbf{j}}+4\hat{\mathbf{k}}$ and $-9\hat{\mathbf{i}}+2\hat{\mathbf{k}}$, respectively. Find the position vector of a point P on the line AB and a point Q on the line CD such that **PQ** is perpendicular to **AB** and **CD** both.

Ans. We have,
$$\mathbf{AB}=3\hat{\mathbf{i}}-\hat{\mathbf{j}}+\hat{\mathbf{k}}$$
and
$$\mathbf{CD}=-3\hat{\mathbf{i}}+2\hat{\mathbf{j}}+4\hat{\mathbf{k}}$$

Also, the position vectors of A and C are $6\hat{\mathbf{i}}+7\hat{\mathbf{j}}+4\hat{\mathbf{k}}$ and $-9\hat{\mathbf{j}}+2\hat{\mathbf{k}}$, respectively.
Since, **PQ** is perpendicular to both **AB** and **CD**.

So, P and Q will be foot of perpendicular to both the lines through A and C.

Now, equation of the line through A and parallel to the vector **AB** is,

$$\mathbf{r}=(6\hat{\mathbf{i}}+7\hat{\mathbf{j}}+4\hat{\mathbf{k}})+\lambda(3\hat{\mathbf{i}}-\hat{\mathbf{j}}+\hat{\mathbf{k}}) \quad \ldots\text{(i)}$$

and the line through C and parallel to the vector **CD** is given by

$$\mathbf{r}=-9\hat{\mathbf{j}}+2\hat{\mathbf{k}}+\mu(-3\hat{\mathbf{i}}+2\hat{\mathbf{j}}+4\hat{\mathbf{k}}) \quad \ldots\text{(ii)}$$

Let $P(6+3\lambda, 7-\lambda, 4+\lambda)$ is any point on the first line and Q be any point on second line is given by $(-3\mu, -9+2\mu, 2+4\mu)$.

$$\therefore \quad \mathbf{PQ}=(-3\mu-6-3\lambda)\hat{\mathbf{i}}+(-9+2\mu-7+\lambda)\hat{\mathbf{j}}+(2+4\mu-4-\lambda)\hat{\mathbf{k}}$$
$$=(-3\mu-6-3\lambda)\hat{\mathbf{i}}+(2\mu+\lambda-16)\hat{\mathbf{j}}+(4\mu-\lambda-2)\hat{\mathbf{k}}$$

Note (✱) Topics related to these question have been removed from latest NCERT.

If **PQ** is perpendicular to the first line, then

$$3(-3\mu-6-3\lambda)-(2\mu+\lambda-16)+(4\mu-\lambda-2)=0$$

$$\Rightarrow \quad -9\mu-18-9\lambda-2\mu-\lambda+16+4\mu-\lambda-2=0$$

$$\Rightarrow \quad -7\mu-11\lambda-4=0 \qquad \text{...(iii)}$$

If **PQ** is perpendicular to the second line, then

$$-3(-3\mu-6-3\lambda)+2(2\mu+\lambda-16)+4(4\mu-\lambda-2)=0$$

$$\Rightarrow \quad 9\mu+18+9\lambda+4\mu+2\lambda-32+16\mu-4\lambda-8=0$$

$$\Rightarrow \quad 29\mu+7\lambda-22=0 \qquad \text{...(iv)}$$

On solving Eqs. (iii) and (iv), we get

$$-49\mu-77\lambda-28=0$$

$$\Rightarrow \quad 319\mu+77\lambda-242=0$$

$$\Rightarrow \quad 270\mu-270=0$$

$$\Rightarrow \quad \mu=1$$

On using μ in Eq. (iii), we get

$$-7(1)-11\lambda-4=0$$

$$\Rightarrow \quad -7-11\lambda-4=0$$

$$\Rightarrow \quad -11-11\lambda=0$$

$$\Rightarrow \quad \lambda=-1$$

$$\therefore \quad \mathbf{PQ}=[-3(1)-6-3(-1)]\,\hat{\mathbf{i}}+[2(1)+(-1)-16]\,\hat{\mathbf{j}}+[4(1)-(-1)-2]\,\hat{\mathbf{k}}$$

$$=-6\hat{\mathbf{i}}-15\hat{\mathbf{j}}+3\hat{\mathbf{k}}$$

Q.27 Show that the straight lines whose direction cosines are given by $2l+2m-n=0$ and $mn+nl+lm=0$ are at right angles.

Ans. Given, $\quad 2l+2m-n=0 \qquad$ ...(i)

and $\quad mn+nl+lm=0 \qquad$ ...(ii)

From Eq. (i), we get $\quad m=\dfrac{n-2l}{2}$

$$\Rightarrow \quad \left(\frac{n-2l}{2}\right)n+nl+l\left(\frac{n-2l}{2}\right)=0 \qquad \text{[From Eq. (ii)]}$$

$$\Rightarrow \quad \frac{n^2-2nl+2nl+nl-2l^2}{2}=0$$

$$\Rightarrow \quad n^2+nl-2l^2=0$$

$$\Rightarrow \quad n^2+2nl-nl-2l^2=0$$

$$\Rightarrow \quad (n+2l)(n-l)=0$$

$$\Rightarrow \quad n=-2l \text{ and } n=l$$

$$\therefore \quad m=\frac{-2l-2l}{2},\, m=\frac{l-2l}{2}$$

$$\Rightarrow \quad m=-2l,\, m=\frac{-l}{2}$$

Thus, the direction ratios of two lines are proportional to $l, -2l, -2l$ and $l, \frac{-l}{2}, l$.

$$\Rightarrow \quad 1, -2, -2 \text{ and } 1, \frac{-1}{2}, 1$$

$$\Rightarrow \quad 1, -2, -2 \text{ and } 2, -1, 2$$

Also, the vectors parallel to these lines are $\mathbf{a} = \hat{\mathbf{i}} - 2\hat{\mathbf{j}} - 2\hat{\mathbf{k}}$ and $\mathbf{b} = 2\hat{\mathbf{i}} - \hat{\mathbf{j}} + 2\hat{\mathbf{k}}$, respectively.

$$\therefore \quad \cos\theta = \frac{\mathbf{a}\cdot\mathbf{b}}{|\mathbf{a}||\mathbf{b}|} = \frac{(\hat{\mathbf{i}} - 2\hat{\mathbf{j}} - 2\hat{\mathbf{k}})\cdot(2\hat{\mathbf{i}} - \hat{\mathbf{j}} + 2\hat{\mathbf{k}})}{3\cdot 3}$$

$$\Rightarrow \quad \cos\theta = \frac{2+2-4}{9} = 0 = \cos\frac{\pi}{2} \qquad \left[\because \cos\frac{\pi}{2} = 0\right]$$

$$\therefore \quad \theta = \frac{\pi}{2}$$

Q.28 If $l_1, m_1, n_1, l_2, m_2, n_2$ and l_3, m_3, n_3 are the direction cosines of three mutually perpendicular lines, then prove that the line whose direction cosines are proportional to $l_1 + l_2 + l_3, m_1 + m_2 + m_3$ and $n_1 + n_2 + n_3$ makes equal angles with them.

Ans. Let
$$\mathbf{a} = l_1\hat{\mathbf{i}} + m_1\hat{\mathbf{j}} + n_1\hat{\mathbf{k}},$$
$$\mathbf{b} = l_2\hat{\mathbf{i}} + m_2\hat{\mathbf{j}} + n_2\hat{\mathbf{k}},$$
$$\mathbf{c} = l_3\hat{\mathbf{i}} + m_3\hat{\mathbf{j}} + n_3\hat{\mathbf{k}}$$
and
$$\mathbf{d} = (l_1 + l_2 + l_3)\hat{\mathbf{i}} + (m_1 + m_2 + m_2)\hat{\mathbf{j}} + (n_1 + n_2 + n_3)\hat{\mathbf{k}}$$

Also, let α, β and γ are the angles between **a** and **d**, **b** and **d**, **c** and **d**.

$$\therefore \quad \cos\alpha = l_1(l_1 + l_2 + l_3) + m_1(m_1 + m_2 + m_3) + n_1(n_1 + n_2 + n_3)$$
$$= l_1^2 + l_1 l_2 + l_1 l_3 + m_1^2 + m_1 m_2 + m_1 m_3 + n_1^2 + n_1 n_2 + n_1 n_3$$
$$= (l_1^2 + m_1^2 + n_1^2) + (l_1 l_2 + l_1 l_3 + m_1 m_2 + m_1 m_3 + n_1 n_2 + n_1 n_3)$$
$$= 1 + 0 = 1$$
$$[\because l_1^2 + m_1^2 + n_1^2 = 1 \text{ and } l_1 \perp l_2, l_1 \perp l_3, m_1 \perp m_2, m_1 \perp m_3, n_1 \perp n_2, n_1 \perp n_3]$$

Similarly, $\cos\beta = l_2(l_1 + l_2 + l_3) + m_2(m_1 + m_2 + m_3) + n_2(n_1 + n_2 + n_3)$
$= 1 + 0$ and $\cos\gamma = 1 + 0$

$$\Rightarrow \quad \cos\alpha = \cos\beta = \cos\gamma$$
$$\Rightarrow \quad \alpha = \beta = \gamma$$

So, the line whose direction cosines are proportional to $l_1 + l_2 + l_3, m_1 + m_2 + m_3, n_1 + n_2 + n_3$ makes equal angles with the three mutually perpendicular lines whose direction cosines are $l_1, m_1, n_1, l_2, m_2, n_2$ and l_3, m_3, n_3 respectively.

Objective Type Questions

Choose the correct answer out of the given four options in each of the exercises from 29 to 36.

Q.29 Distance of the point (α, β, γ) from y-axis is

(a) β
(b) $|\beta|$
(c) $|\beta| + |\gamma|$
(d) $\sqrt{\alpha^2 + \gamma^2}$

Ans. ***(d)*** $\because$ The given point is (α, β, γ)

Any point of y-axis $= (0, \beta, 0)$

Required distance $= \sqrt{(\alpha - 0)^2 + (\beta - \beta)^2 + (\gamma - 0)^2} = \sqrt{\alpha^2 + \gamma^2}$

Q.30 If the direction cosines of a line are k, k and k, then

(a) $k > 0$ (b) $0 < k < 1$

(c) $k = 1$ (d) $k = \frac{1}{\sqrt{3}}$ or $-\frac{1}{\sqrt{3}}$

Ans. *(d)* Since, direction cosines of a line are k, k and k.

$$\therefore \quad l = k, m = k \text{ and } n = k$$

We know that, $l^2 + m^2 + n^2 = 1$

$$\Rightarrow \quad k^2 + k^2 + k^2 = 1 \Rightarrow k^2 = \frac{1}{3}$$

$$\therefore \quad k = \pm \frac{1}{\sqrt{3}}$$

***Q.31** The distance of the plane $\mathbf{r}\left(\frac{2}{7}\hat{\mathbf{i}} + \frac{3}{7}\hat{\mathbf{j}} - \frac{6}{7}\hat{\mathbf{k}}\right) = 1$ from the origin is

(a) 1 (b) 7 (c) $\frac{1}{7}$ (d) None of these

Ans. *(a)* The distance of the plane $\mathbf{r}\left(\frac{2}{7}\hat{\mathbf{i}} + \frac{3}{7}\hat{\mathbf{j}} - \frac{6}{7}\hat{\mathbf{k}}\right) = 1$ from the origin is 1.

[since, $\mathbf{r} \cdot \mathbf{n} = d$ is the form of above equation, where d represents the distance of plane from the origin *i.e.*, $d = 1$]

***Q.32** The sine of the angle between the straight line $\frac{x-2}{3} = \frac{y-3}{4} = \frac{z-4}{5}$ and the plane $2x - 2y + z = 5$ is

(a) $\frac{10}{6\sqrt{5}}$ (b) $\frac{4}{5\sqrt{2}}$ (c) $\frac{2\sqrt{3}}{5}$ (d) $\frac{\sqrt{2}}{10}$

Ans. *(d)* We have, the equation of line as

$$\frac{x-2}{3} = \frac{y-3}{4} = \frac{z-4}{5}$$

Now, the line passes through point (2, 3, 4) and having direction ratios (3, 4, 5).

Since, the line passes through point (2, 3, 4) and parallel to the vector $(3\hat{\mathbf{i}} + 4\hat{\mathbf{j}} + 5\hat{\mathbf{k}})$.

$$\therefore \quad \mathbf{b} = 3\hat{\mathbf{i}} + 4\hat{\mathbf{j}} + 5\hat{\mathbf{k}}$$

Also, the cartesian form of the given plane is $2x - 2y + z = 5$.

$$\Rightarrow \quad (x\hat{\mathbf{i}} + y\hat{\mathbf{j}} + z\hat{\mathbf{k}})(2\hat{\mathbf{i}} - 2\hat{\mathbf{j}} + \hat{\mathbf{k}}) = 5$$

$$\therefore \quad \mathbf{n} = (2\hat{\mathbf{i}} - 2\hat{\mathbf{j}} + \hat{\mathbf{k}})$$

We know that,

$$\sin\theta = \frac{|\mathbf{b} \cdot \mathbf{n}|}{|\mathbf{b}| \cdot |\mathbf{n}|} = \frac{|(3\hat{\mathbf{i}} + 4\hat{\mathbf{j}} + 5\hat{\mathbf{k}}) \cdot (2\hat{\mathbf{i}} - 2\hat{\mathbf{j}} + \hat{\mathbf{k}})|}{\sqrt{3^2 + 4^2 + 5^2} \cdot \sqrt{4+4+1}}$$

$$= \frac{|6 - 8 + 5|}{\sqrt{50} \cdot 3} = \frac{3}{15\sqrt{2}} = \frac{1}{5\sqrt{2}} = \frac{1}{5\sqrt{2}} \times \frac{\sqrt{2}}{\sqrt{2}}$$

$$\Rightarrow \quad \sin\theta = \frac{\sqrt{2}}{10}$$

Note (✱) Topics related to these question have been removed from latest NCERT.

***Q.33** The reflection of the point (α, β, γ) in the XY-plane is

(a) $(\alpha, \beta, 0)$
(b) $(0, 0, \gamma)$
(c) $(-\alpha, -\beta, \gamma)$
(d) $(\alpha, \beta, -\gamma)$

Ans. ***(d)*** In xy-plane, the reflection of the point (α, β, γ) is $(\alpha, \beta, -\gamma)$.

Q.34 The area of the quadrilateral $ABCD$ where $A(0,4,1)$, $B(2, 3, -1)$, $C(4, 5, 0)$, and $D(2, 6, 2)$ is equal to

(a) 9 sq units
(b) 18 sq units
(c) 27 sq units
(d) 81 sq units

Ans. ***(a)*** Given points are $A(0,4,1)$, $B(2, 3, -1)$, $C(4, 5, 0)$ and $D(2, 6, 2)$

We have, $\quad \mathbf{AB} = (2-0)\hat{\mathbf{i}} + (3-4)\hat{\mathbf{j}} + (-1-1)\hat{\mathbf{k}} = 2\hat{\mathbf{i}} - \hat{\mathbf{j}} - 2\hat{\mathbf{k}}$

and $\quad \mathbf{BC} = (4-2)\hat{\mathbf{i}} + (5-3)\hat{\mathbf{j}} + (0+1)\hat{\mathbf{k}} = 2\hat{\mathbf{i}} + 2\hat{\mathbf{j}} + \hat{\mathbf{k}}$

$$\therefore \text{ Area of quadrilateral } ABCD = |\mathbf{AB} \times \mathbf{BC}| = \begin{vmatrix} \hat{\mathbf{i}} & \hat{\mathbf{j}} & \hat{\mathbf{k}} \\ 2 & -1 & -2 \\ 2 & 2 & 1 \end{vmatrix}$$

$$= \left|\hat{\mathbf{i}}(-1+4) - \hat{\mathbf{j}}(2+4) + \hat{\mathbf{k}}(4+2)\right|$$

$$= \left|3\hat{\mathbf{i}} - 6\hat{\mathbf{j}} + 6\hat{\mathbf{k}}\right|$$

$$= \sqrt{9+36+36} = 9 \text{ sq. units}$$

***Q.35** The locus represented by $xy + yz = 0$ is

(a) a pair of perpendicular lines
(b) a pair of parallel lines
(c) a pair of parallel planes
(d) a pair of perpendicular planes

Ans. ***(d)*** We have, $\quad xy + yz = 0$

$\Rightarrow \quad xy = -yz$

Hence, a pair of perpendicular planes.

***Q.36** If the plane $2x - 3y + 6z - 11 = 0$ makes an angle $\sin^{-1} \alpha$ with x-axis, then the value of α is

(a) $\frac{\sqrt{3}}{2}$
(b) $\frac{\sqrt{2}}{3}$
(c) $\frac{2}{7}$
(d) $\frac{3}{7}$

Ans. ***(c)*** Since, $2x - 3y + 6z - 11 = 0$ makes an angle $\sin^{-1}\alpha$ with x-axis.

$$\mathbf{b} = (1\hat{\mathbf{i}} + 0\hat{\mathbf{j}} + 0\hat{\mathbf{k}}) \text{ and } \mathbf{n} = 2\hat{\mathbf{i}} - 3\hat{\mathbf{j}} + 6\hat{\mathbf{k}}$$

We know that, $\quad \sin\theta = \dfrac{|\mathbf{b}\cdot\mathbf{n}|}{|\mathbf{b}|\cdot|\mathbf{n}|}$

$$= \frac{|(1\hat{\mathbf{i}})\cdot(2\hat{\mathbf{i}} - 3\hat{\mathbf{j}} + 6\hat{\mathbf{k}})|}{\sqrt{1}\sqrt{4+9+36}} = \frac{2}{7}$$

Note (✱) Topics related to these question have been removed from latest NCERT.

Fill in the blanks in each of the exercises 37 to 41.

***Q.37** A plane passes through the points (2, 0,0) (0, 3, 0) and (0, 0, 4) the equation of plane is ....

Ans. We know that, equation of a the plane that cut the coordinate axes at $(a, 0, 0)$ $(0, b, 0)$ and $(0, 0, c)$ is $\frac{x}{a}+\frac{y}{b}+\frac{z}{c}=1.$

Hence, the equation of plane passes through the points (2, 0, 0), (0, 3, 0) and (0, 0, 4) is $\frac{x}{2}+\frac{y}{3}+\frac{z}{4}=1.$

Q.38 The direction cosines of the vector $(2\hat{i}+2\hat{j}-\hat{k})$ are ....

Ans. Direction cosines of $(2\hat{i}+2\hat{j}-\hat{k})$ are $\frac{2}{\sqrt{4+4+1}}, \frac{2}{\sqrt{4+4+1}}, \frac{-1}{\sqrt{4+4+1}}$ *i.e.*,

$\frac{2}{3}, \frac{2}{3}, \frac{-1}{3}.$ $\left[\because l=\frac{a}{\sqrt{a^2+b^2+c^2}}, m=\frac{b}{\sqrt{a^2+b^2+c^2}}, n=\frac{c}{\sqrt{a^2+b^2+c^2}}\right]$

Q.39 The vector equation of the line $\frac{x-5}{3}=\frac{y+4}{7}=\frac{z-6}{2}$ is ....

Ans. We have, $\mathbf{a}=5\hat{i}-4\hat{j}+6\hat{k}$ and $\mathbf{b}=3\hat{i}+7\hat{j}+2\hat{k}$

So, the vector equation will be $\mathbf{r}=\mathbf{a}+\lambda\mathbf{b}$

$$\Rightarrow \quad \mathbf{r}=(5\hat{i}-4\hat{j}+6\hat{k})+\lambda(3\hat{i}+7\hat{j}+2\hat{k}$$

$$\Rightarrow \quad (x\hat{i}+y\hat{j}+z\hat{k})-(5\hat{i}-4\hat{j}+6\hat{k})=\lambda(3\hat{i}+7\hat{j}+2\hat{k})$$

$$\Rightarrow \quad (x-5)\hat{i}+(y+4)\hat{j}+(z-6)\hat{k}=\lambda(3\hat{i}+7\hat{j}+2\hat{k})$$

Q.40 The vector equation of the line through the points (3, 4, −7) and (1, −1, 6) is ....

Ans. We know that, vector equation of a line passes through two points is represented by $\mathbf{r}=\mathbf{a}+\lambda(\mathbf{b}-\mathbf{a})$

Here, $\mathbf{r}=x\hat{i}+y\hat{j}+z\hat{k}$, $\mathbf{a}=3\hat{i}+4\hat{j}-7\hat{k}$

and $\mathbf{b}=\hat{i}-\hat{j}+6\hat{k}$

$$\Rightarrow \quad (\mathbf{b}-\mathbf{a})=-2\hat{i}-5\hat{j}+13\hat{k}$$

So, the required equation is

$$x\hat{i}+y\hat{j}+z\hat{k}=3\hat{i}+4\hat{j}-7\hat{k}+\lambda(-2\hat{i}-5\hat{j}+13\hat{k})$$

$$\Rightarrow (x-3)\hat{i}+(y-4)\hat{j}+(z+7)\hat{k}=\lambda(-2\hat{i}-5\hat{j}+13\hat{k})$$

***Q.41** The cartesian equation of the plane $\mathbf{r}\cdot(\hat{i}+\hat{j}-\hat{k})=2$ is ....

Ans. We have, $\mathbf{r}\cdot(\hat{i}+\hat{j}-\hat{k})=2$

$$\Rightarrow \quad (x\hat{i}+y\hat{j}+z\hat{k})\cdot(\hat{i}+\hat{j}-\hat{k})=2$$

$$\Rightarrow \quad x+y-z=2$$

which is the required form

Note (✱) Topics related to these question have been removed from latest NCERT.

State True or False for the statements in each of the exercises 42 to 49.

***Q.42** The unit vector normal to the plane $x + 2y + 3z - 6 = 0$ is $\frac{1}{\sqrt{14}}\hat{\mathbf{i}} + \frac{2}{\sqrt{14}}\hat{\mathbf{j}} + \frac{3}{\sqrt{14}}\hat{\mathbf{k}}$.

Ans. *True*

We have, $\mathbf{n} = \hat{\mathbf{i}} + 2\hat{\mathbf{j}} + 3\hat{\mathbf{k}}$

$$\therefore \quad = \frac{\mathbf{n}}{|\mathbf{n}|}\mathbf{n} = \frac{\hat{\mathbf{i}} + 2\hat{\mathbf{j}} + 3\hat{\mathbf{k}}}{\sqrt{1^2 + 2^2 + 3^2}} = \frac{\hat{\mathbf{i}}}{\sqrt{14}} + \frac{2\hat{\mathbf{j}}}{\sqrt{14}} + \frac{3\hat{\mathbf{k}}}{\sqrt{14}}$$

***Q.43** The intercepts made by the plane $2x - 3y + 5z + 4 = 0$ on the coordinate axis are $-2, \frac{4}{3}$ and $-\frac{4}{5}$.

Ans. *True*

We have, $2x - 3y + 5z + 4 = 0$

$$\Rightarrow \quad 2x - 3y + 5z = -4$$

$$\Rightarrow \quad \frac{2x}{-4} - \frac{3y}{-4} + \frac{5z}{-4} = 1$$

$$\Rightarrow \quad \frac{x}{-2} + \frac{y}{\frac{4}{3}} - \frac{z}{\frac{4}{5}} = 1$$

$$\Rightarrow \quad \frac{x}{-2} + \frac{y}{\frac{4}{3}} + \frac{z}{\left(-\frac{4}{5}\right)} = 1$$

So, the intercepts are $-2, \frac{4}{3}$ and $-\frac{4}{5}$.

***Q.44** The angle between the line $\mathbf{r} = (5\hat{\mathbf{i}} - \hat{\mathbf{j}} - 4\hat{\mathbf{k}}) + \lambda(2\hat{\mathbf{i}} - \hat{\mathbf{j}} + \hat{\mathbf{k}})$ and the plane $\mathbf{r}.(3\hat{\mathbf{i}} - 4\hat{\mathbf{j}} - \hat{\mathbf{k}}) + 5 = 0$ is $\sin^{-1}\left(\frac{5}{2\sqrt{91}}\right)$.

Ans. *False*

We have, $\mathbf{b} = 2\hat{\mathbf{i}} - \hat{\mathbf{j}} + \hat{\mathbf{k}}$ and $\mathbf{n} = 3\hat{\mathbf{i}} - 4\hat{\mathbf{j}} - \hat{\mathbf{k}}$

Let θ is the angle between line and plane.

$$\text{Then,} \quad \sin\theta = \frac{|\mathbf{b}\cdot\mathbf{n}|}{|\mathbf{b}|\cdot|\mathbf{n}|}$$

$$= \frac{|(2\hat{\mathbf{i}} - \hat{\mathbf{j}} + \hat{\mathbf{k}})\cdot(3\hat{\mathbf{i}} - 4\hat{\mathbf{j}} - \hat{\mathbf{k}})|}{\sqrt{6}\cdot\sqrt{26}}$$

$$= \frac{|6 + 4 - 1|}{\sqrt{156}} = \frac{9}{2\sqrt{39}}$$

$$\therefore \quad \theta = \sin^{-1}\frac{9}{2\sqrt{39}}$$

Note (✱) Topics related to these question have been removed from latest NCERT.

***Q.45** The angle between the planes $\mathbf{r}.(2\hat{\mathbf{i}}-3\hat{\mathbf{j}}+\hat{\mathbf{k}})=1$ and $\mathbf{r}.(\hat{\mathbf{i}}-\hat{\mathbf{j}})=4$ is $\cos^{-1}\left(\dfrac{-5}{\sqrt{58}}\right)$.

Ans. *False*

We know that, the angle between two planes is given by $\cos\theta=\dfrac{|\mathbf{n}_1\cdot\mathbf{n}_2|}{|\mathbf{n}_1||\mathbf{n}_2|}$

Here, $\mathbf{n}_1=(2\hat{\mathbf{i}}-3\hat{\mathbf{j}}+\hat{\mathbf{k}})$ and $\mathbf{n}_2=(\hat{\mathbf{i}}-\hat{\mathbf{j}})$

$$\therefore \quad \cos\theta=\frac{|(2\hat{\mathbf{i}}-3\hat{\mathbf{j}}+\hat{\mathbf{k}})\ (\hat{\mathbf{i}}-\hat{\mathbf{j}})|}{\sqrt{4+9+1}\ \sqrt{1+1}}$$

$$\Rightarrow \quad \cos\theta=\frac{|2+3|}{\sqrt{14}\cdot\sqrt{2}}=\frac{5}{\sqrt{28}}$$

$$\therefore \quad \theta=\cos^{-1}\left(\frac{5}{\sqrt{28}}\right)$$

***Q.46** The line $\mathbf{r}=2\hat{\mathbf{i}}-3\hat{\mathbf{j}}-\hat{\mathbf{k}}+\lambda(\hat{\mathbf{i}}-\hat{\mathbf{j}}+2\hat{\mathbf{k}})$ lies in the plane $\mathbf{r}.(3\hat{\mathbf{i}}+\hat{\mathbf{j}}-\hat{\mathbf{k}})+2=0$.

Ans. *False*

We have, $\mathbf{r}=2\hat{\mathbf{i}}-3\hat{\mathbf{j}}-\hat{\mathbf{k}}+\lambda(\hat{\mathbf{i}}-\hat{\mathbf{j}}+2\hat{\mathbf{k}})$

$\Rightarrow \quad (x\hat{\mathbf{i}}+y\hat{\mathbf{j}}+z\hat{\mathbf{k}})=\hat{\mathbf{i}}(2+\lambda)+\hat{\mathbf{j}}(-3-\lambda)+\hat{\mathbf{k}}(-1+2\lambda)$

Since, $x=(2+\lambda)$, $y=(-3-\lambda)$ and $z=(-1+2\lambda)$ are coordinates of general point which should satisfy the equation of the given plane.

$$\mathbf{r}\cdot(3\hat{\mathbf{i}}+\hat{\mathbf{j}}-\hat{\mathbf{k}})+2=0$$

$$\Rightarrow \quad (x\hat{\mathbf{j}}+y\hat{\mathbf{j}}+z\hat{\mathbf{k}})\cdot(3\hat{\mathbf{i}}+\hat{\mathbf{j}}-\hat{\mathbf{k}})+2=0$$

$$\Rightarrow \quad 3x+y-z+2=0$$

$$\Rightarrow \quad 3(2+\lambda)+(-3-\lambda)-(-1+2\lambda)+2=0$$

$$\Rightarrow \quad 6+3\lambda-3-\lambda+1-2\lambda+1-2\lambda+2=0$$

$\Rightarrow 6=0$, which is not

So, the line $r=2\hat{\mathbf{i}}-3\hat{\mathbf{j}}-\hat{\mathbf{k}}+\lambda(\hat{\mathbf{i}}-\hat{\mathbf{j}}+2\hat{\mathbf{k}})$ does not lie in a plane.

Q.47 The vector equation of the line $\dfrac{x-5}{3}=\dfrac{y+4}{7}=\dfrac{z-6}{2}$ is

$$\mathbf{r}=5\hat{\mathbf{i}}-4\hat{\mathbf{j}}+6\hat{\mathbf{k}}+\lambda(3\hat{\mathbf{i}}+7\hat{\mathbf{j}}+2\hat{\mathbf{k}})$$

Ans. *True*

We have, $\mathbf{a}=5\hat{\mathbf{i}}-4\hat{\mathbf{j}}+6\hat{\mathbf{k}}$ and $\mathbf{b}=3\hat{\mathbf{i}}+7\hat{\mathbf{j}}+2\hat{\mathbf{k}}$

$\therefore$ vector equation of given line is

$$\mathbf{r}=\mathbf{a}+\lambda\mathbf{b}$$

$$=(5\hat{\mathbf{i}}-4\hat{\mathbf{j}}+6\hat{\mathbf{k}})+\lambda(3\hat{\mathbf{i}}+7\hat{\mathbf{j}}+2\hat{\mathbf{k}})$$

Q.48 The equation of a line, which is parallel to $2\hat{\mathbf{i}}+\hat{\mathbf{j}}+3\hat{\mathbf{k}}$ and which passes through the point $(5,-2,4)$ is $\dfrac{x-5}{2}=\dfrac{y+2}{-1}=\dfrac{z-4}{3}$.

Note (✱) Topics related to these question have been removed from latest NCERT.

Ans. *False*

Here, $x_1 = 5, y_1 = -2, z_1 = 4$

and $a = 2, b = 1, c = 3$

$\therefore$ Required equation of line is $\dfrac{x-5}{2} = \dfrac{y+2}{1} = \dfrac{z-4}{3}$

Q.49 If the foot of perpendicular drawn from the origin to a plane is $(5, -3, -2)$, then the equation of plane is $\mathbf{r}.(5\hat{\mathbf{i}} - 3\hat{\mathbf{j}} - 2\hat{\mathbf{k}}) = 38$.

Ans. *True*

Let the required plane passes through the point $P(5, -3, -2)$ and is perpendicular to **OP**.

$\therefore$ $\mathbf{a} = 5\hat{\mathbf{i}} - 3\hat{\mathbf{j}} - 2\hat{\mathbf{k}}$

and $\mathbf{n} = \mathbf{OP} = 5\hat{\mathbf{i}} - 3\hat{\mathbf{j}} - 2\hat{\mathbf{k}}$

Now, the equation of the plane is

$$(\mathbf{r} - \mathbf{a}) \cdot \mathbf{n} = 0$$

$$\Rightarrow \quad \mathbf{r} \cdot \mathbf{n} = \mathbf{a} \cdot \mathbf{n}$$

$$\Rightarrow \quad \mathbf{r} \cdot (5\hat{\mathbf{i}} - 3\hat{\mathbf{j}} - 2\hat{\mathbf{k}}) = (5\hat{\mathbf{i}} - 3\hat{\mathbf{j}} - 2\hat{\mathbf{k}}) \cdot (5\hat{\mathbf{i}} - 3\hat{\mathbf{j}} - 2\hat{\mathbf{k}})$$

$$\Rightarrow \quad \mathbf{r}.(5\hat{\mathbf{i}} - 3\hat{\mathbf{j}} - 2\hat{\mathbf{k}}) = 25 + 9 + 4$$

$$\Rightarrow \quad \mathbf{r} \cdot (5\hat{\mathbf{i}} - 3\hat{\mathbf{j}} - 2\hat{\mathbf{k}}) = 38$$

Chapter 12

Linear Programming

Short Answer Type Questions

Q.1 Determine the maximum value of $Z = 11x + 7y$ subject to the constraints: $2x + y \leq 6$, $x \leq 2$, $x \geq 0$, $y \geq 0$.

Thinking Process Using constraints, get the corner points for the bounded region and then for each corner point check the corresponding value of Z.

Ans. We have, maximise $Z = 11x + 7y$ ...(i)

Subject to the constraints

$$2x + y \leq 6 \quad ...(ii)$$

$$x \leq 2 \quad ...(iii)$$

$$x \geq 0, y \geq 0 \quad ...(iv)$$

We see that, the feasible region as shaded determined by the system of constraints (ii) to (iv) is $OABC$ and is bounded. As $x \geq 0$ and $y \geq 0$, then we consider the function only in first quadrant.

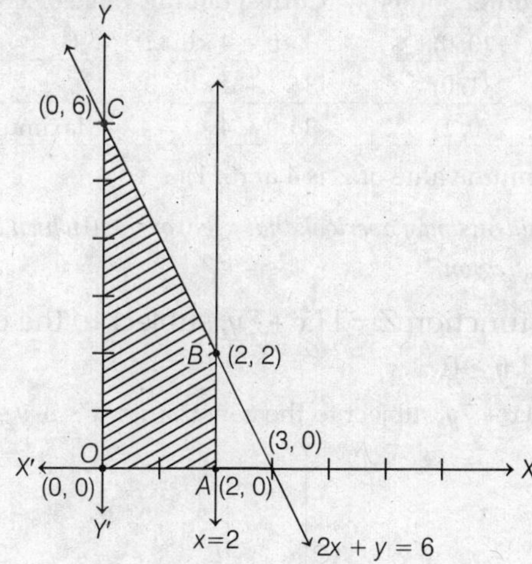

So, now we shall use corner point method to determine the maximum value of Z.

Corner points are $O(0, 0)$, $A(2, 0)$, $B(2, 2)$ and $C(0, 6)$ and value of Z at corner points are given below.

Corner points	Corresponding value of Z
(0, 0)	$11\times0+7\times0=0$
(2, 0)	$11\times2+7\times0=22$
(2, 2)	$11\times2+7\times2=36$
(0, 6)	$11\times0+7\times6=42$ ← Maximum

Hence, the maximum value of Z is 42 at (0, 6).

Q.2 Maximise $Z = 3x + 4y$, subject to the constraints $x + y \le 1$, $x \ge 0$, $y \ge 0$.

Ans. Maximise $Z = 3x + 4y$,

Subject to the constraints $x + y \le 1, x \ge 0, y \ge 0$.

The shaded region shown in the figure as OAB is bounded and the coordinates of corner points O, A and B are (0, 0), (1, 0) and (0, 1), respectively.

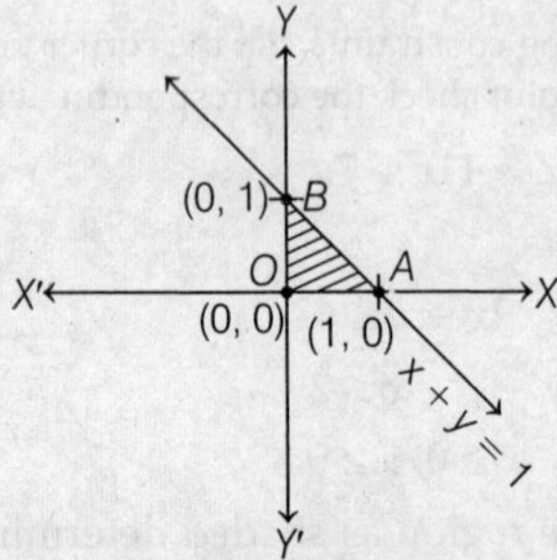

So, now we shall use corner point method to determine the maximum value Z.

Corner points	Corresponding value of Z
(0, 0)	$3\times0+4\times0=0$
(1, 0)	$3\times1+4\times0=3$
(0, 1)	$3\times0+4\times1=4$ ← Maximum

Hence, the maximum value of Z is 4 at (0, 1).

Mistake Alert! *Students may overlook that $x \ge 0$ or $y \ge 0$ which could cause the solution to be in the invalid region.*

Q.3 Maximise the function $Z = 11x + 7y$, subject to the constraints: $x \le 3$, $y \le 2$, $x \ge 0$ and $y \ge 0$.

Ans. Maximise $Z = 11x + 7y$, subject to the constraints $x \le 3, y \le 2, x \ge 0, y \ge 0$.

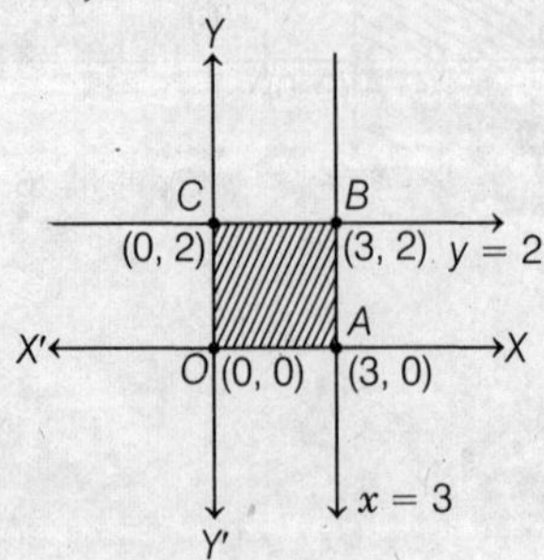

The shaded region as shown in the figure as $OABC$ is bounded and the coordinates of corner points are $O(0, 0)$, $A(3, 0)$, $B(3, 2)$ and $C(0, 2)$, respectively.

So, now we shall use corner point method to determine the maximum value of Z.

Corner points	Corresponding value of Z
(0, 0)	$11 \times 0 + 7 \times 0 = 0$
(3, 0)	$11 \times 3 + 7 \times 0 = 33$
(3, 2)	$11 \times 3 + 7 \times 2 = 47 \leftarrow$ Maximum
(0, 2)	$11 \times 0 + 7 \times 2 = 14$

Hence, Z is maximum at (3, 2) and its maximum value is 47.

Q.4 Minimise $Z = 13x - 15y$ subject to the constraints: $x + y \le 7$, $2x - 3y + 6 \ge 0$, $x \ge 0$ and $y \ge 0$.

Ans. Minimise $Z = 13x - 15y$, subject to the constraints, $x + y \le 7$, $2x - 3y + 6 \ge 0$, $x \ge 0$, $y \ge 0$.

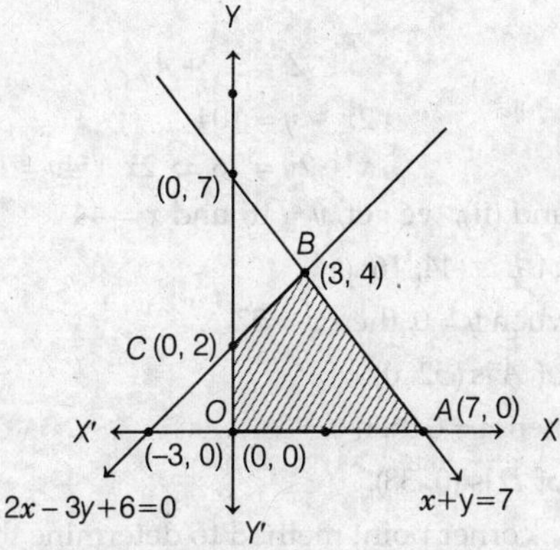

The feasible (shaded) region shown as $OABC$ is bounded and coordinates of its corner points are O (0, 0), A (7, 0), B (3, 4) and C(0, 2), respectively. So, now we shall use corner point method to determine the minimum value of Z.

Corner points	Corresponding value of Z
(0, 0)	$13 \times 0 - 15 \times 0 = 0$
(7, 0)	$13 \times 7 - 15 \times 0 = 91$
(3, 4)	$13 \times 3 - 15 \times 4 = -21$
(0, 2)	$13 \times 0 - 15 \times 2 = -30 \leftarrow$ Minimum

Hence, the minimum value of Z is –30 at (0, 2).

Q.5 Determine the maximum value of $Z = 3x + 4y$, if the feasible region (shaded) for a LPP is shown in following figure.

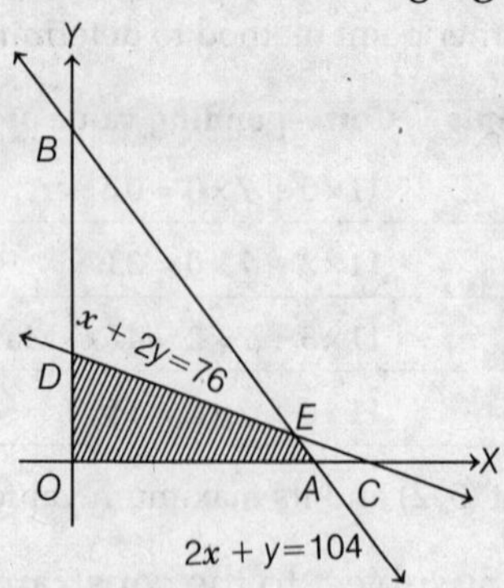

Thinking Process First evaluate the intersection coordinates by the given two equation after that find the coordinates of the bounded region.

Ans. As clear from the graph, corner points are O, A, E and D. Also, given region is bounded.

Here, $Z = 3x + 4y$

$\because$ $2x + y = 104$ ...(i)

and $x + 2y = 76 \Rightarrow 2x + 4y = 152$ ...(ii)

On solving Eqs. (i) and (ii), we get, $y = 16$ and $x = 44$

So, the co-ordinate of E is (44, 16).

Now, from Eq. (i), when $y = 0$, then $x = 52$

So, the co-ordinate of A is (52, 0)

and from Eq. (ii), when $x = 0$ then $y = 38$.

So, the co-ordinate of D is (0, 38).

So, now we shall use corner point method to determine the maximum value of Z.

Corner points	Corresponding value of Z
(0, 0)	$3 \times 0 + 4 \times 0 = 0$
(52, 0)	$3 \times 52 + 4 \times 0 = 156$
(44, 16)	$3 \times 44 + 4 \times 16 = 196 \leftarrow$ Maximum
(0, 38)	$3 \times 0 + 4 \times 38 = 152$

Hence, Z maximum is at (44, 16) and its maximum value is 196.

Q.6 Feasible region (shaded) for a LPP is shown in following figure. Maximise $Z = 5x + 7y$.

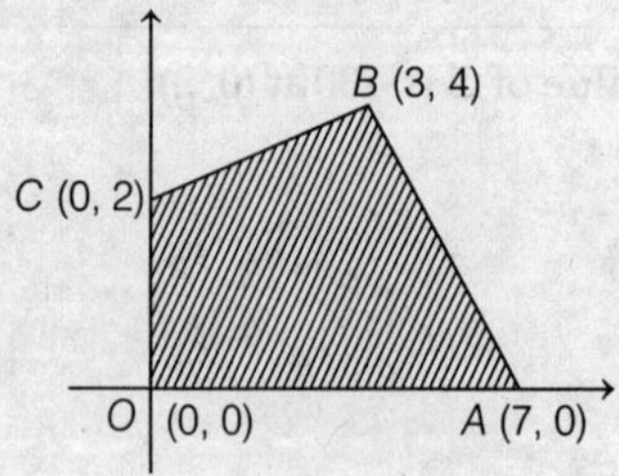

Ans. The shaded region is bounded and has coordinates of corner points as $O(0, 0)$, $A(7, 0)$, $B(3, 4)$ and $C(0, 2)$. Also, $Z = 5x + 7y$.

Now, let us find the corresponding value of Z at the corner points.

Corner points	Corresponding value of Z
(0, 0)	$5 \times 0 + 7 \times 0 = 0$
(7, 0)	$5 \times 7 + 7 \times 0 = 35$
(3, 4)	$5 \times 3 + 7 \times 4 = 43 \leftarrow$ Maximum
(0, 2)	$5 \times 0 + 7 \times 2 = 14$

Hence, the maximum value of Z is 43 at (3, 4).

Q.7 The feasible region for a LPP is shown in following figure. Find the minimum value of $Z = 11x + 7y$.

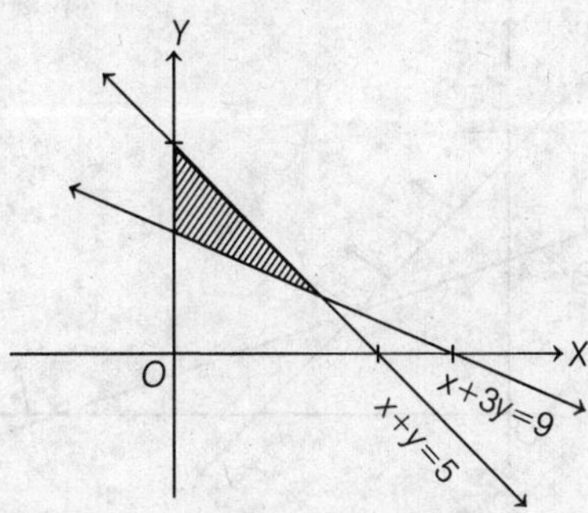

Ans. From the figure, it is clear that feasible region is bounded.

$\because \quad x + 3y = 9 \quad$ ...(i)

and $\quad x + y = 5 \quad$ ...(ii)

On solving Eqs. (i) and (ii), we get

$\therefore \quad y = 2$ and $x = 3$

So, intersection points of $x + y = 5$ and $x + 3y = 9$ is (3, 2).

Now, from Eq. (i) if $x = 0$, then $y = 3$ and from Eq. (ii) if $x = 0$, then $y = 5$

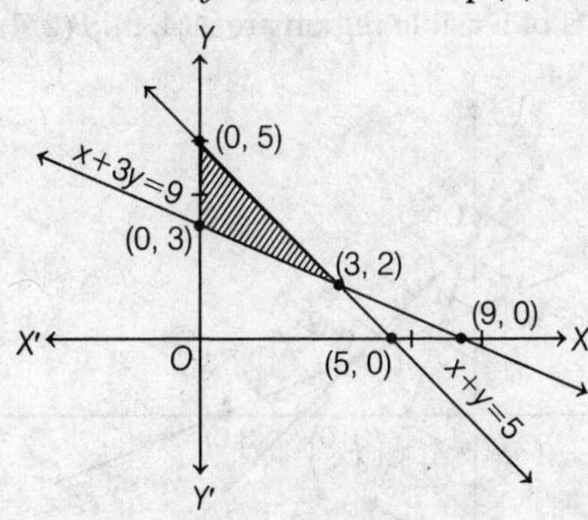

So, the coordinates of corner points of feasible region are (3, 2), (0, 3) and (0, 5).

Now, let us find the corresponding value of Z at the corner points (3, 2), (0, 3) and (0, 5).

Corner points	Corresponding value of Z
(0, 3)	$11 \times 0 + 7 \times 3 = 21 \leftarrow$ Minimum
(3, 2)	$11 \times 3 + 7 \times 2 = 47$
(0, 5)	$11 \times 0 + 7 \times 5 = 35$

Hence, the minimum value of Z is 21 at (0, 3).

Q.8 Refer to question 7 above. Find the maximum value of Z.

Ans. From question 7, above, it is clear that Z is maximum at (3, 2) and its maximum value is 47.

Q.9 The feasible region for a LPP is shown in the following figure. Evaluate $Z = 4x + y$ at each of the corner points of this region. Find the minimum value of Z, if it exists.

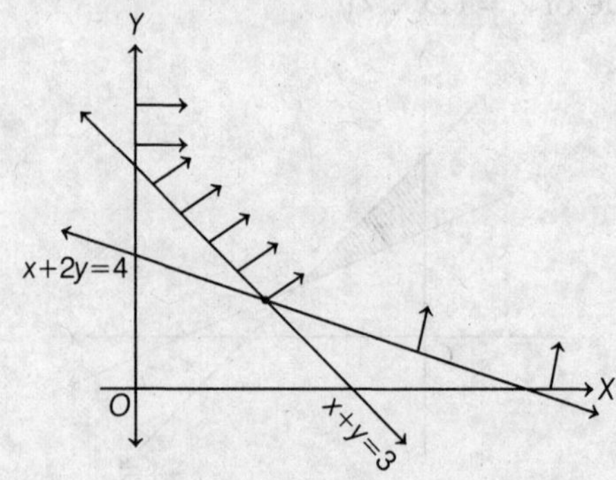

Ans. From the shaded region, it is clear that feasible region is unbounded.

Also, we have $Z = 4x + y.$

$$x + 2y = 4 \quad \text{...(i)}$$

$$x + y = 3 \quad \text{...(ii)}$$

On solving Eqs. (i) and (ii), we get

$$x = 2 \text{ and } y = 1$$

Now, solving Eq. (i) and $y = 0$, we get $x = 4$, $y = 0$

Also, solving Eq. (ii) and $x = 0$, we get $x = 0$ and $y = 3$

Hence, corner points of feasible region are $A(4, 0)$, $B(2, 1)$ and $C(0, 3)$

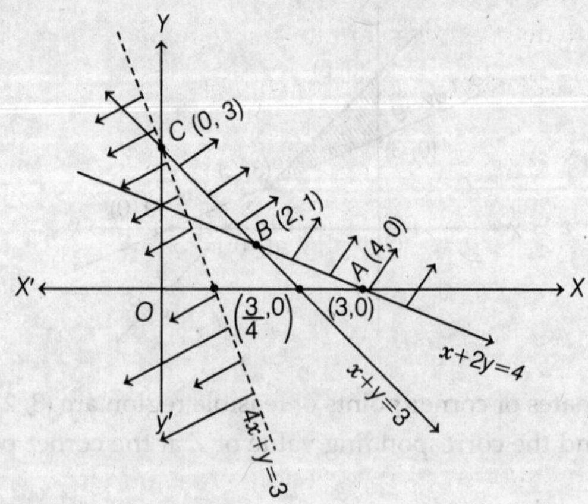

Now, we shall used corner point method to determine the minimum value of Z.

Corner points	Corresponding value of Z
(4, 0)	$4 \times 4 + 0 = 16$
(2, 1)	$4 \times 2 + 1 = 9$
(0, 3)	$4 \times 0 + 3 = 3 \leftarrow$ Minimum

Now, we see that 3 is the smallest value of Z at the corner point (0, 3). Note that here we see that, the region is unbounded, therefore 3 may or may not be the minimum value of Z.

To decide this issue, we graph the inequality $4x + y < 3$ and check whether the resulting open half plan has no point in common with feasible region otherwise, Z has no minimum value.

From the shown graph above, it is clear that there is no point in common with feasible region and hence Z has minimum value 3 at (0, 3).

Note Maximum or minimum value of the objective function may or may not exist if the feasible region is unbounded.

Q.10 In following figure, the feasible region (shaded) for a LPP is shown. Determine the maximum and minimum value of $Z = x + 2y$.

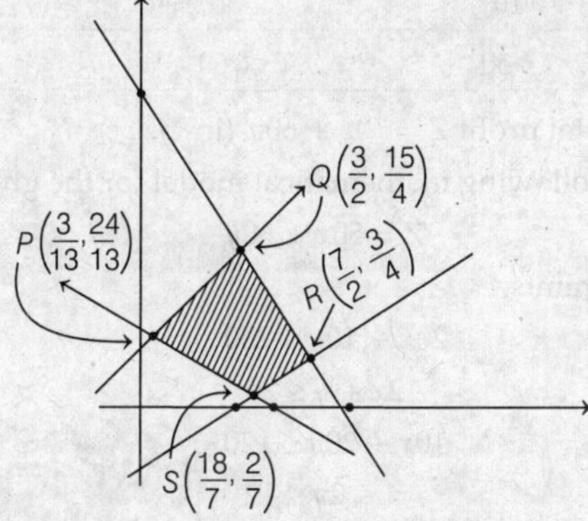

Ans. From the shaded bounded region, it is clear that the coordinates of corner points are $P\left(\frac{3}{13}, \frac{24}{13}\right)$, $S\left(\frac{18}{7}, \frac{2}{7}\right)$, $R\left(\frac{7}{2}, \frac{3}{4}\right)$ and $Q\left(\frac{3}{2}, \frac{15}{4}\right)$.

Also, we have to determine maximum and minimum value of $Z = x + 2y$.

Corner points	Corresponding value of Z
$\left(\frac{3}{13}, \frac{24}{13}\right)$	$\frac{3}{13} + \frac{48}{13} = \frac{51}{13} = 3\frac{12}{13}$
$\left(\frac{18}{7}, \frac{2}{7}\right)$	$\frac{18}{7} + \frac{4}{7} = \frac{22}{7} = 3\frac{1}{7} \leftarrow$ Minimum
$\left(\frac{7}{2}, \frac{3}{4}\right)$	$\frac{7}{2} + \frac{6}{4} = \frac{20}{4} = 5$
$\left(\frac{3}{2}, \frac{15}{4}\right)$	$\frac{3}{2} + \frac{30}{4} = \frac{36}{4} = 9 \leftarrow$ Maximum

Hence, the maximum value is 9 and minimum value is $3\frac{1}{7}$.

Q.11 A manufacturer of electronic circuits has a stock of 200 resistors, 120 transistors and 150 capacitors and is required to produce two types of circuits A and B. Type A requires 20 resistors, 10 transistors and 10 capacitors. Type B requires 10 resistors, 20 transistors and 30 capacitors. If the profit on type A circuit is ₹ 50 and that on type B circuit is ₹ 60, formulate this problem as a LPP, so that the manufacturer can maximise his profit.

Thinking Process For maximising the profit, use resistor constraint, transistor constraint, capacitor constraint and non-negative constraint.

Ans. Let the manufacturer produces x units of type A circuits and y units of type B circuits.

Form the given information, we have following corresponding constraint table.

	Type A (x)	Type B (y)	Maximum stock
Resistors	20	10	200
Transistors	10	20	120
Capacitors	10	30	150
Profit	₹ 50	₹ 60	

Thus, we see that total profit $Z = 50x + 60y$ (in ₹).

Now, we have the following mathematical model for the given problem.

Maximise $\quad Z = 50x + 60y \quad$ ...(i)

Subject to the constraints.

$20x + 10y \le 200 \quad$ [resistors constraint]

$\Rightarrow \quad 2x + y \le 20 \quad$ ...(ii)

and $\quad 10x + 20y \le 120 \quad$ [transistor constraint]

$\Rightarrow \quad x + 2y \le 12 \quad$ ...(iii)

and $\quad 10x + 30y \le 150 \quad$ [capacitor constraint]

$\Rightarrow \quad x + 3y \le 15 \quad$ ...(iv)

and $\quad x \ge 0, y \ge 0 \quad$ [non-negative constraint] ...(v)

So, maximise $Z = 50x + 60y$, subject to $2x + y \le 20$, $x + 2y \le 12$, $x + 3y \le 15$, $x \ge 0$, $y \ge 0$.

Q.12 A firm has to transport 1200 packages using large vans which can carry 200 packages each and small vans which can take 80 packages each. The cost for engaging each large van is ₹ 400 and each small van is ₹ 200. Not more than ₹ 3000 is to be spent on the job and the number of large vans cannot exceed the number of small vans. Formulate this problem as a LPP given that the objective is to minimise cost.

Ans. Let the firm has x number of large vans and y number of small vans. From the given information, we have following corresponding constraint table.

	Large vans (x)	Small vans (y)	Maximum / Minimum
Packages	200	80	1200
Cost	400	200	3000

Thus, we see that objective function for minimum cost is $Z = 400x + 200y$.

Subject to constraints

$$200x + 80y \geq 1200 \quad \text{[package constraint]}$$

$$\Rightarrow \quad 5x + 2y \geq 30 \quad \text{...(i)}$$

$$\text{and} \quad 400x + 200y \leq 3000 \quad \text{[cost constraint]}$$

$$\Rightarrow \quad 2x + y \leq 15 \quad \text{...(ii)}$$

$$\text{and} \quad x \leq y \quad \text{[van constraint] ...(iii)}$$

$$\text{and} \quad x \geq 0, y \geq 0 \quad \text{[non-negative constraints] ...(iv)}$$

Thus, required LPP to minimise cost is minimise $Z = 400x + 200y$, subject to

$$5x + 2y \geq 30$$

$$2x + y \leq 15$$

$$x \leq y \Rightarrow x \geq 0, y \geq 0$$

Q.13 A company manufactures two types of screws A and B. All the screws have to pass through a threading machine and a slotting machine. A box of type A screws requires 2 minutes on the threading machine and 3 min on the slotting machine. A box of type B screws requires 8 minutes of threading on the threading machine and 2 minutes on the slotting machine. In a week, each machine is available for 60 hours. On selling these screws, the company gets a profit of ₹ 100 per box on type A screws and ₹ 170 per box on type B screws. Formulate this problem as a LPP given that the objective is to maximise profit.

Ans. Let the company manufactures x boxes of type A screws and y boxes of type B screws. From the given information, we have following corresponding constraint table

	Type A (x)	Type B (y)	Maximum time available on each machine in a week
Time required for screws on threading machine	2 mins	8 mins	60 × 60 (mins)
Time required for screws on slotting machine	3 mins	2 mins	60 × 60 (mins)
Profit	₹ 100	₹ 170	

Thus, we see that objective function for maximum profit is $Z = 100x + 170y$.

Subject to constraints

$$2x + 8y \leq 60 \times 60$$

[time constraint for threading machine]

$\Rightarrow \quad x + 4y \leq 1800$ ...(i)

and $\quad 3x + 2y \leq 60 \times 60$ [time constraint for slotting machine]

$\Rightarrow \quad 3x + 2y \leq 3600$ ... (ii)

Also, $\quad x \geq 0, y \geq 0$ [non-negative constraints] ...(iii)

$\therefore$ Required LPP is,

Maximise $\quad Z = 100x + 170y$

Subject to constraints, $x + 4y \leq 1800$, $3x + 2y \leq 3600$, $x \geq 0, y \geq 0$.

Q.14 A company manufactures two types of sweaters type A and type B. It costs ₹ 360 to make a type A sweater and ₹ 120 to make a type B sweater. The company can make atmost 300 sweaters and spend atmost ₹ 72000 a day. The number of sweaters of type B cannot exceed the number of sweaters of type A by more than 100. The company makes a profit of ₹ 200 for each sweater of type A and ₹ 120 for every sweater of type B.
Formulate this problem as a LPP to maximise the profit to the company.

Ans. Let the company manufactures x number of type A sweaters and y number of type B sweaters.

From the given information we see that cost to make a type A sweater is ₹360 and cost to make a type B sweater is ₹ 120.

Also, the company spend atmost ₹ 72000 a day.

$\therefore \quad 360x + 120y \leq 72000$

$\Rightarrow \quad 3x + y \leq 600$ ...(i)

Also, company can make atmost 300 sweaters in a day.

$\therefore \quad x + y \leq 300$ ...(ii)

Further, the number of sweaters of type B cannot exceed the number of sweaters of type A by more than 100 *i.e.*,

$$x + 100 \geq y$$

$\Rightarrow \quad x - y \geq -100$ ...(iii)

Also, we have non-negative constraints for x and y *i.e.*, $x \geq 0, y \geq 0$ ...(iv)

Hence, the company makes a profit of ₹200 for each sweater of type A and ₹120 for each sweater of type B.

So, objective function for maximize profit $(Z) = 200x + 120y$

Thus, the required LPP to maximise the profit is

Maximise $Z = 200x + 120y$, subject to constraints,

$$3x + y \leq 600$$
$$x + y \leq 300$$
$$x - y \geq -100$$
$$x \geq 0, y \geq 0$$

Q.15 A man rides his motorcycle at the speed of 50 km/h. He has to spend ₹ 2 per km on petrol. If he rides it at a faster speed of 80 km/h, the petrol cost increases to ₹ 3 per km. He has atmost ₹ 120 to spend on petrol and one hour's time. He wishes to find the maximum distance that he can travel. Express this problem as a linear programming problem.

- **Concept Insight** Used formula, Time $= \dfrac{\text{Distance}}{\text{Speed}}$

Ans. Let the man rides to his motorcycle to a distance x km at the speed of 50 km/h and to a distance y km at the speed of 80 km/h.

Therefore, cost on petrol is $2x + 3y$.

Since, he has to spend ₹ 120 atmost on petrol.

$$\therefore \quad 2x + 3y \le 120 \quad \text{...(i)}$$

Also, he has atmost one hour's time.

$$\therefore \quad \frac{x}{50} + \frac{y}{80} \le 1$$

$$\Rightarrow \quad 8x + 5y \le 400 \quad \text{...(ii)}$$

Also, we have $x \ge 0, y \ge 0$ [non-negative constraints]

Thus, required LPP to travel maximum distance by him is

Maximise $Z = x + y$, subject to $2x + 3y \le 120$, $8x + 5y \le 400$, $x \ge 0$, $y \ge 0$

Mistake Alert! Students may incorrectly take $x - y \le 100$ instead of $x - y \ge -100$.

Long Answer Type Questions

Q.16 Refer to question 11. How many of circuits of type A and of type B, should be produced by the manufacturer, so as to maximise his profit? Determine the maximum profit.

Thinking Process Using the constraints, draw the graph to get the corner points and find the maximum value of possible corner points (as asked).

Ans. Referring to solution 11, we have

Maximise $Z = 50x + 60y$, subject to $2x + y \le 20$, $x + 2y \le 12$, $x + 3y \le 15$

i.e.,

$$2x + y = 20 \quad \text{...(i)}$$

$$x + 2y = 12 \quad \text{...(ii)}$$

$$x + 3y = 15 \quad \text{...(iii)}$$

Now, solving Eqs. (i) and (ii), we get

$$x = \frac{28}{3} \text{ and } y = \frac{4}{3}$$

$\therefore$ Coordinate of B is $\left(\dfrac{28}{3}, \dfrac{4}{3}\right)$

Solving Eqs. (ii) and (iii), we get

$$x = 6, y = 3$$

$\therefore$ Coordinate of C is (6, 3)

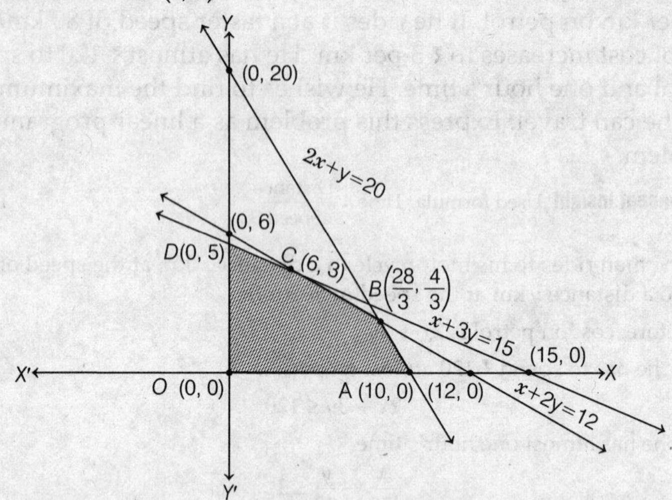

From the shaded region it is clear that the feasible region $OABCD$ determined by the system of constraints is $OABCD$ and the coordinates of corner points are (0, 0), A(10, 0), $B\left(\frac{28}{3}, \frac{4}{3}\right)$, C(6, 3) and D(0, 5), respectively.

Now, we shall use corner point method to determine the maximum value of Z.

Corner points	Corresponding value of $Z = 50x + 60y$
(0, 0)	$50 \times 0 + 60 \times 0 = 0$
(10, 0)	$50 \times 10 + 60 \times 0 = 500$
$\left(\frac{28}{3}, \frac{4}{3}\right)$	$\frac{1400}{3} + \frac{240}{3} = \frac{1640}{3} = 546.66 \leftarrow$ Maximum
(6, 3)	$50 \times 6 + 60 \times 3 = 480$
(0, 5)	$50 \times 0 + 60 \times 5 = 300$

Since, the manufacturer is required to produce two types of circuits A and B and it is clear that parts of resistor, transistor and capacitor cannot be in fraction, so the required maximum profit is ₹ 480 where circuits of type A is 6 and circuits of type B is 3.

Mistake Alert! *Students may forget that the number of products cannot be fractions or negative.*

Q.17 Refer to question 12. What will be the minimum cost?

Ans. Referring to solution 12, we have minimise $Z = 400x + 200y$,

subject to the constraints $5x + 2y \geq 30$, $2x + y \leq 15$, $x \leq y$, $x \geq 0$, $y \geq 0$.

On solving $x - y = 0$ and $5x + 2y = 30$, we get

$$y = \frac{30}{7}, x = \frac{30}{7}$$

$\therefore$ Coordinate of A is $\left(\frac{30}{7}, \frac{30}{7}\right)$

On solving $x - y = 0$ and $2x + y = 15$, we get $x = 5, y = 5$

$\therefore$ Coordinate of B is (5, 5)

On, solving $5x + 2xy = 30$ and $2x + y = 15$, we get

$$x = 0, y = 15$$

$\therefore$ Coordinate of C is (0, 15).

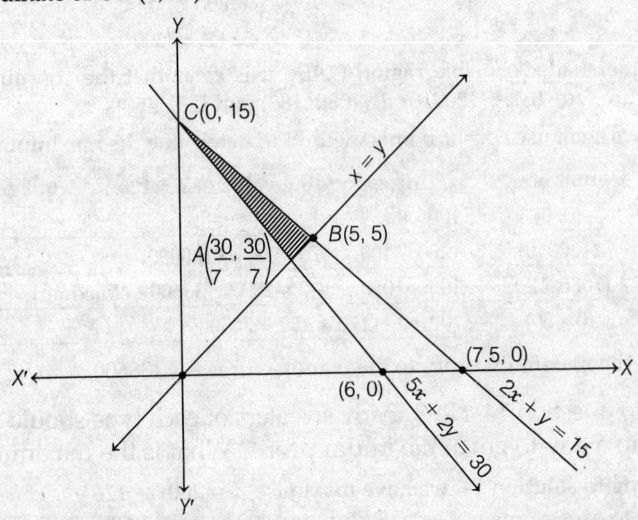

So, from the shaded feasible region ABC it is clear that coordinates of corner points are

$C(0, 15)$, $B(5, 5)$ and $A\left(\frac{30}{7}, \frac{30}{7}\right)$.

Now, we shall use corner point method to determine the minimum cost.

Corner points	Corresponding value of $Z = 400x + 200y$
(0, 15)	$400 \times 0 + 200 \times 15 = 3000$
(5, 5)	$400 \times 5 + 200 \times 5 = 3000$
$\left(\frac{30}{7}, \frac{30}{7}\right)$	$400 \times \frac{30}{7} + 200 \times \frac{30}{7} = \frac{18000}{7}$ $= 2571.43 \leftarrow$ Minimum

Hence, the minimum cost is ₹ 2571.43.

Q.18 Refer to question 13. Solve the linear programming problem and determine the maximum profit to the manufacturer.

Ans. Referring to solution 13, we have

Maximise $Z = 100x + 170y$, subject to the constraints

$$3x + 2y \leq 3600, x + 4y \leq 1800, x \geq 0, y \geq 0$$

On solving $x + 4y = 1800$ and $3x + 2y = 3600$, we get $x = 1080$ and $y = 180$

$\therefore$ Corrdinates of B is (1080, 180).

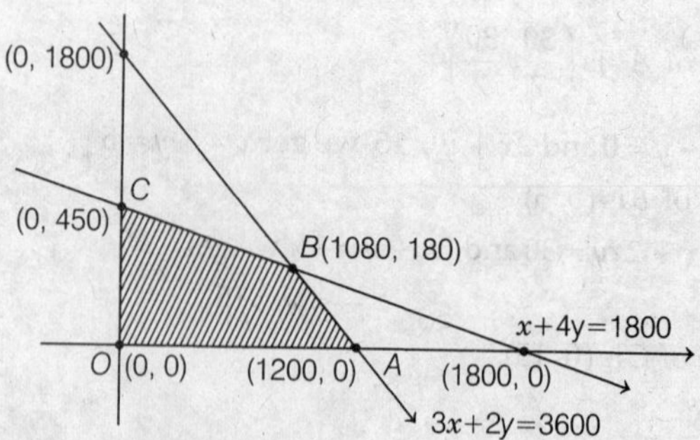

From the shaded feasible region $OABC$ it is clear that the coordinates of corner points are $O(0, 0)$, $A(1200, 0)$, $B(1080, 180)$ and $C(0, 450)$.

Now, we shall use corner points method to determine the maximum profit.

Corner points	Corresponding value of $Z = 100x + 170y$
(0, 0)	$100 \times 0 + 17 \times 0 = 0$
(1200, 0)	$1200 \times 100 + 170 \times 0 = = 120000$
(1080, 180)	$100 \times 1080 + 170 \times 180 = 138600$ ← Maximum
(0, 450)	$100 \times 0 + 170 \times 450 = 76500$

Hence, the maximum profit to the manufacturer is ₹ 138600.

Q.19 Refer to question 14. How many sweaters of each type should the company make in a day to get a maximum profit? What is the maximum profit?

Ans. Referring to solution 14, we have maximise $Z = 200x + 120y$

Subject to constraints $x + y \le 300$, $3x + y \le 600$, $x - y \ge -100$, $x \ge 0$, $y \ge 0$.

On solving $x + y = 300$ and $3x + y = 600$, we get

$$x = 150,\ y = 150$$

∴ The coordinate of B is (150, 150).

On solving $x - y = -100$ and $x + y = 300$, we get $x = 100$, $y = 200$

∴ The coordinate of C is (100, 200)

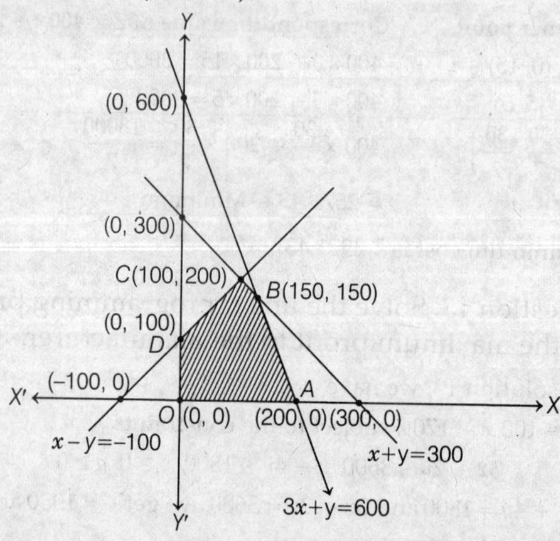

From the shaded feasible region $OABCD$ it is clear that coordinates of corner points are $O(0, 0)$. A (200, 0), B(150, 150), C(100, 200) and D(0, 100).

Now, we shall use corner point method to determine the maximum profit.

Corner points	Corresponding value of $Z = 200x + 120y$
(0, 0)	$200 \times 0 + 120 \times 0 = 0$
(200, 0)	$200 \times 200 + 120 \times 0 = 40000$
(150, 150)	$150 \times 200 + 120 \times 150 = 48000 \leftarrow$ Maximum
(100, 200)	$100 \times 200 + 120 \times 200 = 44000$
(0, 100)	$200 \times 0 + 120 \times 100 = 12000$

Hence, 150 sweaters of each type made by company and maximum profit = ₹ 48000.

Q.20 Refer to question 15. Determine the maximum distance that the man can travel.

Ans. Referring to solution 15, we have

Maximise $Z = x + y$, subject to the constraints.

$$2x + 3y \leq 120,\ 8x + 5y \leq 400,\ x \geq 0, y \geq 0$$

On solving,

$8x + 5y = 400$ and $2x + 3y = 120$, we get

$$x = \frac{300}{7}, y = \frac{80}{7}$$

$\therefore$ The coordinates of B is $\left(\frac{300}{7}, \frac{80}{7}\right)$

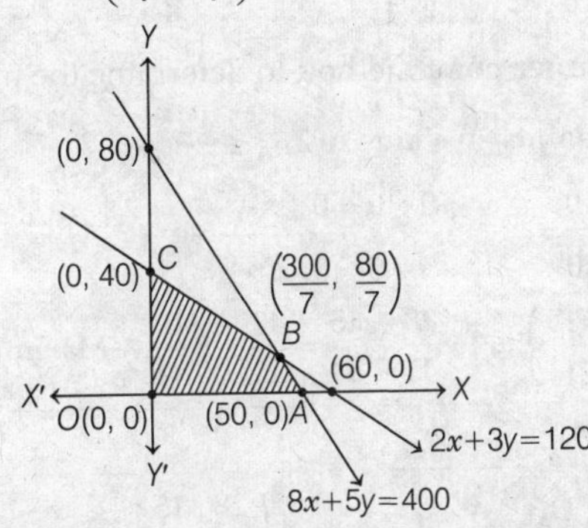

From the shaded feasible region $OABC$, it is clear that coordinates of corner points are $O(0, 0)$, $A(50, 0)$, $B\left(\frac{300}{7}, \frac{80}{7}\right)$ and $C(0, 40)$.

Now, we shall use corner point method to determine the maximum distance.

Corner points	Corresponding value of $Z = x + y$
(0, 0)	$0 + 0 = 0$
(50, 0)	$50 + 0 = 50$
$\left(\frac{300}{7}, \frac{80}{7}\right)$	$\frac{300}{7} + \frac{80}{7} = \frac{380}{7} = 54\frac{2}{7}$ km $\leftarrow$ Maximum
(0, 40)	$0 + 40 = 40$

Hence, the maximum distance that the man can travel is $54\frac{2}{7}$ km.

Q.21 Maximise $Z = x + y$ subject to $x + 4y \leq 8$, $2x + 3y \leq 12$, $3x + y \leq 9$, $x \geq 0$ and $y \geq 0$.

Ans. Here, the given LPP is,

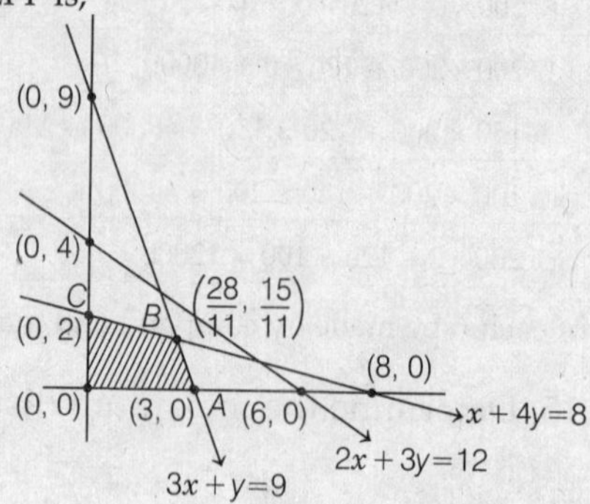

Maximise $Z = x + y$

Subject to, $x + 4y \leq 8$, $2x + 3y \leq 12$, $3x + y \leq 9$, $x \geq 0$, $y \geq 0$.

On solving $x + 4y = 8$ and $3x + y = 9$, we get, $x = \frac{28}{11}, y = \frac{15}{11}$.

$\therefore$ Coordinate of B is $\left(\frac{28}{11}, \frac{15}{11}\right)$

From the graph feasible region $OABC$ is bounded and clear that coordinates of corner

points are $O(0, 0)$, $A(3, 0)$, $B\left(\frac{28}{11}, \frac{15}{11}\right)$ and $C(0, 2)$.

Now, we shall use corner point method to determine the maximum value.

Corner points	Value of $Z = x + y$
$O(0, 0)$	$0 + 0 = 0$
$A(3, 0)$	$3 + 0 = 3$
$B\left(\frac{28}{11}, \frac{15}{11}\right)$	$\frac{28}{11} + \frac{15}{11} = \frac{43}{11} = 3\frac{10}{11}$ ← Maximum
$C(0, 2)$	$0 + 2 = 2$

Hence, the maximum value is of Z $3\frac{10}{11}$ at $\left(\frac{28}{11}, \frac{15}{11}\right)$.

Q.22 A manufacturer produces two models of bikes-model X and model Y. Model X takes a 6 man-hours to make per unit, while model Y takes 10 man-hours per unit. There is a total of 450 man-hour available per week. Handling and marketing costs are ₹ 2000 and ₹ 1000 per unit for models X and Y, respectively. The total funds available for these purposes are ₹ 80000 per week. Profits per unit for models X and Y are ₹ 1000 and ₹ 500, respectively. How many bikes of each model should the manufacturer produce, so as to yield a maximum profit? Find the maximum profit.

Thinking Process First check whether the drawn graph (by using constraints) gives bounded region or not and then get the maximised profit on corresponding corner points.

Ans. Let the manufacturer produces x number of model X bikes and y number of model Y bikes.

Model X takes a 6 man-hours to make per unit and model Y takes a 10 man-hours to make per unit.

There is total of 450 man-hour available per week.

$\therefore \quad 6x + 10y \le 450$

$\Rightarrow \quad 3x + 5y \le 225 \quad \ldots\text{(i)}$

For models X and Y, handling and marketing costs are ₹ 2000 and ₹1000, respectively and total funds available for these purposes are ₹ 80000 per week.

$\therefore \quad 2000x + 1000y \le 80000$

$\Rightarrow \quad 2x + y \le 80 \quad \ldots\text{(ii)}$

Also, $\quad x \ge 0, y \ge 0$

Hence, the profits per unit for models X and Y are ₹ 1000 and ₹ 500, respectively.

$\therefore$Required LPP is,

Maximise $Z = 1000x + 500y$

Subject to, $3x + 5y \le 225$, $2x + y \le 80$, $x \ge 0$, $y \ge 0$

On solving $3x + 5y = 225$ and $2x + y = 80$, we get

$$x = 25, \; y = 30$$

$\therefore$ Coordinates of B is (25, 30).

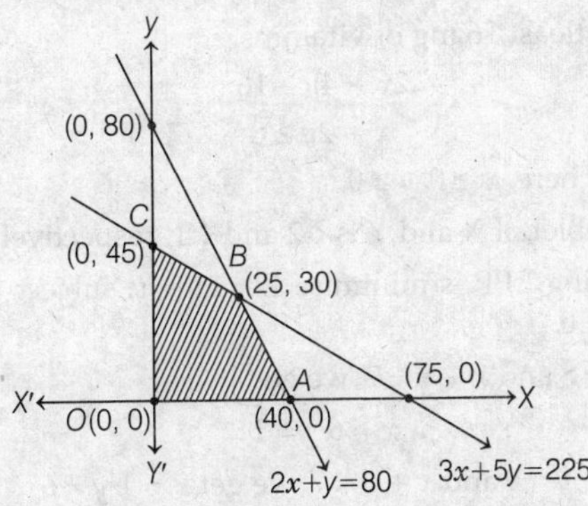

From the shaded feasible region $OABC$, it is clear that coordinates of corner points are O (0, 0), A(40, 0), B(25, 30) and C(0, 45).

Now, we shall use corner point method to determine the maximum profit.

Corner points	Value of $Z = 1000x + 500y$
O (0, 0)	$1000 \times 0 + 500 \times 0 = 0$
A (40, 0)	$1000 \times 40 + 500 \times 0 = 40000$ ← Maximum
B (25, 30)	$1000 \times 25 + 500 \times 30 = 40000$ ← Maximum
C (0, 45)	$1000 \times 0 + 500 \times 45 = 22500$

So, the manufacturer should produce 25 bikes of model X and 30 bikes of model Y to get a maximum profit of ₹ 40000.

Since, in question it is asked that each model bikes should be produced.

Q.23 In order to supplement daily diet, a person wishes to take some X and some wishes Y tablets. The contents of iron, calcium and vitamins in X and Y (in mg/tablet) are given as below

Tablets	Iron	Calcium	Vitamin
X	6	3	2
Y	2	3	4

The person needs atleast 18 mg of iron, 21 mg of calcium and 16 mg of vitamins. The price of each tablet of X and Y is ₹ 2 and ₹1, respectively. How many tablets of each should the person take in order to satisfy the above requirement at the minimum cost?

Ans. Let the person takes x units of tablet X and y units of tablet Y.

So, from the given information, we have

The person needs atleast 18 mg of iron,

$$6x + 2y \geq 18$$

$$\Rightarrow \quad 3x + y \geq 9 \quad \ldots\text{(i)}$$

The person needs atleast 21 mg of calcium.

$$3x + 3y \geq 21$$

$$\Rightarrow \quad x + y \geq 7 \quad \ldots\text{(ii)}$$

The person needs atleast 16 mg of vitamins.

$$2x + 4y \geq 16$$

$$\Rightarrow \quad x + 2y \geq 8 \quad \ldots\text{(iii)}$$

Also, we know that here, $x \geq 0, y \geq 0$ ...(iv)

The price of each tablet of X and Y is ₹ 2 and ₹ 1, respectively.

So, the corresponding LPP is minimise $Z = 2x + y$, subject to, $3x + y \geq 9$, $x + y \geq 7$, $x + 2y \geq 8$, $x \geq 0$, $y \geq 0$.

On solving $x + 2y = 8$ and $x + y = 7$, we get

$$x = 6, y = 1$$

and on solving $3x + y = 9$ and $x + y = 7$, we get $x = 1, y = 6$

So, coordinates of C is (1, 6) and coordinates of B is (6, 1)

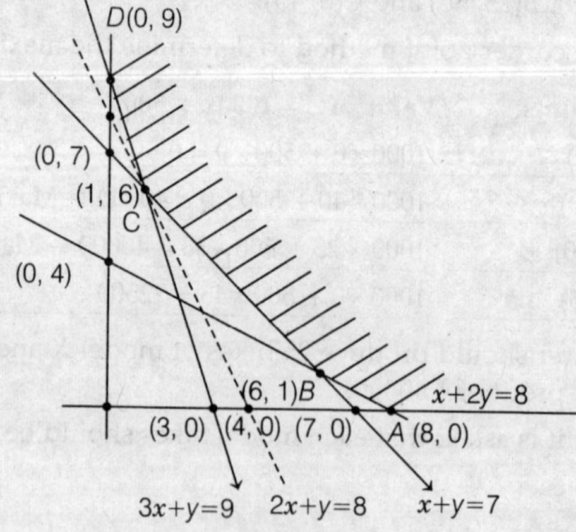

From the shaded graph, we see that for the shown unbounded region, we have coordinates of corner points A, B, C and D as (8, 0), (6, 1), (1, 6), and (0, 9), respectively.

Now, we shall use corner point method to determine the minimum cost.

Corner points	Value of $Z = 2x + y$
(8, 0)	$2 \times 8 + 0 = 16$
(6, 1)	$2 \times 6 + 1 = 13$
(1, 6)	$2 \times 1 + 6 = 8 \leftarrow$ Minimum
(0, 9)	$2 \times 0 + 9 = 9$

Thus, we see that 8 is the minimum value of Z at the corner point (1, 6). Here, we see that the feasible region is unbounded. Therefore, 8 may or may not be the minimum value of Z. To decide this issue, we graph the inequality

$$2x + y < 8 \quad \text{...(v)}$$

and check whether the resulting open half has points in common with feasible region or not. If it has common point, then 8 will not be the minimum value of Z, otherwise 8 will be the minimum value of Z.

Thus, from the graph it is clear that, it has no common point.

Therefore, $Z = 2x + y$ has 8 as minimum value.

Hence, the person should take 1 unit of X tablet and 6 units of Y tablets to satisfy the given requirements and at the minimum cost of ₹ 8.

Mistake Alert! *Assuming that unbounded feasible region always gives no solution, without checking whether the objective function has a minimum or not.*

Q.24 A company makes 3 model of calculators; A, B and C at factory I and factory II. The company has orders for atleast 6400 calculators of model A, 4000 calculators of model B and 4800 calculators of model C. At factory I, 50 calculators of model A, 50 of model B and 30 of model C are made everyday; at factory II, 40 calculators of model A, 20 of model B and 40 of model C are made everyday. It costs ₹ 12000 and ₹ 15000 each day to operate factory I and II, respectively. Find the number of days each factory should operate to minimise the operating costs and still meet the demand.

Ans. Let the factory I operate for x days and the factory II operate for y days.

At factory I, 50 calculators of model A and at factory II, 40 calculators of model A are made everyday. Also, company has ordered for atleast 6400 calculators of model A.

$$\therefore \quad 50x + 40y \geq 6400$$

$$\Rightarrow \quad 5x + 4y \geq 640 \quad \text{...(i)}$$

Also, at factory I, 50 calculators of model B and at factory II, 20 calculators of modal B are made everyday.

Since, the company has ordered atleast 4000 calculators of model B.

$$\therefore \quad 50x + 20y \geq 4000$$

$$\Rightarrow \quad 5x + 2y \geq 400 \quad \text{...(ii)}$$

Similarly, for model C, $30x + 40y \geq 4800$

$\Rightarrow$ $3x + 4y \geq 480$ ...(iii)

Also, $x \geq 0, y \geq 0$ ...(iv)

[since, x and y are non-negative]

It costs ₹ 12000 and ₹ 15000 each day to operate factories I and II, respectively.

$\therefore$ Corresponding LPP is,

Minimise $Z = 12000x + 15000y$,

subject to,

$$5x + 4y \geq 640$$
$$5x + 2y \geq 400$$
$$3x + 4y \geq 480$$

and $x \geq 0, y \geq 0$

On solving $3x + 4y = 480$ and $5x + 4y = 640$, we get $x = 80, y = 60$.

On solving $5x + 4y = 640$ and $5x + 2y = 400$, we get $x = 32, y = 120$.

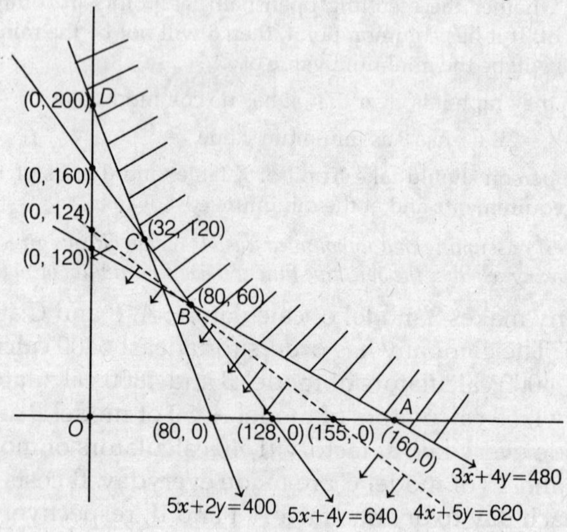

Thus, from the graph, it is clear that feasible region is unbounded and the coordinates of corner points A, B, C and D are (160, 0), (80, 60), (32, 120) and (0, 200), respectively.

Now, we shall use corner point method to determine the minimum value.

Corner points	Value of $Z = 12000x + 15000y$
(160, 0)	$160 \times 12000 + 15000 \times 0 = 1920000$
(80, 60)	$(80 \times 12 + 60 \times 15) \times 1000 = 1860000 \leftarrow$ Minimum
(32, 120)	$(32 \times 12 + 120 \times 15) \times 1000 = 2184000$
(0, 200)	$12000 \times 0 + 200 \times 15000 = 3000000$

From the above table, it is clear that for given unbounded region the minimum value of Z may or may not be 1860000.

Now, for deciding this, we graph the inequality

$$12000x + 15000y < 1860000$$

$$\Rightarrow \qquad 4x + 5y < 620$$

and check whether the resulting open half plane has points in common with feasible region or not.

Thus, as shown in the figure, it has no common points so, Z = 12000*x* + 15000*y* has minimum value 1860000.

So, number of days factory I should be operated is 80 and number of days factory II should be operated is 60 for the minimum cost and satisfying the given constraints.

Q.25 Maximise and minimise $Z = 3x - 4y$, subject to $x - 2y \le 0, -3x + y \le 4$, $x - y \le 6$ and $x, y \ge 0$.

Thinking Process First evaluate the feasible region and their corner points. After that substitute the coordinates of corner points in $Z = 3x - 4y$ to figure out the maximum and minimum values.

Ans. Given LPP is,

maximise and minimise $Z = 3x - 4y$, subject to $x - 2y \le 0$, $-3x + y \le 4$, $x - y \le 6$, $x, y \ge 0$.

On solving $x - y = 6$ and $x - 2y = 0$, we get $x = 12$, $y = 6$

$\therefore$ Coordinate of B is (12, 6).

Now, solving $x - 2y = 0$ and $-3x + y = 4$, we get

$$x = \frac{-8}{5} \text{ and } y = \frac{-4}{5}$$

$\therefore$ Coordinate of A is $\left(-\frac{8}{5}, \frac{-4}{5}\right)$.

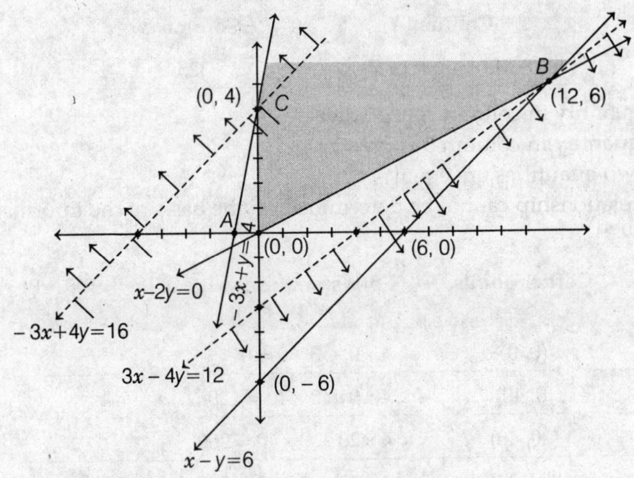

From the shown graph, for the feasible region, we see that it is unbounded and coordinates of corner points are $A\left(-\frac{8}{5}, -\frac{4}{5}\right)$, B (12, 6) and C (0, 4)

Now, we shall use corner point method the determine the maximum and minimum value.

Corner points	Corresponding value of $Z = 3x - 4y$
$\left(-\frac{8}{5}, -\frac{4}{5}\right)$	$3 \times -\frac{8}{5} - 4 \times \left(-\frac{4}{5}\right) = -\frac{8}{5}$
(0, 4)	$3 \times 0 - 4 \times 4 = -16 \leftarrow$ Minimum
(12, 6)	$3 \times 12 - 4 \times 6 = 12 \leftarrow$ Maximum

For given unbounded region the minimum value of Z may or may not be –16. So, for deciding this, we graph the inequality.

$$3x - 4y < -16$$

and check whether the resulting open half plane has common points with feasible region or not.

Thus, from the figure it shows it has common points with feasible region, so it does not have any minimum value.

Also, similarly for maximum value, we graph the inequality $3x - 4y > 12$ and see that resulting open half plane has no common points with the feasible region and hence maximum value 12 exist for $Z = 3x - 4y$.

Objective Type Questions

Choose the correct answer out of the given four options in each of the exercises from 26 to 34.

Q.26 The corner points of the feasible region determined by the system of linear constraints are (0, 0), (0, 40), (20, 40), (60, 20), (60, 0). The objective function is $Z = 4x + 3y$. Compare the quantity in column A and column B.

Column A	Column B
Maximum of Z	325

(a) The quantity in column A is greater

(b) The quantity in column B is greater

(c) The two quantities are equal

(d) The relationship cannot be determined on the basis of the information supplied.

Ans. ***(b)***

Corner points	Corresponding value of $Z = 4x + 3y$
(0, 0)	$4 \times 0 + 3 \times 0 = 0$
(0, 40)	$4 \times 0 + 3 \times 40 = 120$
(20, 40)	$4 \times 20 + 3 \times 40 = 200$
(60, 20)	$4 \times 60 + 3 \times 20 = 300 \leftarrow$ Maximum
(60, 0)	$4 \times 60 + 3 \times 0 = 240$

Hence, maximum value of Z = 300 which is less than 325.

So, the quantity in column B is greater.

Q.27 The feasible solution for a LPP is shown in following figure. Let $Z = 3x - 4y$ be the objective function. Minimum of Z occurs at

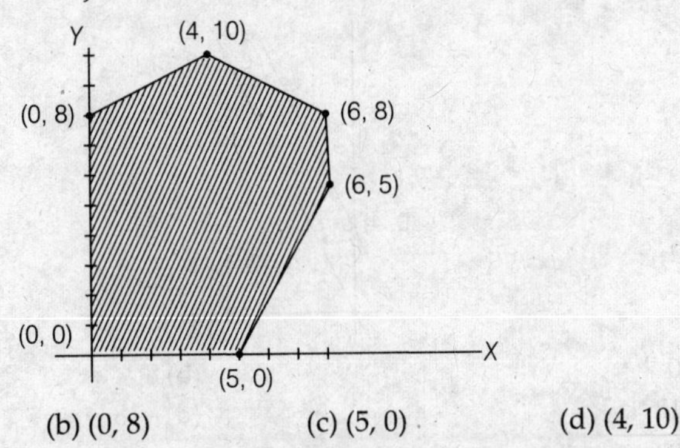

(a) (0, 0) (b) (0, 8) (c) (5, 0) (d) (4, 10)

Ans. ***(b)***

Corner points	Corresponding value of $Z = 3x - 4y$
(0, 0)	$3 \times 0 - 4 \times 0 = 0$
(5, 0)	$3 \times 5 - 4 \times 0 = 15 \leftarrow$ Maximum
(6, 5)	$3 \times 6 - 4 \times 5 = -2$
(6, 8)	$3 \times 6 - 4 \times 8 = -14$
(4, 10)	$3 \times 4 - 4 \times 10 = -28$
(0, 8)	$3 \times 0 - 4 \times 8 = -32 \leftarrow$ Minimum

Hence, the minimum of Z occurs at (0, 8) and its minimum value is (–32).

Q.28 Refer to question 27. Maximum of Z occurs at

(a) (5, 0) (b) (6, 5)
(c) (6, 8) (d) (4, 10)

Ans. ***(a)*** Refering to solution 27, maximum of Z occurs at (5, 0).

Q.29 Refer to question 27. Maximum value of Z + Minimum value of Z is equal to

(a) 13 (b) 1
(c) –13 (d) –17

Ans. ***(d)*** Refering to solution 27, maximum value of Z + minimum value of Z

$$= 15 - 32 = -17$$

Q.30 The feasible region for an LPP is shown in the following figure. Let $F = 3x - 4y$ be the objective function. Maximum value of F is

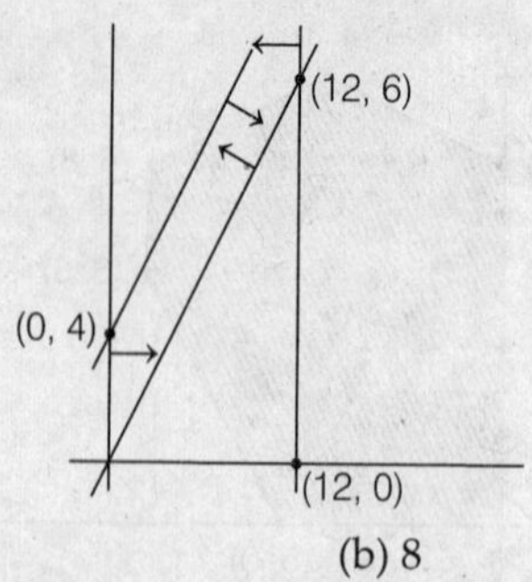

(a) 0 (b) 8
(c) 12 (d) −18

Ans. *(c)* The feasible region as shown in the figure, has objective function $F = 3x - 4y$.

Corner points	Corresponding value of $F = 3x - 4y$
(0, 0)	$3 \times 0 - 4 \times 0 = 0$
(12, 6)	$3 \times 12 - 4 \times 6 = 12 \leftarrow$ Maximum
(0, 4)	$3 \times 0 - 4 \times 4 = -16 \leftarrow$ Minimum

Hence, the maximum value of F is 12.

Q.31 Refer to question 30. Minimum value of F is

(a) 0 (b) −16
(c) 12 (d) Does not exist

Ans. *(b)* Referring to solution 30, minimum value of F is −16 at (0, 4).

Q.32 Corner points of the feasible region for an LPP are (0, 2), (3, 0), (6, 0), (6, 8) and (0, 5). Let $F = 4x + 6y$ be the objective function. The minimum value of F occurs at

(a) only (0, 2)
(b) only (3, 0)
(c) the mid-point of the line segment joining the points (0, 2) and (3, 0)
(d) any point on the line segment joining the points (0, 2) and (3, 0)

Ans. *(d)*

Corner points	Corresponding value of $F = 4x + 6y$
(0, 2)	$4 \times 0 + 6 \times 2 = 12 \leftarrow$ Minimum
(3, 0)	$4 \times 3 + 6 \times 0 = 12 \leftarrow$ Minimum
(6, 0)	$4 \times 6 + 6 \times 0 = 24$
(6, 8)	$4 \times 6 + 6 \times 8 = 72 \leftarrow$ Maximum
(0, 5)	$4 \times 0 + 6 \times 5 = 30$

Hence, minimum value of F occurs at any points on the line segment joining the points (0, 2) and (3, 0).

Note If two corner points of the feasible region are both optimal solutions of the same type *i.e.*, both produce the same maximum or minimum, then any point on the line segment joining these two points is also an optimal solution of the same type.

Q.33 Refer to question 32, maximum of $F-$ minimum of F is

(a) 60 (b) 48

(c) 42 (d) 18

Ans. ***(a)*** Referring to the solution 32, maximum of $F-$ minimum of $F = 72 - 12 = 60$

Q.34 Corner points of the feasible region determined by the system of linear constraints are (0, 3), (1, 1) and (3, 0). Let $Z = px + qy$, where $p, q > 0$. Condition on p and q, so that the minimum of Z occurs at (3, 0) and (1, 1) is

(a) $p = 2q$ (b) $p = \frac{q}{2}$

(c) $p = 3q$ (d) $p = q$

Ans. ***(b)***

Corner points	Corresponding value of $Z = px + qy; p, q > 0$
(0, 3)	$p \times 0 + q \times 3 = 3q$
(1, 1)	$p \times 1 + q \times 1 = p + q$
(3, 0)	$p \times 3 + q \times 0 = 3p$

So, condition of p and q, so that the minimum of Z occurs at (3, 0) and (1, 1) is

$$p + q = 3p$$

$$\Rightarrow \quad 2p = q$$

$$\therefore \quad p = \frac{q}{2}$$

Fill in the blanks in each of the exercises from 35 to 41.

Q.35 In a LPP, the linear inequalities or restrictions on the variables are called ....

Ans. In a LPP, the linear inequalities or restrictions on the variables are called linear constraints.

Q.36 In a LPP, the objective function is always...... .

Ans. In a LPP, objective function is always linear.

Q.37 In the feasible region for a LPP is ......, then the optimal value of the objective function $Z = ax + by$ may or may not exist.

Ans. If the feasible region for a LPP is unbounded, then the optimal value of the objective function $Z = ax + by$ may or may not exist.

Q.38 In a LPP, if the objective function $Z = ax + by$ has the same maximum value on two corner points of the feasible region, then every point on the line segment joining these two points give the same ...... value.

Ans. In a LPP, if the objective function $Z = ax + by$ has the same maximum value on two corner points of the feasible region, then every point on the line segment joining these two points give the same maximum value.

Q.39 A feasible region of a system of linear inequalities is said to be ......, if it can be enclosed within a circle.

Ans. A feasible region of a system of linear inequalities is said to be bounded, if it can be enclosed within a circle.

Q.40 A corner point of a feasible region is a point in the region which is the ...... of two boundary lines.

Ans. A corner point of a feasible region is a point in the region which is the intersection of two boundary lines.

Q.41 The feasible region for an LPP is always a ...... polygon.

Ans. The feasible region for an LPP is always a convex polygon.

State True or False for the statements in each of the exercises 42 to 45.

Q.42 If the feasible region for a LPP is unbounded, maximum or minimum of the objective function $Z = ax + by$ may or may not exist.

Ans. *True*

Q.43 Maximum value of the objective function $Z = ax + by$ in a LPP always occurs at only one corner point of the feasible region.

Ans. *False*

Q.44 In a LPP, the minimum value of the objective function $Z = ax + by$ is always 0, if origin is one of the corner point of the feasible region.

Ans. *False*

Q.45 In a LPP, the maximum value of the objective function $Z = ax + by$ is always finite.

Ans. *True*

Chapter

13 Probability

Short Answer Type Questions

Q.1 For a loaded die, the probabilities of outcomes are given as under

$$P(1) = P(2) = 0.2, P(3) = P(5) = P(6) = 0.1 \text{ and } P(4) = 0.3.$$

The die is thrown two times. Let A and B be the events, 'same number each time' and 'a total score is 10 or more', respectively. Determine whether or not A and B are independent.

Thinking Process First, find $P(A)$, $P(B)$ and $P(A \cap B)$ and then use the concept that two events A and B are called independent events, if $P(A \cap B) = P(A) \cdot P(B)$.

Ans. For a loaded die, it is given that

$$P(1) = P(2) = 0.2,$$
$$P(3) = P(5) = P(6) = 0.1 \text{ and } P(4) = 0.3$$

Also, die is thrown two times.

Here, $A =$ Same number each time and $B =$ Total score is 10 or more

$\therefore \quad A = \{(1, 1), (2, 2), (3, 3), (4, 4), (5, 5), (6, 6)\}$

So, $P(A) = P(1, 1) + P(2, 2) + P(3, 3) + P(4, 4) + P(5, 5) + P(6, 6)$

$= P(1) \cdot P(1) + P(2) \cdot P(2) + P(3) \cdot P(3) + P(4) \cdot P(4) + P(5) \cdot P(5) + P(6) \cdot P(6)$

$= 0.2 \times 0.2 + 0.2 \times 0.2 + 0.1 \times 0.1 + 0.3 \times 0.3 + 0.1 \times 0.1 + 0.1 \times 0.1$

$= 0.04 + 0.04 + 0.01 + 0.09 + 0.01 + 0.01 = 0.20$

and $B = \{(4, 6), (6, 4), (5, 5), (5, 6), (6, 5), (6, 6)\}$

$\therefore \quad P(B) = P(4, 6) + P(6, 4) + P(5, 5) + P(5, 6) + P(6, 5) + P(6, 6)$

$= P(4) \cdot P(6) + P(6) \cdot P(4) + P(5) \cdot P(5) + P(5) \cdot P(6) + P(6) \cdot P(5) + P(6) \cdot P(6)$

$= 0.3 \times 0.1 + 0.1 \times 0.3 + 0.1 \times 0.1 + 0.1 \times 0.1 + 0.1 \times 0.1 + 0.1 \times 0.1$

$= 0.03 + 0.03 + 0.01 + 0.01 + 0.01 + 0.01 = 0.10$

Also, $A \cap B = \{(5, 5), (6, 6)\}$

$\therefore P(A \cap B) = P(5, 5) + P(6, 6) = P(5) \cdot P(5) + P(6) \cdot P(6)$

$= 0.1 \times 0.1 + 0.1 \times 0.1 = 0.01 + 0.01 = 0.02$

We know that, for two events A and B, if $P(A \cap B) = P(A) \cdot P(B)$, then both are independent events.

Here, $P(A \cap B) = 0.02$ and $P(A) \cdot P(B) = 0.20 \times 0.10 = 0.02$

Thus, $P(A \cap B) = P(A) \cdot P(B) = 0.02$

Hence, A and B are independent events.

Q.2 Refer to question 1 above. If the die were fair, determine whether or not the events A and B are independent.

Thinking Process In a fair die, we have equally likely outcomes. So, with the given events A and B, we first find $P(A)$, $P(B)$ and $P(A \cap B)$ and then check whether they are dependent or independent.

Ans. Referring to the above solution, we have

$$A = \{(1, 1), (2, 2), (3, 3), (4, 4), (5, 5), (6, 6)\}$$

$\Rightarrow \quad n(A) = 6$ and $n(S) = 6^2 = 36$ [where, S is sample space]

$$\therefore \quad P(A) = \frac{n(A)}{n(S)} = \frac{6}{36} = \frac{1}{6}$$

and $\quad B = \{(4, 6), (6, 4), (5, 5), (6, 5), (5, 6), (6, 6)\}$

$\Rightarrow \quad n(B) = 6$ and $n(S) = 6^2 = 36$

$$\therefore \quad P(B) = \frac{n(B)}{n(S)} = \frac{6}{36} = \frac{1}{6}$$

Also, $\quad A \cap B = \{(5, 5), (6, 6)\}$

$\Rightarrow \quad n(A \cap B) = 2$ and $n(S) = 36$

$$\therefore \quad P(A \cap B) = \frac{2}{36} = \frac{1}{18}$$

Also, $\quad P(A) \cdot P(B) = \dfrac{1}{36}$

Thus, $\quad P(A \cap B) \neq P(A) \cdot P(B)$ $\quad \left[\because \dfrac{1}{18} \neq \dfrac{1}{36}\right]$

So, we can say that both A and B are not independent events.

Note If A and B are dependent events, the $P(A \cap B) \neq P(A).P(B)$

Q.3 The probability that atleast one of the two events A and B occurs is 0.6. If A and B occur simultaneously with probability 0.3, evaluate $P(\overline{A}) + P(\overline{B})$.

- **Concept Insight** If E be an event, then $P(E) + P(\overline{E}) = 1$.

Ans. We know that, $A \cup B$ denotes the occurrence of atleast one of A and B and $A \cap B$ denotes the occurrence of both A and B, simultaneously.

Thus, $\quad P(A \cup B) = 0.6$ and $P(A \cap B) = 0.3$

Also, $\quad P(A \cup B) = P(A) + P(B) - P(A \cap B)$

$\Rightarrow \quad 0.6 = P(A) + P(B) - 0.3$

$\Rightarrow \quad P(A) + P(B) = 0.9$

$\Rightarrow \quad [1 - P(\overline{A})] + [1 - P(\overline{B})] = 0.9$ $\quad [\because P(A) = 1 - P(\overline{A})$ and $P(B) = 1 - P(\overline{B})]$

$\Rightarrow \quad P(\overline{A}) + P(\overline{B}) = 2 - 0.9 = 1.1$

Q. 4 A bag contains 5 red marbles and 3 black marbles. Three marbles are drawn one by one without replacement. What is the probability that atleast one of the three marbles drawn be black, if the first marble is red?

Ans. Let $R = \{5 \text{ red marbles}\}$ and $B = \{3 \text{ black marbles}\}$

For atleast one of the three marbles drawn be black, if the first marble is red, then the following three conditions will be followed

(i) Second ball is black and third is red (E_1).

(ii) Second ball is black and third is also black (E_2).

(iii) Second ball is red and third is black (E_3).

$$\therefore \quad P(E_1) = P(R_1) \cdot P(B_1|R_1) \cdot P(R_2|R_1B_1) = \frac{5}{8} \cdot \frac{3}{7} \cdot \frac{4}{6} = \frac{60}{336} = \frac{5}{28}$$

$$P(E_2) = P(R_1) \cdot P(B_1|R_1) \cdot P(B_2|R_1B_1) = \frac{5}{8} \cdot \frac{3}{7} \cdot \frac{2}{6} = \frac{30}{336} = \frac{5}{56}$$

$$\text{and} \quad P(E_3) = P(R_1) \cdot P(R_2|R_1) \cdot P(B_1|R_1R_2) = \frac{5}{8} \cdot \frac{4}{7} \cdot \frac{3}{6} = \frac{60}{336} = \frac{5}{28}$$

Left E be the event that atleast one of the three marbles drawn be black, if the first marble is red.

$$\text{Then} \quad P(E) = P(E_1) + P(E_2) + P(E_3) = \frac{5}{28} + \frac{5}{56} + \frac{5}{28}$$

$$= \frac{10 + 5 + 10}{56} = \frac{25}{56}$$

Q.5 Two dice are thrown together and the total score is noted. The events E, F and G are 'a total of 4', 'a total of 9 or more' and 'a total divisible by 5', respectively. Calculate $P(E)$, $P(F)$ and $P(G)$ and decide which pairs of events, if any, are independent.

Ans. Two dice are thrown together

$\therefore \quad n(S) = 36$ [$\because$ here S is sample space]

$E = \text{A total of } 4 = \{(2, 2), (3, 1), (1, 3)\}$

$\Rightarrow \quad n(E) = 3$

$F = \text{A total of 9 or more}$

$= \{(3, 6), (6, 3), (4, 5), (4, 6), (5, 4), (6, 4), (5, 5), (5, 6), (6, 5), (6, 6)\}$

$\Rightarrow \quad n(F) = 10$

$G = \text{a total divisible by } 5 = \{(1, 4), (4, 1), (2, 3), (3, 2), (4, 6), (6, 4), (5, 5)\}$

$\Rightarrow \quad n(G) = 7$

Here, $\quad (E \cap F) = \phi$ and $(E \cap G) = \phi$

$\therefore \quad P(E \cap F) = 0, P(F \cap G) = 0$

Also, $\quad (F \cap G) = \{(4, 6), (6, 4), (5, 5)\}$

$\Rightarrow \quad n(F \cap G) = 3$ and $(E \cap F \cap G) = \phi$

$$\therefore \quad P(E) = \frac{n(E)}{n(S)} = \frac{3}{36} = \frac{1}{12}$$

$$P(F) = \frac{n(F)}{n(S)} = \frac{10}{36} = \frac{5}{18}$$

and $$P(G) = \frac{n(G)}{n(S)} = \frac{7}{36}$$

Also, $$P(F \cap G) = \frac{3}{36} = \frac{1}{12}$$

and $$P(F) \cdot P(G) = \frac{5}{18} \cdot \frac{7}{36} = \frac{35}{648}$$

Here, we see that $P(F \cap G) \neq P(F) \cdot P(G)$

[since, only F and G have common events, so only F and G are used here]

Hence, there is no pair, which is independent.

***Q.6** Explain why the experiment of tossing a coin three times is said to have Binomial distribution.

Ans. We know that, a random variable X taking values 0, 1, 2, ..., n is said to have a binomial distribution with parameters n and p, if its probability distribution is given by

$$P(X = r) = {}^nC_r p^r q^{n-r}$$

where, $$q = 1 - p \qquad [p + q = 1]$$

and $$r = 0, 1, 2, ..., n$$

Similarly, in an experiment of tossing a coin three times, we have $n = 3$ and random variable X can take values $r = 0, 1, 2$ and 3 with $p = \frac{1}{2}$ and $q = \frac{1}{2}$

X	0	1	2	3
P(X)	${}^3C_0\, q^3$	${}^3C_1\, pq^2$	${}^3C_2\, p^2q$	${}^3C_3\, p^3$

So, we see that in the experiment of tossing a coin three times, we have random variable X which can take values 0, 1, 2 and 3 with parameters $n = 3$ and $p = \frac{1}{2}$.

Therefore, it is said to have a binomial distribution.

Q.7 A and B are two events such that

$P(A) = \frac{1}{2}$, $P(B) = \frac{1}{3}$ and $P(A \cap B) = \frac{1}{4}$. Find

(i) $P(A|B)$. (ii) $P(B|A)$. (iii) $P(A'|B)$. (iv) $P(A'|B')$.

Ans. Here, $P(A) = \frac{1}{2}$, $P(B) = \frac{1}{3}$ and $P(A \cap B) = \frac{1}{4}$

(i) $$P(A|B) = \frac{P(A \cap B)}{P(B)} = \frac{1/4}{1/3} = \frac{3}{4}$$

(ii) $$P(B|A) = \frac{P(A \cap B)}{P(A)} = \frac{1/4}{1/2} = \frac{1}{2}$$

(iii) $$P(A'|B) = 1 - P(A/B) = 1 - \frac{3}{4} = \frac{1}{4}$$

Alternate of part (iii)

$$P(A'|B) = \frac{P(A' \cap B)}{P(B)} = \frac{P(B) - P(A \cap B)}{P(B)} = \frac{\frac{1}{3} - \frac{1}{4}}{\frac{1}{3}} = \frac{\frac{1}{12}}{\frac{1}{3}} = \frac{1}{4}$$

Note (✱) Topics related to these question have been removed from latest NCERT.

(iv) $P(A'|B') = \frac{P(A' \cap B')}{P(B')} = \frac{P(A \cup B)'}{P(B')} = \frac{1 - P(A \cup B)}{1 - P(B)}$

$= \frac{1 - [P(A) + P(B) - P(A \cap B)]}{1 - P(B)}$

$= \frac{1 - \left[\frac{1}{2} + \frac{1}{3} - \frac{1}{4}\right]}{1 - \frac{1}{3}} = \frac{1 - \left(\frac{5}{6} - \frac{1}{4}\right)}{\frac{2}{3}} = \frac{1 - 14/24}{2/3} = \frac{10/24}{2/3} = \frac{30}{48} = \frac{5}{8}$

Mistake Alert! *Students may incorrectly use the conditional probability formula* $P(A|B) = \frac{P(A \cap B)}{P(A)}$ *instead of* $P(A|B) = \frac{P(A \cap B)}{P(B)}$

Q.8 Three events A, B and C have probabilities $\frac{2}{5}, \frac{1}{3}$ and $\frac{1}{2}$, respectively. Given that $P(A \cap C) = \frac{1}{5}$ and $P(B \cap C) = \frac{1}{4}$, find the values of $P(C|B)$ and $P(A' \cap C')$.

- **Concept Insight** In this question, we will use $P(C|B) = \frac{P(B \cap C)}{P(B)}$ and $P(A' \cap C') = P(A \cup C)'$ to get the required value.

Ans. Here, $P(A) = \frac{2}{5}, P(B) = \frac{1}{3}, P(C) = \frac{1}{2}, P(A \cap C) = \frac{1}{5}$ and $P(B \cap C) = \frac{1}{4}$

$\therefore \quad P(C|B) = \frac{P(B \cap C)}{P(B)} = \frac{1/4}{1/3} = \frac{3}{4}$

and $P(A' \cap C') = P(A \cup C)' = 1 - P(A \cup C) = 1 - [P(A) + P(C) - P(A \cap C)]$

$= 1 - \left[\frac{2}{5} + \frac{1}{2} - \frac{1}{5}\right] = 1 - \left[\frac{4 + 5 - 2}{10}\right] = 1 - \frac{7}{10} = \frac{3}{10}$

Q.9 Let E_1 and E_2 be two independent events such that $P(E_1) = p_1$ and $P(E_2) = p_2$. Describe in words of the events whose probabilities are

(i) $p_1 p_2$ (ii) $(1 - p_1)p_2$
(iii) $1 - (1 - p_1)(1 - p_2)$ (iv) $p_1 + p_2 - 2p_1 p_2$

Ans. We have, $P(E_1) = p_1$ and $P(E_2) = p_2$

(i) $p_1 p_2 \Rightarrow P(E_1) \cdot P(E_2) = P(E_1 \cap E_2)$

So, E_1 and E_2 occur.

(ii) $(1 - p_1) p_2 = P(E_1') \cdot P(E_2) = P(E_1' \cap E_2)$

So, E_1 does not occur but E_2 occurs.

(iii) $1 - (1 - p_1)(1 - p_2) = 1 - P(E_1')P(E_2') = 1 - P(E_1' \cap E_2')$

$= 1 - [1 - P(E_1 \cup E_2)] = P(E_1 \cup E_2)$

So, either E_1 or E_2 or both E_1 and E_2 occurs.

(iv) $p_1 + p_2 - 2p_1 p_2 = P(E_1) + P(E_2) - 2P(E_1) \cdot P(E_2)$

$= P(E_1) + P(E_2) - 2P(E_1 \cap E_2) = P(E_1 \cup E_2) - P(E_1 \cap E_2)$

So, either E_1 or E_2 occurs but not both.

***Q.10** A discrete random variable X has the probability distribution as given below

X	0.5	1	1.5	2
P(X)	k	k^2	$2k^2$	k

(i) Find the value of k.

(ii) Determine the mean of the distribution.

Ans. We have,

X	0.5	1	1.5	2
$P(X)$	k	k^2	$2k^2$	k

(i) We know that, $\sum_{i=1}^{n} P_i = 1$, where $P_i \geq 0$

$$\Rightarrow \quad P_1 + P_2 + P_3 + P_4 = 1$$
$$\Rightarrow \quad k + k^2 + 2k^2 + k = 1$$
$$\Rightarrow \quad 3k^2 + 2k - 1 = 0$$
$$\Rightarrow \quad 3k^2 + 3k - k - 1 = 0$$
$$\Rightarrow \quad 3k(k+1) - 1(k+1) = 0$$
$$\Rightarrow \quad (3k-1)(k+1) = 0$$
$$\Rightarrow \quad k = 1/3 \text{ or } k = -1$$

Since, $\quad k \geq 0$ or $k = 1/3$

(ii) Mean of the distribution $(\mu) = E(X) = \sum_{i=1}^{n} x_i P_i$

$$= 0.5(k) + 1(k^2) + 1.5(2k^2) + 2(k) = 4k^2 + 2.5k$$
$$= 4 \cdot \frac{1}{9} + 2.5 \times \frac{1}{3} \qquad \left[\because k = \frac{1}{3}\right]$$
$$= \frac{4 + 7.5}{9} = \frac{11.5}{9} = \frac{23}{18}$$

Q.11 Prove that

(i) $P(A) = P(A \cap B) + P(A \cap \overline{B})$

(ii) $P(A \cup B) = P(A \cap B) + P(A \cap \overline{B}) + P(\overline{A} \cap B)$

Ans. (i) We have to prove that $P(A) = P(A \cap B) + P(A \cap \overline{B})$

$\therefore$ RHS $= P(A \cap B) + P(A \cap \overline{B})$
$$= P(A \cap B) + P(A) - P(A \cap B)$$
$$= P(A) = \text{LHS}$$

(ii) We have to prove that $P(A \cup B) = P(A \cap B) + P(A \cap \overline{B}) + P(\overline{A} \cap B)$

$\therefore$ RHS $= P(A \cap B) + P(A \cap \overline{B}) + P(\overline{A} \cap B)$
$$= P(A \cap B) + P(A) - P(A \cap B) + P(B) - P(A \cap B)$$
$$= P(A) + P(B) - P(A \cap B)$$
$$= P(A \cup B) = \text{LHS}$$

Note (✱) Topics related to these question have been removed from latest NCERT.

***Q.12** If X is the number of tails in three tosses of a coin, then determine the standard deviation of X.

Thinking Process First get the values of $P(X)$ at $x = 0, 1, 2, 3$ and then use the formula of standard deviation of $X = \sqrt{\text{Var}(X)}$, where $\text{Var}(X) = E(X^2) - [E(X)]^2 = \Sigma X^2 P(X) - [\Sigma XP(X)]^2$

Ans. Given that, random variable X is the number of tails in three tosses of a coin.

So, $X = 0, 1, 2, 3$.

$$\Rightarrow \quad P(X = x) = {}^nC_x (p)^x q^{n-x},$$

where $n = 3$, $p = 1/2$, $q = 1/2$ and $x = 0, 1, 2, 3$

X	0	1	2	3
$P(X)$	$\frac{1}{8}$	$\frac{3}{8}$	$\frac{3}{8}$	$\frac{1}{8}$
$XP(X)$	0	$\frac{3}{8}$	$\frac{3}{4}$	$\frac{3}{8}$
$X^2P(X)$	0	$\frac{3}{8}$	$\frac{3}{2}$	$\frac{9}{8}$

We know that, $\text{Var}(X) = E(X^2) - [E(X)]^2$ ...(i)

where, $E(X^2) = \sum_{i=1}^{n} x_i^2 P(x_i)$ and $E(X) = \sum_{i=1}^{n} x_i \ P(x_i)$

$$\therefore \quad E(X^2) = \sum_{i=1}^{n} x_i^2 \ P(X_i) = 0 + \frac{3}{8} + \frac{3}{2} + \frac{9}{8} = \frac{24}{8} = 3$$

$$\text{and} \quad [E(X)]^2 = \left[\sum_{i=0}^{n} x_i \ P(x_i)\right]^2 = \left[0 + \frac{3}{8} + \frac{3}{4} + \frac{3}{8}\right]^2 = \left[\frac{12}{8}\right]^2 = \frac{9}{4}$$

$$\therefore \quad \text{Var}(X) = 3 - \frac{9}{4} = \frac{3}{4} \quad \text{[using Eq. (i)]}$$

and standard deviation of $X = \sqrt{\text{Var}(X)} = \sqrt{\frac{3}{4}} = \frac{\sqrt{3}}{2}$

***Q.13** In a dice game, a player pays a stake of ₹ 1 for each throw of a die. She receives ₹ 5, if the die shows a 3, ₹ 2, if the die shows a 1 or 6 and nothing otherwise. What is the player's expected profit per throw over a long series of throws?

Thinking Process Take X as the random variable of profit per throw and at $X = -1, 1$ and 4 get the values of $P(X)$ and use the formula expected profit $E(X) = \Sigma X \ P(X)$ to get the desired result.

Ans. Let X is the random variable of profit per throw.

X	-1	1	4
$P(X)$	$\frac{1}{2}$	$\frac{1}{3}$	$\frac{1}{6}$

Since, she loss ₹ 1 on getting any of 2, 4 or 5.

So, at $X = -1$, $\quad P(X) = \frac{1}{6} + \frac{1}{6} + \frac{1}{6} = \frac{3}{6} = \frac{1}{2}$

Note (✱) Topics related to these question have been removed from latest NCERT.

Similarly, at $X = 1$, $P(X) = \frac{1}{6} + \frac{1}{6} = \frac{1}{3}$ [if die shows of either 1 or 6]

and at $X = 4$, $P(X) = \frac{1}{6}$ [if die shows a 3]

$\therefore$ Player's expected profit $= E(X) = \Sigma X\, P(X)$

$$= \left(-1 \times \frac{1}{2}\right) + \left(1 \times \frac{1}{3}\right) + \left(4 \times \frac{1}{6}\right)$$

$$= \frac{-3 + 2 + 4}{6} = \frac{3}{6}$$

$$= \frac{1}{2} = ₹\, 0.50$$

Q.14 Three dice are thrown at the same time. Find the probability of getting three two's, if it is known that the sum of the numbers on the dice was six.

Ans. On a throw of three dice, we have the of sample space *i.e.*, $[n(S)] = 6^3 = 216$

Let E_1 is the event when the sum of numbers on the dice was six and E_2 is the event when three two's occurs.

$\Rightarrow$ $E_1 = \{(1, 1, 4), (1, 2, 3), (1, 3, 2), (1, 4, 1), (2, 1, 3), (2, 2, 2), (2, 3, 1), (3, 1, 2), (3, 2, 1), (4, 1, 1)\}$

$\Rightarrow$ $n(E_1) = 10$ and $E_2 = \{2, 2, 2\}$

$\Rightarrow$ $n(E_2) = 1$

Also, $(E_1 \cap E_2) = 1$

$\therefore$ $P(E_2 \mid E_1) = \frac{P(E_1 \cap E_2)}{P(E_1)} = \frac{1/216}{10/216} = \frac{1}{10}$

***Q.15** Suppose 10000 tickets are sold in a lottery each for ₹ 1. First prize is of ₹ 3000 and the second prize is of ₹ 2000. There are three third prizes of ₹ 500 each. If you buy one ticket, what is your expectation?

Thinking Process Take X is the random variable for the prize, so at $X = 0, 500, 2000$ and 3000, get $P(X)$ for each X and then use the formula of $E(X) = \Sigma X\, P(X)$ to get the answer.

Ans. Let X be the random variable for the prize.

X	0	500	2000	3000
$P(X)$	$\frac{9995}{10000}$	$\frac{3}{10000}$	$\frac{1}{10000}$	$\frac{1}{10000}$

Since, $E(X) = \Sigma X\, P(X)$

$\therefore$ $E(X) = 0 \times \frac{9995}{10000} + \frac{1500}{10000} + \frac{2000}{10000} + \frac{3000}{10000}$

$$= \frac{1500 + 2000 + 3000}{10000}$$

$$= \frac{6500}{10000} = \frac{13}{20} = ₹\, 0.65$$

Note (*) Topics related to these question have been removed from latest NCERT.

Q.16 A bag contains 4 white and 5 black balls. Another bag contains 9 white and 7 black balls. A ball is transferred from the first bag to the second and then a ball is drawn at random from the second bag. Find the probability that the ball drawn is white.

Ans. Here, $W_1 = \{4 \text{ white balls}\}$ and $B_1 = \{5 \text{ black balls}\}$

and $W_2 = \{9 \text{ white balls}\}$ and $B_2 = \{7 \text{ black balls}\}$

Let E_1 is the event that ball transferred from the first bag is white and E_2 is the event that the ball transferred from the first bag is black.

And let E is the event that the ball drawn from the second bag is white.

$$\therefore \quad P(E \mid E_1) = \frac{10}{17}, \; P(E \mid E_2) = \frac{9}{17}$$

$$\text{and} \quad P(E_1) = \frac{4}{9} \text{ and } P(E_2) = \frac{5}{9}$$

$$\therefore \quad P(E) = P(E_1) \cdot P(E \mid E_1) + P(E_2) \cdot P(E \mid E_2)$$

$$= \frac{4}{9} \cdot \frac{10}{17} + \frac{5}{9} \cdot \frac{9}{17} = \frac{40 + 45}{153} = \frac{85}{153} = \frac{5}{9}$$

Q.17 Bag I contains 3 black and 2 white balls, bag II contains 2 black and 4 white balls. A bag and a ball is selected at random. Determine the probability of selecting a black ball.

Ans. Bag I = $\{3B, 2W\}$, Bag II = $\{2B, 4W\}$

Let E_1 = Event that bag I is selected

E_2 = Event that bag II is selected

and E = Event that a black ball is selected

$$\Rightarrow \quad P(E_1) = 1/2, \; P(E_2) = \frac{1}{2}, \; P(E \mid E_1) = \frac{3}{5}, \; P(E \mid E_2) = \frac{2}{6} = \frac{1}{3}$$

$$\therefore \quad P(E) = P(E_1) \cdot P(E \mid E_1) + P(E_2) \cdot P(E \mid E_2)$$

$$= \frac{1}{2} \cdot \frac{3}{5} + \frac{1}{2} \cdot \frac{2}{6} = \frac{3}{10} + \frac{2}{12} = \frac{18 + 10}{60} = \frac{28}{60} = \frac{7}{15}$$

Q.18 A box has 5 blue and 4 red balls. One ball is drawn at random and not replaced. Its colour is also not noted. Then, another ball is drawn at random. What is the probability of second ball being blue?

Ans. A box = $\{5 \text{ blue}, 4 \text{ red}\}$

Let E_1 be the event that first ball drawn is blue, E_2 be the event that first ball drawn is red and E be the event that second ball drawn is blue.

$$\therefore \quad P(E) = P(E_1) \cdot P(E \mid E_1) + P(E_2) \cdot P(E \mid E_2)$$

$$= \frac{5}{9} \cdot \frac{4}{8} + \frac{4}{9} \cdot \frac{5}{8} = \frac{20}{72} + \frac{20}{72} = \frac{40}{72} = \frac{5}{9}$$

Q.19 Four cards are successively drawn without replacement from a deck of 52 playing cards. What is the probability that all the four cards are king?

Ans. Let E_1, E_2, E_3 and E_4 are the events that the first, second, third and fourth card is king, respectively.

$$\therefore P(E_1 \cap E_2 \cap E_3 \cap E_4) = P(E_1) \cdot P(E_2 \mid E_1) \cdot P(E_3 \mid E_1 \cap E_2) \cdot P[E_4 \mid (E_1 \cap E_2 \cap E_3)]$$

$$= \frac{4}{52} \cdot \frac{3}{51} \cdot \frac{2}{50} \cdot \frac{1}{49} = \frac{24}{52 \cdot 51 \cdot 50 \cdot 49}$$

$$= \frac{1}{13 \cdot 17 \cdot 25 \cdot 49} = \frac{1}{270725}$$

***Q.20 If a die is thrown 5 times. Find the probability that an odd number will come up exactly three times.**

Ans. Here, $n = 5,\ p = \left(\frac{1}{6} + \frac{1}{6} + \frac{1}{6}\right) = \frac{1}{2}$ and $q = 1 - p = 1 - \frac{1}{2} = \frac{1}{2}$

Also, $r = 3$ (when odd number i.e. 1, 3, 5 come)

$$\therefore \quad P(X = r) = {}^nC_r (p)^r (q)^{n-r} = {}^5C_3 \left(\frac{1}{2}\right)^3 \left(\frac{1}{2}\right)^{5-3}$$

$$= \frac{5!}{3!2!} \cdot \frac{1}{8} \cdot \frac{1}{4} = \frac{10}{32} = \frac{5}{16}$$

***Q.21 If ten coins are tossed. What is the probability of getting atleast 8 heads?**

Thinking Process For getting atleast 8 heads, take random variable X for getting head on tossing a coin. So, get sum of $P(8)$, $P(9)$ and $P(10)$ to get the answer.

Ans. In this case, we have to find out the probability of getting atleast 8 heads. Let X is the random variable for getting a head.

Here, $n = 10,\ r \geq 8,$

i.e., $r = 8, 9, 10,\ p = \frac{1}{2}, q = \frac{1}{2}$

We know that, $P(X = r) = {}^nC_r p^r q^{n-r}$

$$\therefore \quad P(X = r) = P(r = 8) + P(r = 9) + P(r = 10)$$

$$= {}^{10}C_8 \left(\frac{1}{2}\right)^8 \left(\frac{1}{2}\right)^{10-8} + {}^{10}C_9 \left(\frac{1}{2}\right)^9 \left(\frac{1}{2}\right)^{10-9} + {}^{10}C_{10} \left(\frac{1}{2}\right)^{10} \left(\frac{1}{2}\right)^{10-10}$$

$$= \frac{10!}{8!2!}\left(\frac{1}{2}\right)^{10} + \frac{10!}{9!1!}\left(\frac{1}{2}\right)^{10} + \frac{10!}{0!10!}\left(\frac{1}{2}\right)^{10}$$

$$= \left(\frac{1}{2}\right)^{10} \left[\frac{10 \times 9}{2} + 10 + 1\right]$$

$$= \left(\frac{1}{2}\right)^{10} \cdot 56 = \frac{1}{2^7 \times 2^3} \cdot 56 = \frac{7}{128}$$

***Q.22 The probability of a man hitting a target is 0.25. He shoots 7 times. Then what is the probability of his hitting atleast twice?**

Thinking Process In this question, we shall first find out the probability of his hitting atmost once (*i.e.,* $r = 0, 1$) and then subtract this probability from 1 to get the desired probability by using binomial distribution formula

$$P(X = r) = 1 - [P(r = 0) + P(r = 1)]$$

Note (✱) Topics related to these question have been removed from latest NCERT.

Ans. Here, $n = 7\ p = 0.25 = \frac{1}{4}, q = 1 - \frac{1}{4} = \frac{3}{4}, r \geq 2$

where, $P(X) = {}^nC_r(p)^r(q)^{n-r}$

$\therefore$ $P(X = r)\ P(2 \leq X \leq 7) = 1 - [P(r = 0) + P(r = 1)]$

$$= 1 - \left[{}^7C_0\left(\frac{1}{4}\right)^0\left(\frac{3}{4}\right)^{7-0} + {}^7C_1\left(\frac{1}{4}\right)^1\left(\frac{3}{4}\right)^{7-1}\right]$$

$$= 1 - \left[\frac{7!}{0!7!}\left(\frac{3}{4}\right)^7 + \frac{7!}{1!6!}\left(\frac{1}{4}\right)\left(\frac{3}{4}\right)^6\right] = 1 - \left[\left(\frac{3}{4}\right)^6\left(\frac{3}{4}\cdot 1 + \frac{1}{4}\cdot 7\right)\right]$$

$$= 1 - \left[\frac{3^6}{4^6}\left(\frac{10}{4}\right)\right] = 1 - \left[\frac{3^6 \times 10}{4^7}\right] = 1 - \left[\frac{27 \times 27 \times 10}{64 \times 256}\right]$$

$$= 1 - \left[\frac{7290}{16384}\right] = 1 - \frac{3645}{8192} = \frac{4547}{8192}$$

***Q.23** A lot of 100 watches is known to have 10 defective watches. If 8 watches are selected (one by one with replacement) at random, what is the probability that there will be atleast one defective watch?

Ans. Probability of defective watch from a lot of 100 watches $= \frac{10}{100} = \frac{1}{10}$

By binomial distribution

$\therefore$ $p = 1/10, q = \frac{9}{10}, n = 8$ and $r \geq 1$

$\therefore$ $P(r \geq 1) = 1 - P(r = 0) = 1 - {}^8C_0\left(\frac{1}{10}\right)^0\left(\frac{9}{10}\right)^{8-0}$

$$= 1 - \frac{8!}{0!8!}\cdot\left(\frac{9}{10}\right)^8 = 1 - \left(\frac{9}{10}\right)^8 \qquad [\because {}^nC_0 = 1]$$

Mistake Alert! *In the statments 'atleast one defective watch', students might incorrectly take $r \leq 1$ instead of $r \geq 1$.*

***Q.24** Consider the probability distribution of a random variable X.

X	0	1	2	3	4
P(X)	0.1	0.25	0.3	0.2	0.15

Calculate

(i) $V\left(\frac{X}{2}\right)$. (ii) Variance of X.

Ans. We have,

X	0	1	2	3	4
$P(X)$	0.1	0.25	0.3	0.2	0.15
$XP(X)$	0	0.25	0.6	0.6	0.60
$X^2P(X)$	0	0.25	1.2	1.8	2.40

Note (✱) Topics related to these question have been removed from latest NCERT.

$$\text{Var}(X) = E(X^2) - [E(X)]^2$$

where, $E(X) = \mu = \sum_{i=1}^{n} x_i P(x_i)$

and $E(X^2) = \sum_{i=1}^{n} x_i^2 P(x_i)$

$\therefore$ $E(X) = 0 + 0.25 + 0.6 + 0.6 + 0.60 = 2.05$

$E(X^2) = 0 + 0.25 + 1.2 + 1.8 + 2.40 = 5.65$

(i) $V\left(\frac{X}{2}\right) = \frac{1}{4}V(X) = \frac{1}{4}[5.65 - (2.05)^2]$

$= \frac{1}{4}[5.65 - 4.2025] = \frac{1}{4} \times 1.4475 = 0.361875$

(ii) $V(X) = 5.65 - (2.05)^2 = 1.4475$

Mistake Alert ! *While calculating variance, students might compute only $E(X^2)$ and forget to subtract $[E(X)]^2$.*

***Q.25** The probability distribution of a random variable X is given below

X	0	1	2	3
$P(X)$	k	$\frac{k}{2}$	$\frac{k}{4}$	$\frac{k}{8}$

(i) Determine the value of k.

(ii) Determine $P(X \le 2)$ and $P(X > 2)$.

(iii) Find $P(X \le 2) + P(X > 2)$.

Ans. We have,

X	0	1	2	3
$P(X)$	k	$\frac{k}{2}$	$\frac{k}{4}$	$\frac{k}{8}$

(i) Since, $\sum_{i=1}^{n} P_i = 1, i = 1, 2, \ldots, n$ and $P_i \ge 0$

$\therefore$ $k + \frac{k}{2} + \frac{k}{4} + \frac{k}{8} = 1 \Rightarrow 8k + 4k + 2k + k = 8$

$\therefore$ $k = \frac{8}{15}$

(ii) $P(X \le 2) = P(0) + P(1) + P(2) = k + \frac{k}{2} + \frac{k}{4}$

$= \frac{(4k + 2k + k)}{4} = \frac{7k}{4} = \frac{7}{4} \cdot \frac{8}{15} = \frac{14}{15}$

and $P(X > 2) = PX = = \frac{k}{8} = \frac{1}{8} \cdot \frac{8}{15} = \frac{1}{15}$

(iii) $P(X \le 2) + P(X > 2) = \frac{14}{15} + \frac{1}{15} = 1$ [substituting above values]

Note (✱) Topics related to these question have been removed from latest NCERT.

***Q.26** For the following probability distribution determine standard deviation of the random variable X.

X	2	3	4
P(X)	0.2	0.5	0.3

Ans. We have,

X	2	3	4
P(X)	0.2	0.5	0.3
XP(X)	0.4	1.5	1.2
$X^2P(X)$	0.8	4.5	4.8

We know that, standard deviation of $X = \sqrt{\text{Var } X}$

where, $\text{Var } X = E(X^2) - [E(X)]^2 = \sum_{i=1}^{n} x_i^2 P(x_i) - \left[\sum_{i=1}^{n} x_i P(x_i)\right]^2$

$\therefore$ $\text{Var } X = [0.8 + 4.5 + 4.8] - [0.4 + 1.5 + 1.2]^2$

$= 10.1 - (3.1)^2 = 10.1 - 9.61 = 0.49$

$\therefore$ Standard deviation of $X = \sqrt{\text{Var } X} = \sqrt{0.49} = 0.7$

***Q.27** A biased die is such that $P(4) = \frac{1}{10}$ and other scores being equally likely. The die is tossed twice. If X is the 'number of fours seen', find the variance of the random variable X.

Ans. Since, X = Number of fours seen

On tossing two die, X = 0, 1, 2.

Also, $P(4) = \frac{1}{10}$ and $P(\text{not } 4)\ 1 - P(4) = \frac{9}{10}$

So, $P(X = 0) = P(\text{not } 4) \times P(\text{not } 4) = \frac{9}{10} \times \frac{9}{10} = \frac{81}{100}$

$P(X = 1) = P(\text{not } 4) \cdot P(4) + P(4) \cdot P(\text{not } 4)$

$= \frac{9}{10} \times \frac{1}{10} + \frac{1}{10} \times \frac{9}{10} = \frac{18}{100}$

$P(X = 2) = P(4) \times P(4) = \frac{1}{10} \times \frac{1}{10} = \frac{1}{100}$

Thus, we get following table

X	0	1	2
P(X)	$\frac{81}{100}$	$\frac{18}{100}$	$\frac{1}{100}$
XP(X)	0	18/100	2/100
$X^2P(X)$	0	18/100	4/100

$\therefore$ $\text{Var}(X) = E(X^2) - [E(X)]^2 = \Sigma X^2 P(X) - [\Sigma XP(X)]^2$

$= \left[0 + \frac{18}{100} + \frac{4}{100}\right] - \left[0 + \frac{18}{100} + \frac{2}{100}\right]^2$

$= \frac{22}{100} - \left(\frac{20}{100}\right)^2 = \frac{11}{50} - \frac{1}{25} = \frac{11-2}{50} = \frac{9}{50} = \frac{18}{100} = 0.18$

Note (✱) Topics related to these question have been removed from latest NCERT.

***Q.28** A die is thrown three times. Let X be the 'number of twos seen', find the expectation of X.

Ans. We have, X = number of twos seen

So, on throwing a die three times, we will have $X = 0, 1, 2, 3$.

$$P(2) = \frac{1}{6}, P(\text{not}^2) = 1 - P(2) = \frac{5}{6}$$

$$\therefore \quad P(X=0) = P(\text{not } 2) \cdot P(\text{not } 2) \cdot P(\text{not } 2) = \frac{5}{6} \times \frac{5}{6} \times \frac{5}{6} = \frac{125}{216}$$

$$P(X=1) = P(\text{not } 2) \cdot P(\text{not } 2) \cdot P(2) + P(\text{not } 2) \cdot P(\text{not } 2) \cdot P(\text{not } 2) + P(2) \cdot P(\text{not } 2) \cdot P(\text{not } 2)$$

$$= \frac{5}{6} \times \frac{5}{6} \ \frac{1}{6} + \frac{5}{6} \times \frac{1}{6} \times \frac{5}{6} + \frac{1}{6} \times \frac{5}{6} \times \frac{5}{6} = \frac{25}{36} \times \frac{3}{6} = \frac{25}{72}$$

$$P(X=2) = P(\text{not } 2) \cdot P(2) \cdot P(2) + P(2) \cdot P(2) \cdot P(\text{not } 2) + P(2) \cdot P(\text{not } 2) + P(2)$$

$$= \frac{5}{6} \times \frac{1}{6} \times \frac{1}{6} + \frac{1}{6} \times \frac{1}{6} \times \frac{5}{6} + \frac{1}{6} \times \frac{5}{6} \times \frac{1}{6}$$

$$= \frac{1}{36} \times \left[\frac{15}{6}\right] = \frac{15}{216}$$

$$P(X=3) = P(2) \cdot P(2) \cdot P(2) = \frac{1}{6} \cdot \frac{1}{6} \cdot \frac{1}{6} = \frac{1}{216}$$

We know that, $E(X) = \Sigma X \ P(X) = 0 \times \frac{125}{216} + 1 \times \frac{25}{72} + 2 \times \frac{15}{216} + 3 \times \frac{1}{216}$

$$= \frac{75 + 30 + 3}{216} = \frac{108}{216} = \frac{1}{2}$$

***Q.29** Two biased dice are thrown together. For the first die $P(6) = \frac{1}{2}$, the other scores being equally likely while for the second die $P(1) = \frac{2}{5}$ and the other scores are equally likely. Find the probability distribution of 'the number of one's seen'.

Ans. For first die, $\quad P_1(6) = \frac{1}{2} \Rightarrow P_1(6') = 1 - \frac{1}{2} = \frac{1}{2}$ [given]

$\Rightarrow \quad P_1(1) + P_1(2) + P(3) + P_1(4) + P_1(5) = \frac{1}{2}$

But $\quad P_1(1) = P_1(2) = P_1(3) = P_1(4) = P_1(5)$

Therefore $\quad 5P_1(1) = \frac{1}{2} \Rightarrow P_1(1) = \frac{1}{10}$

$\therefore \quad P_1(1') = 1 - P_1(1) = 1 - \frac{1}{10} \Rightarrow P_1(1') = \frac{9}{10}$

For second die $\quad P_2(1) = \frac{2}{5}$ [given]

$\Rightarrow \quad P_2(1') = 1 - P_2(1) = 1 - \frac{2}{5} = \frac{3}{5}$

Note (✱) Topics related to these question have been removed from latest NCERT.

Let X be the number of one's seen

$$P(X=0)=P_1(1'),\ P(1')=\frac{9}{10}\times\frac{3}{5}=\frac{27}{50}=0.54$$

For $X=1$, $P(X=1)=P_1(1')\,P_2(1)+P_1(1)\cdot P_2(1')$

$$=\frac{9}{10}\times\frac{2}{5}+\frac{1}{10}\times\frac{3}{5}=\frac{18+3}{50}=\frac{21}{50}=0.42$$

For $X=2$, $P(X=2)=P_1(1)\cdot P_2(1)=\frac{1}{10}\times\frac{2}{5}=\frac{1}{25}=0.04$

Hence, the required probability distribution is as below

X	0	1	2
$P(X)$	0.54	0.42	0.04

***Q.30** Two probability distributions of the discrete random variables X and Y are given below.

X	0	1	2	3
$P(X)$	$\frac{1}{5}$	$\frac{2}{5}$	$\frac{1}{5}$	$\frac{1}{5}$

Y	0	1	2	3
$P(Y)$	$\frac{1}{5}$	$\frac{3}{10}$	$\frac{2}{5}$	$\frac{1}{10}$

Prove that $E(Y^2)=2E(X)$.

Ans.

X	0	1	2	3
$P(X)$	$\frac{1}{5}$	$\frac{2}{5}$	$\frac{1}{5}$	$\frac{1}{5}$

Y	0	1	2	3
$P(Y)$	$\frac{1}{5}$	$\frac{3}{10}$	$\frac{2}{5}$	$\frac{1}{10}$

Since, we have to prove that, $E(Y^2)=2E(X)$

$$\therefore\quad E(X)=\Sigma X\,P(X)=0P(0)+1\cdot P(1)+2\cdot P(2)+3\cdot P(3)$$

$$=0\times\frac{1}{5}+1\times\frac{2}{5}+2\times\frac{1}{5}+3\times\frac{1}{5}=\frac{7}{5}$$

$$\Rightarrow\quad 2E(X)=\frac{14}{5}\qquad \text{...(i)}$$

$$E(Y_2)=\Sigma Y^2P(Y)=0^2P(0)+1^2P(1)+2^2P(2)+3^2P(3)$$

$$=0\times\frac{1}{5}+1\times\frac{3}{10}+4\times\frac{2}{5}+9\times\frac{1}{10}=\frac{3}{10}+\frac{8}{5}+\frac{9}{10}=\frac{28}{10}$$

$$\Rightarrow\quad E(Y^2)=\frac{14}{5}\qquad \text{...(ii)}$$

From Eqs. (i) and (ii), we get

$$E(Y^2)=2E(X)$$

Hence proved.

Note (✱) Topics related to these question have been removed from latest NCERT.

***Q.31** A factory produces bulbs. The probability that any one bulb is defective is $\frac{1}{50}$ and they are packed in 10 boxes. From a single box, find the probability that

(i) none of the bulbs is defective.

(ii) exactly two bulbs are defective.

(iii) more than 8 bulbs work properly.

Ans. Let X be the random variable which denotes that a bulb is defective.

Also, $n = 10$, $p = \frac{1}{50}$ and $q = \frac{49}{50}$ and $P(X = r) = {}^nC_r\, p^r\, q^{n-r}$

(i) None of the bulbs is defective *i.e.*, $r = 0$

$$\therefore \quad P(X = r) = P(0) = {}^{10}C_0\left(\frac{1}{50}\right)^0 \times \left(\frac{49}{50}\right)^{10-0} = \left(\frac{49}{50}\right)^{10}$$

(ii) Exactly two bulbs are defective *i.e.*, $r = 2$

$$\therefore \quad P(X = r) = P(2) = {}^{10}C_2\left(\frac{1}{50}\right)^2\left(\frac{49}{50}\right)^8$$

$$= \frac{10!}{8!2!}\left(\frac{1}{50}\right)^2 \times \left(\frac{49}{50}\right)^8 = 45 \times \left(\frac{1}{50}\right)^{10} \times (49)^8$$

$$= \frac{9}{500}\left(\frac{49}{50}\right)^8$$

(iii) More than 8 bulbs work properly *i.e.*, there is less than 2 bulbs which are defective.

So, $\quad r < 2 \Rightarrow r = 0, 1$

$$\therefore \quad P(X = r) = P(r < 2) = P(0) + P(1)$$

$$= {}^{10}C_0\left(\frac{1}{50}\right)^0\left(\frac{49}{50}\right)^{10-0} + {}^{10}C_1\left(\frac{1}{50}\right)^1\left(\frac{49}{50}\right)^{10-1}$$

$$= \left(\frac{49}{50}\right)^{10} + \frac{10!}{1!9!} \times \frac{1}{50} \times \left(\frac{49}{50}\right)^9$$

$$= \left(\frac{49}{50}\right)^{10} + \frac{1}{5} \times \left(\frac{49}{50}\right)^9 = \left(\frac{49}{50}\right)^9\left(\frac{49}{50} + \frac{1}{5}\right)$$

$$= \left(\frac{49}{50}\right)^9\left(\frac{59}{50}\right) = \frac{59\,(49)^9}{(50)^{10}}$$

Q.32 Suppose you have two coins which appear identical in your pocket. You know that, one is fair and one is 2 headed. If you take one out, toss it and get a head, what is the probability that it was a fair coin?

Ans. Let E_1 = Event that fair coin is drawn, we get

E_2 = Event that 2 headed coin is drawn

E = Event that tossed coin get a head

$$\therefore \quad P(E_1) = 1/2,\ P(E_2) = 1/2,\ P(E \mid E_1) = 1/2 \text{ and } P(E \mid E_2) = 1$$

Note (✱) Topics related to these question have been removed from latest NCERT.

Now, using Baye's theorem,

$$P(E_1 \mid E) = \frac{P(E_1) \cdot P(E \mid E_1)}{P(E_1) \cdot P(E \mid E_1) + P(E_2) \cdot P(E \mid E_2)}$$

$$= \frac{\frac{1}{2} \cdot \frac{1}{2}}{\frac{1}{2} \cdot \frac{1}{2} + \frac{1}{2} \cdot 1}$$

$$= \frac{\frac{1}{4}}{\frac{1}{4} + \frac{1}{2}} = \frac{\frac{1}{4}}{\frac{3}{4}} = \frac{1}{3}$$

Q.33 Suppose that 6% of the people with blood group O are left handed and 10% of those with other blood groups are left handed, 30% of the people have blood group O. If a left handed person is selected at random, what is the probability that he/she will have blood group O?

Ans.

		Blood group 'O'	Other than blood group 'O'
I.	Number of people	30 %	70 %
II.	Percentage of left handed people	6 %	10 %

Let E_1 = Event that the person selected is of blood group O

E_2 = Event that the person selected is of other than blood group O

and E_3 = Event that selected person is left handed.

$$\therefore \quad P(E_1) = 0.30,\ P(E_2) = 0.70$$

$$P(E_3 \mid E_1) = 0.06 \text{ and } P(E_3 \mid E_2) = 0.10$$

By using Baye's theorem, $P(E_1 / E_3) = \dfrac{P(E_1) \cdot P(E_3 \mid E_1)}{P(E_1) \cdot P(E_3 \mid E_1) + P(E_2) \cdot P(E_3 \mid E_2)}$

$$= \frac{0.30 \times 0.06}{(0.30 \times 0.06) + (0.70 \times 0.10)}$$

$$= \frac{0.0180}{0.0180 + 0.0700} = \frac{0.0180}{0.0880} = \frac{180}{880} = \frac{9}{44}$$

Mistake Alert! *In a problem requiring the use of Baye's theorem, students may forget to sum of all the conditional probabilities in the denominator to get the total probability.*

Q.34 Two natural numbers r and s are drawn one at a time, without replacement from the set $S = \{1, 2, 3, \ldots n\}$. Find $P(r \le p \mid s \le p)$, where $p \in S$.

Ans. Given, Set $S = \{1, 2, 3, \ldots, n\}$

$$\therefore \quad P(r \le p \mid s \le p) = \frac{P(p \cap S)}{P(S)}$$

$$= \frac{p-1}{n} \times \frac{n}{n-1} = \frac{p-1}{n-1}$$

***Q.35** Find the probability distribution of the maximum of the two scores obtained when a die is thrown twice. Determine also the mean of the distribution.

Ans. Let X is the random variable score obtained when a die is thrown twice.

$\therefore \quad X = 1, 2, 3, 4, 5, 6$

Here, $\quad S = \{(1, 1), (1, 2), (2, 1), (2, 2), (1, 3), (2, 3), (3, 1), (3, 2), (3, 3), \ldots, (6, 6)\}$

$\therefore \quad P(X = 1) = \frac{1}{6} \times \frac{1}{6} = \frac{1}{36}$

$P(X = 2) = \frac{1}{6} \times \frac{1}{6} + \frac{1}{6} \times \frac{1}{6} + \frac{1}{6} \times \frac{1}{6} = \frac{3}{36}$

$P(X = 3) = \frac{1}{6} \times \frac{1}{6} + \frac{1}{6} \times \frac{1}{6} + \frac{1}{6} \times \frac{1}{6} + \frac{1}{6} \times \frac{1}{6} + \frac{1}{6} \times \frac{1}{6} = \frac{5}{36}$

Similarly, $\quad P(X = 4) = \frac{7}{36}$

$P(X = 5) = \frac{9}{36}$

$P(X = 6) = \frac{11}{36}$

So, the required distribution is,

X	1	2	3	4	5	6
P(x)	1/36	3/36	5/36	7/36	9/36	11/36

Also, we know that, mean $\{E(X)\} = \frac{1}{36} + \frac{6}{36} + \frac{15}{36} + \frac{28}{36} + \frac{45}{36} + \frac{66}{36} = \frac{161}{36}$

***Q.36** The random variable X can take only the values 0, 1, 2. Give that $P(X = 0) = P(X = 1) = p$ and $E(X^2) = E[X]$, then find the value of p.

Ans. Since, X = 0, 1, 2 and $P(X)$ at X = 0 and 1 is p,

Let at X = 2, $P(X)$ is x.

$\therefore \quad p + p + x = 1 \Rightarrow x = 1 - 2p$

We get, the following distribution.

X	0	1	2
P (X)	p	p	$1 - 2p$

$\therefore \quad E[X] = \Sigma X \, P(X)$

$= 0 \cdot p + 1 \cdot p + 2(1 - 2p) = p + 2 - 4p = 2 - 3p$

and $\quad E(X^2) = \Sigma X^2 \, P(X) = 0 \cdot p + 1 \cdot p + 4 \cdot (1 - 2p)$

$= p + 4 - 8p = 4 - 7p$

Also, given that $E(X^2) = E[X]$

$\Rightarrow \quad 4 - 7p = 2 - 3p$

$\Rightarrow \quad 4p = 2 \Rightarrow p = \frac{1}{2}$

Note (✱) Topics related to these question have been removed from latest NCERT.

***Q.37** Find the variance of the following distribution.

X	0	1	2	3	4	5
$P(X)$	$\frac{1}{6}$	$\frac{5}{18}$	$\frac{2}{9}$	$\frac{1}{6}$	$\frac{1}{9}$	$\frac{1}{18}$

Ans. We have,

X	0	1	2	3	4	5
$P(X)$	$\frac{1}{6}$	$\frac{5}{18}$	$\frac{2}{9}$	$\frac{1}{6}$	$\frac{1}{9}$	$\frac{1}{18}$
$XP(X)$	0	$\frac{5}{18}$	$\frac{4}{9}$	$\frac{1}{2}$	$\frac{4}{9}$	$\frac{5}{18}$
$X^2P(X)$	0	$\frac{5}{18}$	$\frac{8}{9}$	$\frac{3}{2}$	$\frac{16}{9}$	$\frac{25}{18}$

$$\therefore \quad \text{Variance} = E(X^2) - [E(X)]^2 = \Sigma X^2 P(X) - [\Sigma X P(X)]^2$$

$$= \left[0 + \frac{5}{18} + \frac{8}{9} + \frac{3}{2} + \frac{16}{9} + \frac{25}{18}\right] - \left[0 + \frac{5}{18} + \frac{4}{9} + \frac{1}{2} + \frac{4}{9} + \frac{5}{18}\right]^2$$

$$= \left[\frac{5 + 16 + 27 + 32 + 25}{18}\right] - \left[\frac{5 + 8 + 9 + 8 + 5}{18}\right]^2$$

$$= \frac{105}{18} - \frac{35 \times 35}{18 \times 18}$$

$$= \frac{18 \times 105 - 35 \times 35}{18 \times 18}$$

$$= \frac{35}{18 \times 18}[54 - 35]$$

$$= \frac{19 \times 35}{324} = \frac{665}{324}$$

Q. 38 A and B throw a pair of dice alternately. A wins the game, if he gets a total of 6 and B wins, if she gets a total of 7. If A starts the game, find the probability of winning the game by A in third throw of the pair of dice.

Ans. Let A_1 = A total of 6 = {(2, 4), (1, 5), (5, 1), (4, 2), (3, 3)}

and B_1 = A total of 7 = {(2, 5), (1, 6), (6, 1), (5, 2), (3, 4), (4, 3)}

Let $P(A_1)$ is the probability, if A wins in a throw $\Rightarrow P(A_1) = \frac{5}{36}$

$$\therefore \quad P(\overline{A_1}) = 1 - P(A_1) = 1 - \frac{5}{36} = \frac{31}{36}$$

and $P(B_1)$ is the probability, if B wins in a throw $\Rightarrow P(B_1) = \frac{1}{6}$

$$\therefore \quad P(\overline{B_1}) = 1 - P(B_1) = 1 - \frac{1}{6} = \frac{5}{6}$$

$\therefore$ Required probability $= P(\overline{A_1}) \cdot P(\overline{B_1}) \cdot P(A_1) = \frac{31}{36} \cdot \frac{5}{6} \cdot \frac{5}{36} = \frac{775}{7776}$

Note (✱) Topics related to these question have been removed from latest NCERT.

Q.39 Two dice are tossed. Find whether the following two events A and B are independent $A = \{(x, y) : x + y = 11\}$ and $B = \{(x, y): x \neq 5\}$, where (x, y) denotes a typical sample point.

Ans. We have, $A = \{(x, y) : x + y = 11\}$ and $B = \{(x, y): x \neq 5\}$

$\therefore \quad A = \{(5, 6), (6, 5)\}$, $B = \{(1,1), (1,2), (1,3), (1,4), (1,5) (1,6), (2,1), (2,2), (2,3), (2,4), (2,5), (2,6), (3,1), (3,2), (3,3), (3,4), (3,5) (3,6), (4,1), (4,2), (4,3), (4,4), (4,5) (4,6), (6,1), (6,2), (6,3), (6,4), (6,5), (6,6)\}$

and $\quad A \cap B = \{6, 5\}$

$\therefore \quad n(S) = 36,\ n(A) = 2,\ n(B) = 30$ and $n(A \cap B) = 1$

$\therefore \quad P(A) = \frac{2}{36} = \frac{1}{18}$ and $P(B) = \frac{30}{36} = \frac{5}{6}$

$\Rightarrow \quad P(A) \cdot P(B) = \frac{5}{108}$

and $\quad P(A \cap B) = \frac{1}{36} \neq P(A) \cdot P(B)$

So, A and B are not independent.

Mistake Alert! *Students often confuse independent events (where to occurence of one event does not affect the other) with dependent events (where one event affects the other).*

Q.40 An urn contains m white and n black balls. A ball is drawn at random and is put back into the urn along with k additional balls of the same colour as that of the ball drawn. A ball is again drawn at random. Show that the probability of drawing a white ball now does not depend on k.

Ans. Let $\quad U = \{m \text{ white}, n \text{ black balls}\}$

$E_1 = \{\text{First ball drawn of white colour}\}$

$E_2 = \{\text{First ball drawn of black colour}\}$

and $\quad E_3 = \{\text{Second ball drawn of white colour}\}$

$\therefore \quad P(E_1) = \frac{m}{m+n}$ and $P(E_2) = \frac{n}{m+n}$

Also, $\quad P(E_3 \mid E_1) = \frac{m+k}{m+n+k}$ and $P(E_3 / E_2) = \frac{m}{m+n+k}$

$$\therefore \quad P(E_3) = P(E_1) \cdot P(E_3 \mid E_1) + P(E_2) \cdot P(E_3 \mid E_2)$$

$$= \frac{m}{m+n} \cdot \frac{m+k}{m+n+k} + \frac{n}{m+n} \cdot \frac{m}{m+n+k}$$

$$= \frac{m(m+k)+nm}{(m+n+k)(m+n)}$$

$$= \frac{m(m+k+n)}{(m+n+k)(m+n)} = \frac{m}{m+n}$$

Hence, the probability of drawing a white ball does not depend on k.

Long Answer Type Questions

Q.41 Three bags contain a number of red and white balls as follows : Bag 1 : 3 red balls, Bag 2 : 2 red balls and 1 white ball and Bag 3 : 3 white balls. The probability that bag i will be chosen and a ball is selected from it is $\frac{i}{6}$, where $i = 1, 2, 3$. What is the probability that

(i) a red ball will be selected? (ii) a white ball is selected?

Ans. Bag 1 : 3 red balls and 0 white ball.

Bag 2 : 2 red balls and 1 white ball.

Beg 3 : 0 red ball and 3 white balls.

Let E_1, E_2 and E_3 be the events that bag 1 , bag 2 and bag 3 is selected respectively and a ball is chosen from it.

$$P(E_1) = \frac{1}{6}, P(E_2) = \frac{2}{6} \text{ and } P(E_3) = \frac{3}{6}$$

(i) Let E be the event that a red ball is selected. Then, probability that red ball will be selected

$$P(E) = P(E_1) \cdot P(E \mid E_1) + P(E_2) \cdot P(E \mid E_2) + P(E_3) \cdot P(E \mid E_3)$$
$$= \left(\frac{1}{6} \times \frac{3}{3}\right) + \left(\frac{2}{6} \times \frac{2}{3}\right) + \left(\frac{3}{6} \times 0\right)$$
$$= \frac{1}{6} + \frac{2}{9} + 0$$
$$= \frac{3+4}{18} = \frac{7}{18}$$

(ii) Let F be the event that a white ball is selected.

$$\therefore \quad P(F) = P(E_1) \cdot P(F \mid E_1) + P(E_2) \cdot P(F / E_2) + P(E_3) \cdot P(F \mid E_3)$$
$$= \left(\frac{1}{6} \times 0\right) + \left(\frac{2}{6} \times \frac{1}{3}\right) + \left(\frac{3}{6} \times 1\right) = \frac{1}{9} + \frac{3}{6} = \frac{11}{18}$$

Note $P(F) = 1 - P(E) = 1 - \frac{7}{18} = \frac{11}{18}$ $\quad [\because P(E) + P(F) = 1]$

Q.42 Refer to question 41 above. If a white ball is selected, what is the probability that it came from

(i) Bag 2? (ii) Bag 3?

Ans. Referring to the previous solution, using Bay's theorem, we have

$$\text{(i) } P(E_2 \mid F) = \frac{P(E_2) \cdot P(F \mid E_2)}{P(E_1) \cdot P(F / E_1) + P(E_2) \cdot P(F \mid E_2) + P(E_3) \cdot P(F \mid E_3)}$$
$$= \frac{\frac{2}{6} \times \frac{1}{3}}{\frac{1}{6} \times 0 + \frac{2}{6} \times \frac{1}{3} + \frac{3}{6} \times 1} = \frac{\frac{2}{18}}{\frac{2}{18} + \frac{3}{6}}$$
$$= \frac{2/18}{\frac{2+9}{18}} = \frac{2}{11}$$

(ii) $$P(E_3 \mid F) = \frac{P(E_3)\cdot P(F \mid E_3)}{P(E_1)\cdot P(F / E_1) + P(E_2)\cdot P(F \mid E_2) + P(E_3)\cdot P(F \mid E_3)}$$

$$= \frac{\frac{3}{6}\times 1}{\frac{1}{6}\times 0 + \frac{2}{6}\times\frac{1}{3} + \frac{3}{6}\times 1}$$

$$= \frac{\frac{3}{6}}{\frac{2}{18}+\frac{3}{6}} = \frac{3/6}{\frac{2}{18}+\frac{9}{18}} = \frac{9}{11}$$

Q.43 A shopkeeper sells three types of flower seeds A_1, A_2 and A_3. They are sold as a mixture, where the proportions are 4 : 4 : 2, respectively. The germination rates of the three types of seeds are 45%, 60% and 35%. Calculate the probability

(i) of a randomly chosen seed to germinate.

(ii) that it will not germinate given that the seed is of type A_3.

(iii) that it is of the type A_2 given that a randomly chosen seed does not germinate.

Ans. We have, $A_1 : A_2 : A_3 = 4:4:2$

Sum of proportions $= 4+4+2 = 10$

$$P(A_1) = \frac{4}{10},\ P(A_2) = \frac{4}{10} \text{ and } P(A_3) = \frac{2}{10}$$

where A_1, A_2 and A_3 denote the three types of flower seeds.

Let E be the event that a seed germinates and $\bar{E}$ be the event that a seed does not germinate.

$\therefore$ $$P(E \mid A_1) = \frac{45}{100},\ P(E \mid A_2) = \frac{60}{100} \text{ and } P(E \mid A_3) = \frac{35}{100}$$

and $$P(\bar{E} \mid A_1) = \frac{55}{100},\ P(\bar{E} \mid A_2) = \frac{40}{100} \text{ and } P(\bar{E} \mid A_3) = \frac{65}{100}$$

(i) $\therefore$ $$P(E) = P(A_1)\cdot P(E \mid A_1) + P(A_2)\cdot P(E \mid A_2) + P(A_3)\cdot P(E \mid A_3)$$

$$= \frac{4}{10}\times\frac{45}{100} + \frac{4}{10}\times\frac{60}{100} + \frac{2}{10}\times\frac{35}{100} \quad \text{[substituting above values]}$$

$$= \frac{180}{1000} + \frac{240}{1000} + \frac{70}{1000} = \frac{490}{1000} = 0.49$$

(ii) $$P(\bar{E} \mid A_3) = 1 - P(E \mid A_3) = 1 - \frac{35}{100} = \frac{65}{100}$$

(iii) $$P(A_2 \mid \bar{E}) = \frac{P(A_2)\cdot P(\bar{E} \mid A_2)}{P(A_1)\cdot P(\bar{E} \mid A_1) + P(A_2)\cdot P(\bar{E} \mid A_2) + P(A_3)\cdot P(\bar{E} \mid A_3)}$$

$$= \frac{\frac{4}{10}\times\frac{40}{100}}{\frac{4}{10}\times\frac{55}{100} + \frac{4}{10}\times\frac{40}{100} + \frac{2}{10}\times\frac{65}{100}}$$

$$= \frac{\frac{160}{1000}}{\frac{220}{1000}+\frac{160}{1000}+\frac{130}{1000}} = \frac{160/1000}{510/1000} = \frac{16}{51} = 0.314$$

Q.44 A letter is known to have come either from 'TATA NAGAR' or from 'CALCUTTA'. On the envelope, just two consecutive letters TA are visible. What is the probability that the letter came from 'TATA NAGAR'?

Ans. Let E_1 be the event that letter is from TATA NAGAR and E_2 be the event that letter is from CALCUTTA.

Also, let E_3 be the event that on the letter, two consecutive letters TA are visible.

$$\therefore \quad P(E_1) = \frac{1}{2} \text{ and } P(E_2) = \frac{1}{2}$$

and
$$P(E_3 \mid E_1) = \frac{2}{8} \text{ and } P(E_3 \mid E_2) = \frac{1}{7}$$

[since, if letter is from TATA NAGAR, we see that the events of two consecutive letters visible are {TA, AT, TA, AN, NA, AG, GA , AR }. So, $P(E_3 / E_1) = \frac{2}{8}$ and if letter is from CALCUTTA, we see that the events of two consecutive letters to visible are {CA, AL, LC, CU, UT, TT, TA}]

So,
$$P(E_3 \mid E_2) = \frac{1}{7}$$

By Baye's theorem using

$$\therefore \quad P(E_1 \mid E_3) = \frac{P(E_1) \cdot P(E_3 \mid E_1)}{P(E_1) \cdot P(E_3 \mid E_1) + P(E_2) \cdot P(E_3 \mid E_2)}$$

$$= \frac{\frac{1}{2} \times \frac{2}{8}}{\frac{1}{2} \times \frac{2}{8} + \frac{1}{2} \times \frac{1}{7}} = \frac{\frac{1}{8}}{\frac{1}{8} + \frac{1}{14}} = \frac{\frac{1}{8}}{\frac{22}{8 \times 14}} = \frac{\frac{1}{8}}{\frac{11}{56}} = \frac{7}{11}$$

Q.45 There are two bags, one of which contains 3 black and 4 white balls while the other contains 4 black and 3 white balls. A die is thrown. If it shows up 1 or 3, a ball is taken from the Ist bag but it shows up any other number, a ball is chosen from the IInd bag. Find the probability of choosing a black ball.

Ans. Since, Bag I = {3 black balls, 4 white balls}, Bag II = {4 black balls, 3 white balls}

Let E_1 be the event that bag I is selected and E_2 be the event that bag II is selected.

Let E_3 be the event that black ball is chosen.

$$\therefore \quad P(E_1) = \frac{1}{6} + \frac{1}{6} = \frac{1}{3}$$

and
$$P(E_2) = 1 - \frac{1}{3} = \frac{2}{3}$$

and
$$P(E_3 \mid E_1) = \frac{3}{7}$$

and
$$P(E_3 \mid E_2) = \frac{4}{7}$$

$$\therefore \quad P(E_3) = P(E_1) \cdot P(E_3 \mid E_1) + P(E_2) \cdot P(E_3 \mid E_2)$$

$$= \frac{1}{3} \times \frac{3}{7} + \frac{2}{3} \times \frac{4}{7} = \frac{11}{21}$$

Q.46 There are three urns containing 2 white and 3 black balls, 3 white and 2 black balls and 4 white and 1 black balls, respectively. There is an equal probability of each urn being chosen. A ball is drawn at random from the chosen urn and it is found to be white. Find the probability that the ball drawn was from the second urn.

Thinking Process First write the event of given data, then use Baye's theroem that is

$$P(E_i \mid E) = \frac{P(E \mid E_i)\,P(E_i)}{\sum_{i=1}^{n} p(E \mid E_i)\,P(E_i)}$$

Ans. Let $U_1 = \{2 \text{ white balls, } 3 \text{ black balls}\}$

$U_2 = \{3 \text{ white balls, } 2 \text{ black balls}\}$

and $U_3 = \{4 \text{ white balls, } 1 \text{ black ball}\}$

$\therefore \quad P(U_1) = P(U_2) = P(U_3) = \frac{1}{3}$

Let E_1 be the event that a ball is chosen from urn U_1, E_2 be the event that a ball is chosen from urn U_2 and E_3 be the event that a ball is chosen from urn U_3.

Also, $\quad P(E_1) = P(E_2) = P(E_3) = 1/3$

Now, let E be the event that white ball is drawn.

$\therefore \quad P(E \mid E_1) = \frac{2}{5},\ P(E \mid E_2) = \frac{3}{5},\ P(E \mid E_3) = \frac{4}{5}$

Now,
$$P(E_2 / E) = \frac{P(E_2)\cdot P(E \mid E_2)}{P(E_1)\cdot P(E \mid E_1) + P(E_2)\cdot P(E \mid E_2) + P(E_3)\cdot P(E \mid E_3)}$$

$$= \frac{\frac{1}{3} \times \frac{3}{5}}{\frac{1}{3} \times \frac{2}{5} + \frac{1}{3} \times \frac{3}{5} + \frac{1}{3} \times \frac{4}{5}}$$

$$= \frac{\frac{3}{15}}{\frac{2}{15} + \frac{3}{15} + \frac{4}{15}} = \frac{3}{9} = \frac{1}{3}$$

Q.47 By examining the chest X-ray, the probability that TB is detected when a person is actually suffering is 0.99. The probability of an healthy person diagnosed to have TB is 0.001. In a certain city, 1 in 1000 people suffers from TB. A person is selected at random and is diagnosed to have TB. What is the probability that he actually has TB?

Ans. Let E_1 = Event that person has TB,

E_2 = Event that person does not have TB

and E = Event that the person is diagnosed to have TB

$\therefore \quad P(E_1) = \frac{1}{1000} = 0.001,\ P(E_2) = 1 - P(E_1) = \frac{999}{1000} = 0.999$

and $\quad P(E \mid E_1) = 0.99$ and $P(E \mid E_2) = 0.001$

$$\therefore \quad P(E\mid E_1)=\frac{P(E_1)\cdot P(E\mid E_1)}{P(E_1)\cdot P(E\mid E_1)+P(E_2)\cdot P(E\mid E_2)}$$

$$=\frac{0.001\times 0.99}{(0.001\times 0.99)+(0.999\times 0.001)}$$

$$=\frac{0.000990}{0.000990+0.000999}=\frac{990}{1989}=\frac{110}{221}$$

Q.48 An item is manufactured by three machines A, B and C. Out of the total number of items manufactured during a specified period, 50% are manufactured on A, 30% on B and 20% on C. 2% of the items produced on A and 2% of items produced on B are defective and 3% of these produced on C are defective. All the items are stored at one godown. One item is drawn at random and is found to be defective. What is the probability that it was manufactured on machine A?

Ans. Let E_1 = Event that item is manufactured on A,

E_2 = Event that an item is manufactured on B,

and E_3 = Event that an item is manufactured on C.

Let E be the event that an item is defective.

$$\therefore \quad P(E_1)=\frac{50}{100}=\frac{1}{2},\ P(E_2)=\frac{30}{100}=\frac{3}{10} \text{ and } P(E_3)=\frac{20}{100}=\frac{1}{5}$$

$$P(E\mid E_1)=\frac{2}{100}=\frac{1}{50},\ P(E\mid E_2)=\frac{2}{100}=\frac{1}{50} \text{ and } P(E\mid E_3)=\frac{3}{100}$$

$$\therefore \quad P(E_1\mid E)=\frac{P(E_1)\cdot P(E\mid E_1)}{P(E_1)\cdot P(E\mid E_1)+P(E_2)\cdot P(E\mid E_2)+P(E_3)\cdot P(E\mid E_3)}$$

$$=\frac{\frac{1}{2}\times\frac{1}{50}}{\frac{1}{2}\times\frac{1}{50}+\frac{3}{10}\times\frac{1}{50}+\frac{1}{5}\times\frac{3}{100}}=\frac{\frac{1}{100}}{\frac{1}{100}+\frac{3}{500}+\frac{3}{500}}=\frac{\frac{1}{100}}{\frac{5+3+3}{500}}=\frac{5}{11}$$

***Q.49** Let X be a discrete random variable whose probability distribution is defined as follows.

$$P(X=x)=\begin{cases} k(x+1), & \text{for } x=1,2,3,4 \\ 2kx, & \text{for } x=5,6,7 \\ 0, & \text{otherwise} \end{cases}$$

where, k is a constant. Calculate

(i) the value of k. (ii) $E(X)$.

(iii) standard deviation of X.

Ans. $P(X=x)=\begin{cases} k(x+1), & \text{for } x=1,2,3,4 \\ 2kx, & \text{for } x=5,6,7 \\ 0, & \text{otherwise} \end{cases}$

Thus, we have following table

Note (*) Topics related to these question have been removed from latest NCERT.

X	1	2	3	4	5	6	7	Otherwise
$P(X)$	$2k$	$3k$	$4k$	$5k$	$10k$	$12k$	$14k$	0
$XP(X)$	$2k$	$6k$	$12k$	$20k$	$50k$	$72k$	$98k$	0
$X^2P(X)$	$2k$	$12k$	$36k$	$80k$	$250k$	$432k$	$686k$	0

(i) Since, $\sum_{i=1}^{n} P'_{x_i} = 1$

$\Rightarrow k(2+3+4+5+10+12+14) = 1$

$\Rightarrow k = \frac{1}{50}$

(ii) $\because E(X) = \Sigma XP(X)$

$\therefore E(X) = 2k + 6k + 12k + 20k + 50k + 72k + 98k + 0 = 260k$

$= 260 \times \frac{1}{50} = \frac{26}{5} = 5.2 \quad \left[\because k = \frac{1}{50}\right]$ ...(i)

(iii) We know that,

$\text{Var}(X) = [E(X^2)] - [E(X)]^2 = \Sigma X^2 P(X) - [\Sigma\{XP(X)\}]^2$

$= [2k + 12k + 36k + 80k + 250k + 432k + 686k + 0] - [5.2]^2$ [using Eq. (i)]

$= [1498k] - 27.04 = \left[1498 \times \frac{1}{50}\right] - 27.04 \quad \left[\because k = \frac{1}{50}\right]$

$= 29.96 - 27.04 = 2.92$

We know that, standard deviation of $X = \sqrt{\text{Var}(X)}$

$= \sqrt{2.92} = 1.7088 = 1.7$ (approx)

***Q.50** The probability distribution of a discrete random variable X is given as under

X	1	2	4	$2A$	$3A$	$5A$
$P(X)$	$\frac{1}{2}$	$\frac{1}{5}$	$\frac{3}{25}$	$\frac{1}{10}$	$\frac{1}{25}$	$\frac{1}{25}$

Calculate

(i) the value of A, if $E(X) = 2.94$. (ii) variance of X.

Ans. (i) We have the following table

X	1	2	4	$2A$	$3A$	$5A$
$P(X)$	$\frac{1}{2}$	$\frac{1}{5}$	$\frac{3}{25}$	$\frac{1}{10}$	$\frac{1}{25}$	$\frac{1}{25}$
$XP(X)$	$\frac{1}{2}$	$\frac{2}{5}$	$\frac{12}{25}$	$\frac{2A}{10}$	$\frac{3A}{25}$	$\frac{5A}{25}$
$X^2P(X)$	$\frac{1}{2}$	$\frac{4}{5}$	$\frac{48}{25}$	$\frac{4A^2}{10}$	$\frac{9A^2}{25}$	$\frac{25A^2}{25}$

$$\Sigma XP(X) = \frac{1}{2} + \frac{2}{5} + \frac{12}{25} + \frac{2A}{10} + \frac{3A}{25} + \frac{5A}{25}$$

$$= \frac{25 + 20 + 24 + 10A + 6A + 10A}{50} = \frac{69 + 26A}{50}$$

Note (✱) Topics related to these question have been removed from latest NCERT.

Since, $E(X) = \Sigma XP(X)$

$\Rightarrow \quad 2.94 = \dfrac{69 + 26A}{50}$

$\Rightarrow \quad 26A = 50 \times 2.94 - 69 \Rightarrow A = \dfrac{147 - 69}{26} = \dfrac{78}{26} = 3$

(ii) We know that,

$$\text{Var}(X) = E(X^2) - [E(X)]^2 = \Sigma X^2 P(X) - [\Sigma XP(X)]^2$$

$$= \frac{1}{2} + \frac{4}{5} + \frac{48}{25} + \frac{4A^2}{10} + \frac{9A^2}{25} + \frac{25A^2}{25} - [E(X)]^2$$

$$= \frac{25 + 40 + 96 + 20A^2 + 18A^2 + 50A^2}{50} - [E(X)]^2$$

$$= \frac{161 + 88A^2}{50} - [E(X)]^2 = \frac{161 + 88 \times (3)^2}{50} - [E(X)]^2 \quad [\because A = 3]$$

$$= \frac{953}{50} - [2.94]^2 \quad [\because E(X) = 2.94]$$

$$= 19.0600 - 8.6436 = 10.4164$$

Mistake Alert! *Students may incorrectly calculate* $Var(X) = E(X^2) - E(X)$ *instead of* $E(X^2) - [(E(X)]^2$.

***Q.51** The probability distribution of a random variable x is given as under

$$P(X = x) = \begin{cases} kx^2, & x = 1, 2, 3 \\ 2kx, & x = 4, 5, 6 \\ 0, & \text{otherwise} \end{cases}$$

where, k is a constant. Calculate

(i) $E(X)$ (ii) $E(3X^2)$ (iii) $P(X \geq 4)$

Ans.

X	1	2	3	4	5	6	Otherwise
$P(X)$	k	$4k$	$9k$	$8k$	$10k$	$12k$	0

We know that, $\Sigma P_{x_i} = 1$

$\Rightarrow \quad k + 4k + 9k + 8k + 10k + 12k + 0 = 1 \Rightarrow 44k = 1 \Rightarrow k = \dfrac{1}{44}$

$\therefore \quad \Sigma XP(X) = k + 8k + 27k + 32k + 50k + 72k + 0 = 190k = 190 \times \dfrac{1}{44} = \dfrac{95}{22}$

(i) So, $E(X) = \Sigma XP(X) = \dfrac{95}{22} = 4.32$

(ii) Also, $E(X^2) = \Sigma X^2 P(X) = k + 16k + 81k + 128k + 250k + 432k$

$$= 908k = 908 \times \frac{1}{44} \quad \left[\because k = \frac{1}{44}\right]$$

$$= 20.636 = 20.64 \text{ (approx)}$$

$\therefore \quad E(3X^2) = 3\,E(X^2) = 3 \times 20.64 = 61.92 = 61.9$

(iii) $P(X \geq 4) = P(X = 4) + P(X = 5) + P(X = 6)$

$$= 8k + 10k + 12k = 30k = 30 \times \frac{1}{44} = \frac{15}{22}$$

Note (*) Topics related to these question have been removed from latest NCERT.

Q.52 A bag contains $(2n+1)$ coins. It is known that n of these coins have a head on both sides whereas the rest of the coins are fair. A coin is picked up at random from the bag and is tossed. If the probability that the toss results in a head is $\frac{31}{42}$, determine the value of n.

Ans. Given, n coins have head on both sides and $(n+1)$ coins are fair coins.

Let E_1 = Event that an unfair coin is selected

E_2 = Event that a fair coin is selected,

and E = Event that the toss results in a head

$$\therefore \quad P(E_1) = \frac{n}{2n+1} \text{ and } P(E_2) = \frac{n+1}{2n+1}$$

Also, $$P(E \mid E_1) = 1 \text{ and } P(E \mid E_2) = \frac{1}{2}$$

$$\therefore \quad P(E) = P(E_1) \cdot P(E \mid E_1) + P(E_2) \cdot P(E \mid E_2)$$

$$= \frac{n}{2n+1} \times 1 + \left(\frac{n+1}{2n+1}\right) \times \frac{1}{2}$$

$$\Rightarrow \quad \frac{31}{42} = \frac{2n+n+1}{2(2n+1)} \Rightarrow \frac{31}{42} = \frac{3n+1}{4n+2}$$

$$\Rightarrow \quad 124n + 62 = 126n + 42$$

$$\Rightarrow \quad 2n = 20 \Rightarrow n = 10$$

***Q.53** Two cards are drawn successively without replacement from a well shuffled deck of cards. Find the mean and standard variation of the random variable X, where X is the number of aces.

Ans. Let X denotes a random variable of number of aces.

$$\therefore \quad X = 0, 1, 2$$

Now, $$P(X=0) = \frac{48}{52} \cdot \frac{47}{51} = \frac{2256}{2652}$$

$$P(X=1) = \frac{48}{52} \cdot \frac{4}{51} + \frac{4}{52} \cdot \frac{48}{51} = \frac{384}{2652}$$

$$P(X=2) = \frac{4}{52} \cdot \frac{3}{51} = \frac{12}{2652}$$

X	0	1	2
$P(X)$	$\frac{2256}{2652}$	$\frac{384}{2652}$	$\frac{12}{2652}$
$XP(X)$	0	$\frac{384}{2652}$	$\frac{24}{2652}$
$X^2P(X)$	0	$\frac{384}{2652}$	$\frac{48}{2652}$

We know that, Mean $(\mu) = E(X) = \Sigma XP(X)$

$$= 0 + \frac{384}{2652} + \frac{24}{2652} = \frac{408}{2652} = \frac{2}{13}$$

Note (✱) Topics related to these question have been removed from latest NCERT.

Also, $\text{Var}(X) = E(X^2) - [E(X)]^2 = \Sigma X^2 P(X) - [E(X)]^2$

$$= \left[0 + \frac{384}{2652} + \frac{48}{2652}\right] - \left(\frac{2}{13}\right)^2 \qquad \left[\because E(X) = \frac{2}{13}\right]$$

$$= \frac{432}{2652} - \frac{4}{169} = 0.1628 - 0.0236 = 0.1391$$

$\therefore$ Standard deviation $= \sqrt{\text{Var}(X)} = \sqrt{0.139} = 0.373$ (approx)

***Q.54** A die is tossed twice. A 'success' is getting an even number on a toss. Then, find the variance of the number of successes.

Ans. Let X be the random variable for a 'success' for getting an even number on a toss.

$$\therefore \quad X = 0, 1, 2, \; n = 2, \; p = \frac{3}{6} = \frac{1}{2} \text{ and } q = \frac{1}{2}$$

By binomial distribution, $P(X = r)\; {}^nC_r = (p)^r (q)^{n-r}$

At $X = 0$, $\qquad P(X = 0) = {}^2C_0 \left(\frac{1}{2}\right)^0 \left(\frac{1}{2}\right)^{2-0} = \frac{1}{4}$

At $X = 1$, $\qquad P(X = 1) = {}^2C_1 \left(\frac{1}{2}\right)^1 \left(\frac{1}{2}\right)^{2-1} = 2 \times \frac{1}{2} \times \frac{1}{2} = \frac{1}{2}$

At $X = 2$, $\qquad P(X = 2) = {}^2C_2 \left(\frac{1}{2}\right)^2 \left(\frac{1}{2}\right)^{2-2} = \frac{1}{4}$

Thus,

X	0	1	2
$P(X)$	$\frac{1}{4}$	$\frac{1}{2}$	$\frac{1}{4}$
$XP(X)$	0	$\frac{1}{2}$	$\frac{1}{2}$
$X^2P(X)$	0	$\frac{1}{2}$	1

$$\therefore \quad \Sigma XP(X) = 0 + \frac{1}{2} + \frac{1}{2} = 1 \qquad \text{...(i)}$$

$$\text{and} \quad \Sigma X^2 P(X) = 0 + \frac{1}{2} + 1 = \frac{3}{2} \qquad \text{...(ii)}$$

$$\because \quad \text{Var}(X) = E(X^2) - [E(X)]^2$$

$$= \Sigma X^2 P(X) - [\Sigma XP(X)]^2 = \frac{3}{2} - (1)^2 = \frac{1}{2}$$

[using Eqs. (i) and (ii)]

***Q.55** There are 5 cards numbered 1 to 5, one number on one card. Two cards are drawn at random without replacement. Let X denotes the sum of the numbers on two cards drawn. Find the mean and variance of X.

Ans. Here, $S = \{(1, 2), (2, 1), (1, 3), (3, 1), (2, 3), (3, 2), (1, 4), (4, 1), (1, 5), (5, 1), (2, 4), (4, 2),$
$(2, 5), (5, 2), (3, 4), (4, 3), (3, 5), (5, 3), (5, 4), (4, 5)\}.$

$$\Rightarrow \quad n(S) = 20$$

Note (✱) Topics related to these question have been removed from latest NCERT.

Let random variable be X which denotes the sum of the numbers on two cards drawn.

$\therefore \quad X = 3, 4, 5, 6, 7, 8, 9$

At $X = 3$, $P(X) = \frac{2}{20} = \frac{1}{10}$ At $X = 4$, $P(X) = \frac{2}{20} = \frac{1}{10}$

At $X = 5$, $P(X) = \frac{4}{20} = \frac{1}{5}$ At $X = 6$, $P(X) = \frac{4}{20} = \frac{1}{5}$

At $X = 7$, $P(X) = \frac{4}{20} = \frac{1}{5}$ At $X = 8$, $P(X) = \frac{2}{20} = \frac{1}{10}$

At $X = 9$, $P(X) = \frac{2}{20} = \frac{1}{10}$

$$\therefore \quad \text{Mean, } E(X) = \Sigma X\, P(X) = \frac{3}{10} + \frac{4}{10} + \frac{5}{5} + \frac{6}{5} + \frac{7}{5} + \frac{8}{10} + \frac{9}{10}$$

$$= \frac{3 + 4 + 10 + 12 + 14 + 8 + 9}{10} = \frac{60}{10} = 6$$

Also,
$$\Sigma X^2 P(X) = \frac{9}{10} + \frac{16}{10} + \frac{25}{5} + \frac{36}{5} + \frac{49}{5} + \frac{64}{10} + \frac{81}{10}$$

$$= \frac{9 + 16 + 50 + 72 + 98 + 64 + 81}{10} = \frac{390}{10} = 39$$

$$\therefore \quad \text{Var}(X) = \Sigma X^2 P(X) - [\Sigma X P(X)]^2$$

$$= 39 - (6)^2 = 39 - 36 = 3$$

Objective Type Questions

Choose the correct answer out of the given four options in each of the exercises from 56 to 93.

Q.56 If $P(A) = \frac{4}{5}$ and $P(A \cap B) = \frac{7}{10}$, then $P(B \mid A)$ is equal to

(a) $\frac{1}{10}$ (b) $\frac{1}{8}$ (c) $\frac{7}{8}$ (d) $\frac{17}{20}$

Thinking Process Use $P(B \mid A) = \frac{P(A \cap B)}{P(A)}$ to get the required value.

Ans. ***(c)*** Since, $P(A) = \frac{4}{5}$, $P(A \cap B) = \frac{7}{10}$

$$\therefore \quad P(B \mid A) = \frac{P(A \cap B)}{P(A)} = \frac{7/10}{4/5} = \frac{7}{8}$$

Q.57 If $P(A \cap B) = \frac{7}{10}$ and $P(B) = \frac{17}{20}$, then $P(A \mid B)$ is equal to

(a) $\frac{14}{17}$ (b) $\frac{17}{20}$ (c) $\frac{7}{8}$ (d) $\frac{1}{8}$

Ans. ***(a)*** Here, $P(A \cap B) = \frac{7}{10}$ and $P(B) = \frac{17}{20}$

$$\therefore \quad P(A \mid B) = \frac{P(A \cap B)}{P(B)} = \frac{7/10}{17/20} = \frac{14}{17}$$

Q.58 If $P(A)=\frac{3}{10}$, $P(B)=\frac{2}{5}$ and $P(A\cup B)=\frac{3}{5}$, then $P(B|A)+P(A|B)$ is equal to

(a) $\frac{1}{4}$ (b) $\frac{1}{3}$ (c) $\frac{5}{12}$ (d) $\frac{7}{12}$

- **Concept Insight** In this question, we will use the formula $P(A\cup B)=P(A)+P(B)-P(A\cap B)$, $P(A|B)=\frac{P(A\cap B)}{P(B)}$ and $P(B|A)=\frac{P(A\cap B)}{P(A)}$.

Ans. (d) Here, $P(A)=\frac{3}{10}$, $P(B)=\frac{2}{5}$ and $P(A\cup B)=\frac{3}{5}$

$$P(B|A)+P(A|B)=\frac{P(B\cap A)}{P(A)}+\frac{P(A\cap B)}{P(B)}$$

$$=\frac{P(A)+P(B)-P(A\cup B)}{P(A)}+\frac{P(A)+P(B)-P(A\cup B)}{P(B)}$$

$$\left[\because P(A\cup B)=P(A)+P(B)-P(A\cap B) \text{ i.e., } P(A\cap B)=P(A)+P(B)-P(A\cup B)\right]$$

$$=\frac{\frac{3}{10}+\frac{2}{5}-\frac{3}{5}}{\frac{3}{10}}+\frac{\frac{3}{10}+\frac{2}{5}-\frac{3}{5}}{\frac{2}{5}}$$

$$=\frac{\frac{1}{10}}{\frac{3}{10}}+\frac{\frac{1}{10}}{\frac{2}{5}}=\frac{1}{3}+\frac{1}{4}=\frac{7}{12}$$

Q.59 If $P(A)=\frac{2}{5}$, $P(B)=\frac{3}{10}$ and $P(A\cap B)=\frac{1}{5}$, then $P(A'|B')\cdot P(B'|A')$ is equal to

(a) $\frac{5}{6}$ (b) $\frac{5}{7}$ (c) $\frac{25}{42}$ (d) 1

Ans. (c) Here, $P(A)=\frac{2}{5}$, $P(B)=\frac{3}{10}$ and $P(A\cap B)=\frac{1}{5}$

$$P(A'|B')=\frac{P(A'\cap B')}{P(B')}=\frac{P(A\cup B)'}{P(B')}=\frac{1-P(A\cup B)}{1-P(B)}$$

$$=\frac{1-[P(A)+P(B)-P(A\cap B)]}{1-P(B)}$$

$$=\frac{1-\left(\frac{2}{5}+\frac{3}{10}-\frac{1}{5}\right)}{1-\frac{3}{10}}$$

$$=\frac{1-\left(\frac{4+3-2}{10}\right)}{\frac{7}{10}}=\frac{1-\frac{1}{2}}{\frac{7}{10}}=\frac{5}{7}$$

and $P(B' \mid A') = \dfrac{P(B' \cap A')}{P(A')} = \dfrac{1 - P(A \cup B)'}{1 - P(A)} = \dfrac{1 - [P(A) + P(B) - P(A \cap B)]}{1 - P(A)}$

$$= \frac{1 - \frac{1}{2}}{1 - \frac{2}{5}} = \frac{1/2}{3/5} = \frac{5}{6} \qquad \left[\because P(A) + P(B) - P(A \cup B) = \frac{1}{2}\right]$$

$$\therefore \; P(A' \mid B') \cdot P(B' \mid A') = \frac{5}{7} \times \frac{5}{6} = \frac{25}{42}$$

Q.60 If A and B are two events such that $P(A) = \frac{1}{2}$, $P(B) = \frac{1}{3}$ and $P(A \mid B) = \frac{1}{4}$, then $P(A' \cap B')$ is equal to

(a) $\frac{1}{12}$ (b) $\frac{3}{4}$ (c) $\frac{1}{4}$ (d) $\frac{3}{16}$

Ans. (c) Here, $P(A) = \frac{1}{2}$, $P(B) = \frac{1}{3}$ and $P(A \mid B) = \frac{1}{4}$

$$\because \quad P(A \mid B) = \frac{P(A \cap B)}{P(B)}$$

$$\Rightarrow \quad P(A \cap B) = P(A \mid B) \cdot P(B) = \frac{1}{4} \times \frac{1}{3} = \frac{1}{12}$$

Now, $P(A' \cap B') = P(A \cup B)' = 1 - P(A \cup B)$

$$= 1 - [P(A) + P(B) - P(A \cap B)]$$

$$= 1 - \left[\frac{1}{2} + \frac{1}{3} - \frac{1}{12}\right] = 1 - \left[\frac{6 + 4 - 1}{12}\right]$$

$$= 1 - \frac{9}{12} = \frac{3}{12} = \frac{1}{4}$$

Q.61 If $P(A) = 0.4$, $P(B) = 0.8$ and $P(B \mid A) = 0.6$, then $P(A \cup B)$ is equal to

(a) 0.24 (b) 0.3
(c) 0.48 (d) 0.96

Ans. (d) Here, $P(A) = 0.4$, $P(B) = 0.8$ and $P(A \mid B) = 0.6$

$$\because \quad P(B \mid A) = \frac{P(B \cap A)}{P(A)}$$

$$\Rightarrow \quad P(B \cap A) = P(B \mid A) \cdot P(A)$$

$$= 0.6 \times 0.4 = 0.24$$

$$\because \quad P(A \cup B) = P(A) + P(B) - P(A \cap B)$$

$$= 0.4 + 0.8 - 0.24$$

$$= 1.2 - 0.24 = 0.96$$

Q.62 If A and B are two events and $A \neq \phi$, $B \neq \phi$, then

(a) $P(A \mid B) = P(A) \cdot P(B)$ (b) $P(A \mid B) = \dfrac{P(A \cap B)}{P(B)}$

(c) $P(A \mid B) \cdot P(B \mid A) = 1$ (d) $P(A \mid B) = P(A) \mid P(B)$

Ans. *(b)* If $A \neq \phi$ and $B \neq \phi$, then $P(A \mid B) = \dfrac{P(A \cap B)}{P(B)}$ [by conditional probability]

Q.63 If A and B are events such that $P(A) = 0.4$, $P(B) = 0.3$ and $P(A \cup B) = 0.5$, then $P(B' \cap A)$ is qual to

(a) $\dfrac{2}{3}$ (b) $\dfrac{1}{2}$ (c) $\dfrac{3}{10}$ (d) $\dfrac{1}{5}$

Ans. *(d)* Here, $P(A) = 0.4$, $P(B) = 0.3$ and $P(A \cup B) = 0.5$

$$\because \quad P(A \cup B) = P(A) + P(B) - P(A \cap B)$$

$$\Rightarrow \quad P(A \cap B) = 0.4 + 0.3 - 0.5 = 0.2$$

$$\therefore \quad P(B' \cap A) = P(A) - P(A \cap B) = 0.4 - 0.2 = 0.2 = \frac{1}{5}$$

Q.64 You are given that A and B are two events such that $P(B) = \dfrac{3}{5}$, $P(A \mid B) = \dfrac{1}{2}$ and $P(A \cup B) = \dfrac{4}{5}$, then $P(A)$ is equal to

(a) $\dfrac{3}{10}$ (b) $\dfrac{1}{5}$ (c) $\dfrac{1}{2}$ (d) $\dfrac{3}{5}$

Ans. *(c)* Here, $P(B) = \dfrac{3}{5}$, $P(A \mid B) = \dfrac{1}{2}$ and $P(A \cup B) = \dfrac{4}{5}$ ...(i)

$$\because \quad P(A \mid B) = \frac{P(A \cap B)}{P(B)} \Rightarrow \frac{1}{2} = \frac{P(A \cap B)}{3/5}$$

$$\therefore \quad P(A \cap B) = \frac{3}{5} \times \frac{1}{2} = \frac{3}{10} \quad \text{...(ii)}$$

and $\quad P(A \cup B) = P(A) + P(B) - P(A \cap B)$

$$\Rightarrow \quad \frac{4}{5} = P(A) + \frac{3}{5} - \frac{3}{10} \quad \text{[using (i) and (ii)]}$$

$$\therefore \quad P(A) = \frac{4}{5} - \frac{3}{5} + \frac{3}{10} = \frac{8 - 6 + 3}{10} = \frac{1}{2}$$

Q.65 In question 64 (above), $P(B \mid A')$ is equal to

(a) $\dfrac{1}{5}$ (b) $\dfrac{3}{10}$ (c) $\dfrac{1}{2}$ (d) $\dfrac{3}{5}$

Ans. *(d)* $P(B \mid A') = \dfrac{P(B \cap A')}{P(A')} = \dfrac{P(B) - P(B \cap A)}{1 - P(A)}$

$$= \frac{\frac{3}{5} - \frac{3}{10}}{1 - \frac{1}{2}} = \frac{\frac{6-3}{10}}{\frac{1}{2}} = \frac{6}{10} = \frac{3}{5}$$

Q.66 If $P(B) = \dfrac{3}{5}$, $P(A \mid B) = \dfrac{1}{2}$ and $P(A \cup B) = \dfrac{4}{5}$, then $P(A \cup B)' + P(A' \cup B)$

(a) $\dfrac{1}{5}$ (b) $\dfrac{4}{5}$ (c) $\dfrac{1}{2}$ (d) 1

Ans. *(d)* Here, $P(B)=\frac{3}{5},\ P(A \mid B)=\frac{1}{2}$

and $P(A \cup B)=\frac{4}{5}$

Since, $P(A \mid B)=\frac{P(A \cap B)}{P(B)}$

$\Rightarrow$ $P(A \cap B)=P(A \mid B)\cdot P(B)$

$=\frac{1}{2}\times\frac{3}{5}=\frac{3}{10}$

Also, $P(A \cup B)=P(A)+P(B)-P(A \cap B)$

$\Rightarrow$ $P(A)=\frac{4}{5}-\frac{3}{5}+\frac{3}{10}=\frac{1}{2}$

$\therefore$ $P(A \cup B)'=1-P(A \cup B)=1-\frac{4}{5}=\frac{1}{5}$

and $P(A' \cup B)=1-P(A-B)=1-P(A \cap B')$

$=1-P(A)\cdot P(B')$

$=1-\frac{1}{2}\times\frac{2}{5}=\frac{4}{5}$ $[\because P(B')=1-P(B)]$

$\Rightarrow$ $P(A \cup B)'+P(A' \cup B)=\frac{1}{5}+\frac{4}{5}=\frac{5}{5}=1$

Q.67 Let $P(A)=\frac{7}{13}$, $P(B)=\frac{9}{13}$ and $P(A \cap B)=\frac{4}{13}$. Then $P(A' \mid B)$ is equal to

(a) $\frac{6}{13}$ (b) $\frac{4}{13}$ (c) $\frac{4}{9}$ (d) $\frac{5}{9}$

Ans. *(d)* Here, $P(A)=\frac{7}{13}$, $P(B)=\frac{9}{13}$ and $P(A \cap B)=\frac{4}{13}$

$\therefore$ $P(A' \mid B)=\frac{P(A' \cap B)}{P(B)}=\frac{P(B)-P(A \cap B)}{P(B)}$

$=\frac{\frac{9}{13}-\frac{4}{13}}{\frac{9}{13}}=\frac{\frac{5}{13}}{\frac{9}{13}}=\frac{5}{9}$

Q.68 If A and B are such events that $P(A)>0$ and $P(B)\neq 1$, then $P(A' \mid B')$ is equal to

(a) $1-P(A \mid B)$ (b) $1-P(A' \mid B)$ (c) $\frac{1-P(A \cup B)}{P(B')}$ (d) $P(A') \mid P(B')$

Ans. *(c)* $\because$ $P(A)>0$ and $P(B)\neq 1$

$\therefore$ $P(A' \mid B')=\frac{P(A' \cap B')}{P(B')}=\frac{P(A \cup B)'}{P(B')}=\frac{1-P(A \cup B)}{P(B')}$

Q.69 If A and B are two independent events with $P(A)=\frac{3}{5}$ and $P(B)=\frac{4}{9}$, then $P(A' \cap B')$ is equal to

(a) $\frac{4}{15}$ (b) $\frac{8}{45}$ (c) $\frac{1}{3}$ (d) $\frac{2}{9}$

Ans. ***(d)*** $P(A' \cap B') = P(A \cup B)' = 1 - P(A \cup B)$

$$= 1 - [P(A) + P(B) - P(A \cap B)]$$

$$= 1 - \left[\frac{3}{5} + \frac{4}{9} - \frac{3}{5} \times \frac{4}{9}\right] \qquad [\because P(A \cap B) = P(A) \cdot P(B)]$$

$$= 1 - \left[\frac{27 + 20 - 12}{45}\right] = 1 - \frac{35}{45} = \frac{10}{45} = \frac{2}{9}$$

Q.70 If two events are independent, then

(a) they must be mutually exclusive
(b) the sum of their probabilities must be equal to 1
(c) Both (a) and (b) are correct
(d) None of the above is correct

Ans. ***(d)*** If two events A and B are independent ,then

$$P(A \cap B) = P(A) \cdot P(B), \ P(A) \neq 0, \ P(B) \neq 0$$

Since, A and B have a common outcome.

Further, mutually exclusive events never have a common outcome.

Note Two independent events having non-zero probabilities of occurrence cannot be mutually exclusive and conversely, *i.e.*, two mutually exclusive events having non-zero probabilities of outcome cannot be independent.

Q.71 Let A and B be two events such that $P(A) = \frac{3}{8}$, $P(B) = \frac{5}{8}$ and $P(A \cup B) = \frac{3}{4}$. Then $P(A | B) \cdot P(A' | B)$ is equal to

(a) $\frac{2}{5}$ (b) $\frac{3}{8}$ (c) $\frac{3}{20}$ (d) $\frac{6}{25}$

Ans. ***(d)*** Here, $P(A) = \frac{3}{8}$, $P(B) = \frac{5}{8}$ and $P(A \cup B) = \frac{3}{4}$

$\because$ $P(A \cup B) = P(A) + P(B) - P(A \cap B)$

$\Rightarrow$ $P(A \cap B) = \frac{3}{8} + \frac{5}{8} - \frac{3}{4} = \frac{3 + 5 - 6}{8} = \frac{2}{8} = \frac{1}{4}$

$\because$ $P(A | B) = \frac{P(A \cap B)}{P(B)} = \frac{1/4}{5/8} = \frac{8}{20} = \frac{2}{5}$

and $P(A' | B) = \frac{P(A' \cap B)}{P(B)} = \frac{P(B) - P(A \cap B)}{P(B)}$

$$= \frac{\frac{5}{8} - \frac{1}{4}}{\frac{5}{8}} = \frac{\frac{5-2}{8}}{\frac{5}{8}} = \frac{3}{5}$$

$\therefore$ $P(A | B) \cdot P(A' | B) = \frac{2}{5} \times \frac{3}{5} = \frac{6}{25}$

Q.72 If the events A and B are independent, then $P(A \cap B)$ is equal to

(a) $P(A) + P(B)$ (b) $P(A) - P(B)$
(c) $P(A) \cdot P(B)$ (d) $P(A) | P(B)$

Ans. ***(c)*** If A and B are independent, then $P(A \cap B) = P(A) \cdot P(B)$

Q.73 Two events E and F are independent. If $P(E) = 0.3$ and $P(E \cup F) = 0.5$, then $P(E \mid F) - P(F \mid E)$ is equal to

(a) $\frac{2}{7}$ (b) $\frac{3}{35}$ (c) $\frac{1}{70}$ (d) $\frac{1}{7}$

Ans. *(c)* Here, $P(E) = 0.3$ and $P(E \cup F) = 0.5$

Let $P(F) = x$

$$\because \quad P(E \cup F) = P(E) + P(F) - P(E \cap F)$$

$$= P(E) + P(F) - P(E) \cdot P(F) \quad [\because P(E \cap F) = P(E) \cdot P(F)]$$

$$\Rightarrow \quad 0.5 = 0.3 + x - 0.3x$$

$$\Rightarrow \quad x = \frac{0.5 - 0.3}{0.7} = \frac{2}{7} = P(F) \quad \text{...(i)}$$

$$\therefore \quad P(E \mid F) - P(F \mid E) = \frac{P(E \cap F)}{P(F)} - \frac{P(F \cap E)}{P(E)}$$

$$= \frac{P(E \cap F) \cdot P(E) - P(F \cap E) \cdot P(F)}{P(E) \cdot P(F)}$$

$$= \frac{P(E \cap F)\,[P(E) - P(F)]}{P(E \cap F)} = P(E) - P(F)$$

$$= \frac{3}{10} - \frac{2}{7} = \frac{21 - 20}{70} = \frac{1}{70} \quad \text{[from Eq. (i)]}$$

Q.74 A bag contains 5 red and 3 blue balls. If 3 balls are drawn at random without replacement, the probability of getting exactly one red ball is

(a) $\frac{45}{196}$ (b) $\frac{135}{392}$ (c) $\frac{15}{56}$ (d) $\frac{15}{29}$

Ans. *(c)* Let R represents red ball.

Probability of getting exactly one red ball

$$= P(R) \cdot P(\overline{R}) \cdot P(\overline{R}) + P(\overline{R}) \cdot P(R) \cdot P(\overline{R}) + P(\overline{R}) \cdot P(\overline{R}) \cdot P(R)$$

$$= \frac{5}{8} \times \frac{3}{7} \times \frac{2}{6} + \frac{3}{8} \times \frac{5}{7} \times \frac{2}{6} + \frac{3}{8} \times \frac{2}{7} \times \frac{5}{6}$$

$$= \frac{15}{4 \times 7 \times 6} + \frac{15}{4 \times 7 \times 6} + \frac{15}{4 \times 7 \times 6}$$

$$= \frac{5}{56} + \frac{5}{56} + \frac{5}{56} = \frac{15}{56}$$

Q.75 Refer to question 74 above. If the probability that exactly two of the three balls were red, the first ball being red, is

(a) $\frac{1}{3}$ (b) $\frac{4}{7}$ (c) $\frac{15}{28}$ (d) $\frac{5}{28}$

Ans. *(b)* Let R and B represent red and blue balls respectively.

Let E_1 = Event that first ball being red

and E_2 = Event that exactly two of the three balls being red

$\therefore \quad E_1 = \{RBR, RRB, RBB, RRR\}$

and $E_2 = \{RRB, RBR, BRR\}$

$$\therefore \quad E_1 \cap E_2 = \{RRB, RBR\}$$

$$\therefore \quad P(E_1) = P(R)\cdot P(\overline{R})\cdot P(R) + P(R)\cdot P(R)\cdot P(\overline{R}) + P(R)\cdot P(\overline{R})\cdot P(\overline{R}) + P(R)\cdot P(R)\cdot P(R)$$

$$= \frac{5}{8}\cdot\frac{3}{7}\cdot\frac{4}{6} + \frac{5}{8}\cdot\frac{4}{7}\cdot\frac{3}{6} + \frac{5}{8}\cdot\frac{2}{6} + \frac{5}{8}\cdot\frac{4}{7}\cdot\frac{3}{6}$$

$$= \frac{60}{336} + \frac{60}{336} + \frac{30}{336} + \frac{60}{336} = \frac{210}{336}$$

$$P(E_1 \cap E_2) = P(R)\cdot P(R)\cdot P(\overline{R}) + P(R)\cdot P(\overline{R})\cdot P(R)$$

$$= \frac{5}{8}\times\frac{4}{7}\times\frac{3}{6} + \frac{5}{8}\times\frac{3}{7}\times\frac{4}{6} = \frac{120}{336} = \frac{120}{336}$$

$$\therefore \quad P(E_2 \mid E_1) = \frac{P(E_1 \cap E_2)}{P(E_1)} = \frac{120/336}{210/336} = \frac{4}{7}$$

Q.76 Three persons A, B and C, fire at a target in turn, starting with A. Their probability of hitting the target are 0.4, 0.3 and 0.2, respectively. The probability of two hits is

(a) 0.024 (b) 0.188 (c) 0.336 (d) 0.452

Ans. (b) Here, $P(A) = 0.4$

$\Rightarrow P(\overline{A}) = 0.6,\ P(B) = 0.3 \Rightarrow P(\overline{B}) = 0.7$ and $P(C) = 0.2 \Rightarrow P(\overline{C}) = 0.8$

$\therefore$ Probability of two hits

$= P(A)\cdot P(B)\cdot P(\overline{C}) + P(A)\cdot P(\overline{B})\cdot P(C) + P(\overline{A})\cdot P(B)\cdot P(C)$

$= 0.4 \times 0.3 \times 0.8 + 0.4 \times 0.7 \times 0.2 + 0.6 \times 0.3 \times 0.2$

$= 0.096 + 0.056 + 0.036 = 0.188$

Q.77 Assume that in a family, each child is equally likely to be a boy or a girl. A family with three children is chosen at random. The probability that the eldest child is a girl given that the family has atleast one girl is

(a) $\frac{1}{2}$ (b) $\frac{1}{3}$ (c) $\frac{2}{3}$ (d) $\frac{4}{7}$

Ans. (d) Here, $S = \{(B, B, B), (G, G, G), (B, G, G), (G, B, G), (G, G, B), (G, B, B), (B, G, B), (B, B, G)\}$ *i.e.*, $n(S) = 8$

Let E_1 = Event that a family has atleast one girl, then

$E_1 = \{(G, B, B), (B, G, B), (B, B, G), (G, G, B), (B, G, G), (G, B, G), (G, G, G)\}$

i.e., $n(E_1) = 7$

E_2 = Event that the eldest child is a girl, then

$E_2 = \{(G, B, B), (G, G, B), (G, B, G), (G, G, G)\}$

$\therefore \quad E_1 \cap E_2 = \{(G, B, B), (G, G, B), (G, B, G), (G, G, G)\}$ *i.e.*, $n(E_1 \cap E_2)$

$$\therefore \quad P(E_2 \mid E_1) = \frac{P(E_1 \cap E_2)}{P(E_1)} = \frac{4/8}{7/8} = \frac{4}{7}$$

Q.78 A die is thrown and a card is selected at random from a deck of 52 playing cards. The probability of getting an even number on the die and a spade card is

(a) $\frac{1}{2}$ (b) $\frac{1}{4}$ (c) $\frac{1}{8}$ (d) $\frac{3}{4}$

Ans. ***(c)*** Let E_1 = Event for getting an even number on the die

and E_2 = Event that a spade card is selected

$$\therefore \quad P(E_1) = \frac{3}{6} = \frac{1}{2} \text{ and } P(E_2) = \frac{13}{52} = \frac{1}{4}$$

Then, $$P(E_1 \cap E_2) = P(E_1) \cdot P(E_2) = \frac{1}{2} \times \frac{1}{4} = \frac{1}{8}$$

Q.79 A box contains 3 orange balls, 3 green balls and 2 blue balls. Three balls are drawn at random from the box without replacement. The probability of drawing 2 green balls and one blue ball is

(a) $\frac{3}{28}$ (b) $\frac{2}{21}$

(c) $\frac{1}{28}$ (d) $\frac{167}{168}$

Ans. ***(a)*** Let G and B represents green and blue balls respectively.

$\therefore$ Probability of drawing 2 green balls and one blue ball

$$= P(G) \cdot P(G) \cdot P(B) + P(B) \cdot P(G) \cdot P(G) + P(G) \cdot P(B) \cdot P(G)$$

$$= \frac{3}{8} \times \frac{2}{7} \times \frac{2}{6} + \frac{2}{8} \times \frac{3}{7} \times \frac{2}{6} + \frac{3}{8} \times \frac{2}{7} \times \frac{2}{6}$$

$$= \frac{1}{28} + \frac{1}{28} + \frac{1}{28} = \frac{3}{28}$$

Q.80 A flashlight has 8 batteries out of which 3 are dead. If two batteries are selected without replacement and tested, then probability that both are dead is

(a) $\frac{33}{56}$ (b) $\frac{9}{64}$ (c) $\frac{1}{14}$ (d) $\frac{3}{28}$

Ans. ***(d)*** Let B be the probability that selected two batteries are dead then

$$\text{Required probability } P(E) = P(D) \cdot P(D) = \frac{3}{8} \times \frac{2}{7} = \frac{3}{28}$$

***Q.81** Eight coins are tossed together. The probability of getting exactly 3 heads is

(a) $\frac{1}{256}$ (b) $\frac{7}{32}$

(c) $\frac{5}{32}$ (d) $\frac{3}{32}$

Ans. ***(b)*** By binomial distribution, $P(X = r) = {}^nC_r\,(p)^r\, q^{n-r}$

Here, $n = 8, r = 3, p = \frac{1}{2}$ and $q = \frac{1}{2}$

$$\therefore \quad P(X = 3) = {}^8C_3 \left(\frac{1}{2}\right)^3 \left(\frac{1}{2}\right)^{8-3} = \frac{8!}{5!3!}\left(\frac{1}{2}\right)^8$$

$$= \frac{8 \times 7 \times 6}{3 \times 2} \times \frac{1}{16 \times 16} = \frac{7}{32}$$

Note (*) Topics related to these question have been removed from latest NCERT.

Q.82 Two dice are thrown. If it is known that the sum of numbers on the dice was less than 6, the probability of getting a sum 3, is

(a) $\frac{1}{18}$ (b) $\frac{5}{18}$ (c) $\frac{1}{5}$ (d) $\frac{2}{5}$

Ans. ***(c)*** Let E_1 = Event that the sum of numbers on the dice was less than 6

and E_2 = Event that the sum of numbers on the dice is 3.

$$\therefore \quad E_1 = \{(1, 4), (4, 1), (2, 3), (3, 2), (2, 2), (1, 3), (3, 1), (1, 2), (2, 1), (1, 1)\}$$

$$\Rightarrow \quad n(E_1) = 10$$

and $\quad E_2 = \{(1, 2), (2, 1)\} \Rightarrow n(E_2) = 2$

$$\therefore \quad E_1 \cap E_2 = \{(1, 2), (2, 1)\} \Rightarrow n(E_1 \cap E_2) = 2$$

$$\therefore \quad P(E_2 \mid E_1) = \frac{P(E_1 \cap E_2)}{P(E_1)} = \frac{\frac{2}{36}}{\frac{10}{36}} = \frac{2}{10} = \frac{1}{5}$$

***Q.83** Which one is not a requirement of a binomial distribution?

(a) There are 2 outcomes for each trial
(b) There is a fixed number of trials
(c) The outcomes must be dependent on each other
(d) The probability of success must be the same for all the trials

Ans. ***(c)*** We know that, in a Binomial distribution,

(i) There are 2 outcomes for each trial.
(ii) There is a fixed number of trials.
(iii) The probability of success must be the same for all the trials.

Q.84 Two cards are drawn from a well shuffled deck of 52 playing cards with replacement. The probability that both cards are queens, is

(a) $\frac{1}{13} \times \frac{1}{13}$ (b) $\frac{1}{13} + \frac{1}{13}$ (c) $\frac{1}{13} \times \frac{1}{17}$ (d) $\frac{1}{13} \times \frac{4}{51}$

Ans. ***(a)*** As, there is replacement.

So, the probability of drawing each card of queen $= \frac{4}{52}$

$$\therefore \text{ Required probability} = \frac{4}{52} \times \frac{4}{52} = \frac{1}{13} \times \frac{1}{13}$$

***Q.85** The probability of guessing correctly atleast 8 out of 10 answers on a true false type examination is

(a) $\frac{7}{64}$ (b) $\frac{7}{128}$ (c) $\frac{45}{1024}$ (d) $\frac{7}{41}$

Ans. ***(b)*** By binomial distribution, $P(X = r) = {}^nC_r\,(p)^r\,(q)^{n-r}$

Here, $\quad n = 10,\ p = \frac{1}{2},\ q = \frac{1}{2}$

and $\quad r \geq 8$ *i.e.*, $r = 8, 9, 10$

$$\Rightarrow \quad P(X = r) = P(r = 8) + P(r = 9) + P(r = 10)$$

Note (✱) Topics related to these question have been removed from latest NCERT.

$$= {}^{10}C_8 \left(\frac{1}{2}\right)^8 \times \left(\frac{1}{2}\right)^{10-8} + {}^{10}C_9 \left(\frac{1}{2}\right)^9 \left(\frac{1}{2}\right) + {}^{10}C_{10} \left(\frac{1}{2}\right)^{10} \times \left(\frac{1}{2}\right)^0$$

$$= \frac{10!}{8!2!}\left(\frac{1}{2}\right)^{10} + \frac{10!}{9!1!}\left(\frac{1}{2}\right)^{10} + \left(\frac{1}{2}\right)^{10}$$

$$= \left(\frac{1}{2}\right)^{10} \times [45 + 10 + 1] = \left(\frac{1}{2}\right)^{10} \times 56$$

$$= \frac{1}{16 \times 64} \times 56 = \frac{7}{128}$$

***Q.86** The probability that a person is not a swimmer is 0.3. The probability that out of 5 persons 4 are swimmers is

(a) ${}^5C_4(0.7)^4\,(0.3)$ (b) ${}^5C_1\,(0.7)\,(0.3)^4$
(c) ${}^5C_4\,(0.7)\,(0.3)^4$ (d) $(0.7)^4\,(0.3)$

Ans. *(a)* By binomial distribution, $P(X = r) = {}^nC_r\,(p)^r(q)^{n-r}$

Here, $\overline{p} = 0.3 \Rightarrow p = 0.7$ and $q = 0.3$, $n = 5$ and $r = 4$

$\therefore$ $P(X = 4) = {}^5C_4\,(0.7)^4\,(0.3)$

***Q.87** The probability distribution of a discrete random variable X is given below

X	2	3	4	5
$P(X)$	$\frac{5}{k}$	$\frac{7}{k}$	$\frac{9}{k}$	$\frac{11}{k}$

The value of k is

(a) 8 (b) 16 (c) 32 (d) 48

Ans.(c) We know that, $\Sigma P(X) = 1$

$\Rightarrow$ $\frac{5}{k} + \frac{7}{k} + \frac{9}{k} + \frac{11}{k} = 1$

$\Rightarrow$ $\frac{32}{k} = 1$

$\therefore$ $k = 32$

***Q.88** For the following probability distribution.

X	– 4	– 3	– 2	– 1	0
$P(X)$	0.1	0.2	0.3	0.2	0.2

$E(X)$ is equal to

(a) 0 (b) –1 (c) –2 (d) –1.8

Ans. *(d)* $\because$ $E(X) = \Sigma X\;P(X)$

$$= -4 \times (0.1) + (-3 \times 0.2) + (-2 \times 0.3) + (-1 \times 0.2) + (0 \times 0.2)$$

$$= -0.4 - 0.6 - 0.6 - 0.2$$

$$= -1.8$$

Note (*) Topics related to these question have been removed from latest NCERT.

***Q.89** For the following probability distribution.

X	1	2	3	4
$P(X)$	$\frac{1}{10}$	$\frac{1}{5}$	$\frac{3}{10}$	$\frac{2}{5}$

$E(X^2)$ is equal to

(a) 3 (b) 5

(c) 7 (d) 10

Ans. *(d)* $E(X^2) = \Sigma X^2 \; P(X)$

$$= 1 \times \frac{1}{10} + 4 \times \frac{1}{5} + 9 \times \frac{3}{10} + 16 \times \frac{2}{5}$$

$$= \frac{1}{10} + \frac{4}{5} + \frac{27}{10} + \frac{32}{5}$$

$$= \frac{1 + 8 + 27 + 64}{10}$$

$$= \frac{100}{10} = 10$$

***Q.90** Suppose a random variable X follows the binomial distribution with parameters n and p, where $0 < p < 1$. If $P(x = r) / P(x = n - r)$ is independent of n and r, then p is equal to

(a) $\frac{1}{2}$ (b) $\frac{1}{3}$

(c) $\frac{1}{5}$ (d) $\frac{1}{7}$

Ans. *(a)* $\because$ $P(X = r) = {}^nC_r \, (p)^r (q)^{n-r} = \frac{n!}{(n-r)!r!} (p)^r (1-p)^{n-r}$ $[\because q = 1 - p]$ ...(i)

$$P(X = 0) = {}^nC_o \, (p)^0 \, (q)^n = q^n = (1-p)^n$$

and $P(X = n - r) = {}^nC_{n-r} \, (p)^{n-r} \, (q)^{n-(n-r)}$

$$= \frac{n!}{(n-r)!r!} (p)^{n-r} (1-p)^r$$ $[\because q = 1 - p]$ $[\because {}^nC_r = {}^nC_{n-r}]$ ...(ii)

Now, $$\frac{P(x = r)}{P(x = n - r)} = \frac{\frac{n!}{(n-r)!r!} p^r (1-p)^{n-r}}{\frac{n!}{(n-r)!\,r!} p^{n-r} (1-p)^r}$$ [using Eqs. (i) and (ii)]

$$= \left(\frac{1-p}{p}\right)^{n-r} \times \frac{1}{\left(\frac{1-p}{p}\right)^r}$$

Above expression is independent of n and r, if $\frac{1-p}{p} = 1$

$$\Rightarrow \qquad \frac{1}{p} = 2 \Rightarrow p = \frac{1}{2}$$

Note (✱) Topics related to these question have been removed from latest NCERT.

Q.91 In a college, 30% students fail in Physics, 25% fail in Mathematics and 10% fail in both. One student is chosen at random. The probability that she fails in Physics, if she has failed in Mathematics is

(a) $\frac{1}{10}$
(b) $\frac{2}{5}$
(c) $\frac{9}{20}$
(d) $\frac{1}{3}$

Ans. ***(b)*** Let *Ph* and *M* represent Physics and Mathematics students resprectively.

$$\therefore \quad P(Ph) = \frac{30}{100} = \frac{3}{10}, \; P(M) = \frac{25}{100} = \frac{1}{4}$$

and $\quad P(M \cap Ph) = \frac{10}{100} = \frac{1}{10}$

Therefore, by using conditional probability

$$\therefore \quad P(Ph \mid M) = \frac{P(Ph \cap M)}{P(M)} = \frac{1/10}{1/4} = \frac{2}{5}$$

Q.92 *A* and *B* are two students. Their chances of solving a problem correctly are $\frac{1}{3}$ and $\frac{1}{4}$, respectively. If the probability of their making a common error is, $\frac{1}{20}$ and they obtain the same answer, then the probability of their answer to be correct is

(a) $\frac{1}{12}$
(b) $\frac{1}{40}$
(c) $\frac{13}{120}$
(d) $\frac{10}{13}$

Ans. ***(d)*** Let E_1 = Event that both *A* and *B* solve the problem

$$\therefore \quad P(E_1) = \frac{1}{3} \times \frac{1}{4} = \frac{1}{12}.$$

Let E_2 = Event that both *A* and *B* got incorrect solution of the problem

$$\therefore \quad P(E_2) = [1 - P(A)]\,[1 - P(B)] = \frac{2}{3} \times \frac{3}{4} = \frac{1}{2}$$

Let E = Event that they got same answer

Here, $\quad P(E \mid E_1) = 1, \; P(E \mid E_2) = \frac{1}{20}$

$$\therefore \quad P(E_1 \mid E) = \frac{P(E_1 \cap E)}{P(E)} = \frac{P(E_1) \cdot P(E \mid E_1)}{P(E_1) \cdot P(E \mid E_1) + P(E_2)\, P(E \mid E_2)}$$

$$= \frac{\frac{1}{12} \times 1}{\frac{1}{12} \times 1 + \frac{1}{2} \times \frac{1}{20}}$$

$$= \frac{1/12}{\frac{10+3}{120}} = \frac{120}{12 \times 13}$$

$$= \frac{10}{13}$$

Note (✱) Topics related to these question have been removed from latest NCERT.

***Q.93** If a box has 100 pens of which 10 are defective. What is the probability that out of a sample of 5 pens drawn one by one with replacement atmost one is defective?

(a) $\left(\frac{9}{10}\right)^5$ (b) $\frac{1}{2}\left(\frac{9}{10}\right)^4$

(c) $\frac{1}{2}\left(\frac{9}{10}\right)^4$ (d) $\left(\frac{9}{10}\right)^5 + \frac{1}{2}\left(\frac{9}{10}\right)^4$

Ans. ***(d)*** Here, $n = 5,\ p = \frac{10}{100} = \frac{1}{10},\ q = \frac{9}{10}$ and $r \le 1$ *i.e.*, $r = 0, 1$

Also, by Binomial, distribution, $P(X = r) = {}^nC_r\, p^r q^{n-r}$

$$\therefore \quad P(X = r) = P(r = 0) + P(r = 1)$$

$$= {}^5C_0\left(\frac{1}{10}\right)^0\left(\frac{9}{10}\right)^5 + {}^5C_1\left(\frac{1}{10}\right)^1\left(\frac{9}{10}\right)^4$$

$$= \left(\frac{9}{10}\right)^5 + 5 \times \frac{1}{10} \times \left(\frac{9}{10}\right)^4$$

$$= \left(\frac{9}{10}\right)^5 + \frac{1}{2}\left(\frac{9}{10}\right)^4$$

State True or False for the statement in each of the exercises 94 to 103.

Q.94 If $P(A) > 0$ and $P(B) > 0$. Then, A and B can be both mutually exclusive and independent.

Ans. *False*

Q.95 If A and B are independent events, then A' and B' are also independent.

Ans. *True*

Since, A and B are independent events.

$$\therefore \quad P(A \cap B) = P(A) \cdot P(B)$$

Now,

$$P(A' \cap B') = P(A \cup B)'$$
$$= 1 - P(A \cup B)$$
$$= 1 - [P(A) + (B) - P(A \cap B)]$$
$$= 1 - [P(A) - P(B) - P(A \cap B)]$$
$$= 1 - P(A) - P(B) + P(A) \cdot P(B)$$
$$= [1 - P(A)] - P(B)\,[1 - P(A)]$$
$$= P(A') - P(B) \cdot P(A')$$
$$= P(A')\,[1 - P(B)]$$
$$= P(A') \cdot P(B')$$

So, A' and B' are also independent events.

Note (✱) Topics related to these question have been removed from latest NCERT.

Q.96 If A and B are mutually exclusive events, then they will be independent also.

Ans. *False*

Q.97 Two independent events are always mutually exclusive.

Ans. *False*

Q.98 If A and B are two independent events, then $P(A \text{ and } B) = P(A) \cdot P(B)$.

Ans. *True*

If A and B are independent events, then $P(A \cap B) = P(A) \cdot P(B)$.

***Q.99** Another name for the mean of a probability distribution is expected value.

Ans. *True*

$$E(X) = \Sigma X \; P(X) = \mu$$

Q.100 If A and B' are independent events, then $P(A' \cup B) = 1 - P(A)\, P(B')$.

Ans. *True*

$$P(A' \cup B) = 1 - P(A - B)(B')$$

$$= 1 - P(A \cap B') = 1 - P(A)\, P(B')$$

[$\because$ A and B' are independent events]

Q.101 If A and B are independent, then P (exactly one of A, B occurs)

$$= P(A)\, P(B') + P(B)\, P(A').$$

Ans. *True*

P(exactly one of A, B occurs)

$$= P(A \cap B') + P(A' \cap B)$$

$$= P(A) \cdot P(B') + P(A')\, P(B)$$ [$\because$ A and B are independent event]

Q.102 If A and B are two events such that $P(A) > 0$ and $P(A) + P(B) > 1$, then $P(B \mid A) \geq 1 - \dfrac{P(B')}{P(A)}$

Ans. *False*

We have, $P(A) > 0$ and $P(A) + P(B) > 1$ ...(i)

$\because$ $$P(B \mid A) = \frac{P(A \cap B)}{P(A)}$$

$$= \frac{P(A) + P(B) - P(A \cup B)}{P(A)} > \frac{1 - P(A \cup B)}{P(A)}$$ [from Eq. (i)]

Note (✱) Topics related to these question have been removed from latest NCERT.

Q.103 If A, B and C are three independent events such that

$$P(A) = P(B) = P(C) = p,$$

then P (atleast two of A, B and C occur) $= 3p^2 - 2p^3$.

Ans. *True*

P (atleast two of A, B and C occur)

$$= P(A)\,P(B)\,P(C') + P(A')\,P(B)\,P(C) + P(A)\,P(B')P(C) + P(A)\,P(B)\,P(C)$$
$$= p \times p \times (1-p) + (1-p) \times p \times p + p(1-p) \times p + p \times p \times p$$
$$= p^2[1 - p + 1 - p + 1 - p + p] = p^2(3 - 2p) = 3p^2 - 2p^3$$

Fill in the blanks in each of the following exercises 104 to 108.

Q.104 If A and B are two events such that $P(A + B) = p$, $P(A) = p$, $P(B) = \frac{1}{3}$ and $P(A \cup B) = \frac{5}{9}$, then $p = \ldots\ldots\ldots\ldots$.

Ans. Here, $P(A) = p$, $P(B) = \frac{1}{3}$ and $P(A \cup B) = \frac{5}{9}$

$$\because \quad P(A \mid B) = \frac{P(A \cap B)}{P(B)} = p \Rightarrow P(A \cap B) = \frac{p}{3}$$

and $P(A \cup B) = P(A) + P(B) - P(A \cap B)$

$$\Rightarrow \quad \frac{5}{9} = p + \frac{1}{3} - \frac{p}{3} \Rightarrow \frac{5}{9} - \frac{1}{3} = \frac{2p}{3}$$

$$\Rightarrow \quad \frac{5-3}{9} = \frac{2p}{3} \Rightarrow p = \frac{2}{9} \times \frac{3}{2} = \frac{1}{3}$$

Q.105 If A and B are such that $P(A' \cup B') = \frac{2}{3}$ and $P(A \cup B) = \frac{5}{9}$, then $P(A') + P(B') = \ldots\ldots\ldots$.

Ans. Here, $P(A' \cup B') = \frac{2}{3}$ and $P(A \cup B) = \frac{5}{9}$

$$\because \quad P(A' \cup B') = P(A \cap B)' = 1 - P(A \cap B)$$

$$\Rightarrow \quad \frac{2}{3} = 1 - P(A \cap B)$$

$$\Rightarrow \quad P(A \cap B) = 1 - \frac{2}{3} = \frac{1}{3}$$

Now, $P(A') + P(B') = 1 - P(A) + 1 - P(B)$

$$= 2 - [P(A) + P(B)]$$
$$= 2 - [P(A \cup B) + P(A \cap B)]$$

$[\because P(A \cup B) = P(A) + P(B) - P(A \cap B)]$

$$= 2 - \left(\frac{5}{9} + \frac{1}{3}\right) = 2 - \left(\frac{5+3}{9}\right) = \frac{18-8}{9} = \frac{10}{9}$$

Note (✱) Topics related to these question have been removed from latest NCERT.

***Q.106** If X follows binomial distribution with parameters $n=5$, p and $P(X=2)=9, P(X=3)$, then $p=$ ....

Ans. $\because$ $$P(X=2)=9\cdot P(X=3)$$

(where, $n=5$ and $q=1-p$)

$$\Rightarrow \quad {}^5C_2\, p^2(1-p)^3 = 9\cdot {}^5C_3 p^3(1-p)^2$$

$$\Rightarrow \quad \frac{5!}{2!3!}p^2(1-p)^3 = 9\cdot\frac{5!}{3!2!}p^3(1-p)^2$$

$$\Rightarrow \quad \frac{p^2(1-p)^3}{p^3(1-p)^2} = 9 \Rightarrow \frac{(1-p)}{p} = 9$$

$$\Rightarrow \quad 9p+p=1 \Rightarrow p=\frac{1}{10}$$

***Q.107** Let X be a random variable taking values $x_1, x_2, x_3, \dots, x_n$ with probabilities $p_1, p_2, p_3, \dots, p_n$, respectively. Then, Var $(x)=$ ....

Ans. $$\text{Var}(X) = E(X)^2 - [E(X)]^2$$

$$= \sum_{i=1}^{n} X^2 P(X) - [\sum_{i=1}^{n} X\, P(X)]^2 = \Sigma\, p_i x_i^2 - (\Sigma\, p_i\, x_i)^2$$

Q.108 Let A and B be two events. If $P(A\mid B)=P(A)$, then A is ... of B.

Ans. $\because$ $$P(A\mid B) = \frac{P(A\cap B)}{P(B)}$$

$$\Rightarrow \quad P(A) = \frac{P(A\cap B)}{P(B)}$$

$$\Rightarrow \quad P(A)\cdot P(B) = P(A\cap B)$$

So, A is independent of B.

Note (✱) Topics related to these question have been removed from latest NCERT.